Everyday Mathematics®

The University of Chicago School Mathematics Project

TEACHER'S LESSON GUIDE
VOLUME 2

McGraw Hill Education

The University of Chicago School Mathematics Project

Max Bell, Director, *Everyday Mathematics* First Edition; James McBride, Director, *Everyday Mathematics* Second Edition; Andy Isaacs, Director, *Everyday Mathematics* Third, CCSS, and Fourth Editions; Amy Dillard, Associate Director, *Everyday Mathematics* Third Edition; Rachel Malpass McCall, Associate Director, *Everyday Mathematics* CCSS and Fourth Editions; Mary Ellen Dairyko, Associate Director, *Everyday Mathematics* Fourth Edition

Authors
Jean Bell, Max Bell, John Bretzlauf, Amy Dillard, Robert Hartfield, Andy Isaacs, James McBride, Rachel Malpass McCall, Kathleen Pitvorec, Peter Saecker

Fourth Edition Grade 1 Team Leader
Rachel Malpass McCall

Writers
Meg Schleppenbach Bates, Kate Berlin, Sarah R. Burns, Gina Garza-Kling, Linda M. Sims

Open Response Team
Catherine R. Kelso, Leader; Kathryn M. Rich

Differentiation Team
Ava Belisle-Chatterjee, Leader; Martin Gartzman; Anne Sommers

Digital Development Team
Carla Agard-Strickland, Leader; John Benson, Gregory Berns-Leone, Juan Camilo Acevedo

Virtual Learning Community
Meg Schleppenbach Bates, Cheryl G. Moran, Margaret Sharkey

Technical Art
Diana Barrie, Senior Artist; Cherry Inthalangsy

UCSMP Editorial
Lila K.S. Goldstein, Senior Editor; Rachel Jacobs, Kristen Pasmore, Delna Weil

Field Test Coordination
Denise A. Porter

Field Test Teachers
Mary Alice Acton, Katrina Brown, Pamela A. Chambers, Erica Emmendorfer, Lara Galicia, Heather A. Hall, Jeewon Kim, Nicole M. Kirby, Vicky Kudwa, Stephanie Merkle, Sarah Orlowski, Jenny Pfeiffer, LeAnita Randolph, Jan Rodgers, Mindy Smith, Kellie Washington

Digital Field Test Teachers
Colleen Girard, Michelle Kutanovski, Gina Cipriani, Retonyar Ringold, Catherine Rollings, Julia Schacht, Christine Molina-Rebecca, Monica Diaz de Leon, Tiffany Barnes, Andrea Bonanno-Lersch, Debra Fields, Kellie Johnson, Elyse D'Andrea, Katie Fielden, Jamie Henry, Jill Parisi, Lauren Wolkhamer, Kenecia Moore, Julie Spaite, Sue White, Damaris Miles, Kelly Fitzgerald

Contributors
William B. Baker, John Benson, Jeanine O'Nan Brownell, Andrea Cocke, Jeanne Mills DiDomenico, Rossita Fernando, James Flanders, Lila K.S. Goldstein, Allison M. Greer, Brooke A. North, Penny Williams

Center for Elementary Mathematics and Science Education Administration
Martin Gartzman, Executive Director; Meri B. Fohran, Jose J. Fragoso, Jr., Regina Littleton, Laurie K. Thrasher

External Reviewers
The *Everyday Mathematics* authors gratefully acknowledge the work of the many scholars and teachers who reviewed plans for this edition. All decisions regarding the content and pedagogy of *Everyday Mathematics* were made by the authors and do not necessarily reflect the views of those listed below.

Elizabeth Babcock, California Academy of Sciences; Arthur J. Baroody, University of Illinois at Urbana-Champaign and University of Denver; Dawn Berk, University of Delaware; Diane J. Briars, Pittsburgh, Pennsylvania; Kathryn B. Chval, University of Missouri-Columbia; Kathleen Cramer, University of Minnesota; Ethan Danahy, Tufts University; Tom de Boor, Grunwald Associates; Louis V. DiBello, University of Illinois at Chicago; Corey Drake, Michigan State University; David Foster, Silicon Valley Mathematics Initiative; Funda Gönülateş, Michigan State University; M. Kathleen Heid, Pennsylvania State University; Natalie Jakucyn, Glenbrook South High School, Glenview, IL; Richard G. Kron, University of Chicago; Richard Lehrer, Vanderbilt University; Susan C. Levine, University of Chicago; Lorraine M. Males, University of Nebraska-Lincoln; Dr. George Mehler, Temple University and Central Bucks School District, Pennsylvania; Kenny Huy Nguyen, North Carolina State University; Mark Oreglia, University of Chicago; Sandra Overcash, Virginia Beach City Public Schools, Virginia; Raedy M. Ping, University of Chicago; Kevin L. Polk, Aveniros LLC; Sarah R. Powell, University of Texas at Austin; Janine T. Remillard, University of Pennsylvania; John P. Smith III, Michigan State University; Mary Kay Stein, University of Pittsburgh; Dale Truding, Arlington Heights District 25, Arlington Heights, Illinois; Judith S. Zawojewski, Illinois Institute of Technology

Note
Many people have contributed to the creation of *Everyday Mathematics*. Visit http://everydaymath.uchicago.edu/authors/ for biographical sketches of *Everyday Mathematics* Fourth Edition staff and copyright pages from earlier editions.

www.everydaymath.com

Send all inquiries to:
McGraw-Hill Education
8787 Orion Place
Columbus, OH 43240

ISBN: 978-0-02-138365-8
MHID: 0-02-138365-0

Printed in the United States of America.

4 5 6 7 8 9 RMN 19 18 17 16 15

Contents

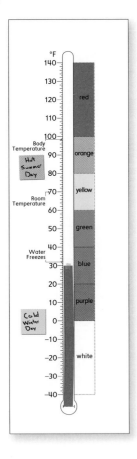

Getting Ready to Teach
First Grade Everyday Mathematics . . . . **xxxvi**

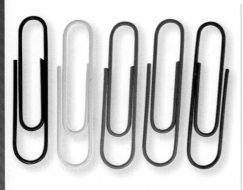

(t)Stockbyte/Getty Images, (b)United States Mint image

Contents   **ix**

Unit 6 Organizer
Addition Fact Strategies

In this unit, children work toward fluency with addition facts. They also explore telling time and solving number stories. Children's learning will focus on five clusters of the Common Core's content standards, as well as in-depth work on two of the Mathematical Practices.

 ## Standards for Mathematical Content

Domain	Cluster
Operations and Algebraic Thinking	Represent and solve problems involving addition and subtraction.
	Add and subtract within 20.
Number and Operations in Base Ten	Understand place value.
	Use place value understanding and properties of operations to add and subtract.
Measurement and Data	Tell and write time.

Because the standards within each domain can be broad, *Everyday Mathematics* has unpacked each standard into Goals for Mathematical Content GMC . For a complete list of Standards and Goals, see page EM1.

For an overview of the CCSS domains, standards, and mastery expectations in this unit, see the **Spiral Trace** on pages 498–499. See the **Mathematical Background** (pages 500–502) for a discussion of the following key topics:

- Clocks and Time
- Number Stories
- Using "Helper" Facts to Add
- Equivalence
- Place Value

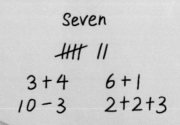

Standards for Mathematical Practice

SMP3 Construct viable arguments and critique the reasoning of others.

SMP8 Look for and express regularity in repeated reasoning.

For a discussion about how *Everyday Mathematics* develops these practices and a list of Goals for Mathematical Practice GMP , see page 503.

 to **vlc.cemseprojects.org** to search for video clips on each practice.

Go Digital with these tools at **connectED.mcgraw-hill.com**

 ePresentations   Student Learning Center   Facts Workshop Game   eToolkit   Professional Development   Home Connections   Spiral Tracker   Assessment and Reporting   English Learners Support   Differentiation Support

Contents

* The standards listed here are addressed in the **Focus** of each lesson. For all the standards in a lesson, see the Lesson Opener.

Unit 6 Materials

VLC Virtual Learning Community

See how *Everyday Mathematics* teachers organize materials. Search "Classroom Tours" at **vlc.cemseprojects.org**.

Lesson	Math Masters	Activity Cards	Manipulative Kit	Other Materials
6-1	pp. 156–158; TA27	70–71	base-10 blocks; number cards 0–12	chart paper; Number Cards; classroom analog clock; brad; colored pencils; prepared hour-hand-only clock; scissors
6-2	p. 159; G42 (optional); G45	72	counters; base-10 blocks	slate; Animal Cards
6-3	pp. 160–163; G30 (optional); G37 (optional); G47	73–75	per partnership: pennies, geoboards; rubber bands; number cards 1–10; 4 sets of number cards 0–9	slate; Number Cards; per partnership: prepared deck of cards, 12 equal signs, numbered egg carton; Two-Dimensional Shapes Poster (optional)
6-4	p. 164; G30 (optional); G47	76	Quick Look Cards 73, 78, 79, 81, 84, 96, 107; counters; per partnership: number cards 1–10	slate; Number Cards
6-5	pp. 165–166; G28 (optional); G29	77	Quick Look Cards 83, 100, 103, 107, 112; doubles dominoes; number cards 2–10	slate; Number Cards; hour-hand-only clock
6-6	pp. 167–168; TA19; G18 (optional); G19–G21; G32 (optional); G33	78	Quick Look Cards 88, 89, 92, 95, 98, 114, 116; 20 counters; 4 sets of number cards 0–10; base-10 longs	slate; Number Cards
6-7	p. 169; TA20 (optional)	79–80	per partnership: 4 each of number cards 1–10, 40 pennies	Number Cards; hour-hand-only clock; *My Reference Book;* stick-on notes (optional)
6-8	pp. 170–172; TA3			slate; Standards for Mathematical Practice Poster; *My Reference Book;* colored pencils; children's work from Day 1
6-9	pp. 173–174; TA28	81–82	counters; base-10 blocks; number grid	slate; prepared name-collection boxes; half-sheet of paper
6-10	pp. 175; TA29; G32 (optional); G33; G35 (optional)		base-10 blocks; number cards	slate; prepared Place-Value Mat; calculator; Number Cards; index cards
6-11	pp. 176; TA29–TA32; G36; G40; G48 (optional)	83	per partnership: pennies, dimes, dot dice; base-10 blocks	slate; prepared one-dollar bills; box; sheet of paper labeled "Bank"; 2 small paper cups
6-12	pp. 177–180; *Assessment Handbook* pp. 37–44			

📖 **Literature Link** 6-7 *Seaweed Soup* (optional), *DK First Encyclopedia* (optional)

Go Online for a complete literature list for Grade 1 and to download all Quick Look Cards.

Problem Solving Professional Development

Everyday Mathematics emphasizes equally all three of the Common Core's dimensions of **rigor:**

- conceptual understanding
- procedural skill and fluency
- applications

Math Messages, other daily work, Explorations, and Open Response tasks provide many opportunities for children to apply what they know to solve problems.

▶ Math Message

Math Messages require children to solve a problem they have not been shown how to solve. Math Messages provide almost daily opportunities for problem solving.

▶ Daily Work

Journal pages, Home Links, Writing/Reasoning prompts, and Differentiation Options often require children to solve problems in mathematical contexts and real-life situations. **Minute Math+** offers number stories and a variety other practice activities for transition times and for spare moments throughout the day. See Routine 6, pages 32–37.

▶ Explorations

In Exploration A, children create number sentences and determine whether they are true or false. In Exploration B, children practice solving doubles facts. And, in Exploration C, children make shapes with given attributes.

▶ Open Response and Reengagement

In Open Response Lesson 6-8, children make sense of and solve a multistep number story and explain their solution strategy. The reengagement discussion on Day 2 may focus on what the answer to each part of the story should look like (for example, is the answer a number or a word?) or different strategies for solving the problem. Making sense of problems is the focus Mathematical Practice for this lesson. Thinking critically about the meaning and possible solution strategies for a problem, either before or after attempting to find an answer, can help children become more efficient and effective problem solvers. GMP1.1

 Virtual Learning Community **Go Online** to watch an Open Response and Reengagement lesson in action. Search "Open Response" at **vlc.cemseprojects.org**.

Look for **GMP1.1-1.6** markers, which indicate opportunities for children to engage in **SMP1:** "Make sense of problems and persevere in solving them." Beyond that, children become better problem solvers as they engage in all of the CCSS Mathematical Practices. The yellow GMP markers throughout the lessons indicate places where you can emphasize the Mathematical Practices and develop children's problem-solving skills.

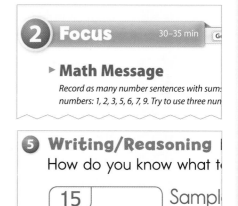

2 Focus 30–35 min

▶ **Math Message**

Record as many number sentences with sum numbers: 1, 2, 3, 5, 6, 7, 9. Try to use three num

5 Writing/Reasoning
How do you know what t

| 15 | Samp|

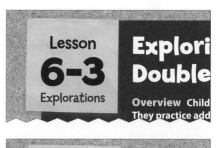

Lesson **6-3** Explorations

Explori Double

Overview Child They practice add

Lesson **6-8** Open Response and Reengagement

Pencils Club

Overview Day disc

Assessment and Differentiation

 Assessment and Reporting

See pages xxii–xxv to learn about this comprehensive online system for recording, monitoring, and reporting children's progress using core program assessments.

 VLC Virtual Learning Community   **Go Online** to **vlc.cemseprojects.org** for tools and ideas related to assessment and differentiation from *Everyday Mathematics* teachers.

✔ Ongoing Assessment

In addition to frequent informal opportunities for "kid watching," every lesson (except Explorations) offers an **Assessment Check-In** to gauge children's performance on one or more of the standards addressed in that lesson.

Lesson	Task Description	**CCSS** Common Core State Standards
6-1	Read time to the hour on an hour-hand-only analog clock.	1.MD.3, SMP5
6-2	Use tools to add 2-digit numbers to 1-digit numbers and multiples of 10.	1.NBT.4, SMP5
6-4	Use doubles facts to solve other addition facts.	1.OA.6
6-5	Represent strategies for addition in writing.	SMP2
6-6	Apply the making-10 strategy to find sums.	1.OA.6
6-7	Use the table of contents to find information.	SMP5
6-8	Make sense of and solve a two-step number story.	1.OA.1, SMP1
6-9	Generate equivalent names for numbers.	1.OA.7, SMP1
6-10	Make place-value exchanges and identify places in numbers.	1.NBT.2, 1.NBT.2a
6-11	Make place-value exchanges.	1.NBT.2, 1.NBT.2a, SMP2

▶ Periodic Assessment

Unit 6 Progress Check This assessment focuses on the CCSS domains of *Operations and Algebraic Thinking, Number and Operations in Base Ten,* and *Measurement and Data.* It also contains a Cumulative Assessment to help monitor children's learning and retention of content that was the focus of previous units.

NOTE Odd-numbered units include an **Open Response Assessment.** Even-numbered units include a **Cumulative Assessment**.

► Unit 6 Differentiation Activities

 Differentiation Support   English Learners Support

Differentiation Options Every regular lesson provides **Readiness, Enrichment, Extra Practice,** and **Beginning English Language Learners Support** activities that address the Focus standards for that lesson.

CCSS 1.OA.6, SMP2	CCSS 1.OA.6, SMP2	CCSS 1.OA.6, SMP6
Readiness 10–15 min	**Enrichment** 10–15 min	**Extra Practice** 10–15 min
Exploring Doubles WHOLE CLASS / SMALL GROUP / PARTNER	**Drawing Doubles Pictures** WHOLE CLASS / SMALL GROUP / PARTNER	**Playing *Roll and Record Doubles*** WHOLE CLASS / SMALL GROUP / PARTNER
number cards 1–10, counters,	Activity Card 76, crayons	

Activity Cards These activities, written to children, enable you to differentiate Part 2 of the lesson through small-group work.

English Language Learners Activities and point-of use support help children at different levels of English language proficiency succeed.

Differentiation Support Two online pages for most lessons provide suggestions for game modifications, ways to scaffold lessons for children who need additional support, and language development suggestions for Beginning, Intermediate, and Advanced English language learners.

Activity Card 78

Making 100

If you split 50 into 20 and 30, then you can add 80 + 20 first to make 100.

What You Need
Making 100 Cards, page 167
base-10 longs
paper

What To Do
Work with a partner.
1. Mix the cards and place them facedown.
2. Draw 2 cards.
3. Use a "making 100" strategy to add the numbers. You can use base-10 longs to help you.

Differentiation Support online pages

Ongoing Practice  Differentiation Support

► Embedded Facts Practice

Basic Facts Practice can be found in every part of the lesson. Look for activities or games labeled with CCSS 1.OA.6; or go online to the Spiral Tracker and search using CCSS 1.OA.6.

For **ongoing distributed practice,** see these activities:
• Mental Math and Fluency
• Differentiated Options: Extra Practice
• Part 3: Journal pages, Math Boxes, *Math Masters,* Home Links
• Print and online games

► Games

Games in *Everyday Mathematics* are an essential tool for practicing skills and developing strategic thinking.

Lesson	Game	Skills and Concepts	**CCSS** Common Core State Standards
6-2	*Stop and Go*	Adding and subtracting 2-digit numbers	1.NBT.4, 1.NBT.6, SMP1
6-3 6-4	*Roll and Record Doubles*	Finding addition doubles	1.OA.6, SMP6
6-3	*Addition Top-It*	Solving addition facts and using relation symbols	1.OA.6, 1.NBT.3, SMP1
6-5	*Domino Top-It*	Finding and comparing sums	1.OA.6, 1.NBT.3, SMP2
6-6 6-10	*Fishing for 10*	Making combinations of 10	1.OA.6, SMP1
6-6	*Ten-Frame Top-It*	Comparing numbers represented on ten frames	1.OA.6, 1.NBT.3, SMP2
6-7	*The Difference Game*	Reinforcing subtraction facts	1.OA.4, 1.OA.6, SMP6
6-10	*The Digit Game with 3-Digit Numbers*	Comparing numbers based on place value	1.NBT.2, 1.NBT.3, SMP6
6-11	*Penny-Dime-Dollar Exchange*	Making place-value exchanges	1.NBT.2, 1.NBT.2a, SMP6
6-11	*Penny-Dime Exchange*	Exchanging ones for tens	1.NBT.2, SMP2
6-11	*Base-10 Exchange*	Exchanging ones for tens	1.NBT.2, 1.NBT.2a, SMP2

VLC Virtual Learning Community — **Go Online** to look for examples of *Everyday Mathematics* games at **vlc.cemseprojects.org.**

CCSS **Spiral Trace:** Skills, Concepts, and Applications

⭐ **Mastery Expectations** This Spiral Trace outlines instructional trajectories for key standards in Unit 6. For each standard, it highlights opportunities for Focus instruction, Warm Up and Practice activities, as well as formative and summative assessment. It describes the **degree of mastery**—as measured against the entire standard—expected at this point in the year.

Operations and Algebraic Thinking

1.OA.3 Apply properties of operations as strategies to add and subtract. *Examples: If 8 + 3 = 11 is known, then 3 + 8 = 11 is also known. (Commutative property of addition.) To add 2 + 6 + 4, the second two numbers can be added to make a ten, so 2 + 6 + 4 = 2 + 10 = 12. (Associative property of addition.)*

⭐ By the end of Unit 6, expect children to **apply the commutative and associative properties of addition to solve problems.**

1.OA.6 Add and subtract within 20, demonstrating fluency for addition and subtraction within 10. Use strategies such as counting on; making ten (e.g., 8 + 6 = 8 + 2 + 4 = 10 + 4 = 14); decomposing a number leading to a ten (e.g., 13 − 4 = 13 − 3 − 1 = 10 − 1 = 9); using the relationship between addition and subtraction (e.g., knowing that 8 + 4 = 12, one knows 12 − 8 = 4); and creating equivalent but easier or known sums (e.g., adding 6 + 7 by creating the known equivalent 6 + 6 + 1 = 12 + 1 = 13).

⭐ By the end of Unit 6, expect children to **use doubles facts and combinations of 10 to help them solve other addition and subtraction facts within 20.**

1.OA.7 Understand the meaning of the equal sign, and determine if equations involving addition and subtraction are true or false. *For example, which of the following equations are true and which are false? 6 = 6, 7 = 8 − 1, 5 + 2 = 2 + 5, 4 + 1 = 5 + 2.*

⭐ By the end of Unit 6, expect children to **find equivalent names for numbers.**

Number and Operations in Base Ten

1.NBT.2 Understand that the two digits of a two-digit number represent amounts of tens and ones.

⭐ By the end of Unit 6, expect children to **tell the value of each digit in a two-digit number.**

Spiral Tracker

Go to **connectED.mcgraw-hill.com** for comprehensive trajectories that show how in-depth mastery develops across the grade.

1.NBT.4 Add within 100, including adding a two-digit number and a one-digit number, and adding a two-digit number and a multiple of 10, using concrete models or drawings and strategies based on place value, properties of operations, and/or the relationship between addition and subtraction; relate the strategy to a written method and explain the reasoning used. Understand that in adding two-digit numbers, one adds tens and tens, ones and ones; and sometimes it is necessary to compose a ten.

| 5-11 Focus Practice | 5-12 Focus Practice | 5-13 Progress Check | 6-2 Focus Practice | 6-7 Focus | 6-8 Warm Up Focus | 6-12 Progress Check | 7-1 Warm Up Practice | 7-5 Practice | 7-9 Warm Up | 8-2 Practice | 8-7 Warm Up |

⭐ By the end of Unit 6, expect children to **add within 100 using tools.**

1.NBT.6 Subtract multiples of 10 in the range 10–90 from multiples of 10 in the range 10–90 (positive or zero differences), using concrete models or drawings and strategies based on place value, properties of operations, and/or the relationship between addition and subtraction; relate the strategy to a written method and explain the reasoning used.

| 5-11 Focus Practice | 6-2 Focus | 6-7 Focus | 6-8 Warm Up | 6-12 Progress Check | 7-1 Practice | 7-5 Practice | 8-7 Warm Up Practice | 8-11 Focus | 9-2 Focus | 9-5 Focus | 9-6 Focus |

⭐ By the end of Unit 6, expect children to **subtract two-digit multiples of 10 from other two-digit multiples of 10 using tools.**

Measurement and Data

1.MD.3 Tell and write time in hours and half-hours using analog and digital clocks.

| 6-1 Focus Practice | 6-5 Warm Up Practice | 6-7 Warm Up | 6-12 Progress Check | 7-11 Focus Practice | 7-12 Progress Check | 8-1 Warm Up | 8-3 Practice | 8-8 Focus Practice | 8-9 Practice | 8-12 Progress Check | 9-1 Warm Up |

⭐ By the end of Unit 6, expect children to **tell time to the hour on an hour-hand-only analog clock.**

Key  = Assessment Check-In  = Progress Check Lesson = Current Unit  = Previous or Upcoming Lessons

Mathematical Background: Content

 The discussion below highlights major content areas and Common Core State Standards addressed in Unit 6. See the online Spiral Tracker for complete information about learning trajectories for all standards.

▶ Clocks and Time (Lesson 6-1)

In Lesson 6-1, children examine the hour hand of an analog clock (a clock with hands) and how it moves as an hour passes. This is important preparation for telling time to the hour and half hour. **1.MD.3** Interpreting the hour hand can be particularly confusing for children when the time shown is almost, but not quite, to the next hour. For example, when an analog clock shows 1:45 with the hour hand closer to the 2 than the 1, children often incorrectly read the time as 2:45. Estimating the time with an hour-hand-only analog clock (one that is missing a minute hand) helps children avoid such errors and familiarizes them with how the hour hand moves as time passes. When the hour hand points exactly at 3, for example, the time is 3 o'clock. When the hour hand is near a number or between two numbers, the time is recorded as "a little before *x* o'clock," "just after *x* o'clock," or "between *x* and *y* o'clock." Children will learn about the minute hand in later units.

▶ Number Stories (Lessons 6-2 and 6-6)

Children work on several different types of number stories—change, parts-and-total, and comparison—throughout *First Grade Everyday Mathematics*. Previous lessons emphasized one type of number story at a time. In Unit 6, children encounter multiple types of number stories within a single lesson, providing additional opportunities to compare strategies for making sense of and solving those problems. **1.OA.1**

Go Online to the *Implementation Guide* for more information about situation diagrams.

▶ Using "Helper" Facts to Add
(Lessons 6-4 through 6-6)

Children develop fact fluency in different ways. While some may simply rely on memory, most children use a variety of strategies to master math facts. In Unit 6, children learn and practice efficient ways of using facts they already know to figure out unknown facts. *Everyday Mathematics* uses the term *helper facts* to refer to facts that can be used to help figure out other facts. **1.OA.6**

Two strategies that use helper facts—*near doubles* and *making 10*—emphasize number relationships that help children develop fact fluency. Initially, these strategies may be too difficult for children to use, but they are introduced in Unit 6 and revisited throughout the remainder of the year to provide children with plenty of practice.

 Standards and Goals for Mathematical Content

Because the standards within each domain can be broad, *Everyday Mathematics* **has unpacked each standard into Goals for Mathematical Content GMC. For a complete list of Standards and Goals, see page EM1.**

An hour-hand-only clock showing almost 3 o'clock

Unit 6 Vocabulary

analog clock	hour hand	*My Reference Book*
equivalent names	hundreds place	name-collection box
flat	making 10	near doubles

Using "Helper" Facts to Add (continued)

The Near-Doubles Strategy

Formally introduced in Unit 4 with Quick Looks of double ten frames, doubles
(1 + 1 = 2; 2 + 2 = 4; etc.) are often among the first facts that children learn fluently.
Children continue to practice doubles as they play *Roll and Record Doubles* and as
they complete the Explorations in Lesson 6-3. By Lesson 6-4, children are ready to
use their knowledge of doubles to help them learn new facts. With the near-doubles
strategy, children use a known double to solve a fact that is close to a double. For
example, 7 + 8 can be solved by thinking, "I know that 7 + 7 = 14, so I'll add 1 more
to get 15" or "I'll start with 8 + 8 = 16 and then subtract 1."

Lesson 6-4 begins with a specific sequence of Quick Looks with double ten frames
to encourage children to notice the relationship between a double and a nearby
fact. For example, children who instantly recognize 4 + 4 = 8 may notice that 4 + 3
is just one less than the familiar double.

The Making-10 Strategy

In Unit 4, children were introduced to a group of facts called combinations of 10
(for example, 7 + 3 and 9 + 1). They continued to practice these facts in *Fishing for
10*. In Lesson 6-6, combinations of 10 are used as helper facts in a strategy called
making 10. To employ this strategy, children break down one addend so that part of
it can be added to the other addend to make 10. For example, 7 + 5 can be solved
by thinking, "What can I add to 7 to get 10?" The needed "3" is obtained by breaking
5 into 3 + 2. Once a 10 is made, the remaining fact (10 + 2) is easy to solve.

> **NOTE** The following set of equations illustrates the example in more
> sophisticated notation. Children are not expected to use this notation.
>
> 7 + 5 =
> 7 + (3 + 2) =
> (7 + 3) + 2 =
> 10 + 2 = 12
>
> Notice that the middle steps of this strategy use the Associative Property of
> Addition to regroup the addends to make addition easier. **1.OA.3**

Lesson 6-6 uses Quick Looks with double ten frames to develop the making-10
strategy. Because children often move dots to fill a ten frame—or *make a 10*—during
Quick Look activities, it makes sense to children that creating a helper fact that is
a combination of 10 can be an effective strategy for solving more difficult facts.

[Go Online] to the *Implementation Guide* for more information about Quick Looks.

These Quick Looks help children
relate 4 + 4 = 8 to 4 + 3 = 7.

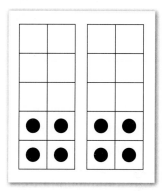

4 + 4 = 8

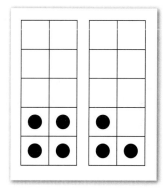

4 + 3 = 7

This Quick Look shows
the making-10 strategy.

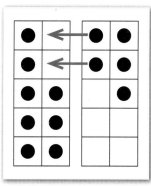

8 + 5 = 8 + 2 + 3 = 10 + 3

▶ Equivalence (Lesson 6-9)

Quantities can be expressed in many ways. In mathematics, we are most accustomed to representing quantities with numerals, but it is important for children to understand that there are many different—or equivalent—ways to represent any quantity. For example, the numeral 5, the name *five*, or ~~HHt~~ all represent the same quantity. Quantities can also be represented in terms of other numbers. For instance, 5 can be seen as a sum of parts ($3 + 2$ or $2 + 2 + 1$), or as part of a larger number ($11 - 6$). In *Everyday Mathematics,* the many ways of representing the same quantity are called "equivalent names." As children learn more about mathematics, they will continue to see numbers in new ways, expanding their knowledge and use of equivalent names.

In *Everyday Mathematics,* children use name-collection boxes to collect equivalent names for a number. Names can include words in any language, tally marks, and drawings, as well as expressions involving addition and subtraction.

Collecting equivalent names for a name-collection box provides an excellent context for reinforcing the meaning of the equal sign and determining if equations are true or false. **1.OA.7** For example, if $5 + 3$ and $4 + 4$ are both in the same name-collection box, then they are both names for the same number. It follows, then, that $5 + 3 = 4 + 4$, $5 + 3 = 8$, and $4 + 4 = 8$ must be true equations. This connection between name-collection boxes and equations is discussed in Lesson 6-9.

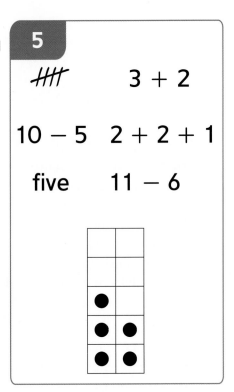

A name-collection box for 5

▶ Place Value (Lessons 6-10 and 6-11)

Place value was formally introduced in Unit 5 when children used base-10 blocks and pennies and dimes to represent 2-digit numbers. This work is reinforced and extended in Unit 6. In Lesson 6-10, children arrange cubes and longs on place-value mats to solve riddles, such as *Show 1 long and 5 cubes. What number is this?* To solve these riddles, children must think beyond $1 + 5 = 6$. They must see "1" as "1 ten." This is not a trivial concept for young children. Use of the place-value mat forces children to arrange the longs and cubes to mimic the placement of digits in a multidigit number, reinforcing the idea that the place of a digit in a number determines the value of the digit. In this lesson, children also use base-10 blocks to demonstrate multiple ways to represent a number (for example, 28 can be shown as 2 longs and 8 cubes, or as 1 long and 18 cubes), and to discuss how and when to make exchanges between place-value groups. **1.NBT.2** To extend children's understanding of the relationship among place-value groups, children explore using a flat (representing 100) to group 10 longs, just as they used a long to group 10 cubes.

In Lesson 6-11, children use dimes and pennies as a familiar context to represent tens and ones and to reinforce place-value concepts. This lesson also extends place-value work to the hundreds place in *Penny-Dime-Dollar Exchange.*

Mathematical Background: Practices

 In Everyday Mathematics, *children learn the **content** of mathematics as they engage in the **practices** of mathematics. As such, the Standards for Mathematical Practice are embedded in children's everyday work, including hands-on activities, problem-solving tasks, discussions, and written work. Read here to see how Mathematical Practices 3 and 8 are emphasized in this unit.*

▶ Standard for Mathematical Practice 3

Unit 6 offers many opportunities for children to "construct viable arguments and critique the reasoning of others," as stated in Mathematical Practice 3. As children respond to prompts such as *How did you get that answer?, Explain your thinking,* or *Do you agree with what Maria just said? Why?,* they are encouraged to demonstrate their thinking and make sense of others' ideas. For young children, clear mathematical communication relies on concrete objects as much as—if not more than—verbal explanations. In Lesson 6-2, children solve number stories with Animal Cards using base-10 blocks, number grids, and drawings to convey their thinking. In Lesson 6-3, they share their understanding of the equal sign through drawings and oral explanations. `GMP3.1` As children explore *My Reference Book* in Lesson 6-7, they are asked to explain sections that their partners found interesting. `GMP3.2`

Of special note are the Quick Look activities used throughout Unit 6. In these activities, children are encouraged to share their processes for solving problems. As children construct their arguments and explain their thinking, others are asked to listen, ask questions, and explain their classmates' thinking in their own words. `GMP3.1` Children are then encouraged to evaluate another child's method by trying it themselves in subsequent Quick Look problems.

▶ Standard for Mathematical Practice 8

As children gain experience with counting and computation, they begin to notice repetition and patterns. Mathematical Practice 8 states that mathematically proficient students make use of these "regularities" to justify rules, create shortcuts, and make generalizations. `GMP8.1` Throughout Unit 6, children explore the relationship between addition facts and use what they notice to create shortcuts that help them develop fact fluency. For example, in Lessons 6-4 and 6-5, children examine addition facts that can be expressed as a "double + 1" or a "double − 1." They apply their knowledge of this pattern as a method to solve other "near doubles." In Lesson 6-6, children examine representations of several teen numbers using double ten frames and are asked to describe how to represent *any* teen number using double ten frames. They later use this generalized representation to develop the useful making-10 strategy for addition.

 Standards and Goals for Mathematical Practice

SMP3 **Construct viable arguments and critique the reasoning of others.**

GMP3.1 Make mathematical conjectures and arguments.

GMP3.2 Make sense of others' mathematical thinking.

SMP8 **Look for and express regularity in repeated reasoning.**

GMP8.1 Create and justify rules, shortcuts, and generalizations.

Go Online ⟩ to the *Implementation Guide* for more information about the Mathematical Practices.

For children's information on the Mathematical Practices, see *My Reference Book,* pages 1–22.

Time and the Hour-Hand-Only Clock

Overview Children practice reading and displaying times on an hour-hand-only clock.

▶ **Before You Begin**
Use *Math Masters,* page 156 to assemble an hour-hand-only clock for use as a demonstration clock. Attach the hour hand with a brad and gather enough brads for each child to make an individual hour-hand-only clock. Be sure your classroom has an analog clock.

▶ **Vocabulary**
analog clock · hour hand

Common Core State Standards

Focus Cluster
Tell and write time.

1 Warm Up 15–20 min

	Materials	
Mental Math and Fluency Children solve number stories.	slate	1.OA.1, 1.OA.6
Daily Routines Children complete daily routines.	See pages 2–37.	See pages xiv–xvii.

2 Focus 30–35 min

Math Message Children write a list of activities that take about an hour to do.		1.MD.3 SMP6
Exploring Hours Children discuss the length of an hour and familiarize themselves with an analog clock.	chart paper, classroom analog clock	1.MD.3 SMP6
Reading Hour-Hand-Only Clocks Children tell time on an hour-hand-only clock. They assemble hour-hand-only clocks and use them to model approximate times.	*Math Journal 2,* Activity Sheet 9; colored pencils; hour-hand-only clock; per child: 1 brad	1.MD.3 SMP5, SMP6
✓ **Assessment Check-In** See page 508.	hour-hand-only clock	1.MD.3, SMP5

CCSS 1.MD.3 Spiral Snapshot

GMC Tell and write time using analog clocks.

 6-1 Focus Practice | 6-5 Warm Up Practice | 6-7 Warm Up | 7-11 Focus Practice | 8-1 Warm Up | 8-3 Practice | 8-8 Focus Practice | 8-9 Practice

 Spiral Tracker   Go Online ▶ to see how mastery develops for all standards within the grade.

3 Practice 15–20 min

Adding and Subtracting 10 on Number Grids Children practice finding 10 more and 10 less than a given number.	*Math Journal 2,* p. 111 and inside back cover	1.NBT.5
Math Boxes 6-1 Children practice and maintain skills.	*Math Journal 2,* p. 112	See page 509.
Home Link 6-1 **Homework** Children draw the hour hand to show times on a clock.	*Math Masters,* p. 158	1.OA.2, 1.OA.3, 1 OA.6, 1.MD.3

 connectED.mcgraw-hill.com ▶

Plan your lessons online with these tools.

 ePresentations   Student Learning Center   Facts Workshop Game   eToolkit   Professional Development   Home Connections   Spiral Tracker   Assessment and Reporting   English Learners Support   Differentiation Support

Differentiation Options RtI

<table>
<tr>
<td>

 CCSS **1.MD.3, SMP6**

Readiness · 10–15 min

Scheduling Daily Events

WHOLE CLASS
SMALL GROUP
PARTNER
INDEPENDENT

hour-hand-only clock

To familiarize children with
an hour-hand-only clock, place an hour-
hand-only clock in the center of a larger
background. Set the hand to the hour
that children tend to wake up. Write "wake
up," or draw a picture near that hour to
represent waking up. Move the hour hand
around the clock, stopping at each hour
to discuss what children typically do at
that time. Use words and pictures around
the clock face to illustrate these activities.
If children bring up activities in which it is
necessary to distinguish "at night" or "in the
morning" (for example, doing homework
at 7 P.M., not 7 A.M.), explain that the hour
hand travels around the clock twice in
a day, so there are A.M. hours and P.M. hours.
GMP6.3

`Go Online` for more information on
A.M and P.M.

</td>
<td>

CCSS **1.MD.3, SMP4**

Enrichment · 10–15 min

Ordering Times of Daily Activities

WHOLE CLASS
SMALL GROUP
PARTNER
INDEPENDENT

Activity Card 70;
Math Masters, p. 157; scissors

To further explore children's understanding
of time, have them illustrate and order
their daily activities according to times
of the day. **GMP4.1**

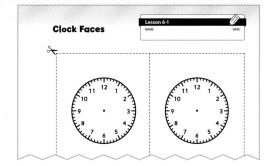

</td>
<td>

CCSS **1.MD.3, SMP2**

Extra Practice · 10–15 min

Setting the Time for Daily Activities

WHOLE CLASS
SMALL GROUP
PARTNER
INDEPENDENT

Activity Card 71;
number cards 1–12;
hour-hand-only clock

To provide practice representing times,
have children show different times on
hour-hand-only clocks and discuss their
daily activities. **GMP2.1**

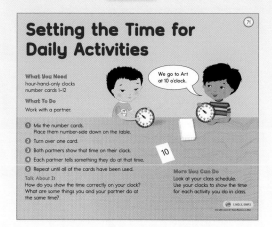

</td>
</tr>
</table>

English Language Learners Support

Beginning ELL Children may be familiar with the body terms *face* and *hand,* but they
may be unfamiliar with the terms when referring to a clock. Point to your face as you say
the word. Repeat the process with hand. Use Total Physical Response prompts to ask
children to show faces and hands. Then use the word *face* and point to the face of an
analog clock, and use the word *hands* as you point to the hands of the clock. Have children
demonstrate their understanding by pointing to their hands and face and the clock hands
and face in response to prompts such as: *Show me your hands. Show me the hands of
the clock. Show me the clock's face.*

`Go Online` **ELL** English Learners
Support

Standards and Goals for
Mathematical Practice

SMP5 **Use appropriate tools strategically.**

GMP5.2 Use tools effectively and make sense of your results.

SMP6 **Attend to precision.**

GMP6.1 Explain your mathematical thinking clearly and precisely.

GMP6.2 Use an appropriate level of precision for your problem.

Beginning a Time Museum

1 Warm Up 15–20 min

▶ Mental Math and Fluency

Have children solve the following number stories and write their answers on their slates.

- ●○○ Cory has 12 pencils. He shares 9 with his friends. How many pencils does Cory have now? 3 pencils
- ●●○ Farah had 16 stickers. She gave 9 stickers to her friend. How many does Farah have left? 7 stickers
- ●●● Brian had 13 pages to read in his book. He read some pages and now has 8 pages left to read. How many pages did Brian read? 5 pages

▶ Daily Routines

Have children complete the Daily Routines. For detailed instructions, see pages 2–37. For specifics on CCSS coverage, see pages xiv–xvii.

2 Focus 30–35 min

▶ Math Message

What are some activities that take about an hour to do? Record your ideas.
GMP6.2

▶ Exploring Hours

| WHOLE CLASS | SMALL GROUP | PARTNER | INDEPENDENT |

Math Message Follow-up Children share their ideas. Expect children to list activities that take exactly an hour, like swim practice. Activities that take approximately an hour, like making dinner, are appropriate as well.
GMP6.2 Discuss the fact that as soon as you say what time it is, it is already a little later. Therefore telling time is always about getting close; it is never exact. This is why we use phrases such as "it's about 2 o'clock," "it's almost 7 o'clock," and "it's just after 8 o'clock." GMP6.2

Begin a Time Museum. Label a piece of chart paper "one hour," and record examples of hour-long activities that children suggested, such as completing a math class, going grocery shopping, practicing a sport, or going out to eat.

NOTE The Time Museum can be expanded as children gain a greater understanding of each time unit. In later lessons, children will label a smaller piece of paper "one half hour" to record activities that take about a half hour, an even smaller piece of paper "one minute" for activities that take about a minute, and a yet smaller piece of paper "one second" for activities that take about a second.

Science Link Children may enjoy discussing how many hours are in a day. Explain that there are 24 hours in a day because it takes 24 hours for Earth to rotate, or spin, once.

Tell children that today they will learn more about telling time to the hour. Direct children's attention to the classroom analog clock. Have them describe what they notice about the clock. GMP6.1 Children may mention the numbers around the clock face, the differences between the hands on the clock, or what they know about telling time. Be sure to address the following points:

An **analog clock** is a clock that shows the time by the positions of the hands. If digital clocks come up, point out that digital clocks have no hour or minute hand. (Digital time notation will be introduced in Lesson 7-11.)

Identify the **hour hand** and the minute hand. Discuss the everyday meaning of *hand,* as well as its meaning in this context. Explain that the hour hand is *shorter* and the minute hand is *longer*. The hour hand also moves more slowly than the minute hand. Tell children that today they will be focusing on the hour hand.

▶ Reading Hour-Hand-Only Clocks

Math Journal 2, Activity Sheet 9

| **WHOLE CLASS** | SMALL GROUP | PARTNER | INDEPENDENT |

Tell children that when the hour hand points to a number, the time is read as that number and the term *o'clock*. Use the demonstration clock made from *Math Masters,* page 156. Starting at 12 o'clock, move the hour hand clockwise and have children call out the hours as the hour hand passes through one hour to the next: 12 o'clock, 1 o'clock, 2 o'clock, and so on. GMP5.2

Compare the hours on a clock to chapters in a familiar book. When reading Chapter 1, you are still in Chapter 1 even as you get closer to Chapter 2. The last page of Chapter 1 is still in Chapter 1, even though you are only one page from Chapter 2. Similarly, emphasize that even when the hand is close to 2 o'clock, it is still moving through the 1 o'clock hour until it reaches 2 o'clock.

Move the hour hand around several more times, stopping at various on-the-hour positions, and let children call out the times. GMP5.2

Have children remove *Math Journal 2,* Activity Sheet 9. Distribute a brad to each child. Modeling with the demonstration clock, guide children to circle the number 12 with a colored pencil and use that pencil to color the section of the clock that is part of that hour. Have children change colors when they continue this process for subsequent numbers. (*See margin.*) Then help children assemble their hour-hand-only clocks.

Position the hand on your clock to show a little past 4 o'clock. *How would you describe the time shown?* Sample answers: after 4 o'clock; between 4 o'clock and 5 o'clock; a little after 4 o'clock

Academic Language Development

Emphasize and practice language for approximating time. For example, with the hour hand between two numbers, we can say the time using phrases such as *about* _____, *almost* _____, *just before* _____, *a little after* _____, and *between* _____ *and* _____. Refer to the classroom analog clock throughout the day, and use this language of approximate time.

A fully-assembled clock from *Math Journal 2,* Activity Sheet 9

Ask the class to watch the hour hand as you move it slowly between the 4 o'clock and 5 o'clock positions. Ask:

- *Is the time getting closer to 4 o'clock or to 5 o'clock?* Sample answer: closer to 5 o'clock **GMP6.2**
- *How do you know it is still within the 4 o'clock hour as we move toward the 5?* Sample answer: because we haven't gotten to 5 yet
- *When will the clock show 5 o'clock?* Sample answers: when the hour hand gets all the way through the 4 o'clock part; when the hour hand gets to the 5

Move the hour hand on the demonstration clock so that it points to 2. *What time is it?* About 2 o'clock

Move the hour hand on a child's clock so that it points about halfway between 2 and 3. Ask children to compare the placement of the hour hands on these two clocks. *How would you describe the time shown?* Sample answers: between 2 o'clock and 3 o'clock; half past 2

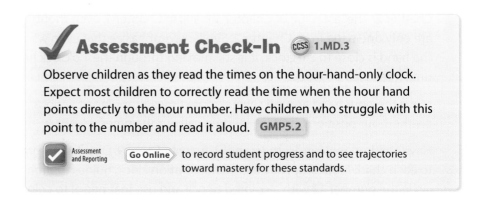

Differentiate **Common Misconception**

Most children will become confused when the hour hand is almost, but not quite, to the next hour. It is common for children to identify the time with the number that is closest to the hour hand. Remind children who make this mistake that an hour is like a chapter. Even if you are close to the end of a chapter, you are not yet in the next chapter.

Go Online Differentiation Support

Move the hour hand to various positions and ask children to follow along and tell the approximate time. Then say a time, and have children model the time on their clocks. Repeat several times.

✔ **Assessment Check-In** **CCSS** 1.MD.3

Observe children as they read the times on the hour-hand-only clock. Expect most children to correctly read the time when the hour hand points directly to the hour number. Have children who struggle with this point to the number and read it aloud. **GMP5.2**

✔ Assessment and Reporting **Go Online** to record student progress and to see trajectories toward mastery for these standards.

Summarize Ask children to look at the hour hand on the classroom clock and tell a partner about what time it is.

NOTE Have children put these clocks in their toolkits for later use.

Math Journal 2, p. 111

Adding and Subtracting 10 on Number Grids Lesson 6-1 DATE

Use the number grid on the inside back cover of your journal.

① Close your eyes and put a finger on your number grid.

Then open your eyes. Answers vary.

I pointed to _____.

What number is 10 more than the number? _____

What number is 10 less than the number? _____

Do the same thing 4 more times.

② I pointed to _____.
What number is 10 more? _____
What number is 10 less? _____

③ I pointed to _____.
What number is 10 more? _____
What number is 10 less? _____

④ I pointed to _____.
What number is 10 more? _____
What number is 10 less? _____

⑤ I pointed to _____.
What number is 10 more? _____
What number is 10 less? _____

1.NBT.5 one hundred eleven 111

3 Practice 15–20 min

 Go Online

ePresentations   eToolkit   Home Connections

▶ Adding and Subtracting 10 on Number Grids

Math Journal 2, p. 111 and inside back cover

| WHOLE CLASS | SMALL GROUP | **PARTNER** | **INDEPENDENT** |

Children practice adding and subtracting 10 using a number grid.

▶ Math Boxes 6-1

Math Journal 2, p. 112

| WHOLE CLASS | SMALL GROUP | **PARTNER** | INDEPENDENT |

Mixed Practice Math Boxes 6-1 are paired with Math Boxes 6-3.

▶ Home Link 6-1

Math Masters, p. 158

Homework Children draw the hour hand to show different times.

Math Journal 2, p. 112

Math Boxes

Lesson 6-1
DATE

1

Brody's Books	
Fairy Tale	///
Adventure	-HIT I
Animal	-HIT II

How many books does Brody have in all?

__16__ books

2 Use your feet (with your shoes on) to measure.

About how long is the path from your desk to your teacher's desk?

_____ feet

Answers vary.

3 Sofia picked 3 daffodils, 5 daisies, and 2 roses. How many flowers did she pick?

$3 + 5 + 2 = \boxed{10}$

Unit
flowers

4 Add using the number grid.
$71 + 10 = ?$
Choose the best answer.
○ 18   ○ 72   ○ 80   ● 81

Unit

5 **Writing/Reasoning** Solve. Exchange if you need to. Write <, >, or = to compare.

▮▮▮▮▮▮ ▭▭▭▭▭ + ▭▭▭▭▭ $\boxed{=}$ 70

How do you know? Sample answer: 6 longs and 10 cubes equals 7 longs, or 70.

① 1.OA.2, 1.NBT.1, 1.MD.4 ② 1.MD.2 ③ 1.OA.2, 1.OA.3, 1.OA.6
④ 1.NBT.4 ⑤ 1.NBT.2, 1.NBT.2a, 1.NBT.2c, 1.NBT.3, 1.NBT.4, SMP2

112   one hundred twelve

Math Masters, p. 158

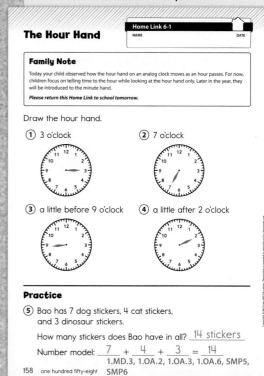

The Hour Hand

Home Link 6-1
NAME   DATE

Family Note

Today your child observed how the hour hand on an analog clock moves as an hour passes. For now, children focus on telling time to the hour while looking at the hour hand only. Later in the year, they will be introduced to the minute hand.

Please return this Home Link to school tomorrow.

Draw the hour hand.

1 3 o'clock   **2** 7 o'clock

3 a little before 9 o'clock   **4** a little after 2 o'clock

Practice

5 Bao has 7 dog stickers, 4 cat stickers, and 3 dinosaur stickers.

How many stickers does Bao have in all? __14 stickers__

Number model: __7__ + __4__ + __3__ = __14__

1.MD.3, 1.OA.2, 1.OA.3, 1.OA.6, SMP5, SMP6

158   one hundred fifty-eight

More 2-Digit Number Stories

Overview Children use tools, strategies, and properties of operations to solve number stories with two or three addends.

▶ **Before You Begin**
Have children locate the Animal Cards (*Math Journal 1*, Activity Sheets 5–8) they used in Unit 5.

Common Core State Standards

Focus Clusters
- Represent and solve problems involving addition and subtraction.
- Understand and apply properties of operations and the relationship between addition and subtraction.
- Add and subtract within 20.
- Use place value understanding and properties of operations to add and subtract.

1 Warm Up 15–20 min

Materials

Mental Math and Fluency Children write and compare numbers.	slate	1.NBT.2, 1.NBT.3
Daily Routines Children complete daily routines.	See pages 2–37.	See pages xiv–xvii.

2 Focus 30–35 min

Math Message Children solve a comparison number story.	Animal Cards	1.OA.1, 1.OA.6
Interpreting and Solving Animal Number Stories Children solve a variety of number stories and discuss the strategies and tools they use.	*Math Journal 2,* inside back cover; Animal Cards; slate; counters; base-10 blocks	1.OA.1, 1.OA.6, 1.NBT.4, 1.NBT.6 SMP1, SMP5
✓ **Assessment Check-In** See page 514.		1.NBT.4, SMP5
Adding Three Animal Acrobats Children solve number stories with three addends.	Animal Cards, slate	1.OA.2, 1.OA.3, 1.OA.6 SMP1, SMP5

CCSS 1.NBT.4 Spiral Snapshot

GMC Understand adding 2-digit numbers and 1-digit numbers.

5-11 Focus Practice	5-12 Focus Practice	6-2 Focus Practice	6-7 Focus	6-8 Warm Up Focus	7-1 Warm Up	7-5 Practice	7-9 Warm Up

Spiral Tracker   **Go Online**   to see how mastery develops for all standards within the grade.

3 Practice 15–20 min

Comparing Numbers of Pennies Children find differences by counting up on a number line.	*Math Journal 2,* p. 113	1.OA.4, 1.OA.6, 1.NBT.3
Math Boxes 6-2 Children practice and maintain skills.	*Math Journal 2,* p. 114	See page 515.
Home Link 6-2 **Homework** Children use a picture to write a number story.	*Math Masters,* p. 159	1.OA.1, 1.OA.6, 1.MD.1

connectED.mcgraw-hill.com

Plan your lessons online with these tools.

ePresentations Student Learning Center Facts Workshop Game eToolkit Professional Development Home Connections Spiral Tracker Assessment and Reporting English Learners Support Differentiation Support

CCSS 1.OA.1, SMP4

Readiness
10–15 min

Modeling Number Stories with Counters

| WHOLE CLASS |
| SMALL GROUP |
| PARTNER |
| INDEPENDENT |

counters, slate

To provide experience with different kinds of number stories, have children model them with counters. **GMP4.1** Tell small-number parts-and-total, change-to-more, change-to-less, and comparison stories. For parts-and-total stories, children draw lines to divide their slates, place counters on each side, and then erase the dividers to find the total. For change stories, they model the starts and the changes with counters to find the answer. For comparison stories, children line up the counters side-by-side to find the answers. You may wish to have children tell and model their own stories.

CCSS 1.OA.1, 1.NBT.4, 1.NBT.6, SMP4

Enrichment
10–15 min

Science and Social Studies Number Stories

| WHOLE CLASS |
| SMALL GROUP |
| PARTNER |
| INDEPENDENT |

Activity Card 72, crayons

Science and Social Studies Link To apply children's understanding of number stories, have them work in partnerships to create, illustrate, solve, and discuss number stories. Select a topic from your science or social studies curriculum as a theme for children's number stories. For example, children might draw two types of clouds and then write a number model for the total number of clouds. Or children might write about schools, libraries, and post offices and find the total number of buildings. **GMP4.1**

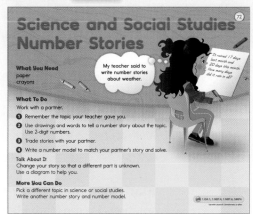

CCSS 1.NBT.4, 1.NBT.6, SMP1

Extra Practice
10–15 min

Playing *Stop and Go*

| WHOLE CLASS |
| SMALL GROUP |
| PARTNER |
| INDEPENDENT |

Math Masters, p. G42 (optional), p. G45; *Stop and Go* Cards; number grid; counters; base-10 blocks

To provide practice adding and subtracting 2-digit numbers, have children play *Stop and Go*. For detailed instructions, see Lesson 5-11.

Observe

• Which children add and subtract multiples of 10 correctly? What strategies do they use?

Discuss

• *How can you check that your answers make sense?* **GMP1.4**

• *If you draw + 6 and then your partner draws − 10, how do you feel? Why?*

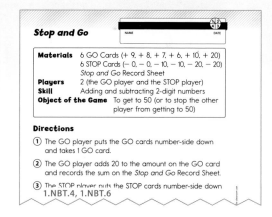

English Language Learners Support

Beginning ELL Build understanding of *same, like,* and *alike*. Show pairs of objects. Point to each object, and model comparing them, pointing to like attributes and thinking aloud: _____ *is like* _____. *They have the same* _____. _____ *and* _____ *are alike because* _____. Give children various objects, and direct them to find two objects that are alike, like each other, or the same. Encourage children to repeat statements in which the terms are used.

Go Online ⟩ **ELL** English Learners Support

Standards and Goals for
Mathematical Practice

SMP1 Make sense of problems and persevere in solving them.
GMP1.6 Compare the strategies you and others use.

SMP5 Use appropriate tools strategically.
GMP5.1 Choose appropriate tools.

1 Warm Up 15–20 min
ePresentations eToolkit

▶ Mental Math and Fluency

Have children record the following pairs of numbers on their slates and compare them using <, >, or =.

● ○ ○ 13, 10 13 > 10 ; 43, 80 43 < 80
● ● ○ 19, 41 19 < 41; 94, 49 94 > 49
● ● ● Write the largest 2-digit number and the smallest 2-digit number. Write <, >, or = to compare. 99 > 10

▶ Daily Routines

Have children complete the Daily Routines. For detailed instructions, see pages 2–37. For specifics on CCSS coverage, see pages xiv–xvii.

NOTE Counting on from larger addends often makes use of the Commutative and Associative Properties of Addition. Children may use these strategies, but should not be expected to know the formal names of the properties.

2 Focus 30–35 min
ePresentations eToolkit

▶ Math Message

Take out your Animal Cards. How much more does the koala weigh than the eagle? Record your answer.

▶ Interpreting and Solving Animal Number Stories

Math Journal 2, inside back cover

| WHOLE CLASS | SMALL GROUP | **PARTNER** | INDEPENDENT |

Math Message Follow-Up Have children share their strategies and solutions for the Math Message. Ask them to compare strategies developed in previous lessons. **GMP1.6** Sample answers: I counted up from the eagle's weight, 15 pounds, to the koala's weight, 19 pounds. My friend counted down from the koala's weight to the eagle's weight. We both found that the koala weighs 4 more pounds than the eagle. Highlight any strategies in which children model the quantities using counters or drawings.

Tell children that today they will use their Animal Cards to create and solve different types of number stories.

> **Differentiate** **Adjusting the Activity**
>
> Provide blank situation diagrams for children to use as needed. Some children may need specific teacher suggestions to get started when solving problems. Go Online Differentiation Support

Tell several types of animal number stories such as those suggested below. Encourage children to begin by writing number models with unknowns on their slates. Some children may wish to represent stories with situation diagrams prior to writing number models. Then encourage children to select tools that can help them efficiently solve the problem and show their thinking, such as base-10 blocks, number grids, or drawings. **GMP5.1** Have children share their thinking with the class and compare their strategies and choices of tools for the problems. **GMP1.6, GMP5.1** Highlight the strategies and tools listed below, or include them for children to analyze if they do not produce them.

Parts-and-total: A cat and a raccoon sit on a scale at the same time. How much do they weigh together?

Possible strategies:

• Counting on from the larger number: I started with the raccoon's weight, which is more, and counted up 7 from 23 to get 30 pounds.
• Composing a 10: I showed the weights with base-10 blocks. I know $3 + 7 = 10$. So I took 3 cubes, or 3 pounds, from the raccoon's weight and added it to the cat's weight to make 10. Then I added the longs, $20 + 10 = 30$, so they weigh 30 pounds all together.

Possible number models:

$7 + 23 = \boxed{30}$, or $23 + 7 = \boxed{30}$

Comparison: The boy stands next to the fox. How much taller is the boy?

Possible strategies:

• Counting up to subtract: I started at 20 on the number grid and counted down 3 rows of 10 to get to 50. The difference is 30, so the boy is 30 inches taller than the fox.
• Subtracting tens: I used my base-10 blocks to show the heights. Then I paired up 2 longs from each group (each long represents 10 inches), which left me with 3 longs. So the boy is 30 inches taller than the fox.

Possible number models:

$20 + \boxed{30} = 50$, or $50 - 20 = \boxed{30}$

Change-to-more: Suppose an eagle picks up an animal in its talons. Now the eagle with the animal in its talons weighs 25 pounds all together. How much does the animal itself weigh? Which animal was it?

Possible strategies:

• Counting up to subtract: I started with the eagle's weight of 15 pounds. I counted up from 15 to 25, which is 10. So the weight that was added is 10 pounds. It's a starfish.
• Counting up by 10s: I marked 15 and 25 on my number grid. I saw that they are one row apart but in the same column; 1 row is 1 ten, so 10 pounds was added to the scale. That's the starfish.

Possible number models:

$15 + \boxed{10} = 25$, or $\boxed{10} = 25 - 15$

First-grade girl	43 in.	41 lb
7-year-old boy	50 in.	50 lb
Cheetah	48 in.	120 lb
Porpoise	72 in.	98 lb
Penguin	36 in.	75 lb
Beaver	30 in.	56 lb
Fox	20 in.	14 lb
Cat	12 in.	7 lb
Raccoon	23 in.	23 lb
Koala	24 in.	19 lb
Eagle	35 in.	15 lb
Rabbit	11 in.	6 lb
Starfish	6 in.	10 lb
Blue crab	4 in.	2 lb
Peacock	60 in.	8 lb
Sun bear	54 in.	110 lb
Toad	5 in.	3 lb
Orangutan	48 in.	165 lb
Parrot	31 in.	3 lb
Squirrel	5 in.	4 lb
Skunk	8 in.	9 lb
Owl	20 in.	5 lb
Flamingo	40 in.	9 lb
Octopus	36 in.	20 lb

Animal Card heights/lengths, and weights

Assessment Check-In 1.NBT.4

As children solve addition problems, observe their strategies and use of tools. GMP5.1 Expect most children to successfully use tools, such as drawings, base-10 blocks, and number grids, to add 2-digit numbers to 1-digit numbers or multiples of 10. If a child is unable to select a tool that will help him or her solve the problem, suggest an appropriate tool and guide usage of the tool to solve the problem accurately.

Assessment and Reporting Go Online to record student progress and to see trajectories toward mastery for these standards.

You may wish to have children work in partnerships, taking turns creating and solving Animal Card number stories. Encourage them to record number models, using situation diagrams as needed.

▶ Adding Three Animal Acrobats

WHOLE CLASS   SMALL GROUP   **PARTNER**   INDEPENDENT

Tell children to imagine that the animals are acting like acrobats in a circus, standing on top of each other in groups of 3. They need to find the total height of these animal acrobats by adding the 3 heights (or lengths) of the animals. Provide the following animal combinations, and ask children to work in partnerships to discuss and compare strategies they can use to combine the lengths. Encourage children to write number models with unknowns on their slates before they solve. Then have them circle the two addends they added first. Highlight strategies such as forming combinations of 10 or looking for doubles. GMP1.6

- Toad, skunk, squirrel 18 inches
- Starfish, blue crab, toad 15 inches
- 2 starfish, skunk 20 inches

Summarize Ask children to think of tools they could use to find the totals if the animal acrobats were taller. GMP5.1 Have children share their suggestions, which are likely to include base-10 blocks or number grids.

Math Journal 2, p. 113

Comparing Numbers of Pennies

Lesson 6-2
DATE

Circle who has more pennies.
Use an addition number model to find the difference.
Example:

Carlos Ⓟ Ⓟ Ⓟ
Ⓛⓨⓝⓝ 0 1 2 3 4 5 6 7 8 9
Ⓟ Ⓟ Ⓟ Ⓟ Ⓟ Ⓟ Ⓟ

Addition Number Model: *3 + ? = 7*
How many more pennies? __4__ pennies

❶ Amy | 13 pennies |

      13 14 15 16 17 18 19 20 21 22
Ⓓⓔⓞⓝ | 21 pennies |

Addition Number Model: 13 + ? = 21
How many more pennies? __8__ pennies

❷ Pete | 7 pennies |

      7 8 9 10 11 12 13 14 15 16
Ⓢⓐⓝⓓⓨ | 15 pennies |

Addition Number Model: 7 + ? = 15
How many more pennies? __8__ pennies

1.NBT.3, 1.OA.4, 1.OA.6   one hundred thirteen   113

③ Practice   15–20 min

Go Online

ePresentations   eToolkit   Home Connections

▶ ## Comparing Numbers of Pennies

Math Journal 2, p. 113

| WHOLE CLASS | SMALL GROUP | PARTNER | **INDEPENDENT** |

Children find differences by counting up on a number line.

▶ ## Math Boxes 6-2

Math Journal 2, p. 114

| WHOLE CLASS | SMALL GROUP | **PARTNER** | INDEPENDENT |

Mixed Practice   Math Boxes 6-2 are paired with Math Boxes 6-5.

▶ ## Home Link 6-2

Math Masters, p. 159

Homework   Children practice writing number stories.

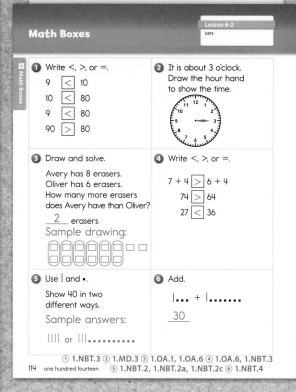

Math Journal 2, p. 114

Math Masters, p. 159

Exploring True and False, Doubles, and Shapes

Overview Children determine whether number sentences are true or false. They practice addition doubles and create shapes with given attributes.

▶ **Before You Begin**

For Exploration A, create one deck of cards per partnership from *Math Masters,* pages 160 and 161. Then cut out 12 equal signs from *Math Masters,* page 162 per partnership (or have children gather the equal sign cards from their number card decks). For Exploration B and the optional Enrichment activity, label one egg carton per partnership with 0–10 in 11 compartments and an extra 5 in the remaining compartment.

Common Core State Standards

Focus Clusters
- Add and subtract within 20.
- Work with addition and subtraction equations.
- Reason with shapes and their attributes.

1 **Warm Up** 15–20 min	**Materials**	
Mental Math and Fluency Children determine whether equations are true or false.		1.OA.6, 1.OA.7
Daily Routines Children complete daily routines.	See pages 2–37.	See pages xiv–xvii.

2 **Focus** 30–35 min		
Math Message Children draw a picture to explain the equal sign.		1.OA.7 SMP2
Explaining the Equal Sign Children share explanations of the equal sign.		1.OA.7
Exploration A: True and False Number Sentences Children create number sentences and determine whether they are true or false.	*Math Masters,* pp. 160–161, 162 (or equal sign cards from number card decks)	1.OA.6, 1.OA.7 SMP1
Exploration B: Adding with Egg Cartons Children practice solving doubles facts.	Activity Card 73; per partnership or small group: numbered egg carton, penny, slate	1.OA.6 SMP6
Exploration C: Geoboard Shapes with Nondefining Attributes Children make shapes with given attributes.	Activity Card 74; *Math Masters,* p. TA16 or TA17; geoboards; rubber bands; Two-Dimensional Shapes Poster (optional)	1.G.1 SMP6

3 **Practice** 15–20 min		
Reviewing the Facts Inventory Record, Parts 1 and 2 Children track which facts they do and do not know.	*Math Journal 2,* pp. 217–221	1.OA.3, 1.OA.6
Math Boxes 6-3 Children practice and maintain skills.	*Math Journal 2,* p. 115	See page 521.
Home Link 6-3 Homework Children draw composite shapes.	*Math Masters,* p. 163	1.NBT.5, 1.G.1, 1.G.2

Differentiation Options RtI

Readiness
10–15 min

| WHOLE CLASS |
| SMALL GROUP |
| **PARTNER** |
| **INDEPENDENT** |

Playing *Roll and Record Doubles*

Math Masters, p. G47, p. G30 (optional); number cards 1–10

To provide experience with addition doubles, have children play a modified version of *Roll and Record Doubles* using number cards 1–10 instead of a die. For detailed instructions, see Lesson 4-7.

Observe
- Which children are automatic with most of their doubles facts?
- With which doubles facts are children not yet automatic?

Discuss
- *Which doubles facts are easiest to remember?* GMP6.4

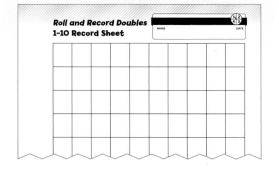

Enrichment
10–15 min

| WHOLE CLASS |
| SMALL GROUP |
| **PARTNER** |
| INDEPENDENT |

Egg Carton Addition with 3 Pennies

Activity Card 75, egg carton, 3 pennies, slate

To further explore finding sums, have children find sums of three addends by shaking 3 pennies inside an egg carton.
GMP6.4

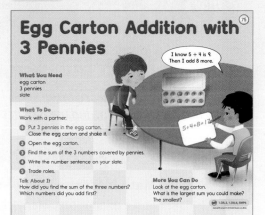

Extra Practice
10–15 min

| WHOLE CLASS |
| SMALL GROUP |
| **PARTNER** |
| INDEPENDENT |

Playing *Addition Top-It*

Math Masters, p. G37 (optional), pp. TA21 and G38; 4 sets of number cards 0–9

Children practice addition facts using relation symbols by playing *Addition Top-It*. For detailed instructions, see Lesson 5-5.

Observe
- What strategies are children using to find the sums? Are they counting all? Counting on? Counting on from the larger addend?

Discuss
- *What can you do if you cannot agree on the sum?* GMP1.4

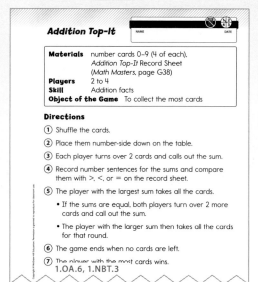

English Language Learners Support

Beginning ELL Provide children with additional experiences hearing and using the words *true* and *false*. Provide them with practice, directing them to use the terms to indicate whether statements are true or false. For example, say: *You are a girl. Juan is ten years old. It is nighttime now. Today is Wednesday.* After each statement, ask: *True or false?* Accept one-word responses, revoicing with complete sentences, such as: *It is false that it is nighttime now.* Ask children to repeat your revoiced statements.

Academic Language Development

Encourage children to articulate their
thinking about the meaning of the
equal sign by using the 4-square
organizer to draw a picture, show an
example and a non-example, and write
a definition in their own words.

---

1 Warm Up   15–20 min   Go Online   ePresentations   eToolkit

▶ Mental Math and Fluency

Have children give thumbs-up if the equation is true and thumbs-down
if it is false.

●○○ $2 = 6 - 3$ False; $8 - 4 = 4$ True
●●○ $19 - 8 = 10$ False; $12 = 3 + 8$ False
●●● $11 - 4 = 7 - 4$ False; $13 + 2 = 10 + 5$ True

▶ Daily Routines

Have children complete the Daily Routines. For detailed instructions,
see pages 2–37. For specifics on CCSS coverage, see pages xiv–xvii.

2 Focus   30–35 min   Go Online   ePresentations   eToolkit

▶ Math Message

*A new child visits our class. He has never seen an equal sign before. Draw an
equal sign, and use pictures or words to explain what it means.* GMP2.1

▶ Explaining the Equal Sign

Math Message Follow-up Allow children to share their drawings
and explanations. Many children will use the equal sign to write true
number sentences. Others may draw pictures to illustrate equivalence.
Review the idea that the equal sign tells us that everything on the left
side "is the same amount as" everything on the right side.

Tell children that they will use their knowledge of the equal sign to
identify true and false number sentences. They will also have
opportunities to practice addition facts and to build shapes.

▶ Exploration A: True and False Number Sentences

Math Masters, pp. 160–162

| WHOLE CLASS | SMALL GROUP | PARTNER | INDEPENDENT |

Model the exploration by drawing the top two cards from the prepared deck (*Math Masters*, pages 160–161) and placing them faceup on either side of an equal-sign card (*Math Masters*, page 162).

Have children read the number sentence and determine whether it is true or false. Encourage them to check whether their answers make sense. **GMP1.4**

Explain that after they make number sentences with their partners, they should separate the true ones and the false ones. When all the cards have been used, children should check each other's work. As time allows, children can shuffle the deck and repeat the activity.

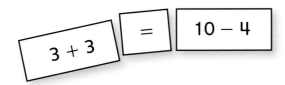

A true number sentence

▶ Exploration B: Adding with Egg Cartons

Activity Card 73

| WHOLE CLASS | SMALL GROUP | PARTNER | INDEPENDENT |

Pairs of children solve doubles facts and record sums on their slates using egg cartons as random-number generators. **GMP6.4**

> **Differentiate** **Adjusting the Activity**
>
> For children who may benefit from initially working with easier sums, use the numbers 0–5. Use larger numbers to provide a challenge for children who are ready to move beyond. **Go Online** 📖 Differentiation Support

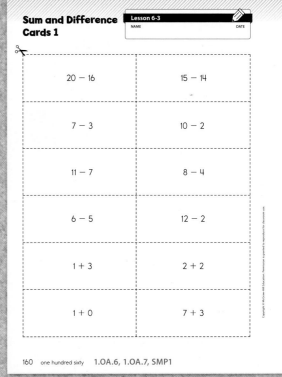

Math Masters, p. 160

Sum and Difference Cards 1 — Lesson 6-3

NAME ___ DATE ___

20 − 16	15 − 14
7 − 3	10 − 2
11 − 7	8 − 4
6 − 5	12 − 2
1 + 3	2 + 2
1 + 0	7 + 3

160   one hundred sixty   1.OA.6, 1.OA.7, SMP1

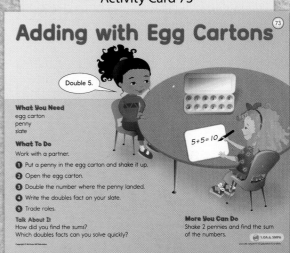

Activity Card 73

Adding with Egg Cartons (73)

"Double 5."

What You Need
egg carton
penny
slate

What To Do
Work with a partner.
1. Put a penny in the egg carton and shake it up.
2. Open the egg carton.
3. Double the number where the penny landed.
4. Write the doubles fact on your slate.
5. Trade roles.

Talk About It
How did you find the sums?
Which doubles facts can you solve quickly?

More You Can Do
Shake 2 pennies and find the sum of the numbers.

5+5=10

Activity Card 74

Geoboard Shapes with Nondefining Attributes

What You Need
5×5 or 7×7 geoboard paper, page TA16 or TA17
a geoboard
rubber bands

What To Do

① Make shapes on your geoboard.

② Carefully record your shapes on geoboard paper.
• Make 2 squares that are different sizes.
• Make 2 triangles that are different sizes.
• Make any shape.
  Make the same shape again with a different color.
• Make a shape with 4 vertices (corners).
• Make a different shape with 4 vertices.
• Make a small shape and record it.
  Turn your geoboard and record what you see now on the same part of the page.

More You Can Do
• Make as many different triangles as you can.
• Make as many differently sized squares as you can.

Talk About It
What is different about the second shape you made with 4 vertices?
Did turning the small shape change what shape it was?

NOTE You may wish to refer children to the Two-Dimensional Shapes Poster as they complete Exploration C.

Math Journal 2, p. 217

My Facts Inventory Record, Part 1			DATE
Addition Fact	Know It	Don't Know It	How I Can Figure It Out...
0 + 1			
7 + 2			
0 + 3			
3 + 2			
0 + 5			
10 + 2			
3 + 1			
2 + 2			
6 + 0			
1 + 1			

1.OA.3, 1.OA.6   two hundred seventeen   217

▶ Exploration C: Geoboard Shapes with Nondefining Attributes

Activity Card 74; *Math Masters,* p. TA16 or TA17

| WHOLE CLASS | **SMALL GROUP** | PARTNER | INDEPENDENT |

Children use rubber bands to make shapes with given attributes on geoboards. They record their shapes on *Math Masters,* page TA16 or TA17. **GMP6.4** Note that, while defining and nondefining attributes are not introduced in this activity, the work children do recognizing that color, position, and size do not define a shape prepares them for Lesson 7-7, where this content is formally introduced.

Summarize Discuss shapes children made. Have children share the different-size squares they made and ask how they knew they were still squares, even though they were different sizes. Sample answer: Both of my squares had 4 equal sides.

③ Practice   15–20 min   Go Online

ePresentations   eToolkit   Home Connections

▶ Reviewing the Facts Inventory Record, Parts 1 and 2

Math Journal 2, pp. 217– 221

| WHOLE CLASS | SMALL GROUP | PARTNER | **INDEPENDENT** |

NOTE After completing them, children should have attached or copied Facts Inventory Record, Parts 1 and 2 from *Math Journal 1,* pages 95–99 into *Math Journal 2,* so they have a complete record of their work in *Math Journal 2.*

Have children review their Facts Inventory Records, Parts 1 and 2 (*Math Journal 2,* pages 217–221). Children update those records as they reassess their facts mastery. Encourage them to revise their strategies for facts they still do not know, using the turn-around rule and other sophisticated strategies more often when applicable. You might expect, for example, that children might change "counting all" to "counting on."

▶ # Math Boxes 6-3 ✏️

Math Journal 2, p. 115

WHOLE CLASS	SMALL GROUP	**PARTNER**	**INDEPENDENT**

Mixed Practice Math Boxes 6-3 are paired with Math Boxes 6-1.

▶ # Home Link 6-3

Math Masters, p. 163

Homework Children practice drawing composite shapes.

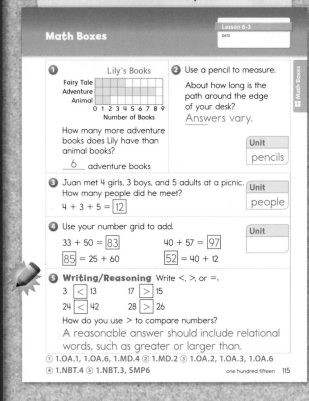

Math Journal 2, p. 115

Math Boxes

Lesson 6-3
DATE

① Lily's Books

Fairy Tale
Adventure
Animal

0 1 2 3 4 5 6 7 8 9
Number of Books

How many more adventure books does Lily have than animal books?

___6___ adventure books

② Use a pencil to measure.

About how long is the path around the edge of your desk?

Answers vary.

Unit
pencils

③ Juan met 4 girls, 3 boys, and 5 adults at a picnic. How many people did he meet?

$4 + 3 + 5 = \boxed{12}$

Unit
people

④ Use your number grid to add.

$33 + 50 = \boxed{83}$ $40 + 57 = \boxed{97}$

$\boxed{85} = 25 + 60$ $\boxed{52} = 40 + 12$

Unit

⑤ **Writing/Reasoning** Write <, >, or =.

3 $\boxed{<}$ 13   17 $\boxed{>}$ 15

24 $\boxed{<}$ 42   28 $\boxed{>}$ 26

How do you use > to compare numbers?

A reasonable answer should include relational words, such as greater or larger than.

① 1.OA.1, 1.OA.6, 1.MD.4 ② 1.MD.2 ③ 1.OA.2, 1.OA.3, 1.OA.6
④ 1.NBT.4 ⑤ 1.NBT.3, SMP6   one hundred fifteen 115

Math Masters, p. 163

Shapes Made from Shapes

Home Link 6-3
NAME   DATE

Family Note

Today your child determined whether number sentences are true or false, practiced addition doubles, and created shapes with given attributes. They will continue this work more formally in future units.

Please return this Home Link to school tomorrow.

① Draw a shape using 2 triangles.
Draw a different shape using 2 triangles.

Sample answers given.

② Draw a shape using a rectangle and a triangle.
Draw another shape using a rectangle and a triangle.

③ How do you know you made a different shape?
Sample answer: My shapes have different numbers of sides.

Practice

④ Find 10 more than each of the numbers.
Think about counting up by 10s to help you.

30 __40__   28 __38__   45 __55__

1.G.1, 1.G.2, 1.NBT.5   one hundred sixty-three 163

Lesson 6-4

Introducing Near Doubles

Overview Children use the near-doubles strategy to solve other addition facts within 20.

▶ **Before You Begin**
For Part 2, select and sequence the following Quick Look Cards: 79, 81, 84, 96, 73, 107, and 78.

▶ **Vocabulary**
near doubles

CCSS Common Core State Standards

Focus Cluster
Add and subtract within 20.

---

1 Warm Up 15–20 min

Materials

Mental Math and Fluency Children solve number stories.		1.OA.1, 1.OA.6
Daily Routines Children complete daily routines.	See pages 2–37.	See pages xiv–xvii.

---

2 Focus 30–35 min

Math Message Children write doubles facts.		1.OA.6
Developing the Near-Doubles Strategy Children use doubles facts to help them find other sums.	Quick Look Cards 73, 78, 79, 81, 84, 96, and 107; ten frame and counters (optional)	1.OA.6 SMP2, SMP3
Using Doubles as Helper Facts Children identify helper facts that they can use to solve near-doubles facts.	Math Journal 2, p. 116	1.OA.6 SMP2, SMP8
✓ **Assessment Check-In** See page 526.	Math Journal 2, p. 116	1.OA.6

CCSS 1.OA.6 Spiral Snapshot

GMC Add and subtract within 20 using strategies.

| 6-1
Warm Up
Practice | 6-2
Focus
Practice | 6-3
Warm Up
Focus
Practice | 6-4
Warm Up
Focus
Practice | 6-5
Focus
Practice | 6-6
Warm Up
Focus
Practice | 6-7
Practice | 6-8
Focus
Practice |

🖩 **Spiral Tracker**   **Go Online**   to see how mastery develops for all standards within the grade.

---

3 Practice 15–20 min

Practicing Subtraction Facts Children solve subtraction facts using strategies.	Math Journal 2, p. 117	1.OA.3, 1.OA.6, 1.OA.8
Math Boxes 6-4 Children practice and maintain skills.	Math Journal 2, p. 118	See page 527.
Home Link 6-4 **Homework** Children solve addition facts and record helper facts that they used.	Math Masters, p. 164	1.OA.6, 1.OA.7 SMP8

---

connectED.mcgraw-hill.com

Plan your lessons online with these tools.

ePresentations   Student Learning Center   Facts Workshop Game   eToolkit   Professional Development   Home Connections   Spiral Tracker   Assessment and Reporting   English Learners Support   Differentiation Support

Differentiation Options RtI

CCSS 1.OA.6, SMP2	**CCSS** 1.OA.6, SMP2	**CCSS** 1.OA.6, SMP6

Readiness   10–15 min

WHOLE CLASS
SMALL GROUP
PARTNER
INDEPENDENT

Exploring Doubles

number cards 1–10,
counters, slate

To explore doubles using a concrete
model, have children represent doubles
facts using counters. **GMP2.1** Children
shuffle the number cards and place
the deck facedown. One child selects
a number card and uses that number of
counters to make a row. The other child
uses the same number of counters to
make a row beneath the first row. Each
child records the doubles fact represented
by the counters. Repeat this activity as
time allows.

Enrichment   10–15 min

WHOLE CLASS
SMALL GROUP
PARTNER
INDEPENDENT

Drawing Doubles Pictures

Activity Card 76, crayons

To further explore doubles
facts, have children illustrate and write
about doubles they see in everyday
situations. Children's doubles may include
a variety of situations, such as "2 front
wheels plus 2 back wheels equals 4 wheels
on a car" and "8 crayons in the front row
and 8 crayons in the back row equals
16 crayons in a box." **GMP2.1**

Go Online for more examples of doubles
in everyday situations.

Extra Practice   10–15 min

WHOLE CLASS
SMALL GROUP
PARTNER
INDEPENDENT

Playing *Roll and Record Doubles*

Math Masters, p. G47,
p. G30 (optional);
per partnership: 1 set of number cards 1–10

Have children practice addition doubles by
playing *Roll and Record Doubles,* this time
with number cards 1–10 instead of a die.
For detailed instructions, see Lesson 4-7.

Observe
- Which children know their doubles
  facts automatically?
- What strategies do children use to
  solve the facts they do not know?

Discuss
- *Why is it helpful to know doubles facts?*
  GMP6.4

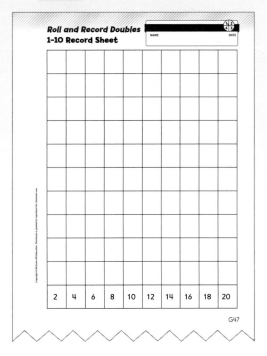

English Language Learners Support

Beginning ELL Use examples and non-examples to demonstrate the meaning of *near.*
Follow with Total Physical Response commands, such as *Stand near the door* and *Put the
pencil near the book* to reinforce children's association of the term with physical position.
Have children identify numbers that are physically near a given number on the number line
in preparation for the discussion about *near doubles.*

Go Online   **ELL** English Learners
Support

Standards and Goals for
Mathematical Practice

SMP2 **Reason abstractly and quantitatively.**
GMP2.2 Make sense of the representations you and others use.

SMP3 **Construct viable arguments and critique the reasoning of others.**
GMP3.2 Make sense of others' mathematical thinking.

SMP8 **Look for and express regularity in repeated reasoning.**
GMP8.1 Create and justify rules, shortcuts, and generalizations.

Professional Development

This lesson continues to allow children to develop their own fact strategies in order to promote basic fact fluency. However, as in other lessons, not all children will spontaneously think of the strategies in this lesson. Using the suggested double ten frames should help elicit such strategies from the class.

---

1 Warm Up 15–20 min | Go Online | ePresentations | eToolkit

▶ Mental Math and Fluency

Pose number stories, and have children share their solutions and strategies orally.

- ●○○ Ben has 5 bottles of orange juice and 5 bottles of apple juice. How many bottles of juice does he have in all? 10 bottles
- ●●○ There are 7 days in the first week of February, and 7 days in the second week of February. How many days are there in the first two weeks of February? 14 days
- ●●● There are 9 players on the home team and 9 players on the visiting team at a baseball game. How many players are there in all? 18 players

▶ Daily Routines

Have children complete the Daily Routines. For detailed instructions, see pages 2–37. For specifics on CCSS coverage, see pages xiv–xvii.

---

2 Focus 30–35 min | Go Online | ePresentations | eToolkit

▶ Math Message

Write 5 doubles facts that you remember from the Adding with Egg Cartons Exploration. Write other doubles you know.

▶ Developing the Near-Doubles Strategy

WHOLE CLASS | SMALL GROUP | PARTNER | INDEPENDENT

Math Message Follow-Up Ask children to name the doubles facts they remembered from the previous Explorations lesson. List doubles facts in a column until you have written all of the facts from $1 + 1 = 2$ through $10 + 10 = 20$. Ask: *What do all doubles facts have in common?* Both addends are the same. Tell children that today they will learn how to use doubles facts to solve other facts.

Display the following two Quick Look Cards in this order: 79, 81. Each time, flash the image for about 2–3 seconds.

NOTE Remind children that Quick Looks are easier to do if they can recognize dot patterns. Tell children to remember or to record how many dots they see all together.

After the second image (Quick Looks Card 81), ask children to share what they saw and how they saw it. GMP2.2 Allow children to share multiple strategies, but highlight starting from the double $4 + 4 = 8$ and adding 1 more dot to make 9. Emphasize this strategy by following up with questions such as *Did everyone understand Damon's strategy?* and *Can someone else explain it again for us?* GMP3.2 Prompt the class to apply it by asking *Can you use Damon's strategy on this next one?* GMP3.2 Repeat this process for the next fact sequence, using Quick Look Cards 84, 96.

Next try displaying images close to doubles *without* first displaying doubles. Use Quick Look Cards 73, 107, and 78. Emphasize using a double and adding or removing a dot to find the new total.

Differentiate **Adjusting the Activity**

Children may need ten frames and counters to relate Quick Looks to the doubles they are near. After they display the doubles on the ten frame, ask them to describe it with the sentence frame "This fact is a double: _____ + _____ = _____." Then have them add or remove a counter to show the Quick Look dot pattern and describe it using the sentence frame "This fact is near a double: _____ + _____ = _____."

Go Online Differentiation Support

▶ # Using Doubles as Helper Facts

Math Journal 2, p. 116

| WHOLE CLASS | SMALL GROUP | PARTNER | INDEPENDENT |

Ask: *What did many of the strategies for the Quick Looks have in common?* They used doubles to help figure out the new sum. Remind children that helper facts are facts that they can use to figure out other facts, such as using doubles facts to figure out facts that are close to doubles. This strategy is called **near doubles** because it uses doubles to help find the sums of nearby facts. GMP8.1 Display Quick Look Card 81 again, and have children identify the helper facts they used. Sample answers: $4 + 4 = 8$; $5 + 5 = 10$

Academic Language Development Have half of the class complete a 4-Square Organizer for the term *near doubles* and the other half complete a 4-Square Organizer for the term *doubles*. Have pairs of children compare and contrast the two different organizers.

Math Journal 2, p. 116

Using Helper Doubles Facts Lesson 6-4
DATE

Record the two doubles facts you could use to find the total. Then write the fact shown on the double ten frame.

1. Helper Doubles Fact: $5 + 5 = 10$ or $6 + 6 = 12$
Fact: $5 + 6 = 11$

2. Helper Doubles Fact: $2 + 2 = 4$ or $3 + 3 = 6$
Fact: $2 + 3 = 5$

3. Helper Doubles Fact: $6 + 6 = 12$ or $7 + 7 = 14$
Fact: $6 + 7 = 13$

4. Helper Doubles Fact: $3 + 3 = 6$ or $4 + 4 = 8$
Fact: $4 + 3 = 7$

5. Explain how you chose a helper fact for Problem 4.
Sample answer: I pretended there were 4 in each ten frame, and I know $4 + 4 = 8$. But one was missing, so this shows $4 + 3 = 7$.

116 one hundred sixteen   1.OA.6, SMP2, SMP8

For each task on journal page 116, have children work in partnerships to record both nearby doubles helper facts that could be used to solve the fact, as well as the fact itself. **GMP2.2** Have children discuss their responses. Have those who started from the double $4 + 4 = 8$ and removed 1 dot in the last task share their thinking, so children see that doubles can be used to solve facts above or below them.

✓ Assessment Check-In CCSS 1.OA.6

Math Journal 2, p. 116

Observe as children identify and record doubles and nearby facts on journal page, 116. Expect most children to use a doubles fact to help them solve each addition fact. Help children who focus on counting individual dots (as opposed to using a double) recall how they found the totals during the Quick Looks activity. Help them recognize a helper double by covering up the extra dot or adding an additional dot.

✅ Assessment and Reporting   (Go Online) to record student progress and to see trajectories toward mastery for these standards.

Summarize Ask children to identify the fact strategies that they learned today. Expect them to suggest using helper facts and the near-doubles strategy. Make sure that using helper facts is on the Strategy Wall, and add an illustration for near doubles.

Math Journal 2, p. 117

Practicing Subtraction Facts

Lesson 6-4
DATE

Solve.

1. $\underline{10} - 5 = 5$ $10 = \underline{12} - 2$ $8 = 16 - \underline{8}$

2. $\underline{14} - 4 = 10$ $10 = 19 - \underline{9}$ $\underline{7} = 14 - 7$

3.
$$\begin{array}{r} 18 \\ -\ 3 \\ \hline 15 \end{array} \qquad \begin{array}{r} 13 \\ -\ 4 \\ \hline 9 \end{array} \qquad \begin{array}{r} 8 \\ -\ 4 \\ \hline 4 \end{array}$$

4. $17 - \underline{7} = 10$ $12 - 6 = \underline{6}$ $8 - \underline{0} = 8$

5. $10 - \underline{0} = 10$ $2 = 11 - \underline{9}$ $19 - \underline{17} = 2$

Try This

6. $999 - 0 = \underline{999}$

7. Explain how you solved Problem 6.
    <u>Sample answer: I know that subtracting 0</u>
    <u>does not change a number. So subtracting 0</u>
    <u>from 999 is 999.</u>

1.OA.3, 1.OA.6, 1.OA.8   one hundred seventeen   117

3 **Practice** 15–20 min | Go Online |

ePresentations eToolkit Home Connections

▶ **Practicing Subtraction Facts**

Math Journal 2, p. 117

| WHOLE CLASS | SMALL GROUP | **PARTNER** | **INDEPENDENT** |

Have children practice solving subtraction facts. Encourage them to use any strategy that makes sense to them.

▶ **Math Boxes 6-4**

Math Journal 2, p. 118

| WHOLE CLASS | SMALL GROUP | **PARTNER** | **INDEPENDENT** |

Mixed Practice Math Boxes 6-4 are paired with Math Boxes 6-6.

▶ **Home Link 6-4**

Math Masters, p. 164

Homework Children practice solving addition facts using helper facts.

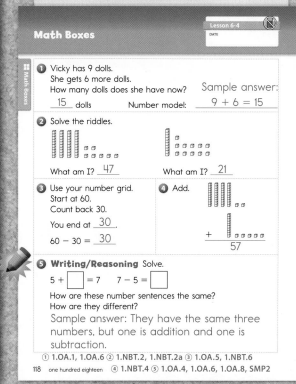

Math Journal 2, p. 118

Math Boxes

Lesson 6-4
DATE

① Vicky has 9 dolls.
She gets 6 more dolls.
How many dolls does she have now?
__15__ dolls   Number model: Sample answer: __9 + 6 = 15__

② Solve the riddles.

What am I? __47__   What am I? __21__

③ Use your number grid.
Start at 60.
Count back 30.
You end at __30__.
60 − 30 = __30__

④ Add.

57

⑤ **Writing/Reasoning** Solve.
5 + ☐ = 7   7 − 5 = ☐
How are these number sentences the same?
How are they different?
Sample answer: They have the same three numbers, but one is addition and one is subtraction.

① 1.OA.1, 1.OA.6 ② 1.NBT.2, 1.NBT.2a ③ 1.OA.5, 1.NBT.6
118 one hundred eighteen ④ 1.NBT.4 ⑤ 1.OA.4, 1.OA.6, 1.OA.8, SMP2

Math Masters, p. 164

Near Doubles

Home Link 6-4
NAME   DATE

Family Note

Today your child learned how to use a doubles fact, such as 8 + 8 = 16, to help solve facts close to doubles, such as 8 + 9 = 17. This strategy is called *near doubles*. Children learned about helper facts (facts that are useful for solving other facts) in Unit 4 and will continue to learn about other helper facts as the year progresses.

Please return this Home Link to school tomorrow.

① Write a helper fact and then the final answer for each number sentence in the table below.

Fact	Helper Fact	Answer
Example: 5 + 6 = ?	5 + 5 = 10 *or* 6 + 6 = 12	5 + 6 = 11
3 + 4 = ?	Sample answer: 3 + 3 = 6	3 + 4 = 7
5 + 4 = ?	Sample answer: 5 + 5 = 10	5 + 4 = 9
7 + 8 = ?	Sample answer: 7 + 7 = 14	7 + 8 = 15

Practice

② Circle the **true** number sentences.

(3 = 3)   (4 = 9 − 5)   6 = 3 + 2
4 = 7 − 2   (10 + 2 = 12)   4 + 9 = 12

164 one hundred sixty-four   1.OA.6, 1.OA.7, SMP8

Recording Near-Doubles Strategies

Overview Children solve facts within 20 and represent their solution strategies with pictures, words, and symbols.

▶ **Before You Begin**
For Part 2, select and sequence the following Quick Look Cards: 100, 107, 83, 112, 103.

Common Core State Standards

Focus Clusters
- Represent and solve problems involving addition and subtraction.
- Add and subtract within 20.

① Warm Up   15–20 min

	Materials	
Mental Math and Fluency Children tell time on an hour-hand-only clock.	hour-hand-only clock	1.MD.3
Daily Routines Children complete daily routines.	See pages 2–37.	See pages xiv–xvii.

② Focus   35–40 min

Math Message Children write two doubles facts that can help them solve a nearby fact.		1.OA.6
Representing Fact Strategies Children practice representing their strategies for solving facts.	Quick Look Cards 83, 100, 107, and 112; slate	1.OA.6 SMP2, SMP6
Assessment Check-In See page 531.		SMP2
Using Near Doubles to Solve Stories Children solve a number story and explain their strategy with pictures, words, or symbols.	*Math Journal 2,* p. 119; Quick Look Card 103 (optional)	1.OA.1, 1.OA.6 SMP2, SMP6

CCSS 1.OA.6   **Spiral Snapshot**

GMC   Add and subtract within 20 using strategies.

6-1 Warm Up Practice	6-2 through 6-4 Warm Up Focus Practice	6-5 Focus Practice	6-6 Warm Up Focus Practice	6-7 Practice	6-8 Focus Practice	6-9 Focus Practice

Spiral Tracker   **Go Online** ▷   to see how mastery develops for all standards within the grade.

③ Practice   10–15 min

Math Boxes 6-5 Children practice and maintain skills.	*Math Journal 2,* p. 120	See page 533.
Home Link 6-5 **Homework** Children solve number stories using a near-doubles strategy.	*Math Masters,* p. 166	1.OA.1, 1.OA.6, 1.OA.8 SMP6

connectED.mcgraw-hill.com ▷

Plan your lessons online with these tools.

 ePresentations   Student Learning Center   Facts Workshop Game   eToolkit   Professional Development   Home Connections   Spiral Tracker   Assessment and Reporting   English Learners Support   Differentiation Support

Differentiation Options RtI

 1.OA.6, SMP1, SMP2

Readiness   5–10 min

Domino Doubles

| WHOLE CLASS |
| SMALL GROUP |
| **PARTNER** |
| INDEPENDENT |

doubles dominoes only

To provide concrete experience with doubles, have children double the dots on dominoes. Display a domino with one side covered. Ask children how many dots are shown. Then have them double the number and identify the total. Reveal the other half of the domino. Have children check their answer by finding the total number of dots. **GMP1.4** Then help them write number sentences to represent each domino. **GMP2.1**

 1.OA.6, SMP1

Enrichment   10–15 min

Recording Near Doubles

| WHOLE CLASS |
| **SMALL GROUP** |
| PARTNER |
| INDEPENDENT |

Activity Card 77;
Math Masters, p. 165;
number cards 2–10

To extend children's understanding of the near-doubles strategy, have them record doubles and nearby facts. Remind children that nearby facts could be 1 or 2 away from a double. For example, $5 + 6$ and $5 + 7$ are near $5 + 5$. Children solve and check their facts. **GMP1.4**

Recording Near Doubles ⑦⑦

What You Need
Near Doubles Record Sheet, page 165
number cards 2–10

What facts are near
$6 + 6 = 12$?

What To Do
Work with a partner.
① Mix the cards and place them number-side down on the table.
② Draw a card.
③ Use the number on the card to make a double. Write down the double.
④ Write 4 facts that are near that doubles fact.
⑤ Have your partner solve the new facts you wrote.
⑥ Take turns until the page is complete.

Talk About It
Check your answers.

More You Can Do
Pick a teen number. Use that number to make a double. Record and solve several addition problems that are near that doubles fact.

1.OA.6, SMP1

Near Doubles Record Sheet   Lesson 6-5
NAME   DATE

Answers vary.

① Double: ____ + ____ = ____
  Nearby Fact: ____ + ____ = ____
  Nearby Fact: ____ + ____ = ____
  Nearby Fact: ____ + ____ = ____
  Nearby Fact: ____ + ____ = ____

② Double: ____ + ____ = ____
  Nearby Fact: ____ + ____ = ____
  Nearby Fact: ____ + ____ = ____
  Nearby Fact: ____ + ____ = ____
  Nearby Fact: ____ + ____ = ____

1.OA.6, SMP1

CCSS **1.OA.6, 1.NBT.3, SMP2**

Extra Practice   10–15 min

Playing *Domino Top-It*

| WHOLE CLASS |
| SMALL GROUP |
| **PARTNER** |
| INDEPENDENT |

Math Masters, p. G28 (optional), p. G29

For practice finding and comparing sums, have children play *Domino Top-It.* For each round, have them record number models using $>$, $<$, or $=$ and check each other's work. For detailed instructions, see Lesson 3-1.

Observe
- Which children are correctly using $>$, $<$, and $=$?
- Which children can easily determine the total on the domino?

Discuss
- *How do your number models represent the dots on your dominoes?* **GMP2.1**

Domino Top-It   NAME   DATE

Materials	1 set of dominoes
Players	2
Skill	Finding and comparing sums
Object of the Game	To collect more dominoes

Directions
① Place all of the dominoes facedown.
② Each player turns over a domino. Each player says the total number of dots.
③ The player with the larger total takes both dominoes. If the totals are the same, each player turns over another domino. The player with the larger total takes all the dominoes from that round.
④ The game is over when all of the dominoes have been turned over.
⑤ The player with more dominoes wins.

1.OA.6, 1.NBT.3

English Language Learners Support

Beginning ELL Help children understand the verb *record* by thinking aloud using the term with children's journals or drawings. Point out children's work displayed in the classroom as ways they recorded lessons, for example: *This is how you recorded your work in Lesson 3. Show me how you recorded some shapes.*

[Go Online] **ELL** English Learners Support

SMP2 Reason abstractly and quantitatively.

GMP2.1 Create mathematical representations using numbers, words, pictures, symbols, gestures, tables, graphs, and concrete objects.

GMP2.3 Make connections between representations.

SMP6 Attend to precision.

GMP6.1 Explain your mathematical thinking clearly and precisely.

1 Warm Up 15–20 min Go Online

ePresentations eToolkit

▶ Mental Math and Fluency

Display the following times on the hour-hand-only demonstration clock. Have children tell the approximate time.

● ○ ○ 1 o'clock; 10 o'clock; 8 o'clock

● ● ○ just after 6 o'clock; just before 3 o'clock

● ● ● between 4 and 5 o'clock; between 11 and 12 o'clock

▶ Daily Routines

Have children complete the Daily Routines. For detailed instructions, see pages 2–37. For specifics on CCSS coverage, see pages xiv–xvii.

2 Focus 35–40 min Go Online

ePresentations eToolkit

▶ Math Message

Record two different helper facts that you could use to solve 5 + 6. How can you use the doubles as helper facts?

▶ Representing Facts Strategies

| WHOLE CLASS | SMALL GROUP | PARTNER | INDEPENDENT |

Math Message Follow-Up As children share their ideas for the Math Message, display Quick Look Card 100 to illustrate how their helper facts relate to 5 + 6. Expect the class to understand that they can use 5 + 5 or 6 + 6 and then add or subtract 1 as appropriate.

Tell children that today they will practice recording their thinking. Display Quick Look Card 107. Ask children to record on slates one way to find the total, using words or numbers. GMP2.1, GMP6.1 Highlight a few representations during the discussion. In particular, elicit a verbal representation, such as *I saw 6 and 6 and then 1 more dot, so that makes 13*, and a more symbolic representation, such as *6 + 6 + 1*. Help children make connections between the verbal and symbolic descriptions of their near-doubles strategies by displaying the representations side-by-side. GMP2.3 It is not necessary for children to develop purely symbolic representations of their thinking in first grade, but they should develop ways to express their thinking using pictures, words, or numbers.

Adjusting the Activity

Differentiate Some children may describe their work more generally, such as saying, "I counted." Prompt them to identify a doubles helper fact within the image and use that fact in their strategies and representations.

Go Online   Differentiation Support

Repeat this process with Quick Look Card 83, having children again record their thinking on their slates. Sample strategies and representations include the following:

- I know 2 rows would be $5 + 5$, but one is missing, so I took away 1 to make 9; $5 + 5 - 1 = 9$.
- I know $4 + 4$ makes 8, but I have one more, so it is 9; $4 + 4 + 1 = 9$.

Have children compare their strategies and representations, noting which helper fact was used and connecting back to the visual image on the Quick Look Card. GMP2.3 Repeat with Quick Look Card 112, helping children recognize that they could add 2 to the double $6 + 6$, subtract 2 from the double $8 + 8$, or move one dot from the 8 image to the 6 image to make $7 + 7$.

 Assessment Check-In CCSS SMP2

Circulate and observe as children work. Expect that most children will be more comfortable describing their thinking in words rather than symbols. GMP2.1 For children who struggle to make any representation, have them describe their thinking verbally to you first, and then help them record what they said.

 Assessment and Reporting ⟨Go Online⟩ to record student progress for this standard.

Academic Language Development

Show children the word *record* in writing. Give examples of the term used as a verb and as a noun, using actions and visuals to accompany statements. For example: *You will record your thinking. Save a written record of your work.* Ask children to think of different words that could be used instead of *record* as a way to help them construct the meanings of the term used as a verb and as a noun. Point out the difference in pronunciation if children don't notice it.

Using Near Doubles to Solve Stories

Lesson 6-5

DATE

Solve two different ways.
Record your strategies using words and numbers.

I have 5 pencils in my desk and 7 pencils in my backpack.
How many pencils do I have all together?

<u>12</u> pencils

First way

Helper fact: _____ + _____ = _____

Tell how you used the helper fact to find the answer
to the number story.

Answers vary. See sample answers in the
Teacher's Lesson Guide.

Second way

Helper fact: _____ + _____ = _____

Tell how you used the helper fact to find the answer
to the number story.

1.OA.1, 1.OA.6, SMP2, SMP6   one hundred nineteen   119

► Using Near Doubles to Solve Stories

Math Journal 2, p. 119

| **WHOLE CLASS** | SMALL GROUP | **PARTNER** | INDEPENDENT |

Tell children that they will now practice recognizing when to use the near-doubles strategy to solve number stories. Remind them that *near* means *close to* a double. Present the following number story:

> *I have 5 pencils in my desk and 7 pencils in my backpack.*
> *How many pencils do I have all together?*

Ask children to solve this problem using two different helper facts. They should record their thinking in words or number sentences on journal page 119. **GMP2.1, GMP6.1** If children are having difficulty recognizing the helper facts, display Quick Look Card 103 or have the class suggest doubles and record those for children before they begin.

Afterward have them compare and discuss responses, such as:

- I know $7 + 7$ is 14, so $7 + 5$ is 2 less, or 12. So, $7 + 7 - 2 = 12$.
- I know $5 + 5 = 10$, so I go up 2 to make 12. So, $5 + 5 + 2 = 12$.
- I took 1 from the 7 and gave it to the 5, so I had $6 + 6$, which is 12.

Make sure children fully explain their strategies by asking questions, such as: *How did you know to add 2 more?* **GMP6.1** Encourage children to revise their work to show their thinking as clearly as they can.

Summarize Display $7 + 8$. Invite children to share two different doubles they can use as helper facts. $7 + 7$ and $8 + 8$ Have them explain how to find the sum using these doubles. Sample answer: I could start with $7 + 7 = 14$, and add 1 to get 15. I could also start with $8 + 8 = 16$, and subtract 1 to get 15. **GMP6.1**

③ **Practice** 10–15 min

Go Online

ePresentations eToolkit Home Connections

▶ **Math Boxes 6-5**

Math Journal 2, p. 120

| WHOLE CLASS | SMALL GROUP | **PARTNER** | **INDEPENDENT** |

Mixed Practice Math Boxes 6-5 are paired with Math Boxes 6-2.

▶ **Home Link 6-5**

Math Masters, p. 166

Homework Children practice solving a number story using near doubles.

Math Journal 2, p. 120

Math Boxes

Lesson 6-5
DATE

① Write <, >, or =.

$24 \boxed{<} 30$

$21 \boxed{>} 12$

$5 + 5 \boxed{=} 10$

$16 \boxed{>} 4 + 6$

② Draw the hour hand to show that the time is a little after 8 o'clock.

③ Draw and solve.

Kristen has 3 stars.
Bella has 10 stars.
How many fewer stars does Kristen have than Bella?

___7___ fewer stars

Sample drawings:
☆☆☆
☆☆⚹⚹⚹⚹⚹⚹⚹
///// //

④ Circle the larger sum.

(11 + 5)

12 + 3

Unit
dolls

Write a number sentence using > or < to show which is larger.

___11 + 5 > 12 + 3___

⑤ Use Ⓓ and Ⓟ.

Show 13 cents two different ways.

Ⓓ Ⓟ Ⓟ Ⓟ;

Ⓟ Ⓟ Ⓟ Ⓟ Ⓟ Ⓟ Ⓟ Ⓟ
Ⓟ Ⓟ Ⓟ

⑥ Add.

📏...... + 📏...... = ___40___

Use | and . to show the sum.

||||

① 1.OA.6, 1.NBT.3 ② 1.MD.3 ③ 1.OA.1, 1.OA.6 ④ 1.NBT.3, 1.NBT.4
120 one hundred twenty ⑤ 1.NBT.2, 1.NBT.2a, 1.NBT.2b ⑥ 1.NBT.2, 1.NBT.2a, 1.NBT.4

Math Masters, p. 166

Recording Near-Doubles Strategies

Home Link 6-5
NAME   DATE

Family Note

Today your child spent more time using doubles facts to help solve nearby facts called *near doubles*. Children focused on explaining their solution strategies using words, pictures, or number sentences. Ask your child to explain how he or she solved the number stories below either with words or with a written number sentence.

Please return this Home Link to school tomorrow.

Solve the number stories.

① Tommy had 4 pretzels.
His mom gave him 4 more pretzels.
How many pretzels does Tommy have now?

___8___ pretzels

② Renee had 4 pretzels.
Her mom gave her 5 more pretzels.
How many pretzels does Renee have now?

___9___ pretzels

③ How can you use the first number story to help you solve the second number story?

Sample answer: I know 4 + 4 = 8, so 1 more is 9.

Practice

④ Complete each number sentence.

$3 + 3 = \underline{6}$ $8 + \underline{2} = 10$ $10 = 1 + \underline{9}$

166 one hundred sixty-six   1.OA.1, 1.OA.6, 1.OA.8, SMP2, SMP6

Lesson 6-6

Introducing Making 10

Overview Children learn the making-10 strategy for adding and subtracting within 20.

▶ **Before You Begin**
For Part 1, select and sequence Quick Look Cards 92, 88, and 89. For Part 2, select and sequence Quick Look Cards 98, 95, 114, and 116. For the optional Enrichment activity, cut out one set of Making 100 Cards (*Math Masters,* page 167) for each partnership.

▶ **Vocabulary**
making 10

 Common Core State Standards

Focus Clusters
- Understand and apply properties of operations and the relationship between addition and subtraction.
- Add and subtract within 20.
- Understand place value.

① Warm Up 15–20 min

	Materials	
Mental Math and Fluency Children identify the total number of dots on ten frames.	Quick Look Cards 88, 89, and 92	1.OA.6 SMP2, SMP6
Daily Routines Children complete daily routines.	See pages 2–37.	See pages xiv–xvii.

② Focus 35–40 min

Math Message Children record how they would show numbers on a double ten frame.	slate	1.OA.6
Developing the Making-10 Strategy Children do Quick Looks with double ten frames, focusing on making 10.	*Math Journal 1,* Activity Sheet 3; *Math Masters,* p. TA19; Quick Look Cards 95, 98, 114, and 116; 20 counters	1.OA.3, 1.OA.6, 1.NBT.2, 1.NBT.2b SMP3, SMP8
Recording Making 10 Children represent the making-10 strategy.	Quick Look Card 116, slate	1.OA.3, 1.OA.6 SMP2
✓ **Assessment Check-In** See page 537.	*Math Journal 1,* Activity Sheet 3 (optional); counters (optional)	1.OA.6

CCSS 1.OA.6 **Spiral Snapshot**

GMC Add and subtract within 20 using strategies.

6-1 Warm Up Practice	6-2 through 6-5 Warm Up Focus Practice	6-6 Warm Up Focus Practice	6-7 Practice	6-8 Focus Practice	6-9 Focus Practice	6-10 Warm Up Practice

 Spiral Tracker ⟨ Go Online ⟩ to see how mastery develops for all standards within the grade.

③ Practice 10–15 min

Facts Inventory Record, Part 3 Children track facts they know and facts they are still learning.	*Math Journal 2,* p. 222	1.OA.3, 1.OA.6
Math Boxes 6-6 Children practice and maintain skills.	*Math Journal 2,* p. 121	See page 539.
Home Link 6-6 **Homework** Children find sums to complete a picture.	*Math Masters,* p. 168	1.OA.6, 1.OA.8

connectED.mcgraw-hill.com ⟩

Plan your lessons online with these tools.

 ePresentations   Student Learning Center   Facts Workshop Game   eToolkit   Professional Development   Home Connections   Spiral Tracker   Assessment and Reporting   English Learners Support   Differentiation Support

CCSS 1.OA.6, SMP1

Readiness   10–15 min

Playing *Fishing for 10*

	WHOLE CLASS
	SMALL GROUP
	PARTNER
	INDEPENDENT

Math Masters, p. G32 (optional),
p. G33; 4 sets of number
cards 0–10

To provide additional exposure to making
combinations of 10, have children play
Fishing for 10. For detailed instructions, see
Lesson 4-9.

Observe

- Which children easily determine
  the number needed to make
  a combination of 10?
- With which combinations of 10 do
  children struggle?

Discuss

- *How did you figure out which card to
  ask for?*
- *How did you and your partner check
  each other's combinations of 10?*
  GMP1.4

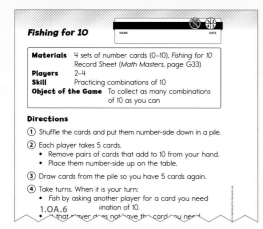

Fishing for 10

Materials	4 sets of number cards (0–10), *Fishing for 10* Record Sheet (*Math Masters*, page G33)
Players	2–4
Skill	Practicing combinations of 10
Object of the Game	To collect as many combinations of 10 as you can

Directions

① Shuffle the cards and put them number-side down in a pile.

② Each player takes 5 cards.
- Remove pairs of cards that add to 10 from your hand.
- Place them number-side up on the table.

③ Draw cards from the pile so you have 5 cards again.

④ Take turns. When it is your turn:
- *Fish* by asking another player for a card you need ... ination of 10.
- ... that player does not have the card you need ...

1.OA.6

CCSS 1.OA.6, 1.NBT.4, SMP2

Enrichment   10–15 min

Making 100

	WHOLE CLASS
	SMALL GROUP
	PARTNER
	INDEPENDENT

Activity Card 78;
Math Masters, p. 167;
base-10 longs

To extend the making-10 strategy, have
children try a "making-100" strategy.
Children decompose one addend to make
a combination of 100, similar to how
they make 10 with addition facts. Provide
base-10 longs as needed, and have
children record their number sentences
on a half-sheet of paper. GMP2.1

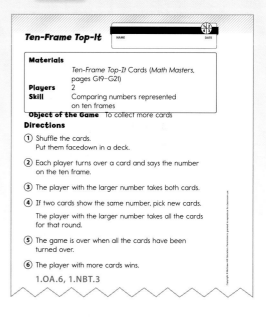

Making 100   78

What You Need
Making 100 Cards, page 167
base-10 longs
paper

If you split 50 into 20 and 30, then you can add 80 + 20 first to make 100.

What To Do
Work with a partner.

① Mix the cards and place them facedown.

② Draw 2 cards.

③ Use a "making 100" strategy to add the numbers. You can use base-10 longs to help you.

④ Write number sentences to match.

⑤ Repeat Steps 2–4.

Talk About It
What other strategies could you use to solve?

More You Can Do
Pick one or two number sentences.
Write how you made 100 to solve.

1.OA.6, 1.NBT.4, SMP2

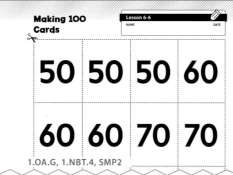

Making 100 Cards	**Lesson 6-6** NAME DATE

50	50	50	60
60	60	70	70

1.OA.G, 1.NBT.4, SMP2

CCSS 1.OA.6, 1.NBT.3, SMP2

Extra Practice   10–15 min

Playing *Ten-Frame Top-It*

	WHOLE CLASS
	SMALL GROUP
	PARTNER
	INDEPENDENT

Math Masters, p. G18 (optional);
pp. G19–G21

To provide practice comparing numbers
represented on ten frames, have children
play *Ten Frame Top-It.* For detailed
instructions, see Lesson 2-2.

Observe

- Which children quickly recognize the
  quantities shown on the ten frames?

Discuss

- *How can you tell what number is shown
  on your ten frame? Can you tell how
  many more you would need to make 10?*
  GMP2.2

Ten-Frame Top-It   NAME   DATE

Materials	*Ten-Frame Top-It* Cards (*Math Masters*, pages G19–G21)
Players	2
Skill	Comparing numbers represented on ten frames

Object of the Game To collect more cards

Directions

① Shuffle the cards.
Put them facedown in a deck.

② Each player turns over a card and says the number on the ten frame.

③ The player with the larger number takes both cards.

④ If two cards show the same number, pick new cards.
The player with the larger number takes all the cards for that round.

⑤ The game is over when all the cards have been turned over.

⑥ The player with more cards wins.

1.OA.6, 1.NBT.3

English Language Learners Support

Beginning ELL Use double ten frames to provide children with oral practice naming
the teen numbers while reinforcing the concept that teen numbers are composed of a 10
and some ones. Model a few examples. Show two ten frames filled in to represent 13, and
say *10 and 3 more* as you point to the individual ten frames. Gesture and bring the two
ten frames together; then say *That's 13.* Repeat with other teen numbers, providing the *first*
statement and prompting children to respond with the number, and vice versa.

Go Online ELL English Learners Support

Standards and Goals for
Mathematical Practice

SMP2 **Reason abstractly and quantitatively.**
  GMP2.3 Make connections between representations.
SMP3 **Construct viable arguments and critique the reasoning of others.**
  GMP3.2 Make sense of others' mathematical thinking.
SMP8 **Look for and express regularity in repeated reasoning.**
  GMP8.1 Create and justify rules, shortcuts, and generalizations.

Math Journal 1, Activity Sheet 3

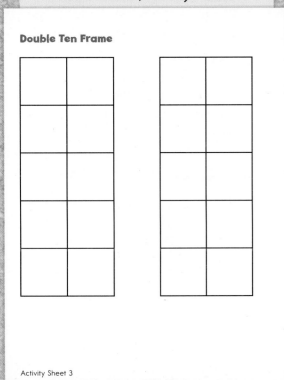

Double Ten Frame

Activity Sheet 3

Adjusting the Activity

Differentiate Provide double ten frames and counters for children who struggle to visualize moving the dots from one ten frame to the other. Children can arrange the counters to form the two addends and then move enough counters from one ten frame to fill the other. Help children connect this model to a written record of the strategy.

Go Online Differentiation Support

---

1 Warm Up 15–20 min

 Go Online ePresentations eToolkit

▶ Mental Math and Fluency

Do Quick Looks with children. Display Quick Look Cards in this order: 92, 88, 89. Show children each image for 2–3 seconds and then remove or cover it. Have children share what they saw and how they saw it.
GMP2.2, GMP6.1 Highlight the strategy of mentally moving dots to complete a ten frame. For detailed instructions, see Lesson 1-1.

▶ Daily Routines

Have children complete the Daily Routines. For detailed instructions, see pages 2–37. For specifics on CCSS coverage, see pages xiv–xvii.

2 Focus 35–40 min

 Go Online ePresentations eToolkit

▶ Math Message

How could you show 15 on a double ten frame? How about 18? 19? Record your thinking on your slate.

▶ Developing the Making-10 Strategy

Math Journal 1, Activity Sheet 3; *Math Masters,* p. TA19

| **WHOLE CLASS** | SMALL GROUP | PARTNER | INDEPENDENT |

Math Message Follow-Up Display a double ten frame (*Math Masters,* p. TA19). Have children use their own double ten frames (*Math Journal 1,* Activity Sheet 3) and counters to share their thinking. Although children may represent the numbers in the Math Message in a variety of ways, emphasize filling one ten frame with ten and then representing the remaining ones on the second ten frame. Generalize this idea by asking: *How can you represent teen numbers?* Represent the tens place with 1 full ten frame and the ones place on the other ten frame **GMP8.1** Tell children to remember this as they see Quick Looks on double ten frames.

Display Quick Look Cards in this order: 98, 95, 114, 116. After each image, ask children to share what they saw and how they saw it. Highlight making 10 when it emerges. Emphasize the strategy by asking questions such as *Did everyone understand Becca's strategy?* and *Can someone else explain it again for us?* **GMP3.2** Sample answer: I moved 2 dots from one frame to the one with 8 to make 10. I had 1 more left to make 11. Record children's thinking for the class. Encourage the class to apply the strategy to the next Quick Look Card.

▶ Recording Making 10

WHOLE CLASS | SMALL GROUP | PARTNER | INDEPENDENT

Ask: *What do many of the strategies for these Quick Looks have in common?* Sample answer: We moved some dots over to fill one ten frame and then added the remaining dots to 10 to find the total. Introduce the phrase **making 10** to describe this strategy, pointing out that filling the ten frame is the same as making 10, which helped make the fact easier to solve.

Display Quick Look Card 116 again, and ask children to record their strategies using words or numbers. Select a few representations to highlight in the discussion, and have children explain their thinking. In particular, elicit a verbal representation, such as "I moved over 2 to the 8, which made 10, and then I had 5 more, which makes 15," and a more symbolic representation, such as $8 + 2 + 5 = 15$.

If children struggle to record their thinking, record their descriptions for the class. Help them make connections between the verbal and symbolic representations of the making-10 strategy by displaying the representations side-by-side. GMP2.3 It is not necessary for children to develop purely symbolic representations of their thinking in first grade, but they should develop ways to express their thinking using words or numbers.

 Differentiate **Adjusting the Activity**

For children who struggle recording their descriptions, provide sentence frames in which they only have to fill in the appropriate numbers. As children become more confident with the statement, provide sentence frames in which they have to fill in key words and phrases, such as *moved over, which made, which makes.*

Go Online 〉 Differentiation Support

Ask children to try applying this thinking to solve $9 + 5$. Continue to highlight various ways of making 10 as children record and discuss their thinking. Sample answers: I moved 1 over from the 5 and added it to the 9 to make 10 which left me with 4. 10 and 4 is 14. $9 + 1 + 4 = 14$; I moved 5 over from 9 and added it to 5 to make 10 which left me with 4. $5 + 5 + 4 = 14$ Repeat with other facts as time permits.

✔ Assessment Check-In (CCSS) 1.OA.6

Observe as children record and share their making-10 strategies. Expect most children to apply the making-10 strategy to find the sum represented by the dots, especially when shown a Quick Look Card. Children who are struggling to visualize moving the dots to make a 10 may benefit from using a double ten frame and counters. Scaffold their use of these tools by modeling, or actually moving the counters to fill one ten frame to make a 10, and by providing a number model.

 Assessment and Reporting **Go Online** 〉 to record student progress and to see trajectories toward mastery for these standards.

My Facts Inventory Record, Part 3

DATE

Addition Fact	Know It	Don't Know It	How I Can Figure It Out...
3 + 4			
7 + 8			
4 + 6			
7 + 9			
6 + 5			
9 + 8			
4 + 5			
7 + 6			
3 + 5			
5 + 7			

222 two hundred twenty-two   1.OA.3, 1.OA.6

Summarize Ask: *What facts do you need to know well in order to make 10? Why?* Sample answer: You need to know combinations of 10 so that you know which parts go together to make 10. **Add an illustration of making 10 to the Strategy Wall.**

3 Practice   10–15 min   Go Online   ePresentations   eToolkit   Home Connections

▶ Facts Inventory Record, Part 3

Math Journal 2, p. 222

| WHOLE CLASS | SMALL GROUP | PARTNER | **INDEPENDENT** |

Remind children that the Facts Inventory Record allows them to keep track of which facts they know well and which ones they are still learning. Have children complete the Facts Inventory Record, Part 3 (*Math Journal 2*, page 222). Have children indicate whether they know each fact. Encourage them to record strategies they could use to figure out facts they do not know well. Many of the facts in Part 3 are near doubles, so encourage children to record near-doubles strategies.

► **Math Boxes 6-6**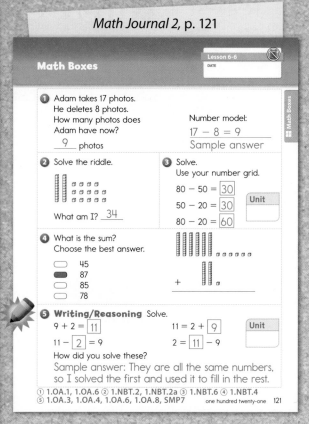

Math Journal 2, p. 121

| WHOLE CLASS | SMALL GROUP | PARTNER | INDEPENDENT |

Mixed Practice Math Boxes 6-6 are paired with Math Boxes 6-4.

► **Home Link 6-6**

Math Masters, p. 168

Homework Children complete a color-by-number picture after finding sums.

Math Journal 2, p. 121

Math Boxes

Lesson 6-6
DATE

① Adam takes 17 photos.
He deletes 8 photos.
How many photos does
Adam have now?
__9__ photos

Number model:
17 − 8 = 9
Sample answer

② Solve the riddle.

What am I? __34__

③ Solve.
Use your number grid.
80 − 50 = [30]
50 − 20 = [30]
80 − 20 = [60]

Unit

④ What is the sum?
Choose the best answer.
◯ 45
▬ 87
◯ 85
◯ 78

+

⑤ **Writing/Reasoning** Solve.
9 + 2 = [11]   11 = 2 + [9]
11 − [2] = 9   2 = [11] − 9

Unit

How did you solve these?
Sample answer: They are all the same numbers,
so I solved the first and used it to fill in the rest.

① 1.OA.1, 1.OA.6 ② 1.NBT.2, 1.NBT.2a ③ 1.NBT.6 ④ 1.NBT.4
⑤ 1.OA.3, 1.OA.4, 1.OA.6, 1.OA.8, SMP7   one hundred twenty-one   121

Math Masters, p. 168

Finding Addition Sums

Home Link 6-6
NAME   DATE

Family Note

Today, your child continued to explore strategies for solving addition facts. Children practiced the making-10 strategy. Ask your child to explain how the making-10 strategy works and how to find each sum on this Home Link.

Please return this Home Link to school tomorrow.

① Solve. Use the color code to color the picture.

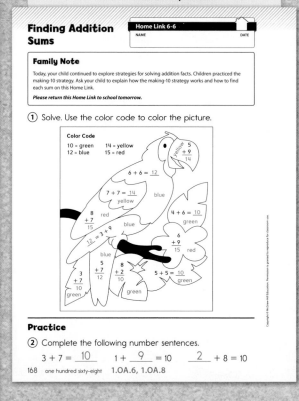

Color Code
10 = green   14 = yellow
12 = blue   15 = red

Practice

② Complete the following number sentences.

3 + 7 = __10__   1 + __9__ = 10   __2__ + 8 = 10

168   one hundred sixty-eight   1.OA.6, 1.OA.8

Lesson 6-6   **539**

Introducing *My Reference Book*

Overview Children learn how to use *My Reference Book* to find mathematical information that will be helpful for solving problems.

▶ **Before You Begin**
If you choose to do the optional Extra Practice activity as a center, read **Seaweed Soup** by Stuart J. Murphy (Harper Collins, 2001) to the class prior to starting the activity.

▶ **Vocabulary**
My Reference Book

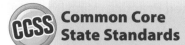

Common Core State Standards

Focus Clusters
• Represent and solve problems involving addition and subtraction.
• Use place value understanding and properties of operations to add and subtract.

1 Warm Up   15–20 min

	Materials	
Mental Math and Fluency Children tell time on an hour-hand-only clock.	hour-hand-only clock	1.MD.3
Daily Routines Children complete daily routines.	See pages 2–37.	See pages xiv–xvii.

2 Focus   35–40 min

Math Message Children share interesting pages in *My Reference Book*.	*My Reference Book*	SMP3
Exploring *My Reference Book* Children use the table of contents to find information in *My Reference Book*.	*My Reference Book*, stick-on notes (optional)	1.OA.1 SMP2, SMP3, SMP5
***My Reference Book* Scavenger Hunt** Children use clues to find specific information in *My Reference Book*.	*Math Journal 2*, p. 122; *My Reference Book*	1.NBT.4, 1.NBT.6
✓ **Assessment Check-In** See page 544.	*My Reference Book*	SMP5

CCSS 1.NBT.4 **Spiral Snapshot**

GMC Understand adding 2-digit numbers and multiples of 10.

5-11 Focus	5-12 Focus Practice	6-2 Focus	6-7 Focus	6-8 Warm Up	7-1 Warm Up	7-5 Practice	8-2 Practice

 **Spiral Tracker**   **Go Online** to see how mastery develops for all standards within the grade.

3 Practice   10–15 min

Playing *The Difference Game* **Game** Children practice finding differences.	*Math Masters*, p. TA20 (optional); per partnership: 4 each of number cards 1–10, 40 toolkit pennies	1.OA.4, 1.OA.6 SMP6
Math Boxes 6-7 Children practice and maintain skills.	*Math Journal 2*, p. 123	See page 545.
Home Link 6-7 **Homework** Children write and solve a comparison number story.	*Math Masters*, p. 169	1.OA.1, 1.OA.6, 1.OA.8

connectED.mcgraw-hill.com

Plan your lessons online with these tools.

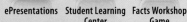

ePresentations   Student Learning Center   Facts Workshop Game   eToolkit   Professional Development   Home Connections   Spiral Tracker   Assessment and Reporting   English Learners Support   Differentiation Support

540   Unit 6 | Addition Fact Strategies

Differentiation Options

  RtI

CCSS SMP5	**CCSS** SMP5

CCSS 1.OA.1, 1.OA.2, 1.OA.6, SMP2

Readiness   10–15 min

WHOLE CLASS
SMALL GROUP
PARTNER
INDEPENDENT

Exploring the Table of Contents

DK First Encyclopedia or another nonfiction book with a table of contents appropriate for children

Literature Link To expose children to using a table of contents, introduce them to **DK First Encyclopedia** edited by Mary Ling [DK Children, 2002] or another nonfiction children's book with a table of contents. Explain that many nonfiction books have a table of contents to help readers find things. Explore the encyclopedia with children. Ask questions about different sections of the table of contents; for example, where they can find information about particular topics, and which subjects appear before, after, or between other subjects. **GMP5.2** If time permits, read some interesting entries to children.

Enrichment   10–15 min

WHOLE CLASS
SMALL GROUP
PARTNER
INDEPENDENT

Playing *My Reference Book I Spy*

Activity Card 79, *My Reference Book*

To challenge children to become more familiar with *My Reference Book,* have them create clues about particular pages or sections. **GMP5.2** Children share clues and try to guess the exact page number(s) described.

Extra Practice   10–15 min

WHOLE CLASS
SMALL GROUP
PARTNER
INDEPENDENT

Creating Number Stories

Activity Card 80, *Seaweed Soup,* slate

Literature Link To provide practice creating number stories, have children tell and solve number stories based on **Seaweed Soup** by Stuart J. Murphy (Harper Collins, 2001). Help children think of addition and subtraction contexts within the story as needed. **GMP2.1**

English Language Learners Support

Beginning ELL Help children understand that the word *table* is used in different ways. Point to tables in the classroom or show pictures. Use drama to illustrate that tables are where people eat or work. Then think aloud while showing examples of data tables that the class may have generated, such as attendance logs, weather charts, or tally charts, to remind children of how they use data tables. Finally, introduce children to tables of contents in various classroom books or on websites. Prepare labels to place next to each example so children will see the terms *data table, table of contents,* and *eating table.* Use Total Physical Response prompts to have them practice identifying different kinds of tables.

Go Online ➤   ELL **English Learners Support**

SMP2 **Reason abstractly and quantitatively.**

GMP2.2 Make sense of the representations you and others use.

SMP3 **Construct viable arguments and critique the reasoning of others.**

GMP3.2 Make sense of others' mathematical thinking.

SMP5 **Use appropriate tools strategically.**

GMP5.2 Use tools effectively and make sense of your results.

NOTE To familiarize children with the Mathematical Practices, see *My Reference Book*, pages 1–22.

My Reference Book, p. 2

Standards for Mathematical Practice

What Are Mathematical Practices?

Read It Together

Mathematical practices are ways that mathematicians think and work as they do mathematics.

When you use mathematical practices, you are thinking and working like a mathematician.

These children are using mathematical practices.

In this section of *My Reference Book*, you will learn about eight mathematical practices.

 two

▶ Mental Math and Fluency

Display the following times on the hour-hand-only demonstration clock. Have children tell the approximate times.

- ●○○ 12 o'clock; 5 o'clock; 9 o'clock
- ●●○ A little past 8 o'clock; just before 8 o'clock
- ●●● Half past 1 o'clock; between 6 and 7 o'clock

▶ Daily Routines

Have children complete the Daily Routines. For detailed instructions, see pages 2–37. For specifics on CCSS coverage, see pages xiv–xvii.

2 **Focus** 35–40 min   Go Online   ePresentations eToolkit

▶ Math Message

Look through My Reference Book. *Find one interesting page, and show it to a partner. Tell your partner something you learned from this page. Then record the page your partner chose. Ask your partner a question to make sure you understand what he or she learned.* **GMP3.2**

▶ Exploring *My Reference Book*

WHOLE CLASS | SMALL GROUP | **PARTNER** | INDEPENDENT

Math Message Follow-Up Display **My Reference Book,** and explain to children that this book may be used during math class in first and second grade. Have children share some of the interesting pages they found. Ask them to share some things their partners learned while exploring *My Reference Book.* **GMP3.2**

NOTE Many teachers integrate *My Reference Book* into nonfiction literacy instruction. Consider using it to build children's understanding of informational text features and reading comprehension, as well as reinforce learning of first-grade mathematics concepts.

Tell children that a reference book is not meant to be read from beginning to end like a story book. Often they may only read one or two pages to find out more about a topic. One way to find topics in a reference book is to use the *table of contents*. GMP5.2   In *My Reference Book,* this is called Contents. Have children turn to the first page of the table of contents on page v and share what they notice. Children should observe the following:

- There are different sections listed.
- Underneath each section is a list of topics.
- Each section and topic has a number next to it. The number is the first page that has information about that section or topic.

Have children use the table of contents to find the first page of Number Stories information. Page 24 Have them turn to this page and share the information they find. Sample answers: The page tells what a number story is. It shows how to use a parts-and-total diagram. It shows a number story and the answer. Read pages 30 and 31 with the class. Have children tell their partners comparison number stories that are not on that page. Then ask them to share how these pages might help them solve their partners' stories. GMP2.2   Sample answers: It reminds me how to use a diagram. It shows how to count up to find an answer.

Next have children use the table of contents to find the section called Measures All Around: Animals and Tools. Pages 117–120 Have children look at the photo essay (on those pages). Explain that this is a photo essay that shows how mathematics is used in the real world. Read aloud the photo essay as children look at the pictures.

Have children find the page containing the first game in the Games section. Page 138 Discuss when this section might be useful. Sample answers: when I have free time in class; if I want to play a game at home

Finally call students' attention to the *My Reference Book* icon in the Math Boxes. Explain that this icon tells them which *My Reference Book* pages to read or review if they need more information to complete the Math Boxes. You may wish to model using a *My Reference Book* page to support solving one of the Math Boxes problems.

> **Differentiate** **Adjusting the Activity**
>
> As the table of contents and Games section are introduced, have children put stick-on notes on the first page of each section to make it easy to find these sections again.
>
>   Differentiation Support

Academic Language Development

Use familiar items to help children generalize the meaning of the term *contents* and its relationship to the term *contain*. Show children items such as a box of cereal, a can of soup, or the packing slip from a shipment. Ask children how they can tell what the boxes or cans *contain*. Restate and ask children how they can tell what the *contents* of the boxes or cans are. Ask them how the *table of contents* of a book tells us what the book contains, or what information we can find in the book.

NOTE In *First Grade Everyday Mathematics* all game directions can be found in the *Math Masters* book. Some game directions can also be found in *My Reference Book.* In some cases, game directions in *My Reference Book* will be more generic, so using the *Math Masters* book is preferable. In other cases, the game directions are identical. In these cases, use what is most convenient for your class.

Math Journal 2, p. 122

My Reference Book
Scavenger Hunt

Lesson 6-7
DATE

Use the table of contents in *My Reference Book* to help you.

1. What is the first page about Adding Larger Numbers?
   page 76
   On this page, what two numbers are being added with base-10 blocks?
   17 and 25

2. Which three pages in the Number Stories section show how to use parts-and-total diagrams?
   pages 24 , 25 , and 26

3. Which section would you like to read?
   Answers vary.
   On what page does that section begin?
   page Answers vary.

4. Where are the directions for the game *Stop and Go*?
   pages 164 and 165
   Read the directions and play the game with your partner.

122   one hundred twenty-two   1.NBT.4, 1.NBT.6

Math Masters, p. G41

The Difference Game

NAME   DATE

Materials	4 each of number cards 1–10, 40 pennies, Bank
Players	2
Skill	Finding differences
Object of the Game	To take more pennies

Directions

1. Shuffle the cards.
   Place the deck facedown on the table.
   Put 40 pennies in the Bank.

2. Each player draws a number card and takes that number of pennies from the Bank.

3. Find out how many more pennies one player has than the other.

4. The player with more pennies keeps the extra pennies, or the difference.

5. The rest of the pennies go back into the bank.

6. The game ends when there are not enough pennies to play another round.

7. The player with more pennies wins.

2 more pennies

1.OA.4, 1.OA.6   G41

▶ # *My Reference Book* Scavenger Hunt

Math Journal 2, p. 122

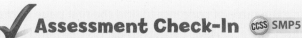

| WHOLE CLASS | SMALL GROUP | PARTNER | INDEPENDENT |

Ask children whether they have ever done a scavenger hunt. Allow a few children to share their experiences. Explain that a scavenger hunt is an activity in which you are given clues about things and have to find them. Tell children that today they will do a scavenger hunt in *My Reference Book.* Partnerships or small groups complete page 122 in their journals, which culminates with children playing the game *Stop and Go.*

✓ **Assessment Check-In**   CCSS SMP5

Observe whether children can use the table of contents to find information in *My Reference Book.*   GMP5.2   Expect most children to find the section opener pages. Some children may have difficulty finding specific information within the sections, but expect children to get better at this as they become accustomed to using the book. Continue to encourage children to use the table of contents to find information in the book so that they become more proficient in using *My Reference Book* as a mathematical tool.

 Assessment and Reporting   Go Online to record student progress for this standard.

Summarize   Ask children about their strategies for finding information in *My Reference Book.* Tell them that they will continue to use *My Reference Book* as a tool for doing mathematics in future lessons.

3 Practice 10–15 min

Go Online
ePresentations   eToolkit   Home Connections

▶ Playing *The Difference Game*

| WHOLE CLASS | SMALL GROUP | **PARTNER** | INDEPENDENT |

Have children practice subtraction by playing *The Difference Game*. For directions for playing the game, use *Math Masters,* page G41. For more information, see Lesson 5-10.

Observe

• How do children find the differences? Do they count up?

Discuss

• *Why is it important to line up the pennies? How does that help you find the difference?* GMP6.1

▶ Math Boxes 6-7

Math Journal 2, p. 123

| WHOLE CLASS | SMALL GROUP | **PARTNER** | **INDEPENDENT** |

Mixed Practice Math Boxes 6-7 are paired with Math Boxes 6-10.

▶ Home Link 6-7

Math Masters, p. 169

Homework Children write and solve a comparison number story.

Math Journal 2, p. 123

Math Boxes Lesson 6-7 DATE

① Solve. Then circle the tens digit.

$6 + 6 = \underline{12}$

② Which number is 10 less than 65?
Use tools if you like.
Choose the best answer.
◯ 70   ◯ 60
⬛ 55   ◯ 5

③ Choose one of these units:
square pattern blocks, crayons, your hands
Use the unit to measure your shoe.
How long is your shoe?
Be sure to write the unit for your answer.
___Answers vary.___

④ What is the number?
Choose the best answer.
◯ 37   ◯ 58
⬛ 62   ◯ 512

⑤ Write <, >, or =.
$28 \boxed{<} 38$ $34 \boxed{<} 43$
$6 + 7 \boxed{>} 12$ $16 \boxed{=} 10 + 6$

⑥ Add. Unit
$9 + 6 = \boxed{15}$ $6 + 9 = \boxed{15}$
$\underline{13} = 5 + 8$ $\underline{13} = 8 + 5$

① 1.NBT.2, 1.NBT.2b, 1.OA.6 ② 1.NBT.5 ③ 1.MD.2 ④ 1.NBT.2, 1.NBT.2a
⑤ 1.NBT.3, 1.OA.6 ⑥ 1.OA.3, 1.OA.6, 1.OA.8   one hundred twenty-three 123

Math Masters, p. 169

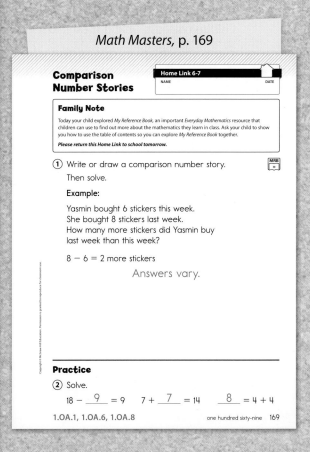

Comparison Number Stories Home Link 6-7
NAME   DATE

Family Note
Today your child explored *My Reference Book,* an important *Everyday Mathematics* resource that children can use to find out more about the mathematics they learn in class. Ask your child to show you how to use the table of contents so you can explore *My Reference Book* together.
Please return this Home Link to school tomorrow.

① Write or draw a comparison number story.
Then solve.

Example:

Yasmin bought 6 stickers this week.
She bought 8 stickers last week.
How many more stickers did Yasmin buy last week than this week?

$8 - 6 = 2$ more stickers
Answers vary.

Practice

② Solve.
$18 - \underline{9} = 9$ $7 + \underline{7} = 14$ $\underline{8} = 4 + 4$

1.OA.1, 1.OA.6, 1.OA.8   one hundred sixty-nine 169

Pencils for the Writing Club

2-Day Lesson

Overview **Day 1:** Children make sense of and solve a multistep number story.
Day 2: The class discusses some solutions, and children revise their work.

Day 1: Open Response

▶ **Before You Begin**
Solve the open response problem and consider how children might interpret the problem. If possible, schedule time to review children's work and plan for Day 2 of this lesson with your grade-level team.

 **Common Core State Standards**

Focus Clusters
- Represent and solve problems involving addition and subtraction.
- Add and subtract within 20.
- Understand place value.
- Use place value understanding and properties of operations to add and subtract.

1 Warm Up   15–20 min

Materials

Mental Math and Fluency Children use a number grid to solve addition and subtraction problems.	*Math Journal 2*, inside back cover; slate	**1.NBT.4, 1.NBT.6**
Daily Routines Children complete daily routines.	See pages 2–37.	See pages xiv–xvii.

2a Focus   45–55 min

Math Message Children decide what the answer to a problem might look like.	*Math Journal 2*, p. 124	**1.NBT.3, 1.NBT.4** SMP1
Making Sense of a Problem Children discuss what it means to make sense of a problem.	*Math Journal 2*, p. 124; *My Reference Book*, p. 3	**1.NBT.3, 1.NBT.4** SMP1
Solving the Open Response Problem Children make sense of and solve a multistep number story.	*Math Masters*, pp. 170–171	**1.OA.1, 1.OA.6** SMP1, SMP4, SMP6

Getting Ready for Day 2 →

Review children's work and plan discussion for reengagement.   *Math Masters*, p. TA3; children's work from Day 1

CCSS 1.OA.1 Spiral Snapshot
GMC Solve number stories by adding and subtracting.

6-4 Warm Up Practice	6-5 Focus Practice	6-7 Focus Practice	6-8 Focus Practice	6-10 Warm Up	6-11 Practice	7-1 Warm Up	7-3 Warm Up Focus

 Spiral Tracker   Go Online   to see how mastery develops for all standards within the grade.

connectED.mcgraw-hill.com

Plan your lessons online with these tools.

 ePresentations   Student Learning Center   Facts Workshop Game   eToolkit   Professional Development   Home Connections   Spiral Tracker   Assessment and Reporting   ELL English Learners Support   Differentiation Support

1 Warm Up 15–20 min

Go Online

ePresentations   eToolkit

▶ Mental Math and Fluency

Have children use the number grid on the inside back cover of their journals to solve. Have them record their answers on their slates.

◉○○ 15 + 7 22; 6 + 29 35

◉◉○ 58 + 10 68; 11 + 75 86

◉◉◉ 70 − 40 30; 100 − 20 80

▶ Daily Routines

Have children complete the Daily Routines. For detailed instructions, see pages 2–37. For specifics on CCSS coverage, see pages xiv–xvii.

2a Focus 45–55 min

Go Online

ePresentations   eToolkit

▶ Math Message

Math Journal 2, p. 124

Look at journal page 124. Work with a partner to try to solve Li's problem and answer the questions. GMP1.1

▶ Making Sense of a Problem

Math Journal 2, p. 124; *My Reference Book,* p. 3

| WHOLE CLASS | SMALL GROUP | PARTNER | INDEPENDENT |

Math Message Follow-Up After children have worked on Li's problem for a few minutes, make sure they stop working on the problem and answer the questions at the bottom of the page. Have them compare their answers with a partner. Make sure children understand that it is fine if they have not completely solved Li's problem yet, but that they can still answer the questions.

CCSS Standards and Goals for Mathematical Practice

SMP1 Make sense of problems and persevere in solving them.
  GMP1.1 Make sense of your problem.

SMP4 Model with mathematics.
  GMP4.1 Model real-world situations using graphs, drawings, tables, symbols, numbers, diagrams, and other representations.

SMP6 Attend to precision.
  GMP6.1 Explain your mathematical thinking clearly and precisely.

Professional Development

The focus mathematical practice for this lesson is **GMP1.1**. To *make sense of a problem* is to think about what the problem is asking and what the answer might look like. Children can start to make sense of a problem before they begin solving it and continue to do so as they work on the problem. For example, even without finding a final answer to the Math Message problem, children should be able to determine that the answer will be a number sentence with a sum of 22, 23, or 24.

Go Online for information about SMP1 in the *Implementation Guide*.

Math Journal 2, p. 124

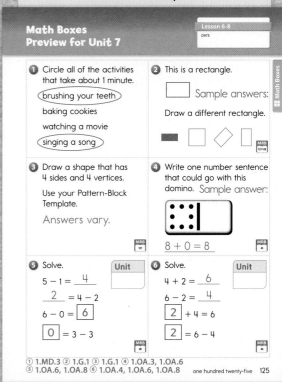

Standards for Mathematical Practice

Problem Solving:
Make Sense and Keep Trying

Write a number sentence that has a sum between 21 and 25.
Use two numbers from the box for the addends.

___ + ___ = A sum between 21 and 25

| 12 | 5 | 15 | 17 | 9 | 14 |

I need to find two numbers
from the box that add up to
a number between 21 and
25. My answer will be an
addition number sentence.

Li

Make sense of your problem.
*Li thinks about what she needs to do and what her
answer will look like.*

three   MRB   **3**

Ask children to share their answers to Problem 1. Ask: *What should Li write?*
GMP1.1 A number sentence *How do you know?* Sample answer: It says
that Li's teacher asked her to write a number sentence. It doesn't say to
write a number or a number story.

Repeat this process for Problem 2. Ask: *Which number could be Li's sum?*
GMP1.1 24 *How do you know?* Sample answer: It says the sum has to be
between 21 and 25. 17 is too small and 31 is too large.

Tell children that when they answered these questions, they were making
sense of Li's problem. Explain that when mathematicians make sense of
a problem, they think about what they need to do and what their answers
might look like.

Have children turn to *My Reference Book,* page 3 and read it together. Ask:
What should Li's answer be, a number sentence or a number story? A number
sentence *What did Li say she needed to do to solve the problem?* Find two
numbers from the box that add up to a number between 21 and 25. *Did
you and your partner agree with Li when you answered the questions in the
Math Message?* Answers vary.

Tell children they will solve a number story where they have to think
carefully about what the answer will look like and what they should do
to solve it. **GMP1.1**

Academic Language Development The term *make sense* can be
used in different ways. When children consider whether $2 + 9 = 20$
makes sense or whether coloring a strawberry blue makes sense, they
are thinking about whether something seems right, logical, or silly. When
children are asked to make sense of the open response problem, they
are being asked to consider what information they know about the
problem, what information they want to find out, and how they will use
the information in the problem to find the answer.

▶ Solving the Open Response Problem

Math Masters, pp. 170–171

| WHOLE CLASS | SMALL GROUP | PARTNER | INDEPENDENT |

Distribute *Math Masters,* pages 170–171. Read the problems aloud. Tell
children that Mr. Khan is not sure if he needs more pencils, and their job
is to figure out whether he has enough. If they do think he has enough,
they should find out how many he has left over. They can use counters,
a number grid, a number line, or any other tools they want to help them
solve the problem.

Remind children that like Li from the Math Message, they will need to make sense of the problem and decide what their answers will look like. Ask: *What are some things we have done before to help us solve number stories?* GMP1.1 Sample answers: Talk to a partner. Act out the problem. Draw a picture. Read the problem again. Explain that these are all ways to make sense of a number story, and children can do these things to help them solve this problem.

> **NOTE** Some children may interpret *enough* in Problem 1 to mean "exactly enough." That is, they may say the answer to the first problem is no and explain that Mr. Khan has too many pencils. Explain to these children that in this problem, Mr. Khan has enough if he does not run out of pencils.

> **Differentiate** **Adjusting the Activity**
>
> If children have difficulty getting started, encourage them to begin by drawing the pencils and the children. A visual representation of the situation may help children see what they need to do to solve the problem.

Allow children time to complete the pages. Note that part of the challenge of this problem is that several pieces of information are presented, and children will not necessarily use them in the order they are given. Encourage children to talk to their partners about the problem and to use drawings or counters to model the situation as needed. Ask: *What will your answer look like for Problem 1? Problem 2? How did you know you had to add those numbers? How did you know when to subtract?* GMP1.1 *Can you use drawings or objects to help you act out the situation?* GMP4.1 *Did you explain your answers in a way that can help someone else solve the problem?* GMP6.1

For children who seem to understand the mathematical concepts but have trouble with the explanation, suggest that they show how they solved the problem using a picture. A sentence frame might help some children get started with a written explanation, for example: "I know Mr. Khan will have enough pencils because _____."

Summarize Ask: *What did you do to help you make sense of the problem?* GMP1.1 Sample answers: acted out the problem; talked to my partner; drew a picture

Collect children's work so that you can evaluate it and prepare for Day 2.

Pencils for the Writing Club

Lesson 6-8

NAME   DATE

Mr. Khan has a pack of 18 pencils.

He wants to give 1 pencil to each child in his writing club.

There are 7 girls and 8 boys in the club.

① Does Mr. Khan have enough to give each child 1 pencil?

   Yes

Use words, numbers, or pictures to show how you know.
Answers vary. See sample children's work on page 555 of the *Teacher's Lesson Guide.*

170 one hundred seventy   1.OA.1, 1.OA.6, SMP1, SMP4, SMP6

Pencils for the Writing Club (continued)

Lesson 6-8

NAME   DATE

② How many pencils will Mr. Khan have left over?

   3 pencils

Show or explain how you found your answer.
Answers vary. See sample children's work on page 555 of the *Teacher's Lesson Guide.*

1.OA.1, 1.OA.6, SMP1, SMP4, SMP6   one hundred seventy-one 171

Sample child's work, Child A

(1) Does Mr. Khan have enough to give each child 1 pencil?

__15__

Use words, numbers, or pictures to show how you know.

$7 + 8 = 15$

Sample child's work, Child B

(1) Does Mr. Khan have enough to give each child 1 pencil?

__yes__

Use words, numbers, or pictures to show how you know.

$18 - 7 - 8 = 3$

Sample child's work, Child C

(1) Does Mr. Khan have enough to give each child 1 pencil?

__yes__

Use words, numbers, or pictures to show how you know.

becuse $7+8=15$ and Mr. Khan
has 18 and 18 is more then
15 so he has enough to give
each child 1 pencil

Math Masters, p. TA3

Planning a Follow-Up Discussion

Review children's work. Use the Reengagement Planning Form (*Math Masters,* page TA3) and the rubric on page 552 to plan ways to help children meet expectations for both the content and practice standards. Look for common misconceptions, such as misinterpreting the problem, as well as interesting ways children represented and solved the problem.

Reengagement Planning Form

Common Core State Standard (CCSS): *1.OA.1 Use addition and subtraction within 20 to solve word problems involving situations of adding to, taking from, putting together, taking apart, and comparing, . . . e.g., by using objects, drawings, and equations with a symbol for the unknown number to represent the problem.*

Goal for Mathematical Practice (GMP): *GMP1.1 Make sense of your problem.*

Strengths and understandings:

Organize the discussion of children's work in one of the ways below or in another way you choose. If children's work is unclear or if you prefer to show work anonymously, rewrite the work for display.

[Go Online] for sample children's work that you can use in your discussion.

1. Show a paper on which a child wrote a number for the answer to Problem 1, as in Child A's work. Have children talk to a partner about whether the answer makes sense for the question. Ask:
   - *What is Problem 1 asking you to do?* GMP1.1 Sample answer: Find out if there are enough pencils.
   - *What did this child do to find the answer to Problem 1?* Sample answer: This child added 7 and 8 to get 15.
   - *Does 15 seem like a good answer? What do you think the answer to Problem 1 should look like? A number? Words?* GMP1.1 Sample answer: 15 doesn't seem right because the answer should be yes or no.
   - *What does the 15 mean here?* Sample answer: 15 is the number of kids in the club.
   - *Did this child need to add 7 and 8 to answer the question? What else should this child do?* GMP1.1 Sample answer: We need to know how many kids are in the club because it helps us figure out if there are enough pencils. Now we need to see if 18 pencils is more than 15.

2. Compare a paper that shows that a child subtracted to find out whether there were enough pencils in Problem 1, as in Child B's work, to a paper that shows that a child compared the number of children to the number of pencils, as in Child C's work. Ask:

- *How did Child B figure out whether there were enough pencils?* GMP1.1 Sample answer: Child B started with 18 pencils and then took away 7 for the girls and 8 for the boys.

- *How did subtracting help Child B figure out whether there were enough pencils?* GMP1.1 Sample answer: After Child B took away the number of pencils for the girls and the number of pencils for the boys, there were still 3 pencils left, so there are enough.

- *How did Child C figure out whether there were enough pencils?* GMP1.1 Sample answer: Child C found out there were 15 kids in the club and then saw that 18 is more than 15, so that's how Child C knew there were enough pencils.

- *Which of these children knew how many pencils would be left over after solving Problem 1?* Child B

- *Did you need to know how many pencils were left over to solve Problem 1?* GMP1.1 Sample answer: You didn't have to know how many pencils were left to solve Problem 1. You can figure it out, but you don't have to. That's Problem 2.

3. Compare two different ways of making sense of the problem. For example, one child might have drawn a picture, as in Child D's work, while another child might have used the problem context to decide when to add or subtract, as in Child E's work. Ask:

- *How did each child make sense of the problem?* GMP1.1 Sample answer: Child D drew a picture. Child E thought about the problem and when numbers should be put together and when they should be subtracted.

- *How did the picture help Child D make sense of the problem?* GMP4.1 Sample answer: Child D drew 18 pencils, then drew 7 girls and 8 boys and crossed out the pencils to see how many were left.

- *Why was it important for Child E to think about whether giving something away means you should subtract?* GMP1.1 Sample answer: Child E needed to know when numbers should be added or taken away.

- *Does everyone always think about problems in the same way? Are these both good ways to make sense of the problem?* GMP1.1 Sample answer: Not everyone has to solve problems in the same way. Yes, both children figured out a way to answer the questions.

- *Was it easy to tell how these children solved the problems? What could these children add to their papers to help others understand what they were thinking?* GMP6.1 Sample answer: Child D could label the picture with words so we could understand it better. Child E could say what numbers were added in Problem 1 to get 15.

Planning for Revisions

Have copies of *Math Masters*, pages 170–171 or extra paper available for children to use in revisions. You might want to ask children to use colored pencils so you can see what they revised.

Sample child's work, Child D

① Does Mr. Khan have enough to give each child 1 pencil?

yes

Use words, numbers, or pictures to show how you know.

② How many pencils will Mr. Khan have left over?

_3_ pencils

Show or explain how you found your answer.

I us 8 boys and 7 Grils then I cros Them out and mr. Khon che have left Over i8 E pencils left

---

Sample child's work, Child E

① Does Mr. Khan have enough to give each child 1 pencil?

yes

Use words, numbers, or pictures to show how you know.

I put the grrls and Boys thgethr so I can figrit out and the answr was 15. mr. han Has 3 more Pencils.

② How many pencils will Mr. Khan have left over?

_3_ pencils

Show or explain how you found your answer.

Because 7+8=15 so mr. Khan Has 3 more Because if you Have Some thing and you Oive Sobode it is Sobchracshin.

Pencils for the Writing Club

Overview   Day 2: The class discusses some solutions, and children revise their work.

Day 2: Reengagement

Common Core State Standards

▶ **Before You Begin**
Have extra copies available of *Math Masters,* pages 170–171 for children to revise their work.

Focus Clusters
- Represent and solve problems involving addition and subtraction.
- Add and subtract within 20.

2b Focus   50–55 min

	Materials	
Setting Expectations Children review the open response problem and discuss what the problem asks them to do and what a good response might include. They also review how to discuss others' work respectfully.	Standards for Mathematical Practice Poster, Guidelines for Discussions Poster	1.OA.1 SMP1, SMP6
Reengaging in the Problem Children discuss how others made sense of the open response problem.	selected samples of children's work	1.OA.1 SMP1, SMP4, SMP6
Revising Work Children revise their work from Day 1.	*Math Masters,* pp. 170–171 (optional); colored pencils (optional); children's work from Day 1	1.OA.1, 1.OA.6 SMP1, SMP4, SMP6
✓ **Assessment Check-In** See page 554 and the rubric below.		1.OA.1 SMP1

Goal for Mathematical Practice GMP1.1 Make sense of your problem.	Not Meeting Expectations	Partially Meeting Expectations	Meeting Expectations	Exceeding Expectations
	May show evidence of recognizing the meaning of some of the numbers in the problem, but does not provide any evidence of relating the number of children to the number of pencils.	Shows evidence (with pictures, number sentences, or written explanations) of recognizing the need to do one, but not both, of the following: • compare the number of pencils to the number of children; • find how many more pencils there are than children.	Shows evidence (with pictures, number sentences, or written explanations) of recognizing the need to do both of the following: • compare the number of pencils to the number of children; • find how many more pencils there are than children.	Meets expectations and shows two different ways to find how many more pencils there are than children (for example, with a picture and by subtracting).

3 Practice   10–15 min

Math Boxes 6-8 Children practice and maintain skills.	*Math Journal 2,* p. 125; Pattern-Block Template	See page 554.
Home Link 6-8 **Homework** Children solve number stories.	*Math Masters,* p. 172	1.OA.1, 1.OA.6 SMP1

2b) Focus   50–55 min   Go Online   ePresentations   eToolkit

NOTE These Day 2 activities will ideally take place within a few days of Day 1. Prior to beginning Day 2, see Planning a Follow-Up Discussion from Day 1.

▶ Setting Expectations

| **WHOLE CLASS** | SMALL GROUP | PARTNER | INDEPENDENT |

Briefly review the open response problem from Day 1. Ask: *What were you asked to do?* GMP1.1 Sample answer: Figure out if Mr. Khan had enough pencils and how many were left over. *What do you think a good response would include?* GMP6.1 Sample answer: It would say if there were enough pencils and how I know. I could use pictures to help me explain or write number models.

After this brief discussion, tell children they are going to look at other children's work to see if everyone thought about the problem in the same way. Refer to **GMP1.1** on the Standards for Mathematical Practice Poster. Explain that the class will try to figure out how some children made sense of the open response problem.

Point out that some children may think differently about the problem or may change their minds during the discussion. Remind children that it is fine to change their minds and that they should listen closely to each other to see if they agree or disagree with what other children say. Refer to your list of discussion guidelines and encourage children to use these sentence frames:

- I like how you _____.
- I wonder why _____.

▶ Reengaging in the Problem

| **WHOLE CLASS** | SMALL GROUP | **PARTNER** | INDEPENDENT |

Children reengage in the problem by analyzing and critiquing other children's work in pairs and in a whole-group discussion. Have children discuss with partners before sharing with the whole group. Guide the discussion based on the decisions you made in Getting Ready for Day 2. GMP1.1, GMP4.1, GMP6.1

▶ Revising Work

| WHOLE CLASS | SMALL GROUP | **PARTNER** | **INDEPENDENT** |

Pass back children's work from Day 1. Before children revise anything, ask them to examine their responses and decide how to improve them. Ask the questions below one at a time. Have partners discuss their responses and then give a thumbs-up or thumbs-down based on their own work.

• *Did you write an answer that makes sense for each problem?* **GMP1.1**

• *Did you use words, numbers, or pictures to show how you got your answers?* **GMP4.1, GMP6.1**

Tell children they now have a chance to revise their work. Those who wrote complete and correct explanations on Day 1 can try to find a different way to check or show their work. Remind children that the explanations presented during the reengagement discussion are not the only correct ones. Tell children to add to their earlier work using colored pencils or to use another sheet of paper, instead of erasing their original work.

Summarize Ask children to reflect on their work and revisions. Ask: *How did you know whether there were enough pencils?* **GMP1.1** Answers vary.

✔ Assessment Check-In CCSS 1.OA.1

Collect and review children's revised work. Expect children to improve their work based on the class discussion. For the content standard, expect most children to correctly conclude that Mr. Khan has enough pencils and has 3 pencils left over. You can use the rubric on page 552 to evaluate children's revised work for **GMP1.1**.

☑ Assessment and Reporting   〈Go Online〉 to record student progress and to see trajectories toward mastery for these standards.

〈Go Online〉 for optional generic rubrics in the *Assessment Handbook* that can be used to assess any additional GMPs addressed in the lesson.

Math Journal 2, p. 125

Math Boxes
Preview for Unit 7

Lesson 6-8
DATE

① Circle all of the activities that take about 1 minute.
(brushing your teeth)
baking cookies
watching a movie
(singing a song)

② This is a rectangle.
Sample answers:
Draw a different rectangle.

③ Draw a shape that has 4 sides and 4 vertices.
Use your Pattern-Block Template.
Answers vary.

④ Write one number sentence that could go with this domino. Sample answer:
8 + 0 = 8

⑤ Solve.
5 − 1 = __4__
__2__ = 4 − 2
6 − 0 = 6
0 = 3 − 3
Unit

⑥ Solve.
4 + 2 = __6__
6 − 2 = __4__
2 + 4 = 6
2 = 6 − 4
Unit

① 1.MD.3 ② 1.G.1 ③ 1.G.1 ④ 1.OA.3, 1.OA.6
⑤ 1.OA.6, 1.OA.8 ⑥ 1.OA.4, 1.OA.6, 1.OA.8   one hundred twenty-five 125

Sample Children's Work—Evaluated

See the sample in the margin. This work meets expectations for the content standard by showing that there were enough pencils with 3 left over. With revision, the work meets expectations for the mathematical practice because the picture shows evidence of comparing the number of pencils to the number of children, as the child drew 18 pencils, crossed 15 out (one for each child), and saw that there were still some left. This child's picture and written description both show evidence of an attempt to find how many more pencils there are than children. The child's written description for Problem 2 makes it clear that the child understood that he or she needed to find how many more pencils there were than children. (Note: The authors interpret this child's answer to Problem 2 to be a description of crossing objects out on the picture. If, instead, the answer to Problem 2 was interpreted as a description of a different method, such as counting back on a number line, this paper could be evaluated as exceeding expectations.) **GMP1.1**

Go Online for other samples of evaluated children's work.

3 Practice 10–15 min

Go Online

ePresentations   eToolkit   Home Connections

▶ Math Boxes 6-8: Preview for Unit 7

Math Journal 2, p. 125

| WHOLE CLASS | SMALL GROUP | **PARTNER** | **INDEPENDENT** |

Mixed Practice Math Boxes 6-8 are paired with Math Boxes 6-12. These problems focus on skills and understandings that are prerequisite for Unit 7. You may want to use information from these Math Boxes to plan instruction and grouping in Unit 7.

▶ Home Link 6-8

Math Masters, p. 172

Homework Children solve number stories and explain to someone at home how they made sense of the problems.

Sample child's work, "Meeting Expectations"

① Does Mr. Khan have enough to give each child 1 pencil?

yes

Use words, numbers, or pictures to show how you know.

② How many pencils will Mr. Khan have left over?

3 pencils

Show or explain how you found your answer.

I strde qt 18 and I - 7
and 3 and I end qt 3

Math Masters, p. 172

Number Stories   Home Link 6-8
NAME   DATE

Family Note

Today your child learned that when solving a problem, it is helpful to think about what the problem is asking. This is called *making sense of a problem.* Making sense of a problem can help your child decide what needs to be done to solve the problem. For example, should the numbers in the problem be added or subtracted?

Throughout the year, encourage your child to explain how he or she knew what to do when solving a problem.

Please return this Home Link to school tomorrow.

Solve.

① You have 12 marbles.
Your friend has 15 marbles.
How many more marbles does your friend have?
3 marbles

② You picked 6 red flowers and 8 blue flowers.
How many flowers did you pick in all? 14 flowers

③ 7 children were at the playground.
2 children went home.
How many children stayed at the playground?
5 children

Tell someone at home how you knew what to do to solve the stories.

Practice

④ Use <, >, or = to make each number sentence true.

13 < 27   44 = 44   80 > 30

172 one hundred seventy-two   1.OA.1, 1.OA.6, 1.NBT.3, SMP1

Understanding Equivalence

Overview Children use addition and subtraction facts to complete name-collection boxes and extend their understanding of equivalence.

▶ **Before You Begin**
For the optional Enrichment activity, prepare blank name-collection boxes from *Math Masters,* page TA28 with different numbers in the tags. There should be as many name-collection boxes as there are children participating. You may wish to attach these to clipboards.

▶ **Vocabulary**
equivalent names • name-collection box

 Common Core State Standards

Focus Clusters
• Understand and apply properties of operations and the relationship between addition and subtraction.
• Add and subtract within 20.
• Work with addition and subtraction equations.

1 Warm Up 15–20 min

	Materials	
Mental Math and Fluency Children find 10 more and 10 less than a given number.	slate	1.NBT.5
Daily Routines Children complete daily routines.	See pages 2–37.	See pages xiv–xvii

2 Focus 30–35 min

Math Message Children list multiple names for themselves.	slate	SMP2
Illustrating Equivalence Children explore the idea of equivalence using addition and subtraction facts.	counters	1.OA.3, 1.OA.6, 1.OA.7 SMP1, SMP2
Introducing Name-Collection Boxes Children learn about name-collection boxes and write equivalent names in them.	*Math Journal 2,* p. 126	1.OA.3, 1.OA.6, 1.OA.7 SMP1
✓ **Assessment Check-In** See page 561.	*Math Journal 2,* p. 126; counters (optional)	1.OA.7 SMP1

CCSS 1.OA.6 **Spiral Snapshot**

GMC Add and subtract within 20 using strategies.

6-2 through 6-6 Warm Up Focus Practice	6-7 Practice	6-8 Focus Practice	6-9 Focus Practice	6-10 Warm Up Practice	6-11 Practice	7-1 Focus Practice

Spiral Tracker   **Go Online** to see how mastery develops for all standards within the grade.

3 Practice 15–20 min

Finding 10 More and 10 Less Children practice finding 10 more and 10 less than a given number.	*Math Journal 2,* p. 127; base-10 blocks; number grid	1.NBT.5 SMP5
Math Boxes 6-9 Children practice and maintain skills.	*Math Journal 2,* p. 128	See page 561.
Home Link 6-9 **Homework** Children find equivalent names for numbers.	*Math Masters,* pp. 173–174	1.OA.6, 1.OA.7, 1.NBT.2, 1.NBT.3, SMP1

connectED.mcgraw-hill.com

Plan your lessons online
with these tools.

ePresentations   Student Learning   Facts Workshop   eToolkit   Professional   Home   Spiral Tracker   Assessment   English Learners   Differentiation
Center   Game   Development   Connections   and Reporting   Support   Support

Differentiation Options RtI

Readiness   10–15 min

WHOLE CLASS
SMALL GROUP
PARTNER
INDEPENDENT

Modeling Equivalence

counters, half-sheet of paper

To explore the concept of equivalence using a concrete model, have children examine combinations of counters for a given number. Model the following process:

1. Write the number 5 on a half-sheet of paper, and circle it.

2. Place 5 counters on the table in a line.

3. Say: *No counters are covered.* Write 0.

4. Say: *5 counters are not covered.* Write + 5.

5. Say: *There are a total of 5 counters.* Write = 5.

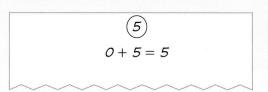

Recording Sheet for Modeling Equivalent Names

Repeat the process, covering 1 counter and writing the number sentence $1 + 4 = 5$ to represent 1 counter covered and 4 uncovered. Have children work in partnerships to write additional number sentences for 5. GMP4.1 Then have them read their number sentences aloud. Repeat the activity using different numbers of counters.

Enrichment   10–15 min

WHOLE CLASS
SMALL GROUP
PARTNER
INDEPENDENT

Musical Name-Collection Boxes

Activity Card 81;
Math Masters, p. TA28

To apply children's understanding of equivalent names, have them complete different name-collection boxes during the activity. GMP2.1 Before children begin, help them choose a short, familiar song to sing.

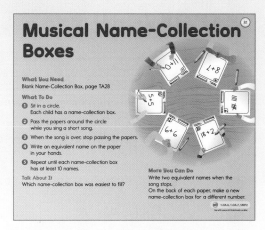

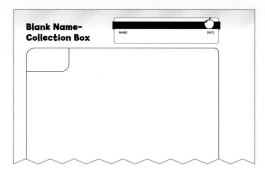

Extra Practice   10–15 min

WHOLE CLASS
SMALL GROUP
PARTNER
INDEPENDENT

Finding Equivalent Names for 10

Activity Card 82;
Math Masters, p. TA28;
10 counters

To provide practice finding equivalent names for 10, children use counters to write as many names as they can for 10. GMP2.1 Although name-collection boxes can include different types of names, for this activity encourage children to write addition and subtraction facts, such as $5 + 5$ and $11 - 1$.

English Language Learners Support

Beginning ELL Role play and think aloud to demonstrate collecting and the meaning of the term *collection*. For example, demonstrate collecting the pencils and say: *I am collecting all the pencils.* Hold all the pencils together and say: *I have a collection of pencils.* Direct children to collect other items and to complete sentence frames, such as: "I am collecting all the _____. I have a collection of _____." Ask questions such as: *Who has a collection of _____?* after several collections have been gathered.

 Go Online ELL English Learners Support

Standards and Goals for
Mathematical Practice

SMP1 Make sense of problems and persevere in solving them.

GMP1.4 Check whether your answer makes sense.

GMP1.5 Solve problems in more than one way.

SMP2 Reason abstractly and quantitatively.

GMP2.1 Create mathematical representations using numbers, words, pictures, symbols, gestures, tables, graphs, and concrete objects.

1 Warm Up 15–20 min [Go Online] ePresentations eToolkit

▶ Mental Math and Fluency

Have children think about number-grid column families to find the answers and record them on their slates.

●○○○ 10 more than 10 20; 10 more than 12 22; 10 less than 20 10

●●○○ 10 more than 24 34; 10 less than 45 35; 10 less than 52 42

●●●○ 10 less than 61 51; 10 more than 11 21; 10 more than 98 108

▶ Daily Routines

Have children complete the Daily Routines. For detailed instructions, see pages 2–37. For specifics on CCSS coverage, see pages xiv–xvii.

2 Focus 30–35 min [Go Online] ePresentations eToolkit

▶ Math Message

Joseph Smith's friends call him Joe. His mom calls him Joey. His grandma calls him sweetie. His baseball coach calls him Smith. Record some names that people call you. GMP2.1

▶ Illustrating Equivalence

| **WHOLE CLASS** | SMALL GROUP | PARTNER | INDEPENDENT |

Math Message Follow-up Discuss the idea that the same person or thing can have different names that refer to the same person or thing, or **equivalent names.** Display the word *equivalent.* Explain that numbers also have different names. True number sentences show two names that are equivalent. For example, when solving 4 + 2 = _____ to make a true number sentence, you are looking for another name for 4 + 2, such as 6. We say that 4 + 2 and 6 are equivalent names because 4 + 2 *is equal to, or equivalent to,* 6. GMP2.1

Ask children to list the addition facts that have sums of 7. Each fact contains an equivalent name for 7: $4 + 3, 2 + 5, 1 + 6, 0 + 7 = 7$ and so on. To illustrate the equivalence, display 7 counters and arrange them into two piles to model each of the facts. (*See margin.*) Write number sentences with equal signs to relate each to 7. You may wish to relate some of them to each other, such as $4 + 3 = 1 + 6$. Discuss how each arrangement still has 7 total counters even though it is organized differently. Ask children if they can think of another way to show 7 with the counters. GMP1.5 If no one suggests it, show how the counters can be arranged into three groups; for example, $2 + 3 + 2$.

Go Online for more ideas on showing equivalencies.

▶ Introducing Name-Collection Boxes

Math Journal 2, p. 126

| WHOLE CLASS | SMALL GROUP | PARTNER | INDEPENDENT |

Draw a name-collection box, write 7 on the tag, and put two addition names for 7 inside the box. Explain that **name-collection boxes** are used to record equivalent names for numbers.

Academic Language Development Point out that a collection is a group of things that are together because they are similar. Collections of objects can include model cars or coins. Extend the meaning to include a collection of different names for a number.

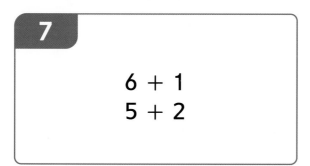

Then add the remaining names from the previous activity, both horizontally and vertically. Ask the class to think of other names to include by thinking about subtraction facts and using the turn-around rule to think of more addition facts. GMP1.5

Demonstrate checking whether a fact belongs in the name-collection box by writing a number sentence with the fact and the number in the tag, such as $9 - 2 = 7$. If the number sentence is true, then the name belongs in the box. GMP1.4 Ask volunteers to write number sentences to check other facts in the box.

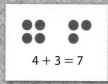

$4 + 3 = 7$

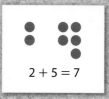

$2 + 5 = 7$

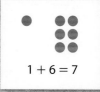

$1 + 6 = 7$

Math Journal 2, p. 126

Name-Collection Boxes

Lesson 6-9
DATE

1 Write other names for 11.

11

8 + 3

Answers vary.

13 − 2

2 Write other names for 12.

12

1 dozen

Answers vary.

|||| |||| ||

3 + 3 + 3 + 3
15 − 3

3 Cross out the names that don't belong in the 10-box.

10

|||| ||||

5 + 5   ~~8 + 3~~
~~0 + 9~~   9 + 1
7 + 3   ~~4 + 4~~

4 Make your own.

Answers vary.

126   one hundred twenty-six   1.OA.6, 1.OA.7, SMP1

Math Journal 2, p. 127

Finding 10 More and 10 Less

Lesson 6-9
DATE

Use base-10 blocks.

1 Use | and ∎ to show the number 31.

Sample answer: |||∎

2 Use | and ∎ to show 10 more than 31.

Sample answer: ||||∎

3 Use | and ∎ to show 10 less than 31.

Sample answer: ||∎

4 Find 76 on your number grid.

What is 10 more than 76? __86__

What is 10 less than 76? __66__

5 Is it easier for you to use blocks or the number grid to find 10 more and 10 less? Explain.

Answers vary.

1.NBT.5, SMP5   one hundred twenty-seven   127

Finally encourage children to provide other names for 7 that are not addition or subtraction facts, such as 7 tally marks.

Examples:

After this discussion, have children complete the name-collection boxes on journal page 126.

Differentiate   **Adjusting the Activity**

Display name-collection boxes in the classroom, and encourage children to fill them with a variety of representations for each number during their free time.

Go Online   Differentiation Support

Assessment Check-In CCSS 1.OA.7

Math Journal 2, p. 126

Observe as children complete journal page 126. Expect most to write at least two other names in Problems 1 and 2. Help children who struggle by providing counters for modeling possible addition facts to include as equivalent names. Encourage children who excel to record names other than numerical expressions, such as representations that use base-10 blocks, money, words, tally marks, and so on. GMP1.5

Assessment and Reporting · Go Online to record student progress and to see trajectories toward mastery for these standards.

Summarize Have children exchange journals with their partners and add one more name to each other's name-collection box in Problem 4.

③ **Practice** 15–20 min · Go Online

ePresentations · eToolkit · Home Connections

▶ **Finding 10 More and 10 Less**

Math Journal 2, p. 127

| WHOLE CLASS | SMALL GROUP | **PARTNER** | **INDEPENDENT** |

Children practice finding 10 more and 10 less than a number using base-10 blocks and the number grid. GMP5.2

▶ **Math Boxes 6-9**

Math Journal 2, p. 128

| WHOLE CLASS | SMALL GROUP | **PARTNER** | **INDEPENDENT** |

Mixed Practice Math Boxes 6-9 are paired with Math Boxes 6-11.

▶ **Home Link 6-9**

Math Masters, pp. 173–174

Homework Children practice finding equivalent names for numbers.

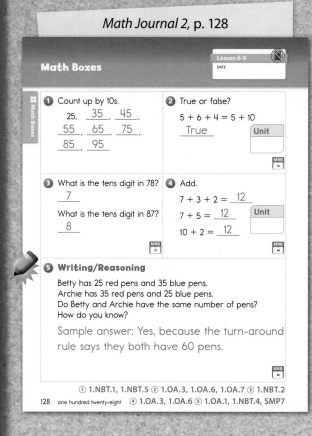
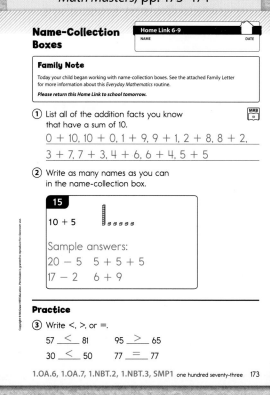

More Place Value

Overview Children use base-10 blocks to solve place-value riddles.

▶ **Before You Begin**
For Part 2, distribute base-10 blocks to children. You may wish to make laminated copies of the Place-Value Mat (*Math Masters,* page TA29) for children to use for the remainder of the year.

▶ **Vocabulary**
flat • hundreds place

Common Core State Standards

Focus Cluster
Understand place value.

---

1 Warm Up 15–20 min

	Materials	
Mental Math and Fluency Children solve number stories by picturing ten frames.		**1.OA.1, 1.OA.6**
Daily Routines Children complete daily routines.	See pages 2–37.	See pages xiv–xvii.

---

2 Focus 35–40 min

Math Message Children show a number with base-10 blocks in two ways.	base-10 blocks	**1.NBT.2, 1.NBT.2a** SMP2
Exploring More Base-10 Block Riddles Children answer riddles and are introduced to the flat.	*Math Masters,* p. TA29; base-10 blocks	**1.NBT.2, 1.NBT.2a, 1.NBT.2b,** **1.NBT.2c, SMP2**
Exchanging More Base-10 Blocks Children exchange base-10 blocks and name the number.	*Math Masters,* p. TA29; base-10 blocks	**1.NBT.2, 1.NBT.2a, 1.NBT.2b,** **1.NBT.2c, SMP2**
✓ **Assessment Check-In** See page 566.	base-10 blocks	**1.NBT.2, 1.NBT.2a**
Extending Base-10 Block Riddles Children solve more place-value riddles.	*Math Journal 2,* p. 129; base-10 blocks	**1.NBT.2, 1.NBT.2a, 1.NBT.2b,** **1.NBT.2c, SMP2**

CCSS 1.NBT.2 **Spiral Snapshot**

GMC Understand exchanging tens and ones.

| 5-7
Warm Up | 5-8
Focus
Practice | 5-9
Practice | 5-11
Focus | 6-10
Focus
Practice | 6-11
Focus | 7-2
Warm Up | 7-7
Practice |

 Spiral Tracker   **Go Online** ▷ to see how mastery develops for all standards within the grade.

---

3 Practice 10–15 min

Playing *Fishing for 10* **Game** Children practice making combinations of ten.	*Math Masters,* p. G32 (optional), p. G33; number cards 0–10	**1.OA.6** SMP1
Math Boxes 6-10 Children practice and maintain skills.	*Math Journal 2,* p. 130	See page 567.
Home Link 6-10 **Homework** Children write numbers represented by base-10 blocks.	*Math Masters,* p. 175	**1.OA.6, 1.OA.7, 1.NBT.2,** **1.NBT.3** SMP2

---

connectED.mcgraw-hill.com

Plan your lessons online with these tools.

 ePresentations   Student Learning Center   Facts Workshop Game   eToolkit   Professional Development   Home Connections   Spiral Tracker   Assessment and Reporting   ELL English Learners Support   Differentiation Support

Differentiation Options RtI

CCSS 1.NBT.2, 1.NBT.2a–1.NBT.2c, SMP2	CCSS 1.NBT.2, 1.NBT.3, SMP6	CCSS 1.NBT.2, 1.NBT.2a, 1.NBT.3, SMP2

Readiness   10–15 min

WHOLE CLASS
SMALL GROUP
PARTNER
INDEPENDENT

Counting Base-10 Blocks with a Calculator

calculator, base-10 blocks

To explore place value, have children count base-10 blocks using calculators. Each partnership takes a small collection of cubes. They program their calculators to count by 1s. One child counts on the calculator as the other child counts each cube. **GMP2.3** Guide children to understand that each time a cube is added, the digit in the ones place changes. Have children notice when the digit in the tens place changes and discuss why it happens. Ask them to predict when the digit in the tens place will change again.

Pause to discuss special cases of numbers, such as count-by-10s numbers and teen numbers, as children reach them. Children continue counting cubes as time permits. Consider having them count longs by programming their calculator to count by 10s. Discuss why the ones place does not change and what they think will happen when they have 10 longs.

Enrichment   10–15 min

WHOLE CLASS
SMALL GROUP
PARTNER
INDEPENDENT

Playing *The Digit Game* with 3-Digit Numbers

Math Masters, p. G35 (optional); number cards

To extend children's ability to compare numbers based on place value, have them play a variation of *The Digit Game.* Each child takes three number cards (instead of two) and creates the largest 3-digit number possible. The partner with the larger number takes all the cards. See Lesson 5-1 for detailed instructions.

Observe

- Which children make the largest possible 3-digit numbers from their cards?
- Which partnerships correctly identify the larger 3-digit numbers?

Discuss

- *How do you know which number is larger?* **GMP6.1**

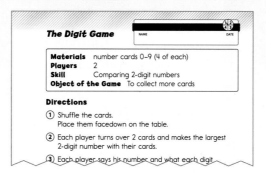

Extra Practice   10–15 min

WHOLE CLASS
SMALL GROUP
PARTNER
INDEPENDENT

Ordering Numbers with Base-10 Representations

index cards, base-10 blocks (optional)

To provide practice with place value, have children work in small groups to order representations of 2-digit numbers. **GMP2.1** Each child in the group takes two index cards and uses | and ■ to represent two 2-digit numbers of their choosing. When all children in the group have completed their cards, they work together to order the cards by determining the numbers represented.

Children may use more than 10 cubes to represent a number, so it may be necessary to trade cubes for longs in order to figure out what number is represented. If necessary, children can model the numbers with base-10 blocks to help them make exchanges.

English Language Learners Support

Beginning ELL   Build on children's familiarity with connecting cubes to demonstrate the meaning of *exchanging for equal value.* Model the expression, *Let's exchange,* as you show children a train of 10 connecting cubes, and demonstrate how two trains of five are exactly the same length. Organize children into pairs, giving one partner trains of connecting cubes and the other partner single cubes to use for making exchanges. Provide oral practice by having children use the expression "Let's exchange" before they make their exchanges.

Go Online ELL English Learners Support

Math Masters, p. TA29

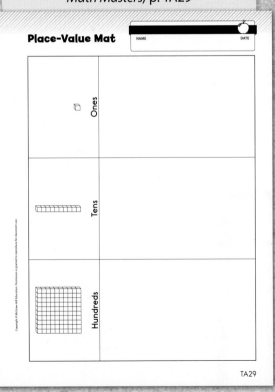

1 Warm Up 15–20 min   Go Online
ePresentations   eToolkit

▶ Mental Math and Fluency

Ask children to solve number stories by visualizing ten frames. Have them briefly share their thinking. *Suggestions:*

◐○○ A ten frame has 10 counters in it. You remove 5. How many counters are there now? 5 counters. Sample answer: I thought of removing one row.

◐◐○ A ten frame has 10 counters in it. You remove 6. How many counters are there now? 4 counters. Sample answer: I thought of removing one full row and one more counter.

◐◐◐ A ten frame has 9 counters in it. You remove 5. How many counters are there now? 4 counters. Sample answer: I thought of starting with a ten frame that is missing one counter and removing a full row of 5 counters. That would leave 4 counters.

▶ Daily Routines

Have children complete the Daily Routines. For detailed instructions, see pages 2–37. For specifics on CCSS coverage, see pages xiv–xvii.

2 Focus 35–40 min   Go Online
ePresentations   eToolkit

▶ Math Message

Show 57 with base-10 blocks in two different ways. **GMP2.1** *Record your ways with* | *and* ■*.*

▶ Exploring More Base-10 Block Riddles

Math Masters, p. TA29

| WHOLE CLASS | SMALL GROUP | PARTNER | INDEPENDENT |

Math Message Follow-Up Have children follow along, placing base-10 blocks on their Place-Value Mats (*Math Masters,* page TA29), as their classmates share various ways to model 57. **GMP2.3** Show 5 longs and 7 cubes, and ask: *What does the tens digit in 57 represent?* 50 *The ones digit?* 7 Tell children that today they will continue to explore numbers with base-10 blocks.

Display the following riddles on your Place-Value Mat, and have children share their answers as they follow along on their own mats:

- Show 6 longs and 4 cubes. *What number is this?* 64
- Show 4 longs and 6 cubes. *Is this number larger or smaller than 64?* Smaller *How do you know?* Sample answer: There are fewer longs.
- Show 7 longs and 0 cubes. *What number is this?* 70 *What does the 0 in the ones place represent?* 0 ones
- Show 1 long and 5 cubes. *What number is this?* 15 *What is the value of the 1 in 15?* 1 ten *The 5?* 5 ones
- Show 3 longs and 7 cubes. *Name a number with the same tens digit and a different ones digit.* Sample answers: 30, 35, 36, 39
- Show 2 longs and 2 cubes. *Name a number with the same ones digit and a different tens digit.* Sample answers: 12, 32, 52, 82
- Show 10 longs and 0 cubes. *What number is this?* 100

Children may struggle with this last problem. Tell children that understanding a new place in numbers may help them find out what number is represented by 10 longs. Distribute flats to children, and invite them to compare their flats with their 10 longs. Tell them that 10 longs can be exchanged for 1 **flat.** A flat represents 100. Have children use longs to completely cover a flat to verify that a flat is equivalent to 10 longs; 100 is the same amount as 10 tens. Guide children to see that 100 cubes is equivalent to 1 flat by asking: *How many cubes are equivalent to 1 long?* 10 *How many longs are equivalent to 1 flat?* 10 *So how many cubes are equivalent to 1 flat?* 100 GMP2.3 Then demonstrate exchanging the 10 longs for 1 flat on your Place-Value Mat.

Write 1, 0, 0 in the appropriate columns on your Place-Value Mat. Ask: *What does the 1 represent?* 1 hundred Note that the 1 is in the **hundreds place.** Then ask what the zeroes in the tens and ones places represent. 0 tens and 0 ones Tell children that 100 is a 3-digit number and that 1, 0, and 0 are the three digits in 100.

Common Misconception

Differentiate **IF** children make an exchange, such as 10 cubes for 1 long, but do not remove the 10 cubes from the total amount of blocks, **THEN** have children count the cubes and longs by 10s and 1s, make the exchange, and count again. Point out that the totals should match after the exchange.

Go Online Differentiation Support

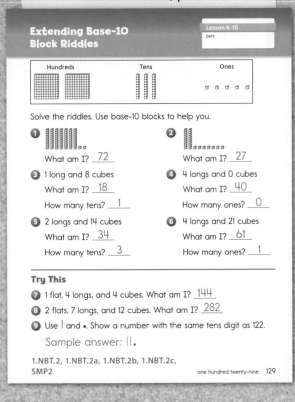

Fishing for 10

Materials	4 sets of number cards (0–10), *Fishing for 10* Record Sheet (*Math Masters*, page G33)
Players	2–4
Skill	Practicing combinations of 10
Object of the Game	To collect as many combinations of 10 as you can

Directions

1. Shuffle the cards and put them number-side down in a pile.
2. Each player takes 5 cards.
   - Remove pairs of cards that add to 10 from your hand.
   - Place them number-side up on the table.
3. Draw cards from the pile so you have 5 cards again.
4. Take turns. When it is your turn:
   - *Fish* by asking another player for a card you need to make a combination of 10.
   - If that player does not have the card you need, *go fish*, or draw a card from the pile.
   - Place any combinations of 10 in your hand number-side up on the table.
5. Make sure everyone's cards add to 10.
6. Play until there are no more cards in the pile and nobody can make another combination of 10.
7. Record 6 of your combinations of 10 on your *Fishing for 10* Record Sheet.

G32   1.OA.6

▶ Exchanging More Base-10 Blocks

Math Masters, p. TA29

Display 2 longs and 16 cubes on your Place-Value Mat. Ask: *What number is shown?* 36 **GMP2.2** Demonstrate the following exchange: Trade 10 of the cubes for 1 long, and place the long in the tens column. There are now 3 longs and 6 cubes. Again ask: *What number is shown?* 36

Academic Language Development Discuss the expression *fair exchange* to extend children's understanding of the term *exchange* as meaning "trading for something of equal value." Show non-examples of *fair exchanges,* and ask children why they would not like to make such exchanges, building on their understanding of what is *fair*.

Repeat this routine with other combinations of longs and cubes that require exchanges:

- 3 longs and 21 cubes 51
- 2 longs and 18 cubes 38
- 4 longs and 10 cubes 50
- 0 longs and 15 cubes 15
- 7 longs and 13 cubes 83
- 10 longs and 4 cubes 104

For this last problem, have children trade 10 longs for a flat. Emphasize that when 10 longs are traded for a flat, there is a 0 in the tens place, just as there is a 0 in the ones place when 10 cubes are traded for a long.

> ✓ **Assessment Check-In**  1.NBT.2, 1.NBT.2a
>
> Observe as children make exchanges between base-10 blocks. Expect most children to exchange cubes for longs and accurately identify the tens digit and ones digit in each number. Have children who struggle count longs and cubes by 10s and 1s to find the numbers represented by the blocks. Have children who easily identify 2-digit numbers build 3-digit numbers with base-10 blocks for partners to identify.
>
> ✓ Assessment and Reporting   [Go Online] to record student progress and to see trajectories toward mastery for these standards.

▶ Extending Base-10-Block Riddles

Math Journal 2, p. 129

| WHOLE CLASS | SMALL GROUP | **PARTNER** | **INDEPENDENT** |

Encourage children to use base-10 blocks to help them solve the riddles on journal page 129. **GMP2.2**

Summarize Discuss how children found their answers to Problem 9 on journal page 129.

3 Practice 10–15 min

Go Online ePresentations eToolkit Home Connections

▶ Playing *Fishing for 10*

Math Masters, p. G33

| WHOLE CLASS | **SMALL GROUP** | **PARTNER** | INDEPENDENT |

Children practice making combinations of 10. For directions for playing *Fishing for 10,* use *Math Masters,* page G32. For more information, see Lesson 4-9.

Observe

• Which combinations of 10 do children know well? Which are most challenging for them?

Discuss

• *How did you and your partner check each other's combinations of 10?*

GMP1.4

▶ Math Boxes 6-10

Math Journal 2, p. 130

| WHOLE CLASS | SMALL GROUP | **PARTNER** | **INDEPENDENT** |

Mixed Practice Math Boxes 6-10 are paired with Math Boxes 6-7.

▶ Home Link 6-10

Math Masters, p. 175

Homework Children practice identifying numbers represented by base-10 blocks.

Math Journal 2, p. 130

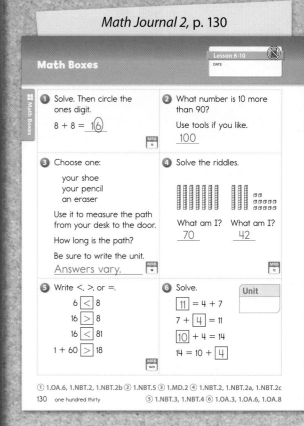

Math Masters, p. 175

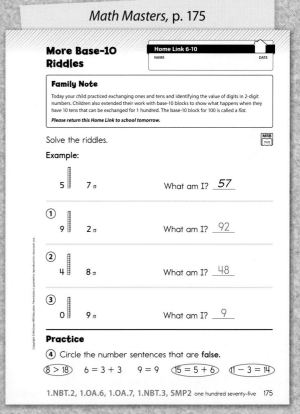

Place-Value Application: Pennies, Dimes, and Dollars

Overview Children apply their understanding of place value to make exchanges between pennies, dimes, and dollars.

▶ **Before You Begin**
Make a 2-sided copy of *Math Masters,* pages TA31 and TA32 for each child, and cut out the bills. For Part 2, each child will need 20 pennies, 10 dimes, and 5 one-dollar bills. For the optional Enrichment activity, provide a box containing about 50 pennies and 50 dimes for each partnership, as well as a sheet of paper labeled "Bank" containing 5 one-dollar bills and 20 dimes.

Common Core State Standards

Focus Cluster
Understand place value.

	Materials	
① Warm Up 15–20 min		
Mental Math and Fluency Children write and compare numbers.	slate	1.NBT.3
Daily Routines Children complete daily routines.	See pages 2–37.	See pages xiv–xvii.

② Focus 35–40 min		
Math Message Children compare two base-10 block representations.		1.NBT.2, 1.NBT.2a, 1.NBT.3 SMP1, SMP2
Making Place-Value Exchanges Children apply place-value understanding to make exchanges between pennies, dimes, and dollars.	*Math Masters,* pp. TA29–TA32; per child: 20 pennies, 20 dimes, 5 one-dollar bills	1.NBT.2, 1.NBT.2a, 1.NBT.2b, 1.NBT.2c, 1.NBT.3 SMP2, SMP6
Introducing *Penny-Dime-Dollar Exchange* **Game** Children practice exchanging pennies, dimes, and dollars.	*Math Masters,* pp. TA30, G48 (optional); per partnership: 2 dice, 5 one-dollar bills, 20 dimes, and 20 pennies	1.NBT.2, 1.NBT.2a SMP6
✓ **Assessment Check-In** See page 573.		1.NBT.2, 1.NBT.2a, SMP2

CCSS 1.NBT.2 **Spiral Snapshot**

GMC Understand exchanging tens and ones.

| 5-8
Focus
Practice | 5-9
Practice | 5-11
Focus | 6-10
Focus
Practice | 6-11
Focus | 7-2
Warm Up | 7-7
Practice | 8-6
Warm Up |

III **Spiral Tracker** 〈 **Go Online** 〉 to see how mastery develops for all standards within the grade.

③ Practice 10–15 min		
Modeling Number Stories with Equations Children solve and match number stories to equations.	*Math Journal 2,* p. 131	1.OA.1, 1.OA.4, 1.OA.6
Math Boxes 6-11 Children practice and maintain skills.	*Math Journal 2,* p. 132	See page 573.
Home Link 6-11 **Homework** Children show how to pay for items.	*Math Masters,* p. 176	1.OA.2, 1.OA.3, 1.NBT.2

connectED.mcgraw-hill.com ▶

Plan your lessons online with these tools.

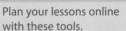

 ePresentations

 Student Learning Center

 Facts Workshop Game

 eToolkit

 Professional Development

 Home Connections

 Spiral Tracker

 Assessment and Reporting

 English Learners Support

 Differentiation Support

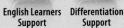

Differentiation Options RtI

CCSS 1.NBT.2, 1.NBT.2a, SMP2

Readiness
10–15 min

WHOLE CLASS
SMALL GROUP
PARTNER
INDEPENDENT

Playing *Penny-Dime Exchange*

Math Masters, p. G36 (optional); 1 dot die; 30 pennies; 5 dimes; 1 sheet of paper labeled "Bank"

To provide additional exposure to exchanging ones for tens, have children play *Penny-Dime Exchange*. For detailed instructions, see Lesson 5-3.

Observe
• Which children make accurate exchanges?

Discuss
• *How could you represent 5 dimes and 3 pennies with base-10 blocks?* GMP2.3

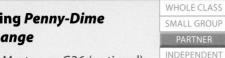

CCSS 1.NBT.2, 1.NBT.2a, SMP6

Enrichment
15–20 min

WHOLE CLASS
SMALL GROUP
PARTNER
INDEPENDENT

Coin Scoop

Activity Card 83; *Math Masters*, p. TA30; slate; 2 small paper cups; box with 50 pennies and 50 dimes; paper labeled "Bank" with 5 one-dollar bills and 20 dimes

To further explore place value using pennies, dimes, and dollars, children scoop up pennies and dimes and determine the value by making exchanges. GMP6.4

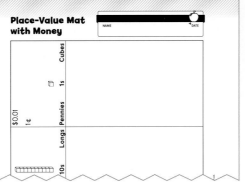

CCSS 1.NBT.2, 1.NBT.2a, SMP2

Extra Practice
15–20 min

WHOLE CLASS
SMALL GROUP
PARTNER
INDEPENDENT

Playing *Base-10 Exchange*

Math Masters, p. G40 (optional), p. TA29; base-10 blocks; 1 dot die

For children who may benefit from practice with place value, have them play a variation of *Base-10 Exchange* in which they exchange 10 longs for 1 flat. The object of this variation is to get 2 flats. If time permits, challenge children to get 3 flats. For detailed instructions, see Lesson 5-8.

Observe
• Which children exchange 10 cubes for 1 long? Which children exchange 10 longs for 1 flat?

Discuss
• *How is exchanging cubes for longs like exchanging pennies for dimes? How is exchanging longs for flats like exchanging dimes for dollars?* GMP2.3

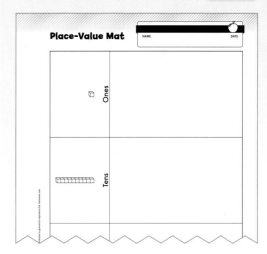

English Language Learners Support

Beginning ELL Provide additional experiences for children to find out more about U.S. money, including exposure to a variety of coins and bills. Let children explore a collection of coins and bills, encouraging them to notice attributes, such as color, thickness, size, weight, texture, design, and illustrations. Name the money and value and have children practice repeating the names. Have them practice identifying the different coins and bills.

Go Online ▶ ELL English Learners Support

1 Warm Up   15–20 min   Go Online
ePresentations   eToolkit

▶ Mental Math and Fluency

Children record pairs of numbers on their slates and compare them using <, >, or =.

●○○ 72, 78   72 < 78; 49, 49   49 = 49; 25, 30   25 < 30

●●○ 77, 87   77 < 87; 90, 89   90 > 89; 38, 83   38 < 83

●●● 54, 62   54 < 62; 100, 93   100 > 93; 121, 121   121 = 121

▶ Daily Routines

Have children complete the Daily Routines. For detailed instructions, see pages 2–37. For specifics on CCSS coverage, see pages xiv–xvii.

2 Focus   35–40 min   Go Online
ePresentations   eToolkit

▶ Math Message

Max has 9 longs and 2 cubes on his mat.
Shawna has 8 longs and 12 cubes on her mat.
Does Max or Shawna have a larger number? How do you know?
Record what you think.   **GMP1.1, GMP2.3**

▶ Making Place-Value Exchanges

Math Masters, pp. TA29–TA33

| WHOLE CLASS | SMALL GROUP | PARTNER | INDEPENDENT |

Math Message Follow-up Have children share their responses to the Math Message. Display Shawna's blocks on a Place-Value Mat (*Math Masters,* page TA29). Ask: *What can I exchange?* Exchange 10 cubes for 1 long. *What is the number?* 92 Then display Max's 9 longs and 2 cubes. Ask: *What is the number?* 92 Help children conclude that both show the same number, 92.

Math Masters, p. TA29

Place-Value Mat

NAME   DATE

Ones   Tens   Hundreds

TA29

Tell children that they will use place-value understandings to exchange coins today. Replace the base-10 blocks with coins that represent the same values in cents. Ask: *Which coin should we use for a cube? Why?* Use a penny because it is a 1-cent coin and a cube represents 1. *Which coin should we use for a long? Why?* Use a dime because it is a 10-cent coin and a long equals 10. Display 8 dimes and 12 pennies and have a child demonstrate exchanging 10 pennies for 1 dime. Have the class count the amount of money represented by 10s and 1s. *How much money is this?* 92 cents GMP2.3

Distribute 20 pennies and 20 dimes to each child. Have children show the quantities shown below. Then have them exchange pennies for dimes and report the total amount of money:

- 7 dimes and 18 pennies 8 dimes, 8 pennies; 88 cents
- 0 dimes and 14 pennies 1 dime, 4 pennies; 14 cents
- 9 dimes, 16 pennies 10 dimes, 6 pennies; 116 cents, or $1.16

Ask whether anyone knows another exchange that can be made in the last problem. Review exchanging 10 longs for 1 flat. Ask: *Can we exchange 10 dimes for something else?* 1 dollar Display a dollar bill, and discuss how 1 dollar has the same value as 10 dimes or 100 pennies, just like a flat has the same value as 10 longs or 100 cubes. Refer to the example with 9 dimes and 26 pennies, and demonstrate exchanging the 10 dimes for 1 dollar. Display 1 dollar, 1 dime, and 6 pennies, and ask: *How much money is this?* 1 dollar and 16 cents

Distribute 5 one-dollar bills (*Math Masters,* pages TA31–TA32) and *Math Masters,* page TA30 to each child. Point out that the Place-Value Mat is just like the one they used before, except that money units have been added in the headings. You may wish to introduce the symbols for dollars ($) and cents (¢) at this time. Explain that the relationships among flats, longs, and cubes are similar to the relationships among dollars, dimes, and pennies.

Display 2 dollars, 4 dimes, and 8 pennies on the Place-Value Mat, and have children do the same. Have them discuss with their partners how much money is represented. Count aloud to find the total amount in cents. Count the dollars first (*100, 200*); then count by 10s as you point to each dime (*210, 220, 230, 240*); finally, count by 1s for each penny (*241, 242, 243, 244, 245, 246, 247, 248*). Explain that 248 cents is equal to 2 dollars and 48 cents and that these are names for the same amount of money. (Briefly review how to write and read dollars-and-cents notation if needed.) Practice counting dollars and cents with other examples.

Suggestions:

- 2 dollars, 1 dime, 8 pennies $2.18
- 4 dollars, 9 dimes, 4 pennies $4.94

Math Masters, p. TA30

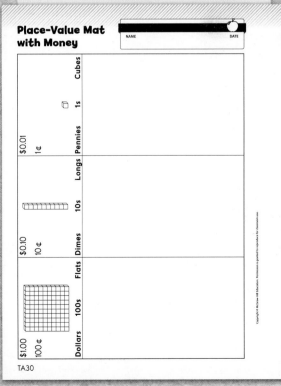

Penny-Dime-Dollar Exchange

NAME   DATE

Materials	1 dollar, 20 dimes, 20 pennies
	1 Place-Value Mat with Money per player
	(Math Masters, page TA30)
	2 dot dice
	1 sheet of paper labeled "Bank"
Players	2
Skill	Making place-value exchanges
Object of the Game	To exchange for a dollar

Directions

① Place all of the money in the bank.

② Players take turns. When it is your turn:
- Roll the dice.
- Take that number of cents from the bank. Place the coins on the mat.
- If you have 10 or more pennies in the Pennies column, exchange 10 pennies for 1 dime.
- If you have 10 or more dimes in the Dimes column, exchange 10 dimes for 1 dollar.

③ The winner is the first player to make the exchange for a dollar.

$1.00	$0.10	$0.01		$1.00	$0.10	$0.01
100¢	10¢	1¢		100¢	10¢	1¢
Dollars 100s Flats	Dimes 10s Longs	Pennies 1s Cubes		Dollars 100s Flats	Dimes 10s Longs	Pennies 1s Cubes

to bank   from bank

G48   1.NBT.2, 1.NBT.2a

Modeling Number Stories with Equations

Lesson 6-11
DATE

Solve the number story. Then draw a line from the number story to the equation that matches it.

❶ Janine ate 7 berries.
Carlos ate 10 berries.
How many more berries did Carlos eat than Janine?

__3__ more berries

$10 - 7 = \square$

❷ Shavon found 3 leaves.
Janelle found 8 leaves.
How many fewer leaves did Shavon find than Janelle?

__5__ fewer leaves

$8 - 3 = \square$

❸ A cat weighs 10 pounds.
A dog weighs 19 pounds.
How much more does the dog weigh than the cat?

__9__ more pounds

$19 - 10 = \square$

$3 + \square = 8$

$10 + \square = 19$

$7 + \square = 10$

1.OA.1, 1.OA.4, 1.OA.6   one hundred thirty-one   131

Display 1 dollar, 3 dimes, and 15 pennies on the Place-Value Mat. Have children do the same. Remind children that they can exchange coins just like they exchange base-10 blocks. Ask: *Which coins can we exchange?* Exchange 10 pennies for a dime. Have children complete the exchange on their mats. Count aloud as a class to confirm the total. Ask: *How much money is shown?* 145 cents; 1 dollar and 45 cents   GMP6.4

Repeat this routine with other combinations of pennies, dimes, and dollars that require exchanges. *Suggestions:*

- 2 dollars, 16 dimes, 4 pennies $3.64
- 3 dollars, 1 dime, 15 pennies $3.25

Display amounts of money using dollars-and-cents notation, and have children show the amounts on their mats using the fewest number of dollars, dimes, and pennies. *Suggestions:* $3.26; $1.10; $1.01; $0.57; $0.08. Emphasize place-value connections with questions such as: *What does the 2 represent in $3.26?* 2 dimes, 2 groups of ten, 20 cents or pennies *What does the 0 represent in $1.10?* 0 cents or pennies, or 0 ones *What does the 5 represent in $5.43?* 5 dollars or 5 groups of 1 hundred cents, 50 dimes or tens, or 500 cents or pennies.   GMP2.3

Differentiate   **Adjusting the Activity**

If children are having difficulty exchanging, ask them to verbally describe their exchanges to a partner during the activity. Scaffold with sentence frames that reflect the content of the problems, such as, "I exchanged, or traded, 10 dimes for _____."

 Go Online   Differentiation Support

▶ Introducing *Penny-Dime-Dollar Exchange*

Math Masters, p. TA30

| WHOLE CLASS | SMALL GROUP | **PARTNER** | INDEPENDENT |

Introduce *Penny-Dime-Dollar Exchange*. For directions, see *Math Masters*, page G48. Demonstrate a round of the game before having children play in partnerships.

Observe

- Which children correctly exchange 10 pennies for a dime, or 10 dimes for a dollar?   GMP6.4

Discuss

- *If you have 3 pennies on your mat and you roll an 8, what should you do? Can you exchange?*

Differentiate   **Game Modifications**   Go Online   Differentiation Support

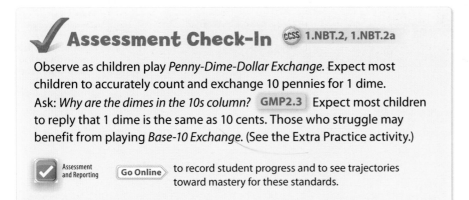

✓ Assessment Check-In   CCSS 1.NBT.2, 1.NBT.2a

Observe as children play *Penny-Dime-Dollar Exchange*. Expect most children to accurately count and exchange 10 pennies for 1 dime. Ask: *Why are the dimes in the 10s column?* GMP2.3 Expect most children to reply that 1 dime is the same as 10 cents. Those who struggle may benefit from playing *Base-10 Exchange*. (See the Extra Practice activity.)

☑ Assessment and Reporting   Go Online to record student progress and to see trajectories toward mastery for these standards.

Summarize Have children work in partnerships. One child shows 1 dollar and 4 dimes, and the other shows 14 dimes on their Place-Value Mats. Have children discuss whether one partner has more money and how they know.

③ Practice   10–15 min   Go Online

ePresentations   eToolkit   Home Connections

▶ Modeling Number Stories with Equations

Math Journal 2, p. 131

| WHOLE CLASS | SMALL GROUP | PARTNER | **INDEPENDENT** |

Children solve number stories and match them with equations.

▶ Math Boxes 6-11 ✏️

Math Journal 2, p. 132

| WHOLE CLASS | SMALL GROUP | **PARTNER** | INDEPENDENT |

Mixed Practice Math Boxes 6-11 are paired with Math Boxes 6-9.

▶ Home Link 6-11

Math Masters, p. 176

Homework Children show how to pay for items using dollar bills and different combinations of coins.

Math Journal 2, p. 132

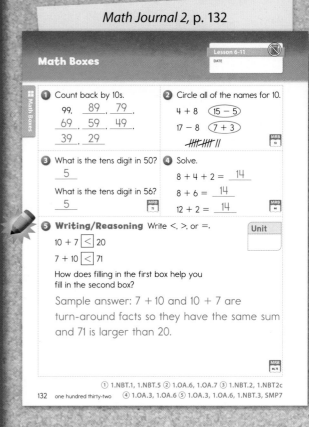

Math Masters, p. 176

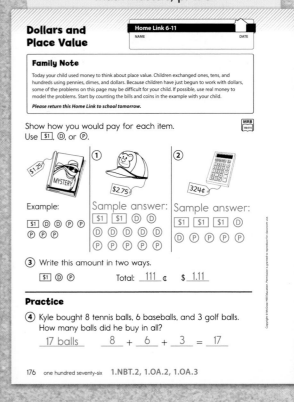

Unit 6 Progress Check

Overview   **Day 1:** Administer the Unit Assessments.
Day 2: Administer the Cumulative Assessment.

2-Day Lesson

 Student Learning Center
Students may take assessments digitally.

 Assessment and Reporting
Record results and track progress toward mastery.

Day 1: Unit Assessments

1 Warm Up   5–10 min

Materials

Self Assessment
Children complete the Self Assessment.

Assessment Handbook, p. 37

2a Assess   35–50 min

Unit 6 Assessment
These items reflect mastery expectations to this point.

Assessment Handbook, pp. 38–40

Unit 6 Challenge (Optional)
Children may demonstrate progress beyond expectations.

Assessment Handbook, p. 41

CCSS Common Core State Standards	**Goals for Mathematical Content (GMC)**	**Lessons**	**Self Assessment**	**Unit 6 Assessment**	**Unit 6 Challenge**
1.OA.1	Solve number stories by adding and subtracting.	6-2, 6-5, 6-7, 6-8		2	
1.OA.3	Apply properties of operations to add or subtract.	6-2, 6-6, 6-9			✔
1.OA.6	Add within 10 fluently.*	6-3 to 6-6, 6-9		5–7	
	Add doubles automatically.*	6-3 to 6-6		4–6	
	Add combinations of 10 automatically.*	6-5, 6-6		6, 7	
	Add and subtract within 20 using strategies.	6-2 to 6-6, 6-8		4, 6	✔
1.OA.7	Understand the meaning of the equal sign.*	6-3, 6-9	3	6, 7	✔
	Determine whether equations involving addition or subtraction are true or false.	6-3, 6-9		6	
1.NBT.2	Understand place value.	6-6, 6-10, 6-11	4	8–10	
	Represent whole numbers as tens and ones.	6-6, 6-10, 6-11		8	
	Understand exchanging tens and ones.	6-10, 6-11		8	
	Understand 10, 20, …, 90 as some tens and no ones.	6-10, 6-11		9, 10	
1.NBT.4	Understand adding 2-digit numbers and 1-digit numbers.	6-2, 6-7, 6-8	2		
	Understand adding 2-digit numbers and multiples of 10.	6-2, 6-7, 6-8	2	2	
	Understand adding 2-digit numbers.	6-8	2		
1.NBT.6	Subtract multiples of 10 from multiples of 10.	6-2, 6-7, 6-8		11	
1.MD.3	Tell and write time using analog clocks.	6-1	1	1	

*Instruction and most practice on this content is complete.

Common Core State Standards	Goals for Mathematical Practice (GMP)	Lessons	Self Assessment	Unit 6 Assessment	Unit 6 Challenge
SMP1	Keep trying when your problem is hard. **GMP1.3**				✓
	Solve problems in more than one way. **GMP1.5**	6-9		7	✓
SMP2	Make sense of the representations you and others use. **GMP2.2**	6-4, 6-7, 6-10		8	
SMP3	Make sense of others' mathematical thinking. **GMP3.2**	6-4, 6-6, 6-7	6		
SMP5	Choose appropriate tools. **GMP5.1**	6-2	5	3	
	Use tools effectively and make sense of your results. **GMP5.2**	6-1, 6-7		2	
SMP6	Explain your mathematical thinking clearly and precisely. **GMP6.1**	6-1, 6-5, 6-6, 6-8		3	
SMP8	Create and justify rules, shortcuts, and generalizations. **GMP8.1**	6-6		4	

Spiral Tracker **Go Online** to see how mastery develops for all standards within the grade.

1 Warm Up 5–10 min

▶ Self Assessment

Assessment Handbook, p. 37

WHOLE CLASS | SMALL GROUP | PARTNER | **INDEPENDENT**

Children complete the Self Assessment to reflect on their progress in Unit 6.

Remind children of a time when they did each type of problem.

Item	Remind children that they . . .
1	told time to the hour on analog clocks. (Lesson 6-1)
2	added animal weights and lengths. (Lesson 6-2)
3	found equivalent names for numbers. (Lesson 6-9)
4	worked with base-10 blocks. (Lesson 6-10)
5	used base-10 blocks and number grids to add and subtract animal weights and lengths. (Lesson 6-2)
6	shared strategies for Quick Looks for near doubles and making 10. (Lessons 6-4 and 6-6)

Assessment Handbook, p. 37

NAME   DATE   Lesson 6-12 ✓

Unit 6 Self Assessment

Put a check in the box that tells how you do each skill.

Skills	I can do this by myself. I can explain how to do this.	I can do this by myself.	I can do this with help.
① Tell time to the hour.			
② Add two-digit numbers.			
③ Fill name-collection boxes.			
④ Tell the value of each digit in a number.			
⑤ Choose a tool to help me solve a problem.			
⑥ Understand how another child solved a problem.			

Assessment Masters **37**

▸ Unit 6 Assessment

Assessment Handbook, pp. 38–40

| WHOLE CLASS | SMALL GROUP | PARTNER | **INDEPENDENT** |

Children complete the Unit 6 Assessment to demonstrate their progress on the Common Core State Standards covered in this unit.

Generic rubrics in the *Assessment Handbook* appendix can be used to evaluate children's progress on the Mathematical Practices.

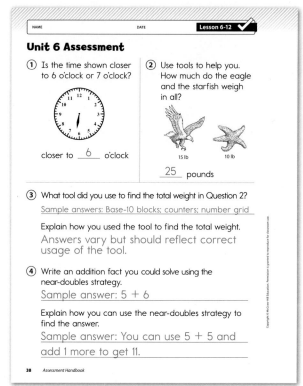

Assessment Handbook, p. 38

Differentiate Adjusting the Assessment

Item(s)	Adjustments
1	To scaffold item 1, have children use an hour-hand-only clock to model the time shown.
2	To extend item 2, have children tell how much 2 eagles and 1 starfish (or 2 eagles and 2 starfish) weigh.
3	To scaffold item 3, give children base-10 blocks, counters, and a number grid. To extend item 3, have children use a different tool to find the total weight.
4	To extend item 4, have children list as many near doubles facts as they can.
5	To scaffold item 5, provide children with counters.
6	To scaffold item 6, provide children with 20 counters.
7	To extend item 7, have children provide at least one name that uses both addition and subtraction.
8–10	To scaffold items 8–10, provide children with base-10 blocks.
11	To extend item 11, have children explain how they could use base-10 blocks to solve the problem.

Advice for Differentiation

All instruction and most practice is complete for the content that is marked with an asterisk (*) on page 574.

Use the online assessment and reporting tools to track children's performance. Differentiation materials are available online to help you address children's needs.

> **NOTE** See the Unit Organizer on pages 498–499 or the online Spiral Tracker for details on Unit 6 focus topics and the spiral.

Assessment Handbook, p. 39

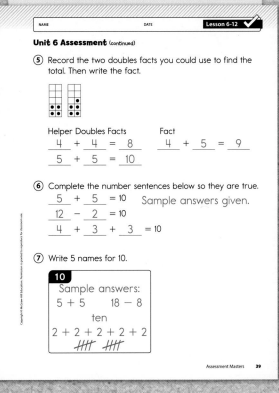

Assessment Handbook, p. 40

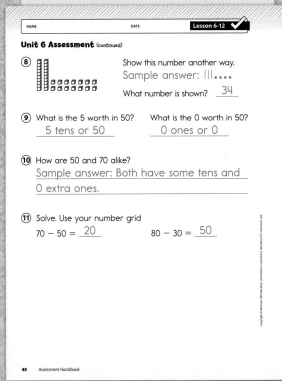

▶ Unit 6 Challenge (Optional)

Assessment Handbook, p. 41

WHOLE CLASS	SMALL GROUP	PARTNER	INDEPENDENT

Children can complete the Unit 6 Challenge after they complete the Unit 6 Assessment.

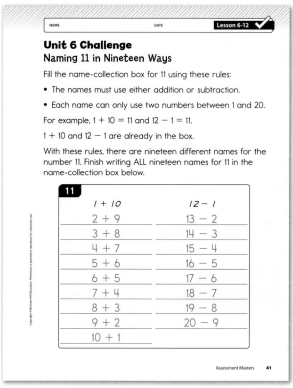

Assessment Handbook, p. 41

Unit 6 Progress Check ✓

Day 2: Cumulative Assessment

2b Assess   30–45 min

Materials

Cumulative Assessment
These items reflect mastery expectations to this point.

Assessment Handbook, pp. 42–44

CCSS Common Core State Standards

	Goals for Mathematical Content (GMC)	**Cumulative Assessment**
1.OA.2	Model and solve number stories involving the addition of 3 addends.*	1
1.OA.3	Apply properties of operations to add or subtract.	1–3, 11
1.OA.6	Add within 10 fluently.*	3
	Add combinations of 10 automatically.*	3
	Add and subtract within 20 using strategies.	1–3
1.OA.7	Understand the meaning of the equal sign.*	3, 7
	Determine whether equations involving addition or subtraction are true or false.	3
1.NBT.1	Count on from any number.*	4–6
	Read and write numbers.*	4–6
1.NBT.3	Compare and order numbers.*	7
	Record comparisons using >, =, or <.	7
1.NBT.4	Understand adding 2-digit numbers and 1-digit numbers.	11
1.MD.1	Order objects by length.*	8
	Compare the lengths of objects indirectly.*	8
1.MD.2	Measure length using same-size units with no gaps or overlaps.*	9
	Express length as a whole number of units.*	9
1.MD.4	Organize and represent data.	10
	Ask questions about data.	12
	Answer questions about data.*	11

	Goals for Mathematical Practice (GMP)	
SMP4	Model real-world situations using graphs, drawings, tables, symbols, numbers, diagrams, and other representations. GMP4.1	10
	Use mathematical models to solve problems and answer questions. GMP4.2	11
SMP6	Think about accuracy and efficiency when you count, measure, and calculate. GMP6.4	1, 9
SMP8	Create and justify rules, shortcuts, and generalizations. GMP8.1	2

*Instruction and most practice on this content is complete.

 Spiral Tracker   **Go Online** ▷ to see how mastery develops for all standards within the grade.

3 Look Ahead   10–15 min

Materials

Math Boxes 6-12
Children preview skills and concepts for Unit 7.

Math Journal 2, p. 133

Home Link 6-12
Children take home the Family Letter that introduces Unit 7.

Math Masters, pp. 177–180

NAME   DATE   Lesson 6-12 ✓

Unit 6 Cumulative Assessment

① Solve.

Lola picked up 5 pink shells, 2 grey shells, and 8 brown shells.

How many shells did Lola pick up in all?

  15 shells

Explain how you found the sum.

Sample answer: I added 8 and 2 to get 10 and then added 10 and 5 to get 15.

② Alice says that if she knows that $8 + 9 = 17$, then she also knows that $9 + 8 = 17$.

Is Alice correct?

Explain why or why not.

Sample answer: Yes, Alice is right. It does not matter which order you add the numbers. The answer will be the same.

③ Write True or False.

$19 = 19$ True $3 + 4 + 5 = 4 + 5 + 3$ True

$8 = 7 + 0$ False $8 + 6 = 6 + 8$ True

$10 = 3 + 6$ False $10 + 2 = 2 + 5 + 5$ True

42 Assessment Handbook

NAME   DATE   Lesson 6-12 ✓

Unit 6 Cumulative Assessment (continued)

Count up by 1s.

④ 66 , 67, 68, 69 , 70 , 71 , 72 , 73

⑤ 88, 89 , 90 , 91, 92 , 93 , 94 , 95

⑥ 113 , 114, 115, 116 , 117 , 118 , 119 , 120

⑦ Write >, <, or =.

$90 \underline{<} 99$ $62 \underline{=} 60 + 2$

$71 \underline{=} 71$ $50 + 7 \underline{<} 70 + 5$

⑧ Ramon compared his pencil to his friends' pencils.
Ramon's pencil is longer than Hugh's pencil.
Ramon's pencil is shorter than Becca's pencil.
Put the three pencils in order from shortest to longest.

  Hugh's pencil   Ramon's pencil   Becca's pencil

⑨ Use one square pattern block.
Measure across your paper.
How many square pattern blocks
across is your paper? Sample answers:
about 8, 9 square pattern blocks

Assessment Masters   43

580   Unit 6 | Addition Fact Strategies

 **Assess**   30–45 min   Go Online

Assessment and Reporting   Differentiation Support

▶ Cumulative Assessment

Assessment Handbook, pp. 42–44

| WHOLE CLASS | SMALL GROUP | PARTNER | **INDEPENDENT** |

Children complete the Cumulative Assessment. The problems in the Cumulative Assessment address content from Units 1–5.

Monitor children's progress on the Common Core State Standards using the online assessment and reporting tools.

Generic rubrics in the *Assessment Handbook* appendix can be used to evaluate children's progress on the Mathematical Practices.

Differentiate	**Adjusting the Assessment**
Item(s)	**Adjustments**
1	To scaffold item 1, provide children with counters. To extend item 1, have children explain another way they could add the same numbers.
2	To scaffold item 2, provide children with counters to justify the turn-around rule. To extend item 2, have children use the turn-around rule to list other facts.
3	To scaffold item 3, provide children with counters. To extend item 3, have children list three additional equations that are true and three that are false.
4–6	To scaffold items 4–6, provide children with number lines or grids. To extend item 6, have children continue counting past 120.
7	To scaffold item 7, provide children with base-10 blocks or number grids.
8	To scaffold item 8, have children draw pencils on their paper and label them with the names.
9	To scaffold item 9, provide children with more than one pattern block, either 2 blocks or 10 blocks, depending on children's abilities.
10	To extend item 10, have children ask the same question of their classmates and create both a tally chart and a bar graph.
11	To scaffold item 11, have children use counters to represent each child who answered the question.
12	To extend item 12, have children write more than one question.

Advice for Differentiation

All instruction and most practice is complete for the content that is marked with an asterisk (*) on page 579.

Use the online assessment and reporting tools to track children's performance. Differentiation materials are available online to help you address children's needs.

③ Look Ahead 10–15 min Go Online 🏠
Home Connections

▶ Math Boxes 6-12: Preview for Unit 7

Math Journal 2, p. 133

| WHOLE CLASS | SMALL GROUP | **PARTNER** | **INDEPENDENT** |

Mixed Practice Math Boxes 6-12 are paired with Math Boxes 6-8. These problems focus on skills and understandings that are prerequisite for Unit 7. You may want to use information from these Math Boxes to plan instruction and grouping in Unit 7.

▶ Home Link 6-12: Unit 7 Family Letter

Math Masters, pp. 177–180

Home Connection The Unit 7 Family Letter provides information and activities related to Unit 7 content.

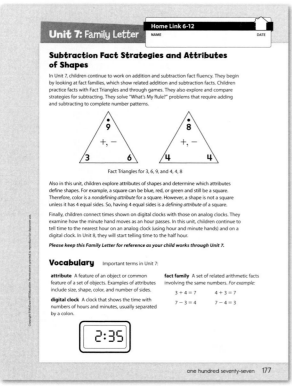

Math Masters, pp. 177–180

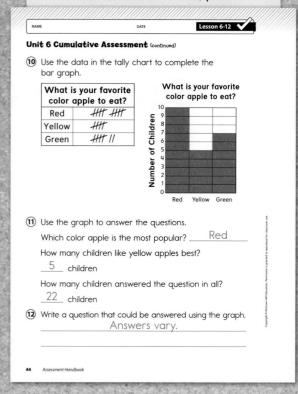

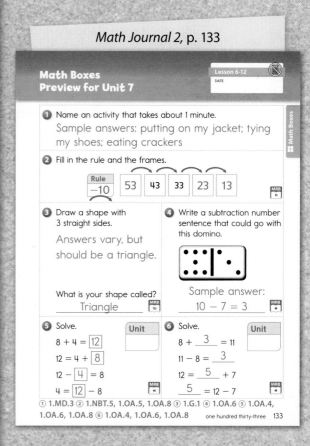

Unit 7 Organizer
Subtraction Fact Strategies and Attributes of Shapes

In this unit, children explore the relationship between addition and subtraction, compare different subtraction strategies, and continue to work on fact fluency. They also explore the defining and nondefining attributes of 2-dimensional shapes and continue their work telling time to the nearest hour, using analog and digital clocks. Children's learning will focus on five clusters of the Common Core's content standards, as well as in-depth work on two of the Mathematical Practices.

 CCSS Standards for Mathematical Content

Domain	Cluster
Operations and Algebraic Thinking	Understand and apply properties of operations and the relationship between addition and subtraction.
	Add and subtract within 20.
	Work with addition and subtraction equations.
Measurement and Data	Tell and write time.
Geometry	Reason with shapes and their attributes.

Because the standards within each domain can be broad, *Everyday Mathematics* has unpacked each standard into Goals for Mathematical Content GMC . For a complete list of Standards and Goals, see page EM1.

For an overview of the CCSS domains, standards, and mastery expectaions in this unit, see the **Spiral Trace** on pages 588–589. See the **Mathematical Background** (pages 590–592) for a discussion of the following key topics:

- Fact Families
- Subtraction Fact Strategies
- Attributes of Shapes
- "What's My Rule?"
- Digital Clocks

 CCSS Standards for Mathematical Practice

SMP7 Look for and make use of structure.

SMP8 Look for and express regularity in repeated reasoning.

For a discussion about how *Everyday Mathematics* develops these practices and a list of Goals for Mathematical Practice GMP , see page 593.

 Virtual Learning Community Go Online to **vlc.cemseprojects.org** to search for video clips on each practice.

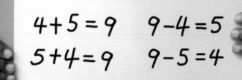

©Joe Polillio/McGraw-Hill Education

Go Digital with these tools at **connectED.mcgraw-hill.com**

 ePresentations
 Student Learning Center
 Facts Workshop Game
 eToolkit
 Professional Development
 Home Connections
 Spiral Tracker
 Assessment and Reporting
 English Learners Support
 Differentiation Support

Contents

*The standards listed here are addressed in the **Focus** of each lesson. For all the standards in a lesson, see the Lesson Opener.

Unit 7 Materials

VLC Virtual Learning Community

See how *Everyday Mathematics* teachers organize materials. Search "Classroom Tours" at **vlc.cemseprojects.org**.

Lesson	*Math Masters*	Activity Cards	Manipulative Kit	Other Materials
7-1	pp. 181–184; TA33–TA34	84	2 dominoes; 10 dominoes with dot sums of 0–9; number cards 0–20; dominoes	slate; Number Cards
7-2	pp. 185–191; TA33; TA35; TA36 (optional); G24–25; G49 (optional)		base-10 blocks; 2 different-colored dice; 4 sets of number cards 0–10; counters	slate; Number Cards; calculator; *My Reference Book* (optional)
7-3	pp. 192–193; G47		Ten Frame and counters (optional); 4 each of number cards 0–10	slate; Ten Frame; Number Cards; *My Reference Book* (optional)
7-4	pp. 194–197; G24–G25; G50 (optional); G51	85	Quick Look Cards 94, 105, 106; per partnership: two dot dice or two 10-sided dice; 20 pennies or other counters; number cards 0–10; 40 toolkit pennies	slate; Number Cards; *My Reference Book* (optional); sheet of paper labeled "Bank;" Class Number Line
7-5	pp. 198–199; G42 (optional); G43–G45	86–87	attribute blocks	number grid (optional); Pattern-Block Template
7-6	pp. 200–202; G52 (optional); G53	88–91	attribute blocks; geoboard; rubber bands; 4 each of number cards 0–10	slate; Number Cards; folder; *My Reference Book* (optional); classroom object
7-7	pp. 203–204; TA30; G48 (optional); G52	92	per partnership: 2 dot dice, 20 dimes, 20 pennies; attribute blocks	Pattern-Block Template; chart paper; per partnership: 5 one-dollar bills; Two-Dimensional Shapes Poster (optional)
7-8	pp. 205–209; TA37; TA38 (optional); G34		craft sticks	slate; paper bag
7-9	pp. 210–211; TA3		square and triangle pattern blocks	Standards for Mathematical Practice Poster; Pattern-Block Template (optional); colored pencils; children's work from Day 1; *My Reference Book*
7-10	pp. 212–214; TA37 (optional); G50 (optional); G51; G54		per partnership: two dot dice, 20 pennies; per group: two dot dice or two 10-sided dice, 20 counters	slate; *My Reference Book* (optional)
7-11	pp. 215–218; TA39; G55–G57	93	toolkit clock; number cards 1–12	slate; digital clock; analog clock; demonstration clock; scissors; brad; hour-hand-only clock; Number Cards
7-12	pp. 219–222; *Assessment Handbook*, pp. 45–51		counters (optional)	crayons or colored pencils

📖 **Literature Link**   7-5   *Windows, Rings, and Grapes—A Look at Different Shapes* (optional); *It's a Shape!* (optional)

Go Online for a complete literature list for Grade 1 and to download all Quick Look Cards.

Problem Solving Professional Development

Everyday Mathematics emphasizes equally all three of the Common Core's dimensions of **rigor:** conceptual understanding, procedural skill and fluency, and applications.

Math Messages, other daily work, Explorations, and Open Response tasks provide many opportunities for children to apply what they know to solve problems.

▶ Math Message

Math Messages require children to solve a problem they have not been shown how to solve. Math Messages provide almost daily opportunities for problem solving.

▶ Daily Work

Journal pages, Home Links, Writing/Reasoning prompts, and Differentiation Options often require children to solve problems in mathematical contexts and real-life situations. *Minute Math+* offers number stories and a variety other practice activities for transition times and for spare moments throughout the day. See Routine 6, pages 32–37.

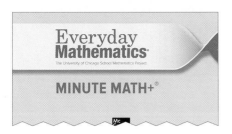

▶ Explorations

In Exploration A, children line up blocks differing in just one attribute as part of the game *Attribute Train.* In Exploration B, children make shapes on geoboards, divide them into two parts, and determine whether the parts are the same size. And in Exploration C, children record fact families for the game *Salute!*

▶ Open Response and Reengagement

In Open Response Lesson 7-9, children use a "What's My Rule?" table and function machine to help them create and justify a rule for a real-world context. The reengagement discussion on Day 2 may focus on why it is important for a rule to always work and how rules can be applied to help solve new problems. Creating and justifying rules, shortcuts, and generalizations is the focus Mathematical Practice for the lesson. Looking for, expressing, and justifying rules can help children find shortcuts for solving complicated problems and make them more efficient problem solvers. GMP8.1

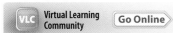 **Virtual Learning Community** Go Online to watch an Open Response and Reengagement lesson in action. Search "Open Response" at **vlc.cemseprojects.org**.

▶ Open Response Assessment

In Progress Check Lesson 7-12, children solve a multistep comparison number story. Making sense of problems is the focus Mathematical Practice for this assessment. As children practice making sense of new and challenging types of problems, they gain confidence in their own problem-solving abilities.

Look for GMP1.1–1.6 markers, which indicate opportunities for children to engage in SMP1: "Make sense of problems and persevere in solving them." Children also become better problem solvers as they engage in all the CCSS Mathematical Practices. The yellow GMP markers throughout the lessons indicate places where you can emphasize the Mathematical Practices and develop children's problem-solving skills.

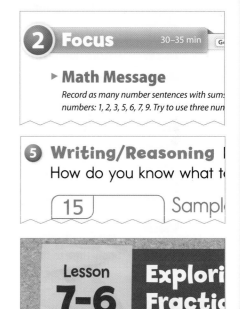

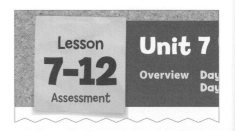

Assessment and Differentiation

Assessment and Reporting

See pages xxii–xxv to learn about this comprehensive online system for recording, monitoring, and reporting children's progress using core program assessments.

VLC Virtual Learning Community   ⟨Go Online⟩ to **vlc.cemseprojects.org** for tools and ideas related to assessment and differentiation from *Everyday Mathematics* teachers.

Ongoing Assessment

In addition to frequent informal opportunities for "kid watching," every lesson (except Explorations) offers an **Assessment Check-In** to gauge children's performance on one or more of the standards addressed in that lesson.

Lesson	Task Description	**CCSS** Common Core State Standards
7-1	Record addition and subtraction facts to represent dominoes.	1.OA.3, 1.OA.4, SMP7
7-2	Record number sentences with an unknown quantity.	1.OA.6, 1.OA.8, SMP2
7-3	Identify the unknown quantity while playing *Salute!*	1.OA.4, 1.OA.6
7-4	Apply a counting strategy to solve subtraction facts.	1.OA.5, 1.OA.6, SMP1
7-5	Identify attributes of triangles.	1.G.1, SMP7
7-7	Recognize defining attributes of squares.	1.G.1, SMP1, SMP6
7-8	Identify the unknown rule in "What's My Rule?" problems.	1.OA.8, SMP8
7-9	Find a rule in a real-world situation.	1.OA.8, SMP8
7-10	Write and solve number sentences with unknowns to check "What's My Rule?" problems.	1.OA.6, 1.OA.8
7-11	Show time to the hour on an analog clock.	1.MD.3, SMP5

▶ Periodic Assessment

Unit 7 Progress Check This assessment focuses on the CCSS domains of *Operations and Algebraic Thinking, Geometry,* and *Measurement and Data.* It also contains an Open Response Assessment to assess children's understanding of how to solve a multistep comparison number story.

> **NOTE** Odd-numbered units include an **Open Response Assessment.** Even-numbered units include a **Cumulative Assessment.**

▶ Unit 7 Differentiation Activities

 Differentiation Support   English Learners Support

Differentiation Options Every regular lesson provides **Readiness, Enrichment, Extra Practice** and **Beginning English Language Learners Support** activities that address the Focus standards for that lesson.

CCSS 1.OA.6, SMP2	CCSS 1.OA.6, SMP2	CCSS 1.OA.6, SMP6
Readiness 10–15 min	**Enrichment** 10–15 min	**Extra Practice** 10–15 min
WHOLE CLASS / SMALL GROUP / PARTNER	WHOLE CLASS / SMALL GROUP / PARTNER	WHOLE CLASS / SMALL GROUP / PARTNER
Exploring Doubles number cards 1–10, counters,	**Drawing Doubles Pictures** Activity Card 76, crayons	**Playing *Roll and Record Doubles***

Activity Cards These activities, written to children, enable you to differentiate Part 2 of the lesson through small-group work.

English Language Learners Activities and point-of-use support help children at different levels of English language proficiency succeed.

Differentiation Support Two online pages for most lessons provide suggestions for game modifications, ways to scaffold lessons for children who need additional support, and language development suggestions for Beginning, Intermediate, and Advanced English language learners.

Differentiation Support online pages

Ongoing Practice

 Differentiation Support

▶ Embedded Facts Practice

Basic Facts Practice can be found in every part of the lesson. Look for activities or games labeled with CCSS 1.OA.6; or go online to the Spiral Tracker and search using CCSS 1.OA.6.

For **ongoing distributed practice,** see these activities:
- Mental Math and Fluency
- Differentiated Options: Extra Practice
- Part 3: Journal pages, Math Boxes, *Math Masters*, Home Links
- Print and online games

▶ Games

Games in *Everyday Mathematics* are an essential tool for practicing skills and developing strategic thinking.

Lesson	Game	Skills and Concepts	CCSS Common Core State Standards
7-2	*Beat the Calculator*	Using mental addition	1.OA.3, 1.OA.4, 1.OA.6, SMP6
7-2 7-4	*Subtraction Bingo*	Subtracting facts	1.OA.6, SMP6
7-3 7-6	*Salute!*	Adding and subtracting facts	1.OA.3, 1.OA.4, 1.OA.6, SMP6, SMP7
7-3	*Top-It* with Subtraction	Subtracting facts using relation symbols	1.OA.6, SMP6
7-3	*Roll and Record Doubles*	Adding doubles facts	1.OA.6, SMP6
7-4 7-10	*Shaker Addition Top-It*	Adding facts	1.OA.6, 1.NBT.3, SMP6
7-4	*The Difference Game*	Subtracting facts	1.OA.4, 1.OA.5, 1.OA.6, SMP6
7-5	*Stop and Go*	Adding and subtracting 2-digit numbers	1.NBT.4, 1.NBT.6, SMP1
7-6 7-7	*Attribute Train*	Comparing attributes	1.G.1, SMP7
7-7	*Penny-Dime-Dollar Exchange*	Making place-value exchanges	1.NBT.2, 1.NBT.2a, SMP6
7-8	*What's Your Way?*	Finding 10 more and 10 less than a number	1.NBT.5, SMP6
7-10	*Tric-Trac*	Adding facts	1.OA.3, 1.OA.6, SMP1
7-11	*Time Match*	Telling time	1.MD.3 ,SMP2

CCSS Spiral Trace: Skills, Concepts, and Applications

⭐ **Mastery Expectations** This Spiral Trace outlines instructional trajectories for key standards in Unit 7. For each standard, it highlights opportunities for Focus instruction, Warm Up and Practice activities, as well as formative and summative assessment. It describes the **degree of mastery**—as measured against the entire standard—expected at this point in the year.

Operations and Algebraic Thinking

1.OA.3 Apply properties of operations as strategies to add and subtract. *Examples: If 8 + 3 = 11 is known, then 3 + 8 = 11 is also known. (Commutative property of addition.) To add 2 + 6 + 4, the second two numbers can be added to make a ten, so 2 + 6 + 4 = 2 + 10 = 12. (Associative property of addition.)*

⭐ By the end of Unit 7, expect children to **use the turn-around rule to generate fact families.**

1.OA.4 Understand subtraction as an unknown-addend problem. *For example, subtract 10 − 8 by finding the number that makes 10 when added to 8.*

⭐ By the end of Unit 7, expect children to **think addition to find the difference between two numbers.**

1.OA.6 Add and subtract within 20, demonstrating fluency for addition and subtraction within 10. Use strategies such as counting on; making ten (e.g., 8 + 6 = 8 + 2 + 4 = 10 + 4 = 14); decomposing a number leading to a ten (e.g., 13 − 4 = 13 − 3 − 1 = 10 − 1 = 9); using the relationship between addition and subtraction (e.g., knowing that 8 + 4 = 12, one knows that 12 − 8 = 4); and creating equivalent but easier or known sums (e.g., adding 6 + 7 by creating the known equivalent 6 + 6 + 1 = 12 + 1 = 13).

⭐ By the end of Unit 7, expect children to **use think addition, counting up, and counting back strategies to solve subtraction facts.**

1.OA.8 Determine the unknown whole number in an addition or subtraction equation relating three whole numbers. *For example, determine the unknown number that makes the equation true in each of the equations 8 + ? = 11, 5 = ☐ − 3, 6 + 6 = ☐.*

⭐ By the end of Unit 7, expect children to **find an unknown rule (including a number and an operation) relating two numbers and describe that relationship with a number sentence.**

Spiral Tracker

Go to **connectED.mcgraw-hill.com** for comprehensive trajectories
that show how in-depth mastery develops across the grade.

Measurement and Data

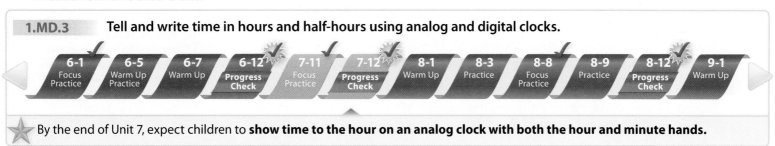

1.MD.3   **Tell and write time in hours and half-hours using analog and digital clocks.**

| 6-1 Focus Practice | 6-5 Warm Up Practice | 6-7 Warm Up | 6-12 Progress Check | 7-11 Focus Practice | 7-12 Progress Check | 8-1 Warm Up | 8-3 Practice | 8-8 Focus Practice | 8-9 Practice | 8-12 Progress Check | 9-1 Warm Up |

By the end of Unit 7, expect children to **show time to the hour on an analog clock with both the hour and minute hands.**

Geometry

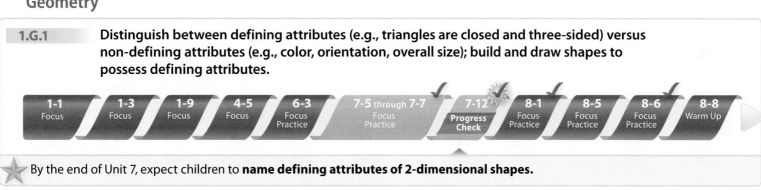

1.G.1   **Distinguish between defining attributes (e.g., triangles are closed and three-sided) versus non-defining attributes (e.g., color, orientation, overall size); build and draw shapes to possess defining attributes.**

| 1-1 Focus | 1-3 Focus | 1-9 Focus | 4-5 Focus | 6-3 Focus Practice | 7-5 through 7-7 Focus Practice | 7-12 Progress Check | 8-1 Focus Practice | 8-5 Focus Practice | 8-6 Focus Practice | 8-8 Warm Up |

By the end of Unit 7, expect children to **name defining attributes of 2-dimensional shapes.**

Key    = Assessment Check-In   = Progress Check Lesson   = Current Unit   = Previous or Upcoming Lessons

Mathematical Background: Content

 The discussion below highlights major content areas and Common Core State Standards addressed in Unit 7. See the online Spiral Tracker for complete information about learning trajectories for all standards.

▶ Fact Families (Lessons 7-1 and 7-2)

In Unit 7, children use fact families to improve fluency with addition and subtraction. **1.OA.6** An addition/subtraction fact family that relates 8, 9, and 17 consists of the facts: $8 + 9 = 17$, $9 + 8 = 17$, $17 - 9 = 8$, and $17 - 8 = 9$. An addition/subtraction fact family that relates 4, 4, and 8 includes only the facts $4 + 4 = 8$ and $8 - 4 = 4$.

Lesson 7-1 uses dominoes to introduce the concept of fact families: the number of dots on each part of the domino and the total number of dots are used to generate the set of four related facts. For example, the 3|6 domino produces two addition and two subtraction facts: $3 + 6 = 9$, $6 + 3 = 9$, $9 - 3 = 6$, and $9 - 6 = 3$.

In Lesson 7-2, children are introduced to addition/subtraction Fact Triangles, the *Everyday Mathematics* version of flash cards. Fact Triangles emphasize fact families and the relationship between addition and subtraction. If given a Fact Triangle that shows 8, 9, and 17, children generate $8 + 9 = 17$, $9 + 8 = 17$, $17 - 8 = 9$, and $17 - 9 = 8$.

Children often work in partnerships as they practice basic facts using Fact Triangles. One child hides a number on the Fact Triangle by covering a corner with a finger. The other child gives an addition or subtraction fact that has the hidden number as its answer. For example, if a child covers the 8 in the 8, 9, 17 Fact Triangle, the other child might say, "$17 - 9 = 8$." With practice, children deepen their understanding of how addition and subtraction facts are related, fostering their development of fact fluency. **1.OA.6**

> **Go Online** to the *Implementation Guide* for more information about Fact Families and Fact Triangles.

▶ Subtraction Fact Strategies

(Lessons 7-3 and 7-4)

The relationship between addition and subtraction is reinforced in Lesson 7-3 as children are encouraged to *think addition* when solving subtraction problems. For instance, to solve $16 - 8 = \boxed{}$, children think of addition to generate a related problem, $8 + \boxed{} = 16$. **1.OA.4** They then use their knowledge of addition facts to find the unknown. **1.OA.8** To help children learn and apply the think-addition strategy, the problems in Lesson 7-3 involve facts that children are most familiar with: doubles and combinations of 10. *Salute!* is a game that provides more practice for the think-addition strategy.

In previous units, children approached most subtraction situations by counting back. In Lesson 7-4, children compare counting back to a new strategy: *counting up to subtract.* Like think addition, this strategy connects subtraction to a related addition

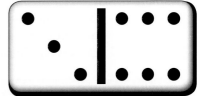

Standards and Goals for Mathematical Content

Because the standards within each domain can be broad, *Everyday Mathematics* has unpacked each standard into Goals for Mathematical Content **GMC**. For a complete list of Standards and Goals, see page EM1.

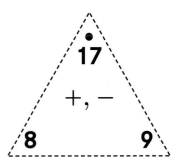

This domino is associated with 3, 6, and 9.

This is an addition/subtraction Fact Triangle. The number under the dot represents the sum of the other two numbers.

Unit 7 Vocabulary

attribute	fact family	polygon
clockwise	Fact Triangle	rule; "What's My Rule?"
closed; open	function machine	subtraction fact
defining attribute	minute hand	think addition
digital clock	nondefining attribute	vertex

▶ Subtraction Fact Strategies *Continued*

problem with an unknown addend. For example, to solve $13 - 9 = \boxed{}$, children notice that it is more efficient to count up from 9 to 13 (in effect, solving $9 + \boxed{} = 13$) than to start at 13 and count back 9. **1.OA.8**

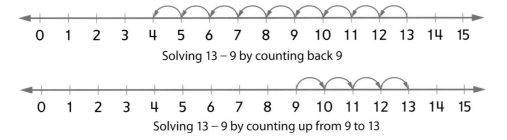

Solving 13 − 9 by counting back 9

Solving 13 − 9 by counting up from 9 to 13

▶ Attributes of Shapes (Lessons 7-5 through 7-7)

In *Kindergarten Everyday Mathematics*, children's work with shapes focused on recognizing, naming, and sorting activities. In Grade 1, this work is extended to include a more formal emphasis on defining and nondefining attributes of shapes. **1.G.1** Children consider questions such as *How can two shapes both be triangles if one is large and green and the other is small and purple?*

The activities in Lesson 7-5 highlight the concept of *attributes*—characteristics such as size, color, shape, and thickness. This work helps children develop a more sophisticated method for observing and describing objects, allowing them to make more detailed statements than, for example, "this shape is a rectangle because it looks like a door." In Lesson 7-6, children practice identifying and comparing attributes of shapes by playing the game *Attribute Train*. Lesson 7-7 formally introduces the terms *defining* and *nondefining attributes*. Children list as many attributes as they can for a given shape (for example: the triangle is green, has 3 sides, has 3 vertices, is big), categorizing each attribute by whether it is a characteristic of *all* examples of that shape (defining) or whether it only applies to *some* examples of that shape (nondefining). For example: All triangles have 3 sides, so having 3 sides is a defining attribute of triangles. But only *some* triangles are green, so color is a nondefining attribute of triangles.

This progression of spatial understanding—from noting only global appearances to focusing on special characteristics and grouping shapes according to these traits—is critical to developing sound geometric concepts.

Children in *First Grade Everyday Mathematics* determine that both shapes are polygons with exactly 3 sides and 3 vertices. Thus, both shapes are triangles.

▶ "What's My Rule?" (Lessons 7-8 and 7-10)

Children are introduced to the "What's My Rule?" routine in Lesson 7-8. This routine provides a format for children to identify relationships between pairs of numbers, to use operations to describe the relationships, and to find unknown numbers. **1.OA.8** The routine also provides practice with basic facts. **1.OA.6**

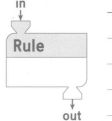

in	out
3	6
7	10
2	5
5	8

Children learn the "What's My Rule?" routine in the context of a function machine. The function machine is a metaphor that helps children understand how the input and output numbers in "What's My Rule?" tables are produced. The machines are "programmed" to take in numbers (input), to change them according to a rule, and then to send out the new numbers (output). The "What's My Rule?" table serves as a place to record each input and output. In some "What's My Rule?" problems, children are asked to find "what comes out" (output numbers). In other instances, the input is the unknown and children must find "what goes in." **1.OA.8** And, in some problems, children must determine what rule is being applied, based on what they notice about the given inputs and outputs.

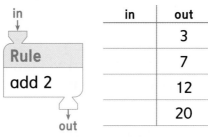

Rule +3

in	out
3	
8	
10	
15	

What comes out?

Rule add 2

in	out
	3
	7
	12
	20

What goes in?

Rule

in	out
10	0
15	5
26	16
34	24

What is the rule?

Lesson 7-10 focuses on "What's My Rule?" problems in which the input numbers (or unknown addends) are missing. For example, if the rule is "add 2" and the output number is 7, children are encouraged to think: "I put a number in the machine. The machine added 2 and then a 7 came out. What number did I put in?" Lesson 7-10 also connects this thinking to the formal number sentence (in this case, ☐ + 2 = 7). **1.OA.8** To ensure ample practice with basic facts, the output numbers used in this lesson do not exceed 20. **1.OA.6**

▶ Digital Clocks (Lesson 7-11)

Up to this point in first grade, children have been telling time using hour-hand-only analog clocks. In Lesson 7-11, children explore the concept of a minute by discussing familiar activities that take "about a minute." Children observe how the minute hand moves in relation to the hour hand. They then begin using analog clocks with minute hands to show "o'clock" times. **1.MD.3** This gradual progression from hour-hand-only clocks to clocks with a minute hand in the "o'clock position" supports children's developing understanding of telling time. Next, children practice representing the time shown on the analog clocks in digital notation. Again, the focus remains on "o'clock" times. In Unit 8, children will learn to tell time to the half hour.

> **NOTE** One option for teaching "What's My Rule?" problems is to create a large function machine so children can act out the problems. Some teachers have children sit under a desk or a table; they receive the "in" numbers on slips of paper or on a slate, change those numbers according to the rule, and then return the output number on another slip of paper or slate. Other teachers use a large box with slots labeled "in" and "out" to represent a function machine. A child sits inside the box, receives "in" numbers, changes them according to the rule, and returns the result through the "out" slot.

Mathematical Background: Practices

 In Everyday Mathematics, *children learn the* **content** *of mathematics as they engage in the* **practices** *of mathematics. As such, the Standards for Mathematical Practice are embedded in children's everyday work, including hands-on activities, problem-solving tasks, discussions, and written work. Read here to see how Mathematical Practices 7 and 8 are emphasized in this unit.*

▶ Standard for Mathematical Practice 7

Standard for Mathematical Practice 7 focuses on the importance of discerning structures. One important structure addressed in Unit 7 is the relationship between addition and subtraction. As children generate fact families through their work with dominoes and Fact Triangles, they begin to notice—and, eventually, use to solve problems—the inverse relationship between addition and subtraction. GMP7.1 For example, in Lessons 7-1 through 7-3, children use the think-addition strategy to solve subtraction problems. They discover that when they encounter a problem such as $14 - 7 = \boxed{}$, they can use a known fact ($7 + 7 = 14$) from the same fact family to help them find the answer. GMP7.2

In Lessons 7-5 through 7-7, children look for structure as they categorize geometric figures according to specific attributes. As children describe characteristics of shapes, sort shapes by attributes, and make decisions about which shapes can join the Attribute Train, they pay close attention to structure. GMP7.1

In Lessons 7-8 and 7-10, children solve "What's My Rule?" problems that require attention to mathematical structure. Each function machine has a rule that is applied to each input number. Children determine the rule and use their understanding of this rule to find unknown inputs or outputs. GMP7.1, GMP7.2

▶ Standard for Mathematical Practice 8

Standard for Mathematical Practice 8 emphasizes the need to generalize strategies to solve a variety of similar mathematical problems. In *First Grade Everyday Mathematics*, children develop this practice as they create simple rules based on patterns that they notice. For example, in Lesson 7-1, children examine fact families and notice that some fact families have only two facts. As children further examine these families, they generate a rule that fact families with doubles have only two facts. GMP8.1

In Lesson 7-4, children solve subtraction problems using two strategies for each problem: counting up and counting back. They note which strategy is more efficient for each problem and are guided to develop general rules about when to use each strategy. For example, children may notice that in problems where the numbers are close together (such as in $11 - 9 = \boxed{}$), it is more efficient to count up than to count back. GMP8.1

Another important component of Mathematical Practice 8 is looking for and making sense of repeated calculations. Many "What's My Rule?" problems in Lessons 7-8 and 7-10 require children to find an unknown rule. To do this, children must notice the pattern—or regularity—between each input and output. By finding the pattern, they can generate a rule. For example, children must notice that for a number to change from 3 to 7 or from 6 to 10, the rule "add 4" applies in each case. GMP8.1

 Standards and Goals for Mathematical Practice

SMP7 **Look for and make use of structure.**
   GMP7.1 Look for mathematical structures such as categories, patterns, and properties.
   GMP7.2 Use structures to solve problems and answer questions.

SMP8 **Look for and express regularity in repeated reasoning.**
   GMP8.1 Create and justify rules, shortcuts, and generalizations.

Go Online to the *Implementation Guide* for more information about the Mathematical Practices.

For children's information on the Mathematical Practices, see *My Reference Book,* pages 1–22.

Fact Families

Overview Children deepen their understanding of the relationship between addition and subtraction by learning about fact families.

▶ **Before You Begin**
For the Math Message, each child will need one half of *Math Masters,* page 181.

▶ **Vocabulary**
subtraction fact · fact family · think addition

Common Core State Standards

Focus Clusters
- Understand and apply properties of operations and the relationship between addition and subtraction.
- Add and subtract within 20.
- Work with addition and subtraction equations.

① Warm Up 15–20 min

	Materials	
Mental Math and Fluency Children solve number stories.	slate	1.OA.1, 1.NBT.4
Daily Routines Children complete daily routines.	See pages 2–37.	See pages xiv–xvii.

② Focus 30–35 min

Math Message Children solve parts-and-total problems on dominoes.	*Math Masters,* p. 181	1.OA.4, 1.OA.6, 1.OA.8 SMP7
Introducing Addition/Subtraction Fact Families Children are introduced to fact families and write fact families for numbers shown on dominoes.	*Math Masters,* pp. 181, TA33–TA34; 2 dominoes	1.OA.3, 1.OA.4, 1.OA.6, 1.OA.8 SMP3, SMP8
Writing Fact Families Children write fact families and use the relationship between addition and subtraction to solve problems.	*Math Journal 2,* pp. 134–135	1.OA.3, 1.OA.4, 1.OA.6, 1.OA.8 SMP3, SMP7, SMP8
✓ **Assessment Check-In** See page 598.	*Math Journal 2,* pp. 134–135	1.OA.3, 1.OA.4, SMP7

CCSS 1.OA.4 **Spiral Snapshot**

GMC Understand subtraction as an unknown-addend problem.

6-11 Practice | 7-1 Focus Practice | 7-2 through 7-4 Focus Practice | 7-6 Focus Practice | 7-7 Warm Up Practice | 8-1 Practice | 8-2 Warm Up

📶 Spiral Tracker   **Go Online**   to see how mastery develops for all standards within the grade.

③ Practice 15–20 min

Subtracting 10s Children subtract multiples of 10 from multiples of 10.	*Math Journal 2,* p. 136	1.NBT.2, 1.NBT.6
Math Boxes 7-1 Children practice and maintain skills.	*Math Journal 2,* p. 137	See page 599.
Home Link 7-1 **Homework** Children write fact families for dominoes.	*Math Masters,* p. 184	1.OA.3, 1.OA.4, 1.OA.6, 1.MD.1, 1.MD.2, SMP7

---

connectED.mcgraw-hill.com

Plan your lessons online with these tools.

ePresentations   Student Learning Center   Facts Workshop Game   eToolkit   Professional Development   Home Connections   Spiral Tracker   Assessment and Reporting   English Learners Support   Differentiation Support

Differentiation Options

 RtI

CCSS 1.OA.6, SMP2

Readiness   10–15 min

Number Card and Domino Concentration

> WHOLE CLASS
> SMALL GROUP
> **PARTNER**
> INDEPENDENT

10 dominoes with dot sums of 0–9, number cards 0–9

To provide experience writing number sentences, have partners do the following activity. Place number cards facedown in 2 rows of 5 cards each. Then arrange the dominoes similarly. Have children take turns flipping over a card and a domino. If the sum of the domino dots matches the number on the card, the child keeps the card-domino pair and flips over another pair. **GMP2.3** If they do not match, the child puts them back in their original places after the partner has seen them. Continue until there are none left. Then have children record addition facts to represent the dominoes they collected. For example, a child might record $3 + 0 = 3$ for a 3|0 domino. **GMP2.3**

CCSS 1.OA.3, 1.OA.4, 1.OA.6, 1.OA.8, SMP7

Enrichment   10–15 min

Fact Family Swap

> WHOLE CLASS
> SMALL GROUP
> **PARTNER**
> INDEPENDENT

Activity Card 84;
Math Masters, p. 182;
number cards 0–20

To further explore finding unknowns in number sentences and recording fact families, have children write facts and complete their partners' fact families. **GMP7.2**

Fact Family Swap

8 4

What You Need
Fact Family Swap Record Sheet, page 182
number cards 0–20

What To Do
Work with a partner.
❶ Take 2 number cards.
❷ Write those numbers in the two squares for Item 1.
❸ Trade papers with your partner. Your partner solves and records the fact family.
❹ Repeat until all 6 items are complete.

More You Can Do
Use number cards 0–9.
Use the numbers drawn to tell the number of tens to record.
For example: If you draw 6 and 2, use the numbers 60 and 20. Solve and record the fact family for those 2 numbers.

Talk About It
Check each other's work.
Which items were easiest to solve? Hardest?

1.OA.4, 1.OA.6, 1.OA.8, SMP7

Fact Family Swap Record Sheet

Lesson 7-1
NAME   DATE

① Fact: ☐ + ☐ = ☐
Fact Family:
_____
_____

② Fact: ☐ + ___ = ☐
Fact Family:
_____
_____

③ Fact: ___ + ☐ = ☐
Fact Family:

④ Fact: ☐ − ___ = ☐
Fact Family:

CCSS 1.OA.3, 1.OA.4, 1.OA.6, SMP2

Extra Practice   10–15 min

Recording Fact Families

> WHOLE CLASS
> SMALL GROUP
> **PARTNER**
> **INDEPENDENT**

Math Masters, p. 183;
dominoes

Have children practice recording fact families by choosing dominoes and filling in *Math Masters,* page 183. Children connect representations of numbers on each domino to the facts in the fact family. **GMP2.3**

Solving Fact Families

Lesson 7-1
NAME   DATE

① Numbers: ___, ___, ___ ☐

Fact Family: ___ + ___ = ___   ___ − ___ = ___
             ___ + ___ = ___   ___ − ___ = ___

② Numbers: ___, ___, ___ ☐

Fact Family: ___ + ___ = ___   ___ − ___ = ___
             ___ + ___ = ___   ___ − ___ = ___

③ Numbers: ___, ___, ___ ☐

Fact Family: ___ + ___ = ___   ___ − ___ = ___
             ___ + ___ = ___   ___ − ___ = ___

④ Numbers: ___, ___, ___ ☐

Fact Family: ___ + ___ = ___   ___ − ___ = ___
             ___ + ___ = ___   ___ − ___ = ___

1.OA.3, 1.OA.4, 1.OA.6   one hundred eighty-three 183

English Language Learners Support

Beginning ELL Introduce the term *fact family* by showing a family portrait. Identify the different people in the picture. For example, say: *This is the mother. This is the father. This is the son.* Emphasize that they are in the same family. Then show the fact family for 5, 3 and 8 and say: *This is a fact family.* Use examples and non-examples, and ask yes/no questions such as: *Is $3 + 5 = 8$ in this fact family? Is $5 + 3 = 8$ in this fact family? Would $5 - 3 = 2$ be in the fact family with $3 + 5 = 8$ and $8 - 5 = 3$?* Provide oral practice with the term *fact family* by showing other fact families and asking: *What is this called?*

 Go Online   **ELL** English Learners Support

① Warm Up   15–20 min   Go Online   ePresentations   eToolkit

▶ Mental Math and Fluency

Have children solve and write the answers on their slates. Encourage them to use number grids. Briefly discuss their strategies.

●○○ Charlotte read 14 books this year. Her sister read 5 books. How many books did Charlotte and her sister read in all? 19 books

●●○ Noah has 66 pennies in his piggy bank. His dad gives him 20 more pennies. How many pennies does he have now? 86 pennies

●●● A teacher has 25 red pens and 13 blue pens. How many pens does he have in all? 38 pens

▶ Daily Routines

Have children complete the Daily Routines. For detailed instructions, see pages 2–37. For specifics on CCSS coverage, see pages xiv–xvii.

② Focus   30–35 min   Go Online   ePresentations   eToolkit

▶ Math Message

Math Masters, p. 181

Complete the problems on the Domino Totals page. GMP7.2

▶ Introducing Addition/Subtraction Fact Families

Math Masters, pp. 181 and TA33–TA34

WHOLE CLASS   SMALL GROUP   PARTNER   INDEPENDENT

Math Message Follow-Up Have children share how they solved the problems on *Math Masters*, page 181.

Problem 2

Possible strategies:

• Counting up: I counted from 7 to 11 on my fingers.

• Guessing and checking: I tried 2. Because 7 + 2 = 9, I knew that 2 was not enough. So I tried 7 + 3 equals 10. But 3 was not enough, so I tried 4. Because 7 + 4 equals 11, I knew there are 4 missing dots.

• Solving an unknown-addend problem: I thought "What do I need to add to 7 to get 11?" Because 7 + 4 is 11, I knew the answer is 4 dots.

Math Masters, p. 181

Domino Totals   Lesson 7-1
NAME   DATE

① Find the total.

Unit
domino
dots

7

Draw the missing dots.

②   11   ③   8

- - - - - - ✂ - - - - - -

Domino Totals   Lesson 7-1
NAME   DATE

① Find the total.

Unit
domino
dots

7

Draw the missing dots.

②   11   ③   8

1.OA.4, 1.OA.6, 1.OA.8, SMP7   one hundred eighty-one 181

Display a 4|6 domino (*Math Masters,* page TA33). Help children see the domino as displaying a fact family by asking questions and eliciting patterns.

- *Which three numbers go with this domino?* 4, 6, 10 Record these for the class.
- *What number sentences can you write to represent this domino?* $4 + 6 = 10$; $6 + 4 = 10$; $10 - 4 = 6$; $10 - 6 = 4$ If children only suggest addition number sentences, ask them for number sentences that use subtraction. Record all four number sentences for the class.
- Circle the two addition facts, and ask: *What do you notice about these facts?* Sample answer: They have the same sum. *Why do they have the same sum?* Sample answer: They have the same sum because we are adding the same two numbers and the order doesn't matter.

Children have been solving subtraction facts since Unit 4, but they might not yet know the term. Underline the two **subtraction facts** and introduce them as such. Explain that these facts are related to or are made from the same three numbers as the addition facts.

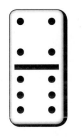

4, 6, 10

$4 + 6 = 10$

$6 + 4 = 10$

$10 - 4 = 6$

$10 - 6 = 4$

Repeat with an 8|2 domino (*Math Masters,* page TA34). Focus on the three numbers associated with this domino and the four number sentences that can be written with these numbers.

Explain that the facts that can be written with the three domino numbers are related and are called a **fact family.** Understanding fact families and the relationships between the facts can help children solve facts they do not already know.

Distribute two dominoes to each child. Be sure to include several dominoes showing doubles. Ask children to record the fact families for each of their dominoes. Facts may be written horizontally or vertically. Have several children share their fact families. Ask: *Do all of you have four number sentences in your fact families?* Encourage children with doubles dominoes to share their number sentences, prompting the class to justify why there are only two facts in those fact families and to consider whether all doubles fact families have only two number sentences. GMP3.1, GMP8.1 Sample answer: For doubles, the turn-around rule does not give a different fact, so I only have 1 addition fact. Because of this, I also have only 1 subtraction fact.

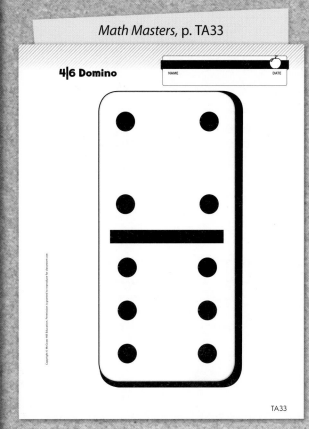

Math Masters, p. TA33

4|6 Domino

NAME   DATE

TA33

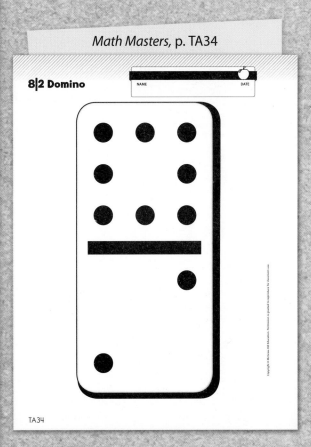

Math Masters, p. TA34

8|2 Domino

NAME   DATE

TA34

Math Journal 2, p. 134

Fact Families

Write the 3 numbers for each domino.
Use the numbers to write the fact family.

1

Numbers: _3_, _5_, _8_

Fact Family: _3_ + _5_ = _8_
_5_ + _3_ = _8_
_8_ − _5_ = _3_
_8_ − _3_ = _5_

2

Numbers: _3_, _3_, _6_

Fact Family: _3_ + _3_ = _6_
_6_ − _3_ = _3_

3

Numbers: _9_, _3_, _12_

Fact Family: _9_ + _3_ = _12_
_3_ + _9_ = _12_
_12_ − _3_ = _9_
_12_ − _9_ = _3_

4

Numbers: _0_, _7_, _7_

Fact Family: _0_ + _7_ = _7_
_7_ + _0_ = _7_
_7_ − _0_ = _7_
_7_ − _7_ = _0_

134 one hundred thirty-four   1.OA.3, 1.OA.4, 1.OA.6, 1.OA.8, SMP3, SMP7, SMP8

Math Journal 2, p. 135

Fact Families (continued)

5 Abbey covers one side of a domino.
You can only see 4 dots.
She says that the total for the domino is 9.
Complete the fact family.

Numbers: 4, _5_, 9

Fact Family: 4 + _5_ = 9
_5_ + _4_ = 9
_9_ − _4_ = _5_
_9_ − _5_ = _4_

6 Juan does not know the answer to 12 − 6 = ☐.
What addition fact could he use to help him find the answer?
Explain.

Sample answer: He could think of the addition
fact in the fact family with 12 − 6. This means he
would think 6 + ☐ = 12. So the answer is 6.

1.OA.3, 1.OA.4, 1.OA.6, 1.OA.8, SMP3,
SMP7, SMP8   one hundred thirty-five 135

▶ Writing Fact Families

Math Journal 2, pp. 134–135

| WHOLE CLASS | SMALL GROUP | **PARTNER** | INDEPENDENT |

Have children complete fact families for dominoes on journal pages 134 and 135. Children extend their thinking about using fact families in Problems 5 and 6 as they work with missing addends and describe how to relate subtraction to addition. GMP3.1, GMP7.2

✔ Assessment Check-In   CCSS 1.OA.3, 1.OA.4

Math Journal 2, pp. 134–135

Observe as children complete the journal pages. Expect children to accurately record the addition and subtraction facts for the dominoes in Problems 1–4. For children who are unable to write the addition number sentences, scaffold by turning the domino to illustrate the turn-around rule. As this is the first exposure to fact families, expect most children will struggle to write subtraction number sentences. You may wish to have children write the domino numbers in parts-and-total diagrams. Then, one at a time, children cover up a part to help write a subtraction fact and uncover the part to find the solution. GMP7.2

✔ Assessment and Reporting   Go Online ▶ to record student progress and to see trajectories toward mastery for these standards.

Display 8 − 5 = ☐ and ask: *Suppose that you do not know this fact. How could a fact family help you solve it?* GMP7.2 Sample answer: I can think of 5 + ☐ = 8 and use that to find the missing number. To help children connect the operations, record the corresponding number sentences side-by-side (8 − 5 = ☐ and 5 + ☐ = 8). Explain that when they do this, they are thinking about addition to help solve a related subtraction problem. GMP7.2, GMP8.1 Add **think addition** to the Strategy Wall.

Go Online ▶ for more information on *think addition*.

Differentiate   **Adjusting the Activity**

Have children who are struggling with subtraction ask themselves "How many more do I need to add to get to _____?" as they try to solve the problems. This question relates addition and subtraction. You may wish to display it for children to reference.

Go Online ▶   Differentiation Support

Summarize Have partnerships apply the think-addition strategy to solve 10 − 2. Tell children they will continue to use think addition to help them subtract in upcoming lessons.

③ Practice 15–20 min

Go Online

ePresentations · eToolkit · Home Connections

▶ Subtracting 10s

Math Journal 2, p. 136

| WHOLE CLASS | SMALL GROUP | **PARTNER** | **INDEPENDENT** |

Have children practice subtracting multiples of 10 from multiples of 10 using drawings.

▶ Math Boxes 7-1

Math Journal 2, p. 137

| WHOLE CLASS | SMALL GROUP | **PARTNER** | **INDEPENDENT** |

Mixed Practice Math Boxes 7-1 are paired with Math Boxes 7-3.

▶ Home Link 7-1

Math Masters, p. 184

Homework Children practice generating fact families for numbers shown on dominoes.

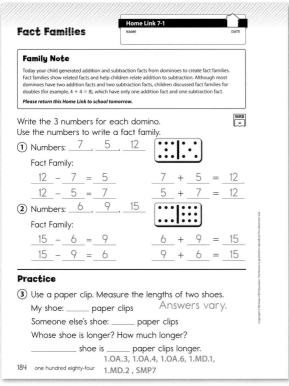

Math Masters, p. 184

Math Journal 2, p. 136

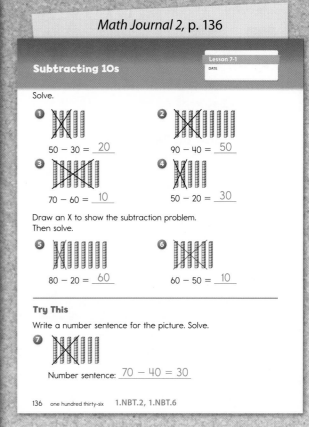

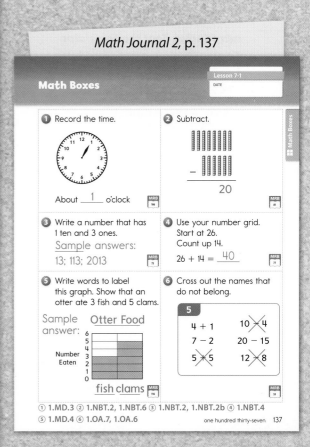

Math Journal 2, p. 137

More Fact Families

Overview Children use Fact Triangles to practice adding and subtracting within 20.

▶ **Before You Begin**

Have children cut apart all the Fact Triangles on Activity Sheets 10–16. On *Math Masters,* page TA35, write 10 near the top vertex of the triangle (just underneath the dot), and 4 and 6 near the bottom two vertices (inside the triangle). Then display it and *Math Masters,* page TA33 with the Math Message.

▶ **Vocabulary**

Fact Triangle

**Common Core
State Standards**

Focus Clusters
- Understand and apply properties of operations and the relationship between addition and subtraction.
- Add and subtract within 20.
- Work with addition and subtraction equations.

1 Warm Up   15–20 min

	Materials	
Mental Math and Fluency Children record the number represented by base-10 blocks.	base-10 blocks, slate	**1.NBT.2, 1.NBT.2a, 1.NBT.2b, 1.NBT.2c**
Daily Routines Children complete daily routines.	See pages 2–37.	See pages xiv–xvii.

2 Focus   30–35 min

Math Message Children compare a domino and a Fact Triangle.	*Math Masters,* pp. TA33 and TA35	**1.OA.3, 1.OA.4, 1.OA.6** **SMP2**
Introducing Fact Triangles Children use Fact Triangles to generate fact families.	*Math Masters,* pp. TA33 and TA35; slate	**1.OA.3, 1.OA.4, 1.OA.6, 1.OA.8, SMP2, SMP7**
✓ **Assessment Check-In** See page 603.		**1.OA.6, 1.OA.8, SMP2**
Exploring Fact Families Children produce number sentences to match Fact Triangles.	*Math Journal 2,* Activity Sheets 10–12; *Math Masters,* p. TA36 (optional)	**1.OA.3, 1.OA.4, 1.OA.6** **SMP2**
Introducing *Beat the Calculator* **Game** Children practice addition facts.	*Math Journal 2,* AS 10–12; *Math Masters,* p. G49 (optional); calculator	**1.OA.3, 1.OA.4, 1.OA.6** **SMP6**

CCSS 1.OA.3   **Spiral Snapshot**

GMC   Apply properties of operations to add or subtract.

6-9 Focus Practice	7-1 Focus Practice	7-2 Focus Practice	7-3 Warm Up Focus Practice	7-4 Practice	7-6 Focus Practice	7-8 Practice	7-9 Warm Up Practice

 **Spiral Tracker**   **Go Online**   to see how mastery develops for all standards within the grade.

3 Practice   15–20 min

Practicing with Name-Collection Boxes Children practice finding equivalent names.	*Math Journal 2,* p. 138	**1.OA.6, 1.OA.7** **SMP2**
Math Boxes 7-2 Children practice and maintain skills.	*Math Journal 2,* p. 139	See page 605.
Home Link 7-2 **Homework** Children practice facts using Fact Triangles.	*Math Masters,* pp. 187–191	**1.OA.3, 1.OA.4, 1.OA.6** **SMP7**

connectED.mcgraw-hill.com

Plan your lessons online with these tools.

 ePresentations   Student Learning Center   Facts Workshop Game   eToolkit   Professional Development   Home Connections   Spiral Tracker   Assessment and Reporting   English Learners Support   Differentiation Support

Differentiation Options

 CCSS 1.OA.3, 1.OA.4, 1.OA.6, SMP7

Readiness · 10–15 min

Constructing Fact Families

	WHOLE CLASS
	SMALL GROUP
	PARTNER
	INDEPENDENT

2 different-colored dice per child

To provide experience with the turn-around rule, have children construct fact families by rolling two different-colored dice. Children roll the dice and record each number using a different crayon color. Children add the two numbers to determine the sum, or third number, and write it in pencil. Then they write the facts for the fact family using the corresponding colors to illustrate the locations of the addends. **GMP7.2**

 CCSS 1.OA.3, 1.OA.4, 1.OA.6, 1.OA.8, 1.NBT.4, 1.NBT.6, SMP7

Enrichment · 10–15 min

Exploring Patterns Using Fact Triangles

	WHOLE CLASS
	SMALL GROUP
	PARTNER
	INDEPENDENT

Math Masters, pp. 185–186

To further explore fact families, have children use Fact Triangles to solve addition facts and fact extensions. Children complete *Math Masters*, pages 185–186 and discuss patterns in the fact families. **GMP7.1**

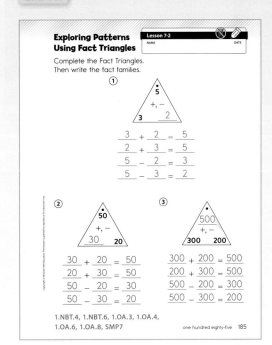

CCSS 1.OA.6, SMP6

Extra Practice · 10–15 min

Playing *Subtraction Bingo*

	WHOLE CLASS
	SMALL GROUP
	PARTNER
	INDEPENDENT

My Reference Book, p. 161 (optional); *Math Masters*, pp. G24–G25; 4 sets of number cards 0–10; counters

To provide practice with subtraction facts, have children play *Subtraction Bingo*. For detailed instructions, see Lesson 2-4. Directions are also included in *My Reference Book*, page 161. Encourage children to think about how fact families and Fact Triangles might help them find differences.

Observe

- What strategies do children use to find the differences?
- Do they apply their strategies accurately? **GMP6.4**

Discuss

- *What did you do to find the differences?*

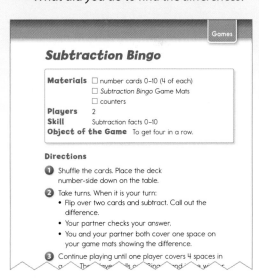

English Language Learners Support

Beginning ELL Scaffold children's understanding of the term *Fact Triangle* by reviewing the terms *triangle, fact,* and *fact family*. Use yes/no questions, such as: *Is this a triangle? Is 6 + 4 = 10 a fact?;* one-word response questions, such as: *What is this called?;* and show-me prompts, such as: *Show me a fact in the 6, 4, and 10 fact family. Show me a triangle.* Think aloud using the review terms to model using a Fact Triangle to make a fact family.

Go Online ELL English Learners Support

SMP2 Reason abstractly and quantitatively.
GMP2.3 Make connections between representations.

SMP6 Attend to precision.
GMP6.4 Think about accuracy and efficiency when you count, measure, and calculate.

SMP7 Look for and make use of structure.
GMP7.2 Use structures to solve problems and answer questions.

1 Warm Up   15–20 min

Go Online   ePresentations   eToolkit

▶ Mental Math and Fluency

Show the following combinations of base-10 blocks. Have children write the numbers represented on their slates.

● ○ ○ 6 cubes 6; 1 long, 2 cubes 12; 15 cubes 15

● ● ○ 1 long 10; 20 cubes 20; 3 longs, 10 cubes 40

● ● ● 3 longs, 5 cubes 35; 7 longs, 12 cubes 82; 25 cubes 25

▶ Daily Routines

Have children complete the Daily Routines. For detailed instructions, see pages 2–37. For specifics on CCSS coverage, see pages xiv–xvii.

2 Focus   30–35 min

Go Online   ePresentations   eToolkit

▶ Math Message

Math Masters, pp. TA33 and TA35

How are the domino and the triangle alike? How are they different? GMP2.3
Record your answers.

▶ Introducing Fact Triangles

Math Masters, pp. TA33 and TA35

| WHOLE CLASS | SMALL GROUP | PARTNER | INDEPENDENT |

Math Message Follow-Up Have children share their responses to the Math Message. Help them compare and connect the two representations by eliciting how each number, 6, 4, and 10, could be found on the domino and the triangle. GMP2.3 Remind children of using dominoes to write fact families in the previous lesson. Ask them to write the fact family for the domino on their slates.

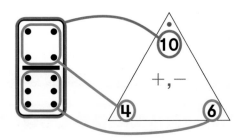

Math Masters, p. TA35

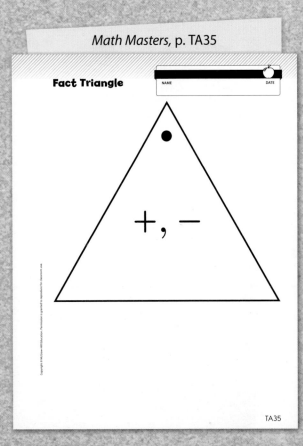

Explain that the triangle is called a **Fact Triangle.** Children will use triangles like it, with the addends in the bottom corners and the sum by the dot at the top, to practice fact families. Discuss using the triangle to write the fact family. Remove the domino, and have children write the fact family for the Fact Triangle on their slates. $4 + 6 = 10; 6 + 4 = 10;$ $10 - 6 = 4; 10 - 4 = 6$

Next display a Fact Triangle with 6 and ? as the addends and 9 as the sum. Ask children to suggest number sentences that they could record using ? to represent the missing number. $? + 6 = 9; 6 + ? = 9; 9 - 6 = ?; 9 - ? = 6$ Be sure to record the entire fact family, and discuss how children might think addition to solve $9 - 6 = ?.$ Sample answer: I can use $6 + ? = 9.$

GMP7.2 Have children solve the subtraction problem and then write the remaining facts in the fact family. Repeat with a few more Fact Triangles that have sums within 10, and have children record their number sentences on slates. Vary the quantity that is unknown.

✓ Assessment Check-In   CCSS 1.OA.6, 1.OA.8

Observe as children record number sentences to match the Fact Triangles with ?s as the missing quantities. Expect most children to record at least one number sentence (most likely an addition fact with the sum unknown) and to solve for the missing quantity. Scaffold by relating the Fact Triangles back to the dominoes from the prior lesson. GMP2.3

☑ Assessment and Reporting   (Go Online) to record student progress and to see trajectories toward mastery for these standards.

▶ # Exploring Fact Families

Math Journal 2, Activity Sheets 10–12

| WHOLE CLASS | SMALL GROUP | PARTNER | INDEPENDENT |

Give children time to examine their Fact Triangles. Ask them to check that the number under the dot on each triangle is the sum of the other two addends. Today children will be working with the Fact Triangles on Activity Sheets 10–12. Those from Activity Sheets 13–16 should be set aside for later use.

Explain that children can practice their addition and subtraction facts using Fact Triangles. Demonstrate practicing with a partner. Start by covering one corner of a Fact Triangle. Have your partner say or write an addition or subtraction fact that has the number you are covering as its answer. For example, if you cover the 3 on a 3, 7, 10 Fact Triangle, your partner might give the subtraction fact $10 - 7 = 3$. If you cover the 10, your partner might give $3 + 7 = 10$ or $7 + 3 = 10$. GMP2.3 Discuss any special facts shown on the triangles, such as doubles or combinations of 10. Cover each of the other numbers to practice the whole fact family.

Math Journal 2, Activity Sheet 10

Fact Triangles 1

Activity Sheet 10

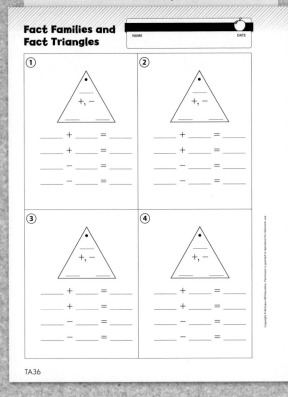

Fact Families and Fact Triangles

NAME   DATE

① ② ③ ④

TA36

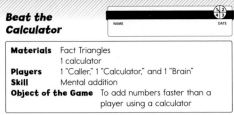

Beat the Calculator

NAME   DATE

Materials	Fact Triangles
	1 calculator
Players	1 "Caller," 1 "Calculator," and 1 "Brain"
Skill	Mental addition
Object of the Game	To add numbers faster than a player using a calculator

Directions

① Mix the Fact Triangles. Place them facedown on the table.

② The Caller:
- takes a Fact Triangle from the pile.
- covers the sum.
- says the fact *without* the sum.

For example, "3 + 4 is equal to what?"

③ The Calculator solves the problem *with* a calculator. The Brain solves it *without* a calculator. The Caller decides who found the sum first.

④ Play about 10 rounds. Then trade roles.

Another Way to Play

The Caller covers other numbers on the Fact Triangles. Players find the covered number. They can subtract to find the missing addend.

1.OA.3, 1.OA.4, 1.OA.6   G49

Differentiate | **Adjusting the Activity**

For children who struggle, provide *Math Masters,* page TA36. These children may benefit from writing the numbers each time before saying the number sentences aloud.

Go Online | Differentiation Support

Circulate and assist as partnerships practice their facts with Fact Triangles. Have children store the Fact Triangles in their toolkits, either clipped together or in envelopes.

▶ Introducing *Beat the Calculator*

Math Journal 2, Activity Sheets 10–12

WHOLE CLASS | **SMALL GROUP** | PARTNER | INDEPENDENT

Introduce *Beat the Calculator* to help children practice addition facts. Prior to playing the game, show children how to find sums using a calculator. Remind them to clear the calculator before and after each use. Display an addition fact, such as $2 + 1 =$ _____. Tell children to press 2 ⊞ 1 ⊟ to find the answer. Ask them to tell what number appears on the screen. Repeat as necessary.

To play as a whole class, take the roll of "Caller" and divide the class into two groups, "Calculators" and "Brains." Model the game by covering the sums on the Fact Triangles. Emphasize doubles and combinations of 10. Have the Brains say their answers aloud, and have the Calculators hold up their calculators with the answers. Groups race to find the correct answer. **GMP6.4**

Be the Caller until children are comfortable with the game. In the future, children will play in groups of three, with one child each as Caller, Calculator, and Brain. Play this game occasionally throughout the remainder of the year. The eventual goal is to beat the calculator on all the addition facts within 10. You may wish to distribute game directions (*Math Masters,* page G49).

Differentiate | **Game Modifications** | Go Online | Differentiation Support

Observe:

Which children can solve most addition facts in their heads?

Discuss:

Why is it sometimes more useful to solve facts in your head than on a calculator?

Summarize Ask children to share a few facts for which they beat the calculator. Discuss that some facts can be solved more easily and quickly in one's head than with a calculator, and explain that as children continue working with facts throughout the year, they will be able to beat the calculator with more and more of the facts. **GMP6.4**

③ Practice 15–20 min

Go Online

ePresentations · eToolkit · Home Connections

▶ Practicing with Name-Collection Boxes

Math Journal 2, p. 138

| WHOLE CLASS | SMALL GROUP | **PARTNER** | INDEPENDENT |

Have children practice finding equivalent names and writing true number sentences. **GMP2.1**

▶ Math Boxes 7-2

Math Journal 2, p. 139

| WHOLE CLASS | SMALL GROUP | **PARTNER** | INDEPENDENT |

Mixed Practice Math Boxes 7-2 are paired with Math Boxes 7-5.

▶ Home Link 7-2

Math Masters, pp. 187–191

Homework The five-page Home Link includes a family letter that explains Fact Triangles and provides four sets of Fact Triangles for use at home when practicing facts.

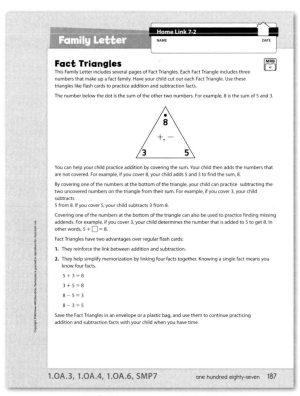

Math Masters, pp. 187–191

Math Journal 2, p. 138

Practicing with Name-Collection Boxes

Lesson 7-2
DATE

Write as many addition and subtraction names as you can for each number.

① 13
Sample answers:
10 + 3   6 + 7
15 − 2   20 − 7

② 18
Sample answers:
18 + 0   20 − 2
10 + 8   9 + 9

③ Cross out the names that DO NOT belong in the box.

20
10 + 10   14 ✗ 5
30 − 10   21 ✗ 1
13 + 7   40 − 20
20 + 0

④ Write true number sentences with names that DO belong in the box for Problem 3.

20 + 0 = 20
Sample answers:
20 = 13 + 7
30 − 10 = 20
13 + 7 = 30 − 10
40 − 20 = 20

138   one hundred thirty-eight   1.OA.6, 1.OA.7, SMP2

Math Journal 2, p. 139

Math Boxes

Lesson 7-2
DATE

① Use a marker to measure.
How tall is your desk or table?
_____ markers
Answers vary.

② Write the missing numbers.

Rule
Add 10 | 50 | 60 | 70 | 80 | 90 | 100 |

③ What is the value of the 3 in 38?
◯ 3
◯ 8
◯ 10
⬤ 30

④ Write 3 names.

17
Sample answers:
~~HHH HHH HHH II~~
7 + 10   20 − 3

⑤ **Writing/Reasoning** Solve. 3 + 5 + 2 = __10__

Which numbers did you add first? Why did you choose those?
Sample answer: I added the 2 and the 3 because I know 2 + 3 = 5. 5 and 5 is 10.

① 1.MD.2 ② 1.NBT.5 ③ 1.NBT.2
④ 1.OA.6, 1.OA.7 ⑤ 1.OA.3, 1.OA.6, SMP3   one hundred thirty-nine   139

Relating Special Addition and Subtraction Facts

Overview Children apply the think-addition strategy to doubles and combinations of 10 facts.

Common Core State Standards

Focus Clusters
- Represent and solve problems involving addition and subtraction.
- Understand and apply properties of operations and the relationship between addition and subtraction.
- Add and subtract within 20.
- Work with addition and subtraction equations.

1 Warm Up 15–20 min

	Materials	
Mental Math and Fluency Children solve number stories and record number models.	slate	1.OA.1, 1.OA.2, 1.OA.3, 1.OA.6, 1.OA.8
Daily Routines Children complete daily routines.	See pages 2–37.	See pages xiv–xvii.

2 Focus 30–35 min

Math Message Children solve doubles number stories.		1.OA.1, 1.OA.6 SMP7
Think Addition to Subtract with Doubles Children think addition to solve subtraction doubles facts.		1.OA.4, 1.OA.6, 1.OA.8 SMP7
Think Addition to Subtract with Combinations of 10 Children think addition to solve subtraction facts.	ten frames and counters (optional)	1.OA.4, 1.OA.6, 1.OA.8 SMP7
Introducing *Salute!* **Game** Children solve related addition and subtraction facts.	*My Reference Book*, pp. 162–163; 4 each of number cards 0–10	1.OA.3, 1.OA.4, 1.OA.6 SMP6
✓ **Assessment Check-In** See page 610.		1.OA.4, 1.OA.6

CCSS 1.OA.6 Spiral Snapshot

GMC Subtract doubles automatically.

| 6-7
Practice | 6-10
Warm Up | 7-1
Focus | 7-2
Focus
Practice | 7-3
Focus
Practice | 7-4
Focus
Practice | 7-6
Focus
Practice | 8-3
Warm Up
Practice |

Spiral Tracker **Go Online** to see how mastery develops for all standards within the grade.

3 Practice 15–20 min

Generating More Fact Families Children generate fact families from Fact Triangles.	*Math Journal 2*, p. 140	1.OA.3, 1.OA.4, 1.OA.6, 1.OA.8
Math Boxes 7-3 Children practice and maintain skills.	*Math Journal 2*, p. 141	See page 611.
Home Link 7-3 **Homework** Children think addition to subtract.	*Math Masters*, p. 193	1.OA.4, 1.OA.6, 1.OA.8, 1.MD.3 SMP7

connectED.mcgraw-hill.com

Plan your lessons online with these tools.

ePresentations Student Learning Center Facts Workshop Game eToolkit Professional Development Home Connections Spiral Tracker Assessment and Reporting English Learners Support Differentiation Support

Differentiation Options

RtI

CCSS 1.OA.6, 1.NBT.3, SMP7

Readiness | 10–15 min

Patterns in *Addition Top-It*

	WHOLE CLASS
	SMALL GROUP
	PARTNER
	INDEPENDENT

My Reference Book, pp. 170–172 (optional); *Math Masters,* p. 192

To prepare children to think about doubles and combinations of 10 as they subtract, have them find sums for an imaginary *Addition Top-It* game. Refer them to *My Reference Book,* pages 170–172 for game directions. Children should recognize and discuss familiar combinations of 10 that emerge from the tasks. **GMP7.1**

Patterns in Addition Top-It

Lesson 7-3
NAME   DATE

Pretend you are playing *Addition Top-It*.
Record a number sentence to match each set of cards.
Then look for patterns in your number sentences.

① Cards: 3 and 7
What is the sum? __10__
Number model:
__3 + 7 = 10__

② Cards: 1 and 9
What is the sum? __10__
Number model:
__1 + 9 = 10__

③ Cards: 6 and 4
What is the sum? __10__
Number model:
__6 + 4 = 10__

④ Cards: 5 and 5
What is the sum? __10__
Number model:
__5 + 5 = 10__

⑤ Cards: 2 and 8
What is the sum? __10__
Number model:
__2 + 8 = 10__

⑥ What patterns do you see in your number models?
__They are__
__combinations of 10.__

Sample number models given.

192   one hundred ninety-two   1.OA.6, SMP7

CCSS 1.OA.6, 1.NBT.3, SMP6

Enrichment | 10–15 min

Playing *Top-It* with Subtraction

	WHOLE CLASS
	SMALL GROUP
	PARTNER
	INDEPENDENT

My Reference Book, pp. 170–172 (optional); 4 each of number cards 0–10

To further explore subtraction fact strategies, children play a version of *Top-It* in which they find the difference between two number cards. Children should look for and apply strategies to efficiently and accurately find the differences. **GMP6.4**

CCSS 1.OA.6, SMP6

Extra Practice | 10–15 min

Playing *Roll and Record Doubles*

	WHOLE CLASS
	SMALL GROUP
	PARTNER
	INDEPENDENT

My Reference Book, p. 160 (optional); *Math Masters,* p. G47; 4 each of number cards 1–10

To provide practice with addition doubles, have children play *Roll and Record Doubles.* Directions are included in *My Reference Book,* page 160.

Observe

- Which doubles do children know automatically?
- What strategies do children use for doubles that they do not know automatically?

Discuss

- *What strategies did you use to find the sums?* **GMP6.4**
- *Are some doubles easier for you to remember than others? Which ones?*

Games

Roll and Record Doubles

Materials   ☐ *Roll and Record Doubles* Record Sheet
                ☐ 1 six-sided die
Players   2
Skill   Finding addition doubles
Object of the Game   To fill one column.

Directions

Work with a partner. When it is your turn:

English Language Learners Support

Beginning ELL   To prepare children for discussions about similarities between facts, review the meaning of *similar* using real objects or pictures of familiar objects that look alike. Think aloud and gesture to indicate how the objects are like each other, using the phrases *like each other* and *similar to each other* interchangeably. Provide oral practice with the terms using sentence frames, such as: "The _____ and _____ are like each other. The _____ and _____ are similar to each other."

Go Online   ELL English Learners Support

1 Warm Up 15–20 min

▶ Mental Math and Fluency

Have children record number models for the following number stories using ☐. Sample number models are given below. Then have children solve.

- ●○○ A necklace has 3 green, 4 yellow, and 6 pink beads. How many beads does it have in all? $3 + 4 + 6 = \square$; 13 beads
- ●●○ The school band played 12 songs. It played 4 slow songs. The rest were fast songs. How many fast songs did the band play? $4 + \square = 12$ or $12 - 4 = \square$; 8 fast songs
- ●●● Some kids started a basketball game. Then 5 kids went home. 3 kids continued to play. How many kids started the basketball game? $\square - 5 = 3$; 8 kids

▶ Daily Routines

Have children complete the Daily Routines. For detailed instructions, see pages 2–37. For specifics on CCSS coverage, see pages xiv–xvii.

2 Focus 30–35 min

▶ Math Message

Solve each problem.

1. *Imagine you are carrying an egg carton with 12 eggs. Oops! You drop the egg carton, and 6 eggs break! How many eggs are not broken?*

2. *Two baseball teams are on the field, for a total of 18 players. One team of 9 players goes into the dugout. How many players are left on the field?*

How are the problems similar? GMP7.1 *Record your answers.*

▶ Think Addition to Subtract with Doubles

| WHOLE CLASS | SMALL GROUP | PARTNER | INDEPENDENT |

Math Message Follow-Up Have children share the strategies they used to solve number stories. 6 eggs broke. 9 players remain. Some might have counted back, and some might have drawn pictures.

Remind children that they put the think-addition strategy on the Strategy Wall in Lesson 7-1. Tell them that today they will be applying this strategy with other facts.

Discuss what the problems had in common. **GMP7.1** Children might suggest that both are subtraction problems or that both can be thought of as "taking away." Ask: *How can you think addition to help solve each problem?* Sample answer: I can figure out an easier addition fact in the same fact family. Record the doubles 6 + 6 = 12 and 9 + 9 = 18 to help children see how thinking of addition doubles could help solve each problem. For example, think aloud: *I don't know what 18 − 9 is. But I can think of 9 +* ☐ *= 18. That makes me think of the double 9 + 9 = 18. So the answer must be 9.*

Next display 8 − 4 = ☐ and 14 − 7 = ☐. Ask children to describe how they could use a double to solve each. Record their thinking. *For example:*

- For 8 − 4 = ☐, I think of 4 + ☐ = 8. I know 4 + 4 = 8, so 8 − 4 = 4.
- For 14 − 7 = ☐, I think of 7 + ☐ = 14. I know 7 + 7 = 14, so 14 − 7 = 7. **GMP7.2**

Ask: *Why might using doubles be helpful for solving some subtraction problems?* Sample answer: I know my doubles well, so using doubles can be faster than counting back or drawing a picture. Add "think addition with doubles" to the Strategy Wall.

▶ Think Addition to Subtract with Combinations of 10

WHOLE CLASS | SMALL GROUP | PARTNER | INDEPENDENT

Ask children how they would solve this subtraction fact: 10 − 8 = ☐. If no one mentions it, remind them that another group of addition facts that they know well are the combinations of 10. Ask children to name combinations of 10 that they could use to solve the problem. 8 + 2 or 2 + 8

Have children list a few other combinations of 10. Then ask them to suggest related subtraction facts that could be solved by thinking about these addition facts. Point out that understanding combinations of 10 can also be useful for games such as *Fishing for 10*. For example, if they are holding a 3, they may think 3 + ☐ = 10 or 10 − 3 = ☐. Therefore, thinking of addition might help them solve problems involving decomposing a 10. **GMP7.2** Add "think addition with combinations of 10" to the Strategy Wall.

> **Differentiate** **Adjusting the Activity**
>
> For children who struggle to decompose 10 with subtraction facts that are related to combinations of 10, place 10 counters on a ten frame, and then model subtracting by removing counters. Have them examine the number of counters left and then record related addition and subtraction number models to match.
>
> **Go Online** Differentiation Support

Academic Language Development

Have children share their thinking using sentence frames, such as:

- "I know that 6 + 6 = 12, so 12 − 6 = 6."
- "I figured out that 12 − 6 = 6."
- "12 − 6 = 6 is similar to 6 + 6 = 12 because both problems have the numbers 6, 6, and 12."

Games

Salute!

Materials ☐ number cards 0–10 (4 of each)
Players 3
Skill Practicing addition and subtraction facts
Object of the Game To solve for the number on your card.

Directions

1. One person begins as the Dealer. The Dealer gives one card to each of the other two Players.

2. Without looking at their cards, the Players hold them on their foreheads with the number facing out.

3. The Dealer looks at both cards and says the sum of the two numbers.

4. Each Player looks at the other Player's card. They use the number they see and the sum said by the Dealer to figure out what the number on their card must be. They say that number out loud.

MRB
162 one hundred sixty-two

NOTE If your class does not divide evenly into groups of 3, the fourth child in a group can be the Checker and check the Dealer's sum before play continues. Alternatively, you can fill in as the third person, rotating through the groups with two children.

Common Misconception

Differentiate **IF** children do not know how to interpret a zero card and that adding zero to any number will always yield it unchanged (because zero is the Additive Identity), **THEN** encourage their groups to discuss adding and subtracting 0 and how that might look when playing *Salute!*

Go Online Differentiation Support

▶ Introducing *Salute!*

WHOLE CLASS | SMALL GROUP | PARTNER | INDEPENDENT

Salute! provides children with practice solving related addition and subtraction facts. Directions can be found on *My Reference Book,* pages 162–163. Children play in groups of 3, taking turns being the dealer. Introduce the game by demonstrating a few rounds with two children and you as the dealer. When children understand the game, have them play several rounds.

Observe

- What strategies do children use to determine the numbers on their cards?
- Do children understand what to do when they see the other player holding a 0 card? When they see a card that equals the sum?

Discuss

- *How did you figure out which card is on your forehead?* GMP6.1
- *When was it easiest to figure out your card? Hardest?* GMP6.1

Differentiate **Game Modifications** Go Online Differentiation Support

✔ Assessment Check-In CCSS 1.OA.4, 1.OA.6

As children play *Salute!* observe and ask about the strategies Dealers used to determine their cards. Although you can expect some inaccuracies the first time this game is played, expect most children to accurately determine their cards when the numbers are small, particularly when doubles are involved. Have children who struggle to determine their cards refer to the Strategy Wall, particularly the think-addition strategy. Help them formulate number sentences from the cards, such as

$5 - 3 = \square$ or $3 + \square = 5$.

 Assessment and Reporting | Go Online ▷ to record student progress and to see trajectories toward mastery for these standards.

Summarize Pose a *Salute!* example, such as: *If the dealer says 14 and you see a 7 on your partner's forehead, which number do you have? How do you know?* Sample answers: I know it is 7 because I counted up 7 from 7 to 14; I used the double $7 + 7 = 14$, so I know the answer is 7. **Have children share their strategies, highlighting counting up and thinking addition with doubles.** GMP6.1

3 Practice 15–20 min

Go Online

ePresentations eToolkit Home Connections

▶ Generating More Fact Families

Math Journal 2, p. 140

| WHOLE CLASS | SMALL GROUP | **PARTNER** | **INDEPENDENT** |

Have children practice generating fact families.

▶ Math Boxes 7-3

Math Journal 2, p. 141

| WHOLE CLASS | SMALL GROUP | **PARTNER** | INDEPENDENT |

Mixed Practice Math Boxes 7-3 are paired with Math Boxes 7-1.

▶ Home Link 7-3

Math Masters, p. 193

Homework Children practice solving subtraction problems using the think-addition strategy.

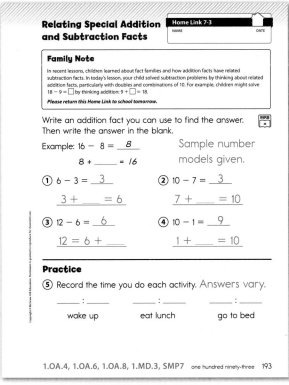

Math Masters, p. 193

Math Journal 2, p. 140

Fact Families and Fact Triangles Lesson 7-3 DATE

Write the fact family for each Fact Triangle.

①

$2 + 3 = 5$
$3 + 2 = 5$
$5 - 3 = 2$
$5 - 2 = 3$

②

$1 + 5 = 6$
$5 + 1 = 6$
$6 - 1 = 5$
$6 - 5 = 1$

③ Write the missing number. Complete the fact family.

$7 + 2 = 9$
$2 + 7 = 9$
$9 - 2 = 7$
$9 - 7 = 2$

④ Explain how you found the missing number in Problem 3.

Sample answer: I subtracted to find the number.
$9 - 7 = 2$

140 one hundred forty 1.OA.3, 1.OA.4, 1.OA.6, 1.OA.8

Math Journal 2, p. 141

Math Boxes Lesson 7-3 DATE

① Record the time.

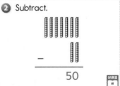

A little after ___12___ o'clock

② Subtract.

___ − ___
50

③ Which number has 5 ones and 6 tens?

○ 11
○ 56
● 65
○ 605

④ Use your number grid. Start at 36. Count up 22.

$36 + 22 = \underline{58}$

⑤ This graph shows the colors of beads on a necklace.

It has 1 red bead, 4 gray beads, and 2 blue beads. Finish labeling the graph.

Beads on a Necklace

Number of Beads

red gray blue

⑥ Write names that equal 10.

10
Answers vary.
$16 - 6$ $2 + 4 + 4$
$12 - 2$ $5 + 5$

① 1.MD.3 ② 1.NBT.2, 1.NBT.6 ③ 1.NBT.2
④ 1.NBT.4 ⑤ 1.MD.4 ⑥ 1.OA.6, 1.OA.7   one hundred forty-one 141

More Subtraction Fact Strategies

Overview Children learn the counting-up and counting-back strategies for subtraction and compare the efficiency of various subtraction strategies.

▶ **Before You Begin**
For Part 1, sequence Quick Look Cards 94, 105, and 106.

 Common Core State Standards

Focus Clusters
- Understand and apply properties of operations and the relationship between addition and subtraction.
- Add and subtract within 20.
- Work with addition and subtraction equations.

1 Warm Up 15–20 min

	Materials	
Mental Math and Fluency Children do Quick Looks with ten frames.	Quick Look Cards 94, 105, and 106	**1.OA.6** **SMP2, SMP6**
Daily Routines Children complete daily routines.	See pages 2–37.	See pages xiv–xvii.

2 Focus 30–35 min

Math Message Children find the unknown card in a round of *Salute!*		**1.OA.4, 1.OA.6**
Comparing Counting Up and Counting Back to Subtract Children count up and back to subtract.	*Math Journal 2,* p. 142; Class Number Line; colored pencils (optional)	**1.OA.4, 1.OA.5, 1.OA.6,** **1.OA.8** **SMP1, SMP7, SMP8**
✓ **Assessment Check-In** See page 615.	*Math Journal 2,* p. 142	**1.OA.5, 1.OA.6, SMP1**
Reviewing Subtraction Fact Strategies Children review and practice the subtraction fact strategies.	slate	**1.OA.4, 1.OA.5, 1.OA.6** **SMP1**

 1.OA.5 Spiral Snapshot

GMC Relate counting to addition and subtraction.

| 4-7
Warm Up | 4-8
Practice | 4-9
Focus | 5-4
Warm Up
Practice | 5-10
Focus
Practice | 5-12
Focus
Practice | 7-4
Focus | 8-2
Warm Up |

Spiral Tracker   **Go Online**   to see how mastery develops for all standards within the grade.

3 Practice 15–20 min

Introducing *Shaker Addition Top-It* **Game** Children practice addition facts.	*Math Masters,* p. G50 (optional), p. G51; per partnership: two dot dice or two 10-sided dice, 20 counters	**1.OA.6, 1.NBT.3** **SMP6**
Math Boxes 7-4 Children practice and maintain skills.	*Math Journal 2,* p. 143	See page 617.
Home Link 7-4 **Homework** Children practice writing fact families and fill in a name-collection box.	*Math Masters,* pp. 195–197	**1.OA.3, 1.OA.4, 1.OA.6,** **1.OA.7, 1.NBT.2**

connectED.mcgraw-hill.com

Plan your lessons online with these tools.

ePresentations Student Learning Center Facts Workshop Game eToolkit Professional Development Home Connections Spiral Tracker Assessment and Reporting English Learners Support Differentiation Support

Differentiation Options RtI

Readiness 10–15 min

Playing *Subtraction Bingo*

| WHOLE CLASS |
| SMALL GROUP |
| **PARTNER** |
| INDEPENDENT |

My Reference Book, p. 161 (optional); *Math Masters,* pp. G24–G25; 4 each of number cards 0–10; counters

For additional experience solving subtraction facts, have children play *Subtraction Bingo.* You may wish to have children share the strategies they used. For detailed instructions, see Lesson 2-4. Directions are also included in *My Reference Book,* page 161.

Observe

- What strategies do children use to find differences?
- Do they apply their strategies accurately? GMP6.4

Discuss

- *What did you do to find differences?*

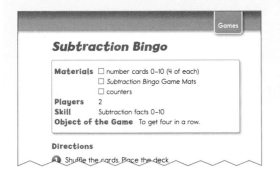

Games

Subtraction Bingo

Materials	☐ number cards 0–10 (4 of each)
	☐ *Subtraction Bingo* Game Mats
	☐ counters
Players	2
Skill	Subtraction facts 0–10
Object of the Game	To get four in a row.

Directions

① Shuffle the cards. Place the deck

Enrichment 10–15 min

Strategy Draw

| WHOLE CLASS |
| SMALL GROUP |
| **PARTNER** |
| INDEPENDENT |

Activity Card 85; *Math Masters,* p. 194

Have children further explore efficient subtraction strategies by writing the facts best suited for particular strategies. GMP6.4, GMP7.2

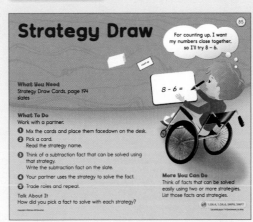

Strategy Draw Cards

Lesson 7-4
NAME   DATE

count up	count up
count back	count back
think addition with doubles	think addition with doubles
think addition with combinations of 10	think addition with combinations of 10
think addition with doubles	count up

1.OA.4, 1.OA.6, SMP6, SMP7

Extra Practice 10–15 min

Playing *The Difference Game*

| WHOLE CLASS |
| SMALL GROUP |
| **PARTNER** |
| INDEPENDENT |

My Reference Book, pp. 142–143 (optional); per partnership: 4 each of number cards 1–10, 40 toolkit pennies

To provide practice with subtraction facts, have children play *The Difference Game.* For detailed instructions, see Lesson 5-10.

Observe

- How do children find differences?
- Do they think addition to subtract?

Discuss

- *How did you find the sums? Did you still need to line up the pennies?* GMP6.1

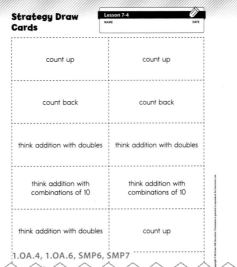

Games

The Difference Game

Materials	☐ number cards 1–10 (4 of each)
	☐ 40 pennies
	☐ 1 sheet of paper labeled "Bank"
Players	2
Skill	Finding differences
Object of the Game	To take more pennies.

Directions

① Shuffle the cards. Place the deck number-side down on the table.

② Put 40 pennies in the bank.

③ To play a round, each player:
- Takes 1 card from the top of the deck.
- Takes the same number of pennies from the bank as the number shown on the card.

④ Find out how many more pennies one player has than the other. Pair as many pennies as you can.

MRB 142 one hundred forty-two

English Language Learners Support

Beginning ELL Demonstrate counting up as you move your hand forward along a number line or on a number grid. Then demonstrate counting back as you gesture in the opposite direction. Check for children's understanding of the expressions *count up* and *count back* by pointing to different numbers on the number line or number grid and giving oral directions and gestures to count up or count back.

Go Online ELL English Learners Support

Standards and Goals for
Mathematical Practice

SMP1 Make sense of problems and persevere
in solving them.
GMP1.6 Compare the strategies you and
others use.

SMP7 Look for and make use of structure.
GMP7.1 Look for mathematical structures such
as categories, patterns, and properties.

SMP8 Look for and express regularity in
repeated reasoning.
GMP8.1 Create and justify rules, shortcuts,
and generalizations.

Professional Development

Children were informally introduced to
counting up as a subtraction strategy in
Unit 2. It is a powerful subtraction fact
strategy because it makes use of the
relationship between addition
and subtraction. **1.OA.4** Counting up
to subtract is an accessible strategy
that improves children's accuracy.
It also transfers well to multidigit
computation.

1 Warm Up 15–20 min   Go Online
ePresentations   eToolkit

▶ Mental Math and Fluency

Do Quick Looks with children. Display Quick Look Cards in this order: 94,
105, 106. Show children each image for 2–3 seconds. Then remove or
cover it. Have children share what they saw and how they saw it.
GMP2.2, GMP6.1 While discussing Card 94, emphasize mentally moving
the dots to make a 10. Encourage children to apply this strategy on the
next two cards.

▶ Daily Routines

Have children complete the Daily Routines. For detailed instructions,
see pages 2–37. For specifics on CCSS coverage, see pages xiv–xvii.

2 Focus 30–35 min   Go Online
ePresentations   eToolkit

▶ Math Message

You are playing Salute! *The Dealer says the sum is 11 and you see an 8 on*
your friend's forehead. Record what card you have.

▶ Comparing Counting Up and
Counting Back to Subtract

Math Journal 2, p. 142

| WHOLE CLASS | SMALL GROUP | PARTNER | INDEPENDENT |

Math Message Follow-Up Have children share how they solved the
problem in the Math Message. In particular, have those who counted up
explain their strategy. I started at 8 and counted up to 11. So, the card I
have is 3. Tell children that today they will be comparing this subtraction
strategy with others.

Ask whether anyone counted back to solve the Math Message. Ask
children to discuss with a partner how they might solve this by counting
back 8 from 11. Have partnerships share and record both strategies while
modeling them on the Class Number Line.

Have children compare these two methods and consider which one is more efficient for solving this problem. GMP1.6 They should notice that it is more efficient to count up 3 from 8 to 11 than to count back 8 from 11 because it requires fewer counts. Record the following number sentences: $11 - 8 = \square$ and $8 + \square = 11$. Discuss how these number sentences illustrate each strategy—the first shows taking away 8 from 11; the second shows counting up from 8 to 11 to find the missing addend. Help children see that thinking addition and then counting up to find the missing addend makes it easier to solve this subtraction problem. Repeat this process with other subtraction problems such as $9 - 7$, $13 - 11$, and $16 - 12$.

Have children complete journal page 142 with a partner. Children solve each problem twice—once counting up and once counting back. Then they record which method was quicker for each. GMP1.6, GMP7.1 Complete the first task together, modeling how children can record their counts to help them determine the more efficient strategy.

Differentiate **Adjusting the Activity**

Have children who struggle with counting up and back to add and subtract use a number line to mark their hops from one number to another. Seeing and comparing the number of hops up or back might allow them to better visualize which strategy—counting up or counting back—is more efficient. Consider also having children use two different colors to indicate hops up and hops back.

Go Online Differentiation Support

✓ **Assessment Check-In** CCSS 1.OA.5, 1.OA.6

Math Journal 2, p. 142

Observe as children complete journal page 142. Expect most children to apply at least one of the counting strategies to accurately solve each problem. Most children will be more comfortable counting back. Refer to the Adjusting the Activity to support children who struggle, and then discuss whether counting back or counting up was quicker. GMP1.6 Expect that some children might struggle to identify the strategy that is more efficient.

Assessment and Reporting **Go Online** to record student progress and to see trajectories toward mastery for these standards.

Have children share their responses for each task. Ask: *What do you notice about the facts you circled as counting up? Counting back?* GMP7.1, GMP8.1 Sample answer: In the counting-up facts, the two numbers were close together. In the counting-back facts, we subtracted a small number. Add notes to the Strategy Wall indicating the types of facts that would be most efficiently solved by counting up and counting back.

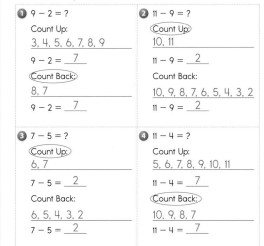

Math Journal 2, p. 142

Counting Up and Counting Back
Lesson 7-4
DATE

Solve each problem by counting up. Record your counts. Then solve each problem by counting back. Record your counts. Circle which way was faster.

① $9 - 2 = ?$
Count Up:
3, 4, 5, 6, 7, 8, 9
$9 - 2 = \underline{7}$
(Count Back:)
8, 7
$9 - 2 = \underline{7}$

② $11 - 9 = ?$
(Count Up:)
10, 11
$11 - 9 = \underline{2}$
Count Back:
10, 9, 8, 7, 6, 5, 4, 3, 2
$11 - 9 = \underline{2}$

③ $7 - 5 = ?$
(Count Up:)
6, 7
$7 - 5 = \underline{2}$
Count Back:
6, 5, 4, 3, 2
$7 - 5 = \underline{2}$

④ $11 - 4 = ?$
Count Up:
5, 6, 7, 8, 9, 10, 11
$11 - 4 = \underline{7}$
(Count Back:)
10, 9, 8, 7
$11 - 4 = \underline{7}$

142 one hundred forty-two   1.OA.4, 1.OA.5, 1.OA.6, SMP1, SMP7

Academic Language Development

To promote children's understanding of the term *efficient strategies,* prompt them with questions such as: *Which way is quicker? Which strategy can we use to do the problem more quickly? Which way helps us save time? Is it faster to start at _____ or _____ ?*

Math Masters, p. G50

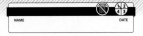

Shaker Addition Top-It

NAME		DATE

Materials	Two dot dice or two 10-sided dice; counters
Players	2 to 5
Skill	Fact review and practice
Object of the Game	To collect the most counters

Directions

1. Make a pile of counters.
2. Take turns shaking and rolling the dice.
   - Add the numbers.
   - Say the sum aloud.
   - If you say the wrong answer, you cannot win the round.
3. The player with the highest correct sum takes a counter from the pile.
   If there is a tie, the tied players each take a counter.
4. Play as long as you have time.
   Record 5 rounds of play.
   Circle the winning fact in each round.
5. The player with the most counters when the game is over wins.

G50   1.OA.6, 1.NBT.3

▶ Reviewing Subtraction Fact Strategies

WHOLE CLASS   SMALL GROUP   **PARTNER**   INDEPENDENT

Use the Strategy Wall to review the various subtraction fact strategies that children have developed, including think addition (particularly with combinations of 10 and with doubles), count up, and count back. Ask children to solve the following four subtraction facts in partnerships using an appropriate strategy from the Strategy Wall. Tell them to try to use a different strategy for each and to record the strategies they used next to each fact on their slates.

- $10 - 4 = \boxed{}$ 6; Sample answer: think addition with combinations of 10
- $11 - 2 = \boxed{}$ 9; Sample answer: count back
- $12 - 6 = \boxed{}$ 6; Sample answer: think addition with doubles
- $8 - 6 = \boxed{}$ 2; Sample answer: count up

Have children share their solutions and discuss and compare the strategies they used for each fact. **GMP1.6** Children may use any strategies they choose, but they should be able to justify their selections.

Summarize Display $10 - 8 = \boxed{}$, and ask children to think of two different strategies they might use to solve it. Sample answers: Count up from 8 to 10. Think addition with a combination of 10. Have children share their ideas.

---

3 Practice   15–20 min

Go Online
ePresentations   eToolkit   Home Connections

▶ Introducing *Shaker Addition Top-It*

Math Masters, p. G51

WHOLE CLASS   **SMALL GROUP**   **PARTNER**   INDEPENDENT

Have children practice addition facts by playing this variation of *Addition Top-It.* At first, limit the game to two players. Model the game with a volunteer. You may wish to distribute the directions (*Math Masters,* page G50) for children to follow as you model the game. Have children record five rounds on the *Shaker Addition Top-It* Record Sheet (*Math Masters,* page G51).

Using dice with more sides will provide practice for more addition facts.

Math Masters, p. G51

Shaker Addition Top-It Record Sheet

NAME		DATE

G51

Observe

- Which children demonstrate fluency with their addition facts?
- How do children check each other's answers?

Discuss

- *When you and your partner do not agree on a sum, how do you check which sum is correct?* GMP6.1
- *What strategies can you use to solve a fact you do not know?*

Differentiate **Game Modifications**   Go Online   Differentiation Support

---

▶ Math Boxes 7-4

Math Journal 2, p. 143

| WHOLE CLASS | SMALL GROUP | **PARTNER** | INDEPENDENT |

Mixed Practice Math Boxes 7-4 are paired with Math Boxes 7-6.

▶ Home Link 7-4

Math Masters, pp. 195–197

Homework Children practice addition and subtraction facts and fact families.

Math Journal 2, p. 143

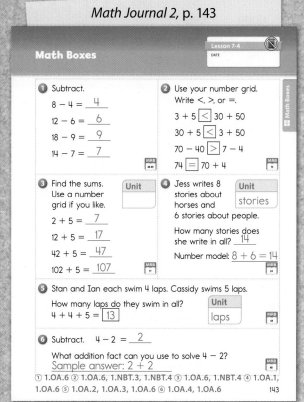

Math Boxes

Lesson 7-4
DATE

① Subtract.

$8 - 4 = \underline{4}$

$12 - 6 = \underline{6}$

$18 - 9 = \underline{9}$

$14 - 7 = \underline{7}$

② Use your number grid. Write <, >, or =.

$3 + 5 \boxed{<} 30 + 50$

$30 + 5 \boxed{<} 3 + 50$

$70 - 40 \boxed{>} 7 - 4$

$74 \boxed{=} 70 + 4$

③ Find the sums. Use a number grid if you like. | Unit |

$2 + 5 = \underline{7}$

$12 + 5 = \underline{17}$

$42 + 5 = \underline{47}$

$102 + 5 = \underline{107}$

④ Jess writes 8 stories about horses and 6 stories about people. | Unit stories |

How many stories does she write in all? $\underline{14}$

Number model: $8 + 6 = 14$

⑤ Stan and Ian each swim 4 laps. Cassidy swims 5 laps.

How many laps do they swim in all? | Unit laps |

$4 + 4 + 5 = \boxed{13}$

⑥ Subtract. $4 - 2 = \underline{2}$

What addition fact can you use to solve $4 - 2$?
Sample answer: $\underline{2 + 2}$

① 1.OA.6 ② 1.OA.6, 1.NBT.3, 1.NBT.4 ③ 1.OA.6, 1.NBT.4 ④ 1.OA.1, 1.OA.6 ⑤ 1.OA.2, 1.OA.3, 1.OA.6 ⑥ 1.OA.4, 1.OA.6   143

Math Masters, pp. 195–197

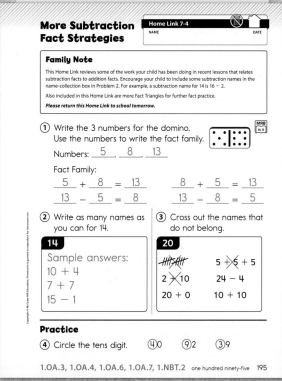

More Subtraction Fact Strategies

Home Link 7-4
NAME
DATE

Family Note

This Home Link reviews some of the work your child has been doing in recent lessons that relates subtraction facts to addition facts. Encourage your child to include some subtraction names in the name-collection box in Problem 2. For example, a subtraction name for 14 is 16 − 2.

Also included in this Home Link are more Fact Triangles for further fact practice.

Please return this Home Link to school tomorrow.

① Write the 3 numbers for the domino. Use the numbers to write the fact family.

Numbers: $\underline{5}$, $\underline{8}$, $\underline{13}$

Fact Family:

$\underline{5} + \underline{8} = \underline{13}$ $\underline{8} + \underline{5} = \underline{13}$

$\underline{13} - \underline{5} = \underline{8}$ $\underline{13} - \underline{8} = \underline{5}$

② Write as many names as you can for 14.

14

Sample answers:
10 + 4
7 + 7
15 − 1

③ Cross out the names that do not belong.

20

~~HH HH~~   5 +̶5̶ + 5

2 ×̶ 10   24 − 4

20 + 0   10 + 10

Practice

④ Circle the tens digit.   ④0   ⑨2   ③9

1.OA.3, 1.OA.4, 1.OA.6, 1.OA.7, 1.NBT.2   one hundred ninety-five   195

Attributes of Shapes

Overview Children identify the attributes of attribute blocks and sort blocks by their attributes.

▶ **Before You Begin**

For Part 2, place a large collection of attribute blocks near the Math Message. You may also wish to make a large demonstration circle. For the optional Extra Practice activity, gather several books that feature shapes, such as *Windows, Rings, and Grapes—A Look at Different Shapes* by Brian P. Cleary (Millbrook Press, 2011) and *It's a Shape!* by Marianne Waering Penn (Capstone Press, 2011).

▶ **Vocabulary**

attribute

Common Core State Standards

Focus Cluster
Reason with shapes and their attributes.

---

1 Warm Up 15–20 min

	Materials	
Mental Math and Fluency Children find 10 more or 10 less than a given number.	number grid (optional)	1.NBT.5
Daily Routines Children complete daily routines.	See pages 2–37.	See pages xiv–xvii.

---

2 Focus 30–35 min

Math Message Children record what they notice about an attribute block.	attribute blocks	1.G.1 SMP7
Introducing Attributes Children learn about attributes of shapes and sort them.	attribute blocks	1.G.1 SMP1, SMP6, SMP7
Sorting Attribute Blocks Children sort attribute blocks using rules.	attribute blocks	1.G.1 SMP7
✓ **Assessment Check-In** See page 622.	3 triangle attribute blocks	1.G.1, SMP7

CCSS 1.G.1 **Spiral Snapshot**

GMC Distinguish between defining and non-defining attributes.

| 4-5
Focus | 6-3
Focus
Practice | 7-5
Focus
Practice | 7-6
Focus
Practice | 7-7
Focus
Practice | 8-1
Focus
Practice | 8-5
Focus
Practice | 8-6
Focus
Practice |

/// Spiral Tracker   **Go Online**   to see how mastery develops for all standards within the grade.

---

3 Practice 15–20 min

Playing *Stop and Go* **Game** Children practice adding and subtracting 2-digit numbers.	*Math Masters,* pp. G42 (optional), G43–G45	1.NBT.4, 1.NBT.6 SMP1
Math Boxes 7-5 Children practice and maintain skills.	*Math Journal 2,* p. 144	See page 623.
Home Link 7-5 **Homework** Children practice identifying shape attributes.	*Math Masters,* p. 199	1.MD.3, 1.G.1 SMP7

Plan your lessons online with these tools.

ePresentations | Student Learning Center | Facts Workshop Game | eToolkit | Professional Development | Home Connections | Spiral Tracker | Assessment and Reporting | English Learners Support | Differentiation Support

Differentiation Options

 RtI

Readiness   10–15 min

| WHOLE CLASS |
| SMALL GROUP |
| PARTNER |
| INDEPENDENT |

Fishing for Attributes

To provide experience with the meaning and use of attributes, have children determine sorting rules. Tell them that you are going fishing for children. Challenge them to determine what kind of children you want by observing what you catch. **GMP7.1** For example, fish for children wearing blue pants. Call children who are wearing blue pants to the front of the class by name, but do not explain why you called them. Let them guess what you are fishing for until someone says "Children wearing blue pants." When children understand the game, let them try fishing. **GMP8.1**

Enrichment   10–15 min

| WHOLE CLASS |
| SMALL GROUP |
| PARTNER |
| INDEPENDENT |

What's My Attribute Rule?

Activity Card 86, attribute blocks

To extend children's understanding of sorting rules, have partners guess each other's rules for sorting attribute blocks. **GMP7.1, GMP8.1**

Extra Practice   15–20 min

| WHOLE CLASS |
| SMALL GROUP |
| PARTNER |
| INDEPENDENT |

Reading about Shapes

Activity Card 87;
Math Masters, p. 198;
books about shapes;
Pattern-Block Template

Literature Link Have children practice identifying shapes with different attributes by searching for shapes in books, such as ***Windows, Rings, and Grapes—A Look at Different Shapes*** by Brian P. Cleary (Millbrook Press, 2011) and ***It's a Shape!*** by Marianne Waering Penn (Capstone Press, 2011). **GMP7.1** Children can use their Pattern-Block Templates for reference.

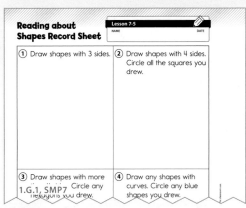

English Language Learners Support

Beginning ELL Use familiar classroom objects to introduce the words *thick* and *thin* and *large* and *small*. For example, have children pick up a thick notebook. Then have them pick up a thin notebook. Have them place them side by side and say: *This notebook is thick* and *this notebook is thin.* Continue asking children to show or point to the thick or thin and large or small object of pairs. Ask yes/no questions, such as: *Is this thin?* or *Is this thick?*

Go Online   English Learners Support

Standards and Goals for
Mathematical Practice

SMP1 Make sense of problems and persevere in solving them.
GMP1.1 Make sense of your problem.

SMP6 Attend to precision.
GMP6.1 Explain your mathematical thinking clearly and precisely.

SMP7 Look for and make use of structure.
GMP7.1 Look for mathematical structures such as categories, patterns, and properties.

NOTE Place a set of attribute blocks in a box or basket near the Math Message. Because of the differences among attribute block sets, the activities in this lesson are generic. Adapt the activities to the blocks that you are using.

1 Warm Up 15–20 min

Go Online
ePresentations eToolkit

▶ Mental Math and Fluency

Encourage children to solve the following problems mentally, but allow them to use number grids, if necessary.

● ○ ○ 10 more than 30. 40 10 less than 30. 20
● ● ○ 10 more than 67. 77 10 less than 67. 57
● ● ● 10 less than 94. 84 10 more than 94. 104

▶ Daily Routines

Have children complete the Daily Routines. For detailed instructions, see pages 2–37. For specifics on CCSS coverage, see pages xiv–xvii.

2 Focus 30–35 min

Go Online
ePresentations eToolkit

▶ Math Message

Choose a block. Record everything you notice about your block. GMP7.1

▶ Introducing Attributes

| **WHOLE CLASS** | SMALL GROUP | PARTNER | INDEPENDENT |

Math Message Follow Up Have children share their observations.
GMP6.1, GMP7.1 Children may share the shapes, colors, thicknesses, sizes, or numbers of sides and vertices (or corners) of the blocks.

Tell children that today they will use observations such as these to sort shapes. Ask children who have circles to go to one area of the room, those with triangles to go to another area, and those with rectangles to go to a third area. Tell children in the rectangle group who have squares (special rectangles) to go to a fourth area. Ask children who remain seated to show their attribute blocks. Ask: *What shape are the blocks?* Hexagons

Elicit observations about the blocks using questions such as the following:

- *What do the blocks in your area have in common?* GMP7.1 Sample answers: They are circles. They don't have corners.
- *Are all the blocks the same color?* Sample answer: No; they are red, yellow, or blue.
- *Are they the same size?* Sample answer: No; some shapes are large, and some shapes are small.
- *Do all the blocks have the same thickness?* Sample answer: No; some are thick, and some are thin.

Next, have children return to their seats and then sort themselves so they are in groups with similar block colors. GMP1.1 Finally, have children sort themselves so they are in groups with similar block sizes—the smaller sizes of each shape in one area, and the larger sizes of each shape in another. GMP1.1

Explain that attribute blocks can be sorted by shape, color, size, or thickness. Shape, color, size, and thickness are **attributes** of the blocks. Explain that people have attributes, too. Traits such as hair color, height, and eye color are attributes of a person. Tell children that the class can be sorted by these attributes. For example, have all children with short hair stand up and all children with long hair sit down.

Ask children with circles to come to the front of the room and display their shapes.

- *Even though these are different colors, are they all circles?* Yes
- *How do you know?* GMP6.1 Sample answers: because they are the same shape; because they are round
- *Even though some are big and some are small, are they still circles?* Yes
- *If we painted some of them orange, would they still be circles?* Yes
- *If we chopped the circle in half, would it still be a circle?* No If children need help visualizing the change in shape, divide the demonstration circle in half and display it next to the group of circles.
- *Why is it not a circle anymore?* Sample answers: because it isn't round now; because the shape changed; because it has a side, and circles don't have sides

---

NOTE In this lesson, children are informally introduced to the concepts of defining and nondefining attributes. In Lesson 7-7, they will explore these attributes further and will be introduced to the terms *defining attributes* and *nondefining attributes*.

Professional Development

In Kindergarten, children learn to distinguish 2-D from 3-D shapes. Representations of 2-D shapes, such as attribute blocks, are not really 2-dimensional because these objects have depth. You may wish to discuss this technicality with children, but do not expect them to fully understand it.

► **Sorting Attribute Blocks**

| **WHOLE CLASS** | SMALL GROUP | PARTNER | INDEPENDENT |

Ask children to stand or remain seated based on rules about attributes. Ask all children who have a blue square to stand up. Say: *All of you fit the rule because you have a blue square.* Continue the activity with other rules. **GMP7.1** For example, say: *Stand up and show me your attribute block if it is*

- *a red circle.*
- *not yellow.* (This is an example of a negative rule.)
- *a thin rectangle.*
- *a large circle.*
- *a small, blue hexagon.*
- *not red and not large.*

Include at least one rule that does not fit any of the blocks, such as *has 10 vertices.*

Have children turn in their attribute blocks according to the rules. For example, say *large blocks, thick blocks, all yellow circles,* and so on, until all the blocks have been collected. Be sure to take this opportunity to practice negative rules.

✔ Assessment Check-In CCSS 1.G.1

Display three triangle attribute blocks of varying sizes and colors. Ask children to put their thumbs up for yes and down for no to answer the following questions:

- *Do all these blocks have the same color?* Thumbs down
- *Do all these blocks have the same shape?* Thumbs up
- *If we made all these blocks very tiny, would they still be the same shape?* Thumbs up
- *If we cut off the corners, would they still be triangles?* Thumbs down

Expect most children to correctly answer the first three questions. Ask children who struggle with the third question to draw a picture of what they think a smaller block would look like. The idea of defining and nondefining attributes will be revisited in Lesson 7-6 and formally introduced in Lesson 7-7. **GMP7.1**

 Assessment and Reporting **Go Online** to record student progress and to see trajectories toward mastery for these standards.

Summarize Ask children to discuss with their partners one attribute that all triangles share and one attribute that they do not all share.

Math Masters, p. G42

Stop and Go

NAME   DATE

Materials	6 GO Cards (+ 9, + 8, + 7, + 6, + 10, + 20)
	6 STOP Cards (− 0, − 0, − 10, − 10, − 20, − 20)
	Stop and Go Record Sheet
Players	2 (the GO player and the STOP player)
Skill	Adding and subtracting 2-digit numbers
Object of the Game	To get to 50 (or to stop the other player from getting to 50)

Directions

① The GO player puts the GO cards number-side down and takes 1 GO card.

② The GO player adds 20 to the amount on the GO card and records the sum on the *Stop and Go* Record Sheet.

③ The STOP player puts the STOP cards number-side down and takes 1 STOP card.

④ The STOP player subtracts the amount on the STOP card from the sum and records it on the Record Sheet.

⑤ The players take turns adding and subtracting.

- If the GO player reaches 50, the GO player wins.
- If the STOP player pushes the GO player back to 0, the STOP player wins.
- If the players run out of cards before the GO player reaches 50, the STOP player wins.

G42 1.NBT.4, 1.NBT.6

(3) Practice 15–20 min

Go Online | ePresentations | eToolkit | Home Connections

► Playing *Stop and Go*

Math Masters, pp. G43– G45

| WHOLE CLASS | SMALL GROUP | **PARTNER** | INDEPENDENT |

Have children play *Stop and Go* to practice addition and subtraction. For directions for playing the game, use *Math Masters,* page G42. For more information, see Lesson 5-11.

Observe

- Which children add and subtract multiples of 10 correctly? What strategies do they use?
- Which children add single-digit numbers to double-digit numbers correctly? What strategies do they use?

Discuss

- *Compare addition and subtraction strategies with your partner. How were your strategies the same and different?* GMP1.6

► Math Boxes 7-5

Math Journal 2, p. 144

| WHOLE CLASS | SMALL GROUP | **PARTNER** | **INDEPENDENT** |

Mixed Practice Math Boxes 7-5 are paired with Math Boxes 7-2.

► Home Link 7-5

Math Masters, p. 199

Homework Children look for and describe attributes of a hexagon and a parallelogram.

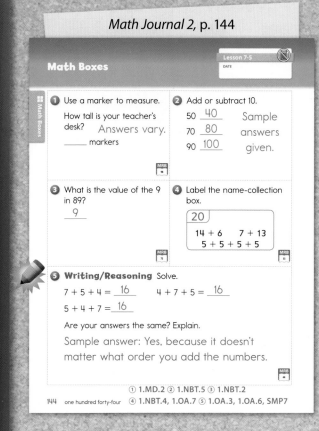

Math Journal 2, p. 144

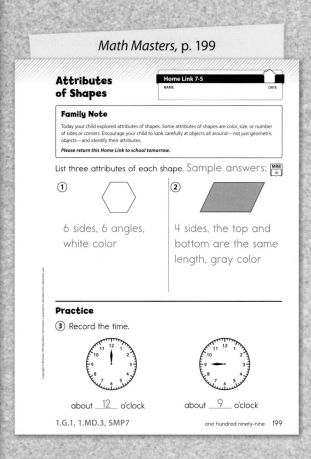

Math Masters, p. 199

Exploring Attributes, Fractions, and *Salute!*

Overview Children sort by attribute rules, explore whether shapes are divided equally, and practice addition and subtraction facts.

▶ **Before You Begin**
For Part 2, place a set of attribute blocks near the Math Message. For the optional Readiness activity, gather classroom objects that children can sort.

Common Core State Standards

Focus Clusters
- Understand and apply properties of operations and the relationship between addition and subtraction.
- Add and subtract within 20.
- Reason with shapes and their attributes.

1 Warm Up 15–20 min	**Materials**	
Mental Math and Fluency Children write fact families for given numbers.	slate	1.OA.3, 1.OA.6
Daily Routines Children complete daily routines.	See pages 2–37.	See pages xiv–xvii.

2 Focus 35–40 min		
Math Message Children list the attributes of an attribute block.	attribute blocks	1.G.1 SMP7
Making an Attribute Train Children review the attributes of blocks and identify blocks differing in just one attribute.	attribute blocks	1.G.1 SMP7
Exploration A: Introducing *Attribute Train* **Game** Children line up blocks differing in just one attribute.	*Math Masters,* p. G52 (optional); attribute blocks	1.G.1 SMP7
Exploration B: Dividing Shapes Children make shapes on geoboards, divide them into two parts, and determine whether the parts are the same size.	Activity Card 88, geoboard, rubber bands	1.G.1, 1.G.3 SMP1
Exploration C: Recording *Salute!* **Game** Children practice solving related addition and subtraction facts.	Activity Card 89; *My Reference Book,* pp. 162–163 (optional); *Math Masters,* p. G53; 4 each of number cards 0–10	1.OA.3, 1.OA.4, 1.OA.6 SMP7

3 Practice 10–15 min		
Solving Garden Number Stories Children practice solving number stories.	*Math Journal 2,* p. 145	1.OA.1, 1.OA.2, 1.OA.3, 1.OA.6, SMP4
Math Boxes 7-6 Children practice and maintain skills.	*Math Journal 2,* p. 146	See page 629.
Home Link 7-6 **Homework** Children find shapes in everyday objects and draw them.	*Math Masters,* pp. 200–202	1.NBT.2, 1.NBT.2a. 1.G.1 SMP7

Differentiation Options RtI

Readiness
10–15 min

Sorting Classroom Objects by Attributes

| WHOLE CLASS |
| SMALL GROUP |
| **PARTNER** |
| INDEPENDENT |

classroom objects

To provide experience sorting by attributes, have children sort various objects. Gather a variety of easy-to-sort classroom objects, such as crayons or books. Partners sort the objects and explain how they sorted. **GMP7.1** Discuss other ways to sort the same set of objects by a different attribute, such as color.

Enrichment
10–15 min

Attribute-Train Puzzles

| WHOLE CLASS |
| SMALL GROUP |
| **PARTNER** |
| INDEPENDENT |

Activity Card 90, attribute blocks, folder

Have children apply their understandings of attributes by making and solving attribute-train puzzles. **GMP7.2**

Attribute-Train Puzzles 90

What You Need
attribute blocks
folder
paper

What To Do
Work with a partner.
1 Sit across from your partner with a folder between you.
2 Both partners make attribute trains.
- Pick any block to start the train.
- Pick the next block that is different from the first block in only one way—in shape, size, thickness, or color.
- Continue until your trains are at least 7 blocks long.
- Take out a few of the blocks and write ?s in their places to make an attribute-train puzzle.
3 Trade puzzles. Place blocks on the ?s that complete the attribute train.
4 Check your work with your partner.

Talk About It
How did you solve your partner's puzzle?

More You Can Do
Try solving your partner's puzzle in a different way.

Extra Practice
10–15 min

Comparing Attributes

| WHOLE CLASS |
| **SMALL GROUP** |
| PARTNER |
| INDEPENDENT |

Activity Card 91, folder

Children practice exploring the attributes of shapes by discussing the attributes that shapes have in common. **GMP6.3**

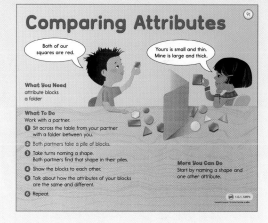

Comparing Attributes 91

Both of our squares are red.

Yours is small and thin. Mine is large and thick.

What You Need
attribute blocks
a folder

What To Do
Work with a partner.
1 Sit across the table from your partner with a folder between you.
2 Both partners take a pile of blocks.
3 Take turns naming a shape. Both partners find that shape in their piles.
4 Show the blocks to each other.
5 Talk about how the attributes of your blocks are the same and different.
6 Repeat.

More You Can Do
Start by naming a shape and one other attribute.

English Language Learners Support

Beginning ELL Use Total Physical Response prompts as you model, and have children practice responding to directions using the attribute blocks, such as: *Show me _____. Point to _____. Pick up the _____. Put all the _____ together.* Use these directions to have children indicate thick, thin, large, small, as well as those with different shapes and colors.

Go Online ELL English Learners Support

Adjusting the Activity

Differentiate Encourage children to explain each choice of block after their turns. If necessary, demonstrate explaining a block choice. For example, *I changed the color to blue* or *I changed the thickness, but I did not change the shape.*

Go Online ▸ Differentiation Support

1 Warm Up 15–20 min

Go Online ▸

ePresentations eToolkit

▸ Mental Math and Fluency

Have children record fact families for the following sets of numbers on their slates.

● ○ ○ **2, 6, 8** $2 + 6 = 8$; $6 + 2 = 8$; $8 - 2 = 6$; $8 - 6 = 2$

● ● ○ **3, 6, 9** $3 + 6 = 9$; $6 + 3 = 9$; $9 - 6 = 3$; $9 - 3 = 6$

● ● ● **5, 9, 14** $5 + 9 = 14$; $9 + 5 = 14$; $14 - 9 = 5$, $14 - 5 = 9$

▸ Daily Routines

Have children complete the Daily Routines. For detailed instructions, see pages 2–37. For specifics on CCSS coverage, see pages xiv–xvii.

2 Focus 35–40 min

Go Online ▸

ePresentations eToolkit

▸ Math Message

Pick up one attribute block to use today. Record as many attributes as you can for your block. GMP7.1

▸ Making an Attribute Train

WHOLE CLASS | SMALL GROUP | PARTNER | INDEPENDENT

Math Message Follow-up Children discuss the attributes of their blocks such as shape, color, size, and thickness. Have all children stand, and ask one child to read the attributes from his or her list, one at a time. As each attribute is read, children whose blocks do not share that attribute sit down. Those children whose blocks share all the attributes that were read continue standing until only children who have identical blocks are left standing. Discuss how all the blocks remaining are the same because they have all the same attributes.

Tell children that today they will play a game using attributes. Have one child come to the front with his or her attribute block and be the "conductor." The conductor chooses a child to join the train. That child's block must differ from the conductor's block in only one way. GMP7.1 For example, a child with a large, thick, blue triangle may choose a child with a large, thin, blue triangle, or a large, thick, blue square. Each child who joins the train chooses the next child. Explain that in Exploration A, children will play a game like this one by making a train of attribute blocks that differ from the previous and next blocks in only one way. Then explain that in Exploration B, they will continue to work with shapes, and in Exploration C, they will play another game.

Academic Language Development To extend children's vocabulary using derivatives of the word *differ*, use questions such as: *How are these two blocks different? How does this block differ from this block?* Provide sentence frames for children to use, such as: "This block differs from this block because _____" or "This block is different from this block because _____."

Math Masters, p. G52

▶ # Exploration A: Introducing
Attribute Train

Review the game directions found on *Math Masters,* page G52 with the class. Be sure children understand what it means to change only one attribute when placing the next block in the train.

Each block differs from the previous
one by just one attribute.

Observe
• Which children place a block that changes in only one attribute?

Discuss
• *Which attribute changed when you chose this block?*
• *Which attributes did not change when you chose this block?* **GMP7.1**

Differentiate **Game Modifications** **Go Online** Differentiation Support

---

▶ # Exploration B: Dividing Shapes

Activity Card 88

WHOLE CLASS | **SMALL GROUP** | **PARTNER** | INDEPENDENT

Children create shapes on geoboards and then divide them into two parts. Partnerships or groups talk about whether the two parts are the same size. Encourage children to find multiple ways to divide the shapes and to use multiple methods to test whether the parts are the same size. **GMP1.5**

Attribute Train   NAME   DATE

Materials	Attribute blocks
Players	2
Skill	Comparing attributes
Object of the Game	To continue the train until all blocks are played

Directions

1. Pick any block to start the train.
2. Your partner picks a block that is different in only one way. The difference can be in the shape, size, thickness, or color.
3. You pick the next block that is different in only one way.
4. Keep taking turns.
5. The game ends when all the blocks have been played.

Each block is different in only one way.

Other Ways to Play
• Pick blocks that change in all but one attribute.
• Follow a pattern for every round. Change color, size, shape, and then thickness.

G52   1.G.1

Activity Card 88

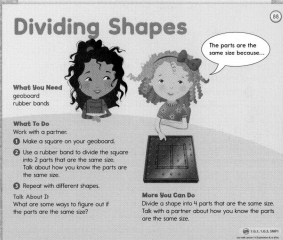

Recording *Salute!*

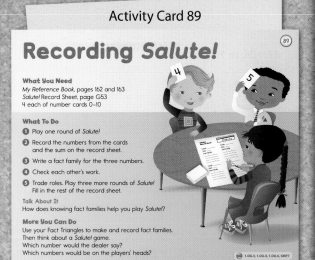

What You Need
My Reference Book, pages 162 and 163
Salute! Record Sheet, page G53
4 each of number cards 0–10

What To Do
1. Play one round of *Salute!*
2. Record the numbers from the cards and the sum on the record sheet.
3. Write a fact family for the three numbers.
4. Check each other's work.
5. Trade roles. Play three more rounds of *Salute!* Fill in the rest of the record sheet.

Talk About It
How does knowing fact families help you play *Salute!*?

More You Can Do
Use your Fact Triangles to make and record fact families. Then think about a *Salute!* game.
Which number would the dealer say?
Which numbers would be on the players' heads?

1.OA.3, 1.OA.4, 1.OA.6, SMP7

Games

Salute!

Materials	☐ number cards 0–10 (4 of each)
Players	3
Skill	Practicing addition and subtraction facts
Object of the Game	To solve for the number on your card.

Directions

1. One person begins as the Dealer. The Dealer gives one card to each of the other two Players.

2. Without looking at their cards, the Players hold them on their foreheads with the number facing out.

3. The Dealer looks at both cards and says the sum of the two numbers.

4. Each Player looks at the other Player's card. They use the number they see and the sum said by the Dealer to figure out what the number on their card must be. They say that number out loud.

MRB
162 one hundred sixty-two

▶ # Exploration C: Recording *Salute!*

Activity Card 89; *Math Masters,* p. G53

WHOLE CLASS	**SMALL GROUP**	PARTNER	INDEPENDENT

Children play *Salute!,* which was introduced in Lesson 7-3. See *My Reference Book,* pages 162 and 163 for directions. Children make a more explicit connection between addition and subtraction as they complete *Math Masters,* page G53. GMP7.2

Differentiate **Game Modifications** Go Online ▶ Differentiation Support

Salute! Record Sheet

NAME _____ DATE _____

Cards: _____, _____   Cards: _____, _____

Sum: _____   Sum: _____

Fact Family: Fact Family:

_____ + _____ = _____   _____ + _____ = _____

_____ + _____ = _____   _____ + _____ = _____

_____ − _____ = _____   _____ − _____ = _____

_____ − _____ = _____   _____ − _____ = _____

Cards: _____, _____   Cards: _____, _____

Sum: _____   Sum: _____

Fact Family: Fact Family:

_____ + _____ = _____   _____ + _____ = _____

_____ + _____ = _____   _____ + _____ = _____

_____ − _____ = _____   _____ − _____ = _____

_____ − _____ = _____   _____ − _____ = _____

1.OA.3, 1.OA.4, 1.OA.6   G53

Math Masters, p. G53

3 Practice 10–15 min

Go Online

ePresentations · eToolkit · Home Connections

▶ Solving Garden Number Stories

Math Journal 2, p. 145

| WHOLE CLASS | SMALL GROUP | **PARTNER** | **INDEPENDENT** |

Have children practice solving number stories. **GMP4.1**

▶ Math Boxes 7-6

Math Journal 2, p. 146

| WHOLE CLASS | SMALL GROUP | **PARTNER** | **INDEPENDENT** |

Mixed Practice Math Boxes 7-6 are paired with Math Boxes 7-4.

▶ Home Link 7-6

Math Masters, pp. 200–202

Homework Children identify shapes in everyday objects.

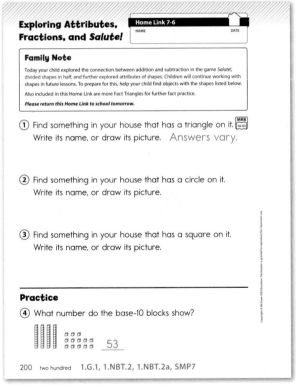

Math Masters, p. 200

Math Journal 2, p. 145

Solving Garden Number Stories Lesson 7-6
DATE

Solve. Then write a number model.

1 Frank grew 7 red flowers.
He grew 5 more flowers.
How many flowers did he grow in all? __12__ flowers

Number model: __7 + 5 = 12__

2 Clark planted 18 seeds.
Ashley planted 15 seeds.
Who planted more seeds? __Clark__
How many more? __3__ seeds

Number model: __18 − 15 = 3__

3 In the garden, there are 6 green peppers,
7 yellow squash, and 4 yellow peppers.
How many vegetables are in the garden? __17__ vegetables

Number model: __6 + 7 + 4 = 17__

Try This

4 For Problem 3, Dan first found the number of peppers.
Then he added the number of squash.
Pam first found the number of yellow vegetables.
Then she added the number of green vegetables.
Do both ways get the same answer? Which is faster for you?
__Sample answer: Yes; I added 6 and 4 first__
__because I know that's 10.__

1.OA.1, 1.OA.2, 1.OA.3, 1.OA.6, SMP4   one hundred forty-five 145

Math Journal 2, p. 146

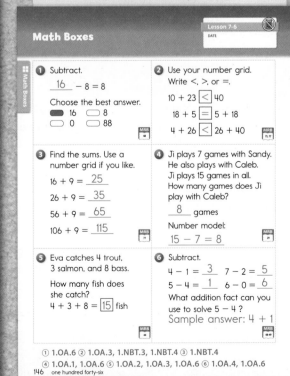

① 1.OA.6 ② 1.OA.3, 1.NBT.3, 1.NBT.4 ③ 1.NBT.4
④ 1.OA.1, 1.OA.6 ⑤ 1.OA.2, 1.OA.3, 1.OA.6 ⑥ 1.OA.4, 1.OA.6
146   one hundred forty-six

Defining and Nondefining Attributes

Overview Children differentiate between defining and nondefining attributes of 2-dimensional shapes.

▶ **Before You Begin**
For Part 2, decide how you will display the posters. These posters will be used again in Lesson 8-1. For the optional Readiness activity, cut out sets of Shape Sorting Cards (*Math Masters*, page 203).

▶ **Vocabulary**
closed • open • vertex • defining attribute • nondefining attribute • polygon

Common Core State Standards

Focus Cluster
Reason with shapes and their attributes.

---

1 Warm Up 15–20 min

	Materials	
Mental Math and Fluency Children solve number stories.	slate	1.OA.1, 1.OA.4, 1.OA.6
Daily Routines Children complete daily routines.	See pages 2–37.	See pages xiv–xvii.

---

2 Focus 30–35 min

Math Message Children compare and contrast two shapes.		1.G.1 SMP6
Identifying Defining and Nondefining Attributes of Shapes Children identify defining and nondefining attributes of triangles. They discuss attributes of other polygons.	Pattern-Block Template, chart paper, Two-Dimensional Shapes Poster (optional)	1.G.1 SMP1, SMP7
Defining a Rectangle Children identify attributes of rectangles.	Pattern-Block Template, chart paper	1.G.1 SMP1
✓ **Assessment Check-In** See page 634.		1.G.1, SMP1, SMP6

 1.G.1 **Spiral Snapshot**

GMC Distinguish between defining and non-defining attributes.

| 4-5
Focus | 6-3
Focus
Practice | 7-5
Focus
Practice | 7-6
Focus
Practice | 7-7
Focus
Practice | 8-1
Focus
Practice | 8-5
Focus
Practice | 8-6
Focus
Practice |

III Spiral Tracker   Go Online   to see how mastery develops for all standards within the grade.

---

3 Practice 15–20 min

Playing *Penny-Dime-Dollar Exchange* **Game** Children practice exchanging pennies for dimes and dimes for a dollar.	*Math Masters*, p. G48 (optional), p. TA30; per partnership: 2 dot dice, 5 one-dollar bills, 20 dimes, 20 pennies	1.NBT.2, 1.NBT.2a SMP6
Math Boxes 7-7 Children practice and maintain skills.	*Math Journal 2*, p. 147	See page 635.
Home Link 7-7 **Homework** Children draw triangles and distinguish between defining and nondefining attributes.	*Math Masters*, p. 204	1.OA.3, 1.OA.6, 1.OA.8, 1.G.1 SMP7

---

connectED.mcgraw-hill.com ▷

Plan your lessons online with these tools.

 ePresentations   Student Learning Center   Facts Workshop Game   eToolkit   Professional Development   Home Connections   Spiral Tracker   Assessment and Reporting   English Learners Support   Differentiation Support

Differentiation Options RtI

| Readiness | 15–20 min | | Enrichment | 10–15 min | | Extra Practice | 10–15 min |

Readiness — 15–20 min

Sorting Shapes by Attributes

	WHOLE CLASS
	SMALL GROUP
	PARTNER
	INDEPENDENT

Math Masters, p. 203

To provide additional experience with attributes of shapes, have children use the Shape Sorting Cards prepared from *Math Masters,* page 203. Ask children to find three shapes that have one or more attributes in common. Then have them determine a rule that tells the commonality. Then have children choose three more shapes with at least one attribute in common and suggest a rule that tells what those shapes have in common. `GMP8.1` Repeat as time permits.

Enrichment — 10–15 min

Growing Polygons

	WHOLE CLASS
	SMALL GROUP
	PARTNER
	INDEPENDENT

Activity Card 92, pattern blocks

To extend children's understandings of defining and nondefining attributes, have them build composite polygons with pattern blocks. Children discuss attributes of polygons and composite polygons. They determine whether shapes are the same after being rotated. `GMP7.1`

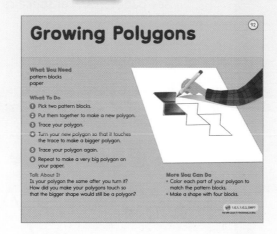

Extra Practice — 10–15 min

Playing *Attribute Train*

	WHOLE CLASS
	SMALL GROUP
	PARTNER
	INDEPENDENT

Math Masters, p. G52; attribute blocks

To provide practice comparing attributes, have children play *Attribute Train.* `GMP7.1` For detailed instructions, see Lesson 7-6.

Observe
- Which children change only one attribute when placing the next shape?

Discuss
- *Which attribute did you change when you chose this shape?*
- *What can you do if you do not know what shape to play next?*

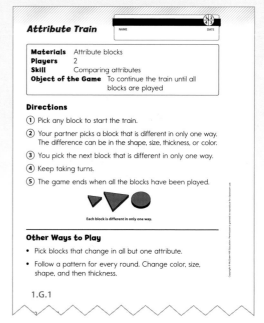

English Language Learners Support

Beginning ELL Prepare children for discussing shapes by looking at examples. Display pairs of shapes with one differing attribute. Ask yes/no questions about color, shape, and size to identify how the shapes are the same and different. For example: *Are all the shapes squares? Are all the shapes the same color? Are they the same shape?* Use the terms *in common* and *same* to summarize answers. Repeat with the words *different* and *not the same.*

  Go Online   ELL English Learners Support

1 Warm Up 15–20 min

Go Online   ePresentations   eToolkit

▶ Mental Math and Fluency

Have children solve number stories on their slates and briefly discuss strategies, including think addition.

● ○ ○ Sun Kim has 16 feathers. He gives some away. Now he has 8 feathers. How many feathers did Sun Kim give away? 8 feathers

● ● ○ Jack has 10 sticks. He loses some sticks, and now he has 8 sticks. How many sticks did Jack lose? 2 sticks

● ● ● Tonya has 15 beads. She used some beads to make a necklace. Now she has 7 beads. How many beads did Tonya use in her necklace? 8 beads

▶ Daily Routines

Have children complete the Daily Routines. For detailed instructions, see pages 2–37. For specifics on CCSS coverage, see pages xiv–xvii.

2 Focus 30–35 min

Go Online ePresentations eToolkit

▶ Math Message

How are these shapes alike? How are they different? Record your ideas. GMP6.1

▶ Identifying Defining and Nondefining Attributes of Shapes

| **WHOLE CLASS** | SMALL GROUP | PARTNER | INDEPENDENT |

Math Message Follow Up Have children share their observations about the shapes. GMP7.1 They may observe that both have 3 sides, both are roughly the same size, and both look like triangles—although the one on the right appears to be "broken." Explain that the sides of a **closed** shape are like a fence with no openings. When two sides of a shape do not meet, the shape is **open.** Explain that only the closed shape is a triangle because being *closed* is one of the attributes that makes it a triangle.

Explain that today children will learn more about how attributes define shapes. If you have not already displayed it, you may wish to display your Two-Dimensional Shapes Poster.

Display three different triangles, similar to those shown below, each colored or shaded and oriented differently from the others.

Be sure children agree that all these shapes are triangles. Ask them to identify one side and one **vertex** (or corner), a point where two sides meet. Label one side and one vertex of a triangle.

> **NOTE** Encourage children to refer to a corner of a polygon as a vertex. The plural of *vertex* is *vertices*. You may wish to display both so children can compare their spellings again. See Lesson 1-3 for a mnemonic illustration.

Display a blank two-column chart. Ask: *What attributes do these three triangles have in common?* Sample answers: They have exactly 3 vertices. They have exactly 3 straight sides. They are closed. Write these common attributes in the first column. GMP7.1

Ask: *What attributes are different about the triangles?* Sample answers: They are shaded differently. They point different ways. They are different sizes. Write these attributes in the second column.

Ask: *Even though they look different, are all three shapes still triangles?* Yes Explain that all three shapes are triangles because each has all the attributes listed in the first column. All triangles have exactly 3 straight sides, have exactly 3 vertices, and are closed. The attributes in the second column could be true of any shape. Label the first column **Defining Attributes** and the second **Nondefining Attributes**. To help children understand these terms, under the Defining Attributes label, explain that a defining attribute of a shape is always true for that shape. Then write "All triangles . . ." and under the Nondefining Attributes label, write "Some triangles and other shapes might . . ." GMP1.1

Ask: *Is there anything else we should add to our list of Defining Attributes?* Sample answers: The sides do not cross. They are flat, so they can't be held.

Invite children to look at their Pattern-Block Templates. Discuss how the shapes other than circles are alike and different.

- Each shape is made up of straight sides.
- Each shape is closed.
- Each shape has vertices.
- Each shape has as many sides as it has vertices.
- The numbers of sides and vertices vary from shape to shape.

For example, hexagons have 6 sides and 6 vertices; squares, trapezoids, and rhombuses have 4 sides and 4 vertices.

Defining Attributes	Nondefining Attributes
All triangles ...	Some triangles and other shapes might ...
• have exactly 3 vertices.	• be different sizes.
• have exactly 3 straight sides.	• be different colors.
• are closed.	• have different patterns.
	• point in different directions.

> **NOTE** If children notice sizes of corners, you may wish to introduce the term *angle*. The angles of squares look like the angles of books. Such angles are called square angles, or right angles. The polygons on the Pattern-Block Templates are mostly examples of regular polygons, so the hexagon angles are larger than the triangle angles. However, there are many irregular hexagons and triangles for which this is not true.

Defining Attributes	Nondefining Attributes
All rectangles ...	Some rectangles and other shapes might ...
• have 4 vertices. • have 4 sides. • are polygons. • have 4 square corners.	• be different sizes. • be different colors. • have different patterns. • be turned in different directions.

Explain that each of these shapes is a **polygon.** A polygon is a closed, flat shape with three or more straight sides meeting only at their vertices. The sides of the polygon cannot cross.

Differentiate   **Adjusting the Activity**

For children who struggle to identify common and differing attributes in shapes, write attributes on stick-on notes and use a fresh two-column chart. Model sorting those attributes into the two columns of the chart. Then have children sort and post to the chart attribute stick-on notes, such as different patterns, have 3 sides, and different colors.

Go Online ▶   Differentiation Support

▶ Defining a Rectangle

WHOLE CLASS   SMALL GROUP   PARTNER   INDEPENDENT

Display several rectangles of different sizes and orientations, including a square. Create another two-column chart to list the defining and nondefining attributes of rectangles. Discuss the defining attributes children see. Ask: *What attributes do all rectangles have?* Sample answers: All rectangles have 4 straight sides. They have 4 vertices. They have square corners. They are polygons. Include these answers in the Defining Attributes column on the chart. *What are some attributes a rectangle might have?* Sample answers: It might be small or large. It might be fat or skinny. It might be in different colors. It might be turned around. Record some examples in the column labeled Nondefining Attributes. Ask children to draw other rectangles that fit the defining attributes they listed. GMP1.1

✔️ **Assessment Check-In**   CCSS 1.G.1

Display a square alongside the list of defining attributes of rectangles. Have children go through the list and tell which defining attributes of rectangles also define squares. Expect children to recognize that all defining attributes of rectangles are also defining attributes of squares. Have children who do not recognize this try to draw a square that does not fit the defining attributes of a rectangle. Ask children who exceed expectations to explain which attributes of squares define them as being different from other rectangles. GMP1.1, GMP6.1

Assessment and Reporting   Go Online   to record student progress and to see trajectories toward mastery for these standards.

Summarize   Ask children whether they think all squares are rectangles based on their list of defining attributes. Remind them that squares are special kinds of rectangles. Have children who listed defining attributes unique to squares share with the class.

Math Masters, p. G48

Penny-Dime-Dollar Exchange

NAME _____ DATE _____

Materials	1 dollar, 20 dimes, 20 pennies 1 Place-Value Mat with Money per player (*Math Masters*, page TA30) 2 dot dice 1 sheet of paper labeled "Bank"
Players	2
Skill	Making place-value exchanges
Object of the Game	To exchange for a dollar

Directions

① Place all of the money in the bank.

② Players take turns. When it is your turn:
   • Roll the dice.
   • Take that number of cents from the bank. Place the coins on the mat.
   • If you have 10 or more pennies in the Pennies column, exchange 10 pennies for 1 dime.
   • If you have 10 or more dimes in the Dimes column, exchange 10 dimes for 1 dollar.

③ The winner is the first player to make the exchange for a dollar.

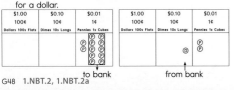

G48   1.NBT.2, 1.NBT.2a

3 Practice 15–20 min

 Go Online   ePresentations   eToolkit   Home Connections

▶ Playing *Penny-Dime-Dollar Exchange*

Math Masters, p. TA30

| WHOLE CLASS | **SMALL GROUP** | **PARTNER** | INDEPENDENT |

Have children practice place-value concepts in the context of money by playing *Penny-Dime-Dollar Exchange.* For directions for playing the game, use *Math Masters,* page G48. For more information, see Lesson 6-11.

Observe

• Which children correctly exchange when they have 10 pennies or 10 dimes?

Discuss

• *If you have 6 pennies on your mat and you roll a 5, can you exchange?*

• *If you have 8 dimes and 1 penny, can you win on your next turn? Explain why or why not.* GMP6.1, GMP6.4

▶ Math Boxes 7-7

Math Journal 2, p. 147

| WHOLE CLASS | SMALL GROUP | **PARTNER** | **INDEPENDENT** |

Mixed Practice Math Boxes 7-7 are paired with Math Boxes 7-10.

▶ Home Link 7-7

Math Masters, p. 204

Homework Children draw triangles with different attributes and describe the attributes of their triangles.

Math Journal 2, p. 147

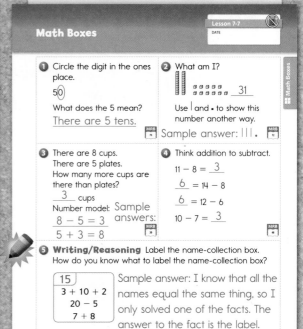

Math Boxes

Lesson 7-7
DATE

① Circle the digit in the ones place.

5⓪

What does the 5 mean?
There are 5 tens.

MRB

② What am I?

▯▯▯ ⚬⚬⚬⚬⚬ / ⚬⚬⚬⚬⚬⚬ ___31___

Use | and • to show this number another way.

Sample answer: ||| .

MRB

③ There are 8 cups.
There are 5 plates.
How many more cups are there than plates?

___3___ cups

Number model: Sample answers:
8 − 5 = 3
5 + 3 = 8

④ Think addition to subtract.

11 − 8 = __3__
__6__ = 14 − 8
__6__ = 12 − 6
10 − 7 = __3__

MRB

⑤ Writing/Reasoning Label the name-collection box. How do you know what to label the name-collection box?

15
3 + 10 + 2
20 − 5
7 + 8

Sample answer: I know that all the names equal the same thing, so I only solved one of the facts. The answer to the fact is the label.

① 1.NBT.2, 1.NBT.2c ② 1.NBT.2, 1.NBT.2a
③ 1.OA.1, 1.OA.6 ④ 1.OA.4, 1.OA.6, 1.OA.8
⑤ 1.OA.7, SMP8

one hundred forty-seven 147

Math Masters, p. 204

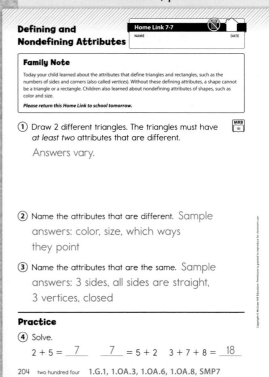

Defining and Nondefining Attributes

Home Link 7-7
NAME
DATE

Family Note

Today your child learned about the attributes that define triangles and rectangles, such as the numbers of sides and corners (also called *vertices*). Without these defining attributes, a shape cannot be a triangle or a rectangle. Children also learned about nondefining attributes of shapes, such as color and size.

Please return this Home Link to school tomorrow.

① Draw 2 different triangles. The triangles must have *at least* two attributes that are different.

MRB

Answers vary.

② Name the attributes that are different. Sample answers: color, size, which ways they point

③ Name the attributes that are the same. Sample answers: 3 sides, all sides are straight, 3 vertices, closed

Practice

④ Solve.

2 + 5 = __7__   __7__ = 5 + 2   3 + 7 + 8 = __18__

204 two hundred four   1.G.1, 1.OA.3, 1.OA.6, 1.OA.8, SMP7

Finding Unknowns: "What's My Rule?"

Overview Children learn the "What's My Rule?" routine and use it to practice finding unknown numbers in number sentences.

▶ **Before You Begin**
For Part 2, prepare a "magic bag" by gathering a paper bag and craft sticks.

▶ **Vocabulary**
rule • "What's My Rule?" • function machine

Common Core State Standards

Focus Clusters
• Add and subtract within 20.
• Work with addition and subtraction equations.
• Use place value understanding and properties of operations to add and subtract.

---

① Warm Up 15–20 min

	Materials	
Mental Math and Fluency Children solve facts within 20.	slate	**1.OA.6**
Daily Routines Children complete daily routines.	See pages 2–37.	See pages xiv–xvii.

---

② Focus 30–35min

Math Message Children solve number stories and determine a pattern.		**1.OA.6** **SMP8**
Finding Rules Children find the rule for the output of a "magic bag."	paper bag, craft sticks	**1.OA.6** **SMP7, SMP8**
Introducing "What's My Rule?" Children are introduced to function machines. They find the rules for function machines, given the inputs and outputs.	*Math Masters*, p. TA37; TA38 (optional)	**1.OA.6, 1.OA.8, 1.NBT.5** **SMP7, SMP8**
Solving "What's My Rule?" Problems Children find and apply rules for function machines.	*Math Journal 2*, p. 148	**1.OA.6, 1.OA.8, 1.NBT.5** **SMP7, SMP8**
✔ **Assessment Check-In** See page 641.	*Math Journal 2*, p. 148	**1.OA.8, SMP8**

CCSS 1.OA.8 Spiral Snapshot

GMC Find the unknown in addition and subtraction equations.

6-4 Practice	7-1 through 7-4 Warm Up Focus Practice	7-8 Focus Practice	7-9 Warm Up Focus Practice	7-10 Focus Practice	8-1 through 8-3 Warm Up Practice

 Spiral Tracker   Go Online ▶ to see how mastery develops for all standards within the grade.

---

③ Practice 15–20 min

Facts Inventory Record, Part 4 Children track facts they know or are still learning.	*Math Journal 2*, p. 223	**1.OA.3, 1.OA.6** **SMP6**
Math Boxes 7-8: Preview for Unit 8 Children practice prerequisite skills for Unit 8.	*Math Journal 2*, p. 149	See page 641.
Home Link 7-8 **Homework** Children solve "What's My Rule?" problems.	*Math Masters*, pp. 205–209	**1.OA.6, 1.OA.8** **SMP7, SMP8**

---

connectED.mcgraw-hill.com

Plan your lessons online with these tools.

ePresentations Student Learning Center Facts Workshop Game eToolkit Professional Development Home Connections Spiral Tracker Assessment and Reporting English Learners Support Differentiation Support

Differentiation Options

RtI

Readiness
5–10 min

	WHOLE CLASS
	SMALL GROUP
	PARTNER
	INDEPENDENT

Identifying Arrow Rules

slate

To provide experience with numeric rules, have children identify arrow rules for counting patterns in Frames and Arrows. For example, display Frames and Arrows with 2, 4, 6, and 8 in the frames. Discuss the number that goes in the next frame and how children know it. Encourage children to state a rule for the pattern, using both counting language ("count up by 2") and operational language ("add 2" or "+ 2"). **GMP8.1** Repeat using other common counting patterns.

Enrichment
10–15 min

	WHOLE CLASS
	SMALL GROUP
	PARTNER
	INDEPENDENT

Making "What's My Rule?" Problems

Math Masters, p. TA37

To further explore finding unknown numbers and rules, have children create four "What's My Rule?" problems for their partners. Encourage them to create some problems in which all the inputs and outputs are given and the rule is unknown and other problems in which the rule is known but some inputs and outputs are missing. Then have children complete each other's problems. **GMP7.2 , GMP8.1**

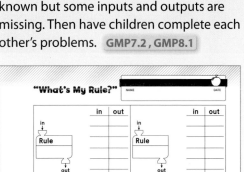

Extra Practice
10–15 min

	WHOLE CLASS
	SMALL GROUP
	PARTNER
	INDEPENDENT

Playing *What's Your Way?*

Math Masters, p. G34

To practice applying a rule and explaining their mathematical thinking, have children play *What's Your Way?* **GMP6.1** It also provides an opportunity to practice finding 10 more and 10 less. For detailed instructions, see Lesson 4-11.

Observe
- Which children mentally find 10 more or 10 less?
- Which children use place value to explain how they found their answers? **GMP6.1**

Discuss
- *Can you make a number sentence for finding 10 more than 38?*

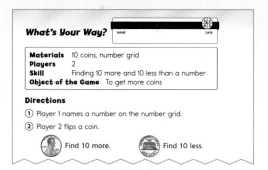

English Language Learners Support

Beginning ELL Help children understand language and procedures for "What's My Rule?" by displaying a large chart of a "What's My Rule?" table labeled with *Rule, in,* and *out.* Use hand gestures to demonstrate the procedure for putting something into and taking something out of the machine. Use the words *in* and *input* as well as *out* and *output* interchangeably. Ask children to point to the table and practice the gestures and phrases. "I put in _____, so the input is _____. I take out _____, so the output is _____."

Go Online ELL English Learners Support

Standards and Goals for
Mathematical Practice

SMP7 Look for and make use of structure.
  GMP7.1 Look for mathematical structures such as categories, patterns, and properties.
  GMP7.2 Use structures to solve problems and answer questions.

SMP8 Look for and express regularity in repeated reasoning.
  GMP8.1 Create and justify rules, shortcuts, and generalizations.

① Warm Up   15–20 min   [Go Online] ePresentations eToolkit

▶ Mental Math and Fluency

Encourage children to use near doubles to solve each fact. Have children briefly explain how they used the strategy.

●○○ $4 + 5$ 9; I know $4 + 4 = 8$, and 1 more is 9. $5 + 6$ 11; I know $5 + 5 = 10$, and 1 more is 11.

●●○ $6 + 7$ 13; I know $6 + 6 = 12$, and 1 more is 13. $5 + 7$ 12; I know $5 + 5 = 10$, and 2 more is 12.

●●● $9 + 8$ 17; I know $9 + 9 = 18$, and 1 less is 17. $8 + 6$ 14; I know $6 + 6 = 12$, and 2 more is 14.

▶ Daily Routines

Have children complete the Daily Routines. For detailed instructions, see pages 2–37. For specifics on CCSS coverage, see pages xiv–xvii.

② Focus   30–35 min   [Go Online] ePresentations eToolkit

▶ Math Message

The library has a rule for how many more books children must check out from one week to the next.

Antoine checked out 5 books one week and 7 books the next week.

Quan checked out 9 books one week and 11 books the next week.

Sooki checked out 2 books one week and 4 books the next week.

What is the library's rule for giving children books from one week to the next?
GMP8.1 *Record your answer.*

▶ Finding Rules

| WHOLE CLASS | SMALL GROUP | PARTNER | INDEPENDENT |

Math Message Follow-Up Discuss children's responses to the Math Message. Emphasize that even though each child started with different numbers of books, they followed the same **rule.** The library's rule is each child must check out 2 more books next week than what he or she got this week. So no matter how many books they check out the first week, children would add 2 to find how many books they can check out next week.

Academic Language Development Build on children's understanding of rules, such as classroom rules. Remind children that rules are instructions that tell us what to do or what not to do. In the same way, each time we follow a mathematical rule, we do things the same way.

Introduce children to your "magic bag," which has the power to change whatever is put in it. Put 1 stick in the bag, say a magic word, and take 2 sticks out of the bag. Then put 4 sticks in and pull 5 sticks out. Repeat several times until children see a pattern and can predict how many sticks will come out. GMP7.1 , GMP7.2 Have children discuss the rule that fits the action of the magic bag and explain how they know the rule is correct. GMP8.1 The magic bag adds 1 stick to the number of sticks put in.

Next tell children that you are going to have the magic bag change the rule. Repeat with *add 2* as the rule, followed by rules such as *subtract 1, minus 3,* or *double;* limit the numbers and rules to focus on addition and subtraction facts that children know.

▶ Introducing "What's My Rule?"

Math Masters, p. TA37

| **WHOLE CLASS** | SMALL GROUP | PARTNER | INDEPENDENT |

Tell children that they are going to learn a new math routine called **"What's My Rule?"** that works like the magic bag. But instead of using craft sticks, "What's My Rule?" uses numbers.

Display a **function machine.** Explain that the function machine is like the magic bag. If you put a number into the machine, the number will follow the rule on the machine, and a different number will come out. Any number you put in will follow the same rule.

Tell children that if you put 3 in this machine, 4 will come out. If you put in 6, 7 will come out. If you put in 9, 10 will come out. Encourage children to look for a pattern. GMP7.1 Ask: *What number do you think will come out if you put in 5?* 6

Have children suggest a number to put in the function machine. *What number will come out if we put that number in?* The number that is 1 more than the number that was put in

Tell children that you want to write number sentences to represent what happens in the function machine. Ask: *Should they be addition or subtraction number sentences? Why?* Addition; because the numbers that come out are larger than the numbers that go in Ask: *How could you write number sentences to show what will happen to each number you put in?* Remind children that they need to use a symbol to represent the rest of the unknown rule.

$3 + \boxed{} = 4$

$6 + \boxed{} = 7$

$9 + \boxed{} = 10$

$5 + \boxed{} = 6$

Professional Development

The *Everyday Mathematics* routine, "What's My Rule?," provides practice with number patterns and arithmetic facts as well as a format for thinking about relationships between pairs of numbers in a function. A function machine is a diagram or metaphor that indicates how the input and output numbers in "What's My Rule?" tables are related.

Go Online to the *Implementation Guide* for more information about "What's My Rule?" and function machines.

Math Masters, p. TA37

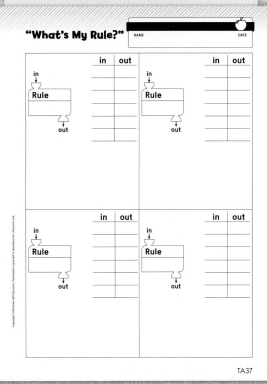

Rule: + 2

in	out
1	3
4	6
7	9
8	10
12	14

Rule: − 3

in	out
4	1
5	2
8	5
7	4
10	7

Rule: + 10

in	out
14	24
27	37
30	40
15	25
56	66

Rule: − 10

in	out
15	5
26	16
49	39
32	22
50	40

Math Journal 2, p. 148

Then ask children what should go in each answer box. This number represents *how much* the machine adds to each number you put in. 1 Help children see that finding the function machine's rule is like finding the unknown number in a number sentence. **GMP8.1** The unknown numbers should be the same for all these number sentences, just as the rule should apply to all the numbers that go in and come out of the function machine.

Ask: *What does this function machine do to the numbers you put in?* It adds 1 to them. *What should you write in the rule box for this function machine?* Sample answers: + 1; add 1 Write + 1 in the rule box.

Explain that other function machines may have different rules. For example, one machine might subtract 2 from any number you put in and another machine might add 10. Tell children that they are going to figure out the rules for different function machines.

Pose several "What's My Rule?" problems for children. (*See margin.*) Record the numbers that you put in the machine using tables such as those on *Math Masters,* page TA37. Note that the numbers you put into the machine are called inputs and that the numbers that come out are called outputs. For each problem, have children look for a pattern to help them find the rule. **GMP7.1, GMP8.1** Then ask them to find additional outputs using the rule. **GMP7.2**

Differentiate   **Adjusting the Activity**

If children have trouble remembering which direction to apply the rules, draw arrows connecting the "in" spaces to the "out" spaces, or you may wish to provide *Math Masters,* TA38.   **Go Online**   Differentiation Support

---

▶ **Solving "What's My Rule?" Problems**

Math Journal 2, p. 148

| WHOLE CLASS | SMALL GROUP | **PARTNER** | **INDEPENDENT** |

Have children complete "What's My Rule?" problems in their journals. Explain that the numbers under the headings "in" and "out" on the journal page show which numbers go into the function machine and which numbers come out. Tell children to find the unknown rule that determines which numbers come out of the machine and to use that rule to find the missing output numbers. **GMP7.2 , GMP8.1** Have children check their rules by writing number sentences for inputs and outputs. You may wish to complete the first problem together.

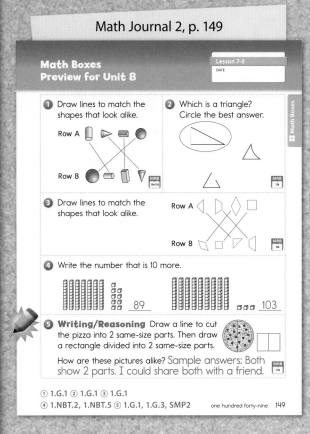

Math Journal 2, p. 149

✓ Assessment Check-In ⓒⓒⓢⓢ 1.OA.8

Math Journal 2, p. 148

Observe children's work on journal page 148. Expect children to find the rules for Problems 1 and 2. **GMP8.1** For children who have difficulty identifying the rule, break down the process into two steps. First have them determine whether the rule involves addition or subtraction by determining whether the outputs are larger or smaller than the inputs. Second have them use this operation, the input, and the output to write a number sentence, leaving an unknown for the number part of the rule. For example, they can write $7 - \square = 5$ for Problem 1. Children then fill in the unknown and use that number and the operation symbol to record a rule for the function machine. If children exceed expectations, they may enjoy the Enrichment activity.

☑ Assessment and Reporting   Go Online to record student progress and to see trajectories toward mastery for these standards.

Summarize Discuss the children's answers to Problem 4 on journal page 148. Have them share how they found the rule and how this rule differs from other rules they found during the lesson.

③ Practice   15–20 min   Go Online

ePresentations   eToolkit   Home Connections

▶ Facts Inventory Record, Part 4

Math Journal 2, p. 223

| WHOLE CLASS | SMALL GROUP | PARTNER | **INDEPENDENT** |

Have children complete the Facts Inventory Record, Part 4 (*Math Journal 2*, page 223). Children indicate whether they know each fact. Encourage them to record strategies that they could use to efficiently figure out the facts that they do not yet know. **GMP6.4** Ask children who know all their facts to record strategies for a few of the most difficult ones. Many of the facts in Part 4 could be solved by making 10, so encourage children to try that strategy.

▶ Math Boxes 7-8: Preview for Unit 8

Math Journal 2, p. 149

| WHOLE CLASS | SMALL GROUP | **PARTNER** | INDEPENDENT |

Mixed Practice Math Boxes 7-8 are paired with Math Boxes 7-12.

▶ Home Link 7-8

Math Masters, pp. 205–209

Homework Children practice solving "What's My Rule?" problems.

Math Masters, pp. 205–209

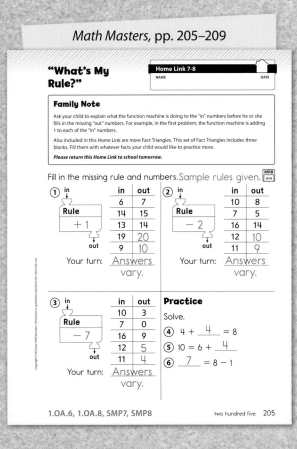

Desks and Chairs

2-Day Lesson

Overview **Day 1:** Children find a rule for a real-world situation and use it to solve a problem. **Day 2:** The class discusses some children's rules and explanations for finding the rules, then children revise their work.

Day 1: Open Response

▶ **Before You Begin**
Make sure you have enough pattern blocks for about 10 squares and 10 triangles per child or partnership. If possible, schedule time to review children's work and plan for Day 2 of this lesson with your grade-level team.

CCSS **Common Core State Standards**

Focus Clusters
- Add and subtract within 20.
- Work with addition and subtraction equations.
- Use place value understanding and properties of operations to add and subtract.

1 Warm Up   15–20 min

	Materials	
Mental Math and Fluency Children solve number stories.	slate	1.OA.1, 1.OA.8, 1.NBT.4
Daily Routines Children complete daily routines.	See pages 2–37.	See pages xiv–xvii.

2a Focus   45–55 min

Math Message Children complete "What's My Rule?" problems.	*Math Journal 2,* p. 150	1.OA.6, 1.OA.8 SMP8
Discussing Rules Children discuss how they found the rule and why rules are useful.	*Math Journal 2,* p. 150	1.OA.6, 1.OA.8 SMP8
Solving the Open Response Problem Children find a rule for a real-world situation and use it to solve a problem.	*Math Masters,* p. 210; square and triangle pattern blocks; Pattern-Block Template (optional)	1.OA.6, 1.OA.8, 1.NBT.4 SMP1, SMP4, SMP8

Getting Ready for Day 2 →

Review children's work and plan discussion for reengagement.   *Math Masters,* p. TA3;
children's work from Day 1

CCSS 1.OA.8 **Spiral Snapshot**

GMC Find the unknown in addition and subtraction equations.

6-4 Practice	7-1 through 7-4 Warm Up Focus Practice	7-8 Focus Practice	7-9 Warm Up Focus Practice	7-10 Focus Practice	8-1 through 8-3 Warm Up Practice

 Spiral Tracker   **Go Online**   to see how mastery develops for all standards within the grade.

connectED.mcgraw-hill.com

Plan your lessons online with these tools.

 ePresentations   Student Learning Center   Facts Workshop Game   eToolkit   Professional Development   Home Connections   Spiral Tracker   Assessment and Reporting   English Learners Support   Differentiation Support

642   Unit 7 | Subtraction Fact Strategies and Attributes of Shapes

1 Warm Up

15–20 min Go Online ePresentations eToolkit

▶ Mental Math and Fluency

Pose number stories. Have children record number models on their slates, share their solutions, and briefly discuss their solution strategies.
Suggestions:

● ○ ○ Elizabeth has 6 blue ribbons and 11 red ribbons. How many ribbons does she have all together? 17 ribbons

● ● ○ Thomas has 17 blue balloons and 10 yellow balloons. How many balloons does he have all together? 27 balloons

● ● ● Christian has 14 comic books in his bag and some on his desk. He has 25 comic books in all. How many comic books are on his desk? 11 comic books

▶ Daily Routines

Have children complete the Daily Routines. For detailed instructions, see pages 2–37. For specifics on CCSS coverage, see pages xiv–xvii.

2a Focus

45–55 min Go Online ePresentations eToolkit

▶ Math Message

Math Journal 2, p. 150

Complete journal page 150. Tell your partner how you found the rule in Problem 2. GMP8.1

▶ Discussing Rules

Math Journal 2, p. 150

| WHOLE CLASS | SMALL GROUP | PARTNER | INDEPENDENT |

Math Message Follow-Up Review function machines by asking children to describe them in their own words. Sample answer: You put a number in a function machine and the function machine does something to it. What it does is the rule. Ask: *How did you find the missing numbers in Problem 1?* Sample answer: I knew the rule was + 5, so I added 5 to all of the *in* numbers. Explain that one way to find the missing numbers is to think, *1 plus 5 equals what?, 6 plus 5 equals what?,* and so on. Display the number sentences below.

$1 + 5 = \square$ $6 + 5 = \square$ $3 + 5 = \square$ $10 + 5 = \square$

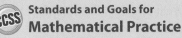
Standards and Goals for Mathematical Practice

SMP1 Make sense of problems and persevere in solving them.
GMP1.4 Check whether your answer makes sense.

SMP4 Model with mathematics.
GMP4.1 Model real-world situations using graphs, drawings, tables, symbols, numbers, diagrams, and other representations.

SMP8 Look for and express regularity in repeated reasoning.
GMP8.1 Create and justify rules, shortcuts, and generalizations.

Professional Development

The focus of this lesson is GMP8.1. When children fill in a missing rule in a "What's My Rule?" problem, they are creating a rule and justifying it based on the information given in the in/out table. In the open response problem, children create a rule for the number of chairs needed for a given number of desks and justify the rule using the models they build for the situation.

Go Online for information about SMP8 in the *Implementation Guide*.

Math Journal 2, p. 150

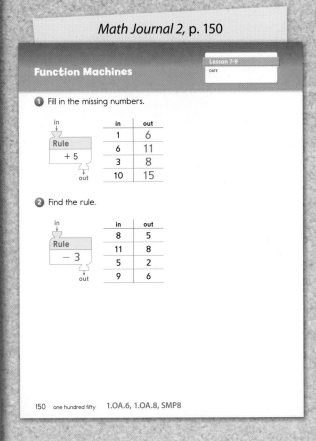

Ask children to share their *out* numbers from Problem 1. Use their answers to fill in the answer boxes in the equations.

Next, have children share their rules from Problem 2. Possible rules children might share are − 3, subtract 3, and count back 3. Discuss different ways of stating the rule. Then ask: *How did you know it would be a subtraction rule and not an addition rule?* GMP8.1 Sample answer: The numbers in the *out* column are smaller than the numbers in the *in* column. Point out that once children knew the function machine was subtracting, they had to figure out *how much* it was subtracting. Explain that one way to find a rule for this function machine is to think, *8 minus what is 5?, 11 minus what is 8?,* and so on. Display the number sentences below.

$$8 - \square = 5 \qquad\qquad 11 - \square = 8$$
$$5 - \square = 2 \qquad\qquad 9 - \square = 6$$

Then model filling in each answer box with a 3 to show that the rule works for all four rows of the table. Ask: *Does our rule need to work for all the rows?* Yes Point out that when children find a rule, they should check that it works for every row in the table.

Remind children that the rule can be used to determine what the function machine will do to other numbers. Display the number sentences $12 - 3 = \square$ and $6 - 3 = \square$. Ask: *If we put 12 in, what will come out?* 9 *If we put 6 in, what will come out?* 3 Fill in the boxes with the *out* numbers as children share answers.

Ask: *Why are rules useful?* Sample answers: We can use one rule to solve lots of problems. Rules tell you to add or subtract and how many to add or subtract. Tell children they will find another rule and use it to solve a problem. GMP8.1

▶ Solving the Open Response Problem

Math Masters, p. 210

WHOLE CLASS   SMALL GROUP   PARTNER   INDEPENDENT

Set up the scenario below. Use square and triangle pattern blocks to model it:

Luka is setting up desks and chairs in the hall to make a craft area. He is pushing the desks against a wall and putting them right next to each other in a long row. After he lines up the desks, he is going to put chairs all the way around them.

First, Luka will line up desks.

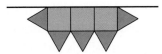

Then Luka will put chairs around the desks.

Explain that you have shown what the craft area would look like with 3 desks. *Luka will need to set up 18 desks in all. He wonders if there is a rule he can use to help him predict how many chairs he will need to go all the way around the 18 desks.* Tell children that their job is to figure out a rule to help Luka.

Ask: *How did we figure out the rule for the function machine in the Math Message?* Sample answer: We looked at the table. Explain that children will use pattern blocks to help them make a table like the ones they have used before to find rules.

Give each child or partnership about 10 square pattern blocks and 10 triangle pattern blocks. Draw a table on the board with columns labeled "Desks" and "Chairs." (*See margin.*) Have children use pattern blocks to model the situation with three desks. GMP4.1 They can use their journals or another object to simulate the wall. As a class, count the number of desks, and write a 3 in the Desks column of your table. Then count the number of chairs, and write a 5 in the Chairs column.

Ask: *How could we find more numbers to write in the table?* Sample answer: We could try a different number of desks. Make sure children understand that to fill in another row of the table, they need to use pattern blocks to show a different number of desks. You may wish to have a volunteer choose a second number of desks to try, then build the model and fill in a second row of the table together.

Tell children they will now have the chance to use pattern blocks to fill in their own tables. Distribute *Math Masters,* page 210. Allow children time to fill in their tables using any numbers they wish in the Desks column. You may want to have children use their Pattern-Block Templates to make drawings of their desk-and-chair models next to the table or on another sheet of paper.

Once children have completed their tables, have them finish the *Math Masters* page by writing a rule and explaining how they found and checked the rule. Then they use the rule to find the number of chairs that Luka will need for 18 desks. Encourage children to explain to their partners how they know that their rules are correct. GMP8.1

Differentiate **Adjusting the Activity**

If children have difficulty getting started, help them think about the problem step-by-step. Model how to complete one or two rows of the table. Choose a number of desks. Use pattern blocks to show that number of desks and the chairs that go around them. Then count the desks and write the number in the Desks column. Finally, count the chairs and write the number in the Chairs column. Ask children to choose a number of desks to try, and have them complete the process.

Math Masters, p. 210

Desks and Chairs — Lesson 7-9

NAME / DATE

① Use pattern blocks to help you fill in the table.

Desks	Chairs
3	5

Example:

Answers vary, but all rows should show two more chairs than desks.

② Write a rule for your table.

Write how you found and checked your rule.

Desks ↓
Rule
+ 2
↓ Chairs

Answers vary. See sample children's work on page 651 of the *Teacher's Lesson Guide.*

③ How many chairs will Luka need for 18 desks?
___20___ chairs

1.OA.6, 1.OA.8, 1.NBT.4, SMP1, SMP4, SMP8

210   two hundred ten

Desks	Chairs
3	5

Sample child's work, Child A

① Use pattern blocks to help you fill in the table.

Desks	Chairs
3	5
5	7
4	6
2	3

Example:

② Write a rule for your table.

Write how you found and checked your rule.

We yousd desks and chairs.

Desks
↓
Rule
+2
↓
Chairs

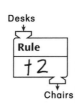

Sample child's work, Child B

① Use pattern blocks to help you fill in the table.

Desks	Chairs
3	5
7	9
4	6
2	4

Example:

② Write a rule for your table.

Write how you found and checked your rule.

I started at 3 and counted to 5 and it was +2

Desks
↓
Rule
+2
↓
Chairs

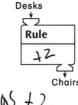

Circulate and observe. Make sure children are using their pattern blocks to model correct scenarios. While it is important for children to create their own models, the focus of this task is to determine and use the rule, so assist them in lining up the blocks as needed.

As children are working, ask: *Are there always more chairs than desks?* Yes *Does the rule use addition or subtraction?* **GMP8.1** Addition *Do all the rows in your table follow your rule?* **GMP1.4** Answers vary. For children who found the rule but have trouble with the explanation in Problem 2, provide sentence frames such as:

- I saw there were _____ (more/fewer) chairs than desks.
- I saw that _____ + _____ = _____.
- I saw that _____ − _____ = _____.

> **Differentiate** **Common Misconception**
>
> Some children may have difficulty connecting the Desks and Chairs table to the idea of a function machine. Have these children write *in* and *out* at the top of the columns to help them make the connection to "What's My Rule?" tables. This may help them use the table to write a rule.

Some children may try to answer Problem 3 by building the scenario with 18 desks. Ask: *Do you need to use blocks to show 18 desks? How could you use your rule to find out how many chairs Luka needs?* Encourage these children to think of 18 as an *in* number and decide how they could find the *out* number.

Summarize Ask: *How did your table help you find a rule?* **GMP8.1**
Sample answer: I could see that there were always 2 more chairs than desks.

Collect children's work so that you can evaluate it and prepare for Day 2.

Getting Ready for Day 2

Math Masters, p. TA3

Planning a Follow-up Discussion

Review children's work. Use the Reengagement Planning Form (*Math Masters,* page TA3) and the rubric on page 648 to plan ways to help children meet expectations for both the content and practice standards. Look for common misconceptions or inconsistencies in children's tables as well as interesting ways children explained and applied their rules.

Organize the discussion in one of the ways below or in another way you choose. If children's work is unclear or if you prefer to show work anonymously, rewrite the work for display.

Go Online for sample children's work that you can use in your discussion.

1. Show work with a rule that is not supported by all four rows of the table, such as in Child A's work. Ask:
   - *What do you notice about this child's paper?* GMP1.4 Sample answer: This child filled out the table and wrote a rule. The rule works for some rows, but the last row doesn't work for + 2.
   - *Should the rule work for all the rows in the table? Why?* GMP8.1 Sample answer: Yes. The rule is supposed to work no matter what numbers you use for the desks.
   - *Is the rule wrong, or is the last row wrong? How can you check?* GMP1.4 Sample answer: It seems like the rule is right because it works for the other rows. We could build the last row with pattern blocks to see if it is wrong. Have a volunteer use pattern blocks to show how to check the last row. GMP4.1

2. Show a paper with an explanation in Problem 2 that refers to only one row of the table, as in Child B's work. Compare it to a paper that refers to more than one row of the table, as in Child C's work. Ask:
   - *How did Child B find the rule?* GMP8.1 Sample answer: Child B used the first row and counted from the number of desks to the number of chairs.
   - *Did Child B check the rule? How can you tell?* GMP1.4 Sample answer: I'm not sure. Child B only talks about checking one row, not all the rows.
   - *How did Child C find the rule and make sure it was correct?* GMP1.4, GMP8.1 Sample answers: Child C noticed that 3 + 2 = 5. Child C found the number that was added each time and made sure it was + 2 for all rows.
   - *Why is it important to make sure your rule works for all the rows?* GMP1.4, GMP8.1 Sample answer: You should make sure your rule always works or else it might not really be the rule. You could find a mistake and fix it if you check all the rows.

3. Show a paper that gives a justification for the rule that goes beyond the numbers in the table, as in Child D's work. Ask:
   - *How did this child explain the rule?* GMP8.1 Sample answer: This child said Luka needed 2 more chairs for the sides.
   - *What do you think this child meant? Does it make sense?* GMP1.4 Sample answer: I think this child is talking about the 2 chairs that are sideways on the ends. This makes sense because if Luka only put chairs in front of the desks, he wouldn't have to add any. But when he puts them on the sides, that's 2 more chairs.

Go Online for four more discussion ideas.

Planning for Revisions

Have copies of *Math Masters,* page 210 or extra paper available for children to use in revisions. You might want to ask children to use colored pencils so you can see what they revised.

Sample child's work, Child C

① Use pattern blocks to help you fill in the table.

Desks	Chairs
3	5
ten	twelve
six	eight
one	three

Example:

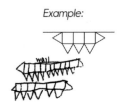

② Write a rule for your table.

Write how you found and checked your rule.

I tonow Because I know 3+2=5 and 10+2=12 and 6+2=8 and 1+2=3 that is how I got my rule!

Desks ↓
Rule
Add two
↓ Chairs

Sample child's work, Child D

① Use pattern blocks to help you fill in the table.

Desks	Chairs
3	5
5	7
4	6
6	8

Example:

② Write a rule for your table.

Write how you found and checked your rule.

Desks ↓
Rule
+2
↓ Chairs

I know the Rule is +2. becase 5+2=7 4+2=6 and 6+2=8 and I k that 6 desks its cust two more chairs for the sides.

③ How many chairs will Luka need for 18 desks?

20 chairs

Desks and Chairs

Overview **Day 2:** The class discusses some children's rules and explanations for finding the rules, then children revise their work.

Day 2: Reengagement

▶ **Before You Begin**
Have extra copies available of *Math Masters,* page 210 for children to revise their work.

CCSS Common Core State Standards

Focus Clusters
• Add and subtract within 20.
• Work with addition and subtraction equations.
• Use place value understanding and properties of operations to add and subtract.

2b Focus   50–55 min

	Materials	
Setting Expectations Children discuss what it means to find and use a rule, and what a good response to the open response problem might include. They also review how to discuss others' work respectfully.	*My Reference Book,* p. 22; Standards for Mathematical Practice Poster; Guidelines for Discussions Poster	1.OA.8 SMP8
Reengaging in the Problem Children discuss various strategies for solving the open response problem by examining others' tables and rules.	selected samples of children's work	1.OA.6, 1.OA.8, 1.NBT.4 SMP1, SMP4, SMP8
Revising Work Children revise their work from Day 1.	*Math Masters,* p. 210 (optional); square and triangle pattern blocks; Pattern-Block Template (optional); colored pencils (optional); children's work from Day 1	1.OA.6, 1.OA.8, 1.NBT.4 SMP1, SMP4, SMP8
✓ **Assessment Check-In** See page 650 and the rubric below.		1.OA.8 SMP8

Goal for Mathematical Practice **GMP8.1** Create and justify rules, shortcuts, and generalizations.	Not Meeting Expectations	Partially Meeting Expectations	Meeting Expectations	Exceeding Expectations
	Does not write a rule, or writes a rule that is not supported by at least three rows of the table.	Writes a rule that is supported by at least three rows of the table, but does not justify the rule in Problem 2 using at least two rows of the table.	Writes a rule that is supported by at least three rows of the table, and justifies the rule using at least two rows of the table.	Meets expectations and further justifies the rule by explaining that the two extra chairs are the ones at the ends.

3 Practice   10–15 min

Math Boxes 7-9 Children practice and maintain skills.	*Math Journal 2,* p. 151	See page 650.
Home Link 7-9 **Homework** Children find rules in "What's My Rule?" problems.	*Math Masters,* p. 211	1.OA.6, 1.OA.8 SMP8

 **Focus** 50–55 min | Go Online | ePresentations | eToolkit

NOTE These Day 2 activities will ideally take place within a few days of Day 1. Prior to beginning Day 2, see Planning a Follow-Up Discussion from Day 1.

▶ Setting Expectations

My Reference Book, p. 22

| **WHOLE CLASS** | SMALL GROUP | PARTNER | INDEPENDENT |

Remind children that on Day 1 they solved a problem about rules. Ask: *Is a rule something that works just one time?* No. Rules work for lots of problems. Read *My Reference Book,* page 22 as a class. Ask: *Did Rosa use just one example to explain how she uses doubles to help her figure out other facts?* No. Rosa gave two examples. Refer to **GMP8.1** on the Standards for Mathematical Practice Poster. Explain to children that to create and justify a rule, you have to say what the rule is and also show that it works for more than one problem. **GMP8.1**

Briefly review the open response problem from Day 1. Ask: *What were you asked to do?* Sample answer: Figure out a rule that can help Luka predict how many chairs he needs. *What do you think a good response would include?* Sample answers: There would be four rows in the table and a rule that works for all the rows. There would be some sentences about finding the rule and checking other rows of the table to make sure the rule works. There would be an answer about how many chairs Luka needs for 18 desks.

After this brief discussion, tell children that they are going to look at other children's work and see if they found the same rule. Point out that some children may have made mistakes or not finished the page. Remind them that they should help each other learn and ask questions when they don't understand something. Refer to your list of discussion guidelines and encourage children to use sentence frames, such as:

- Could you explain _____?
- I'd like to add _____.

▶ Reengaging in the Problem

| **WHOLE CLASS** | SMALL GROUP | **PARTNER** | INDEPENDENT |

Children reengage in the problem by analyzing and critiquing other children's work in pairs and in a whole-group discussion. Have children discuss with partners before sharing with the whole group. Guide the discussion based on the decisions you made in Getting Ready for Day 2. **GMP1.4, GMP4.1, GMP8.1**

My Reference Book, p. 22

Standards for Mathematical Practice

Create Rules and Shortcuts

Rosa knows her doubles facts. She knows that

$7 + 7 = 14$ $8 + 8 = 16$ $9 + 9 = 18$

Rosa figures out how to use doubles to add near doubles:

$7 + 8 = ?$ $8 + 9 = ?$

If $7 + 7 = 14$, then $7 + 8$ is 1 more.
So, $7 + 8 = 15$. I know $8 + 8 = 16$,
so $8 + 9$ is 1 more. So, $8 + 9 = 17$.

Rosa

Create and justify rules, shortcuts, and generalizations.
Rosa justifies, or explains, how she uses doubles to help her figure out other facts.

Mathematical Practice 8: Look for and express regularity in repeated reasoning.

Try It Together

If you know $10 + 10 = 20$, what is $10 + 11$?

If you know $25 + 25 = 50$, what is $25 + 26$?

MRB 22 | twenty-two

▶ Revising Work

WHOLE CLASS | SMALL GROUP | **PARTNER** | **INDEPENDENT**

Pass back children's work from Day 1. Have pattern blocks available. Before children revise anything, ask them to examine their responses and decide how to improve them. Ask the questions below one at a time. Have partners discuss their responses and give a thumbs-up or thumbs-down based on their own work.

- *Did you write a correct rule?* GMP8.1
- *Do all four rows in your table work for the rule?* GMP1.4
- *Did you show how you found your rule and checked it using all the rows of your table?* GMP1.4, GMP8.1
- *Did you use the rule to predict how many chairs Luka will need?*

Tell children they now have a chance to revise their work. Remind them to add to their earlier work using colored pencils or to use another sheet of paper instead of erasing their original work. Encourage children to check their work using pattern blocks and to explain to you or a partner how they know that their rule is correct. GMP4.1 If you observe children checking the rule against the rows of their table but not writing about it in Problem 2, encourage them to add to their explanation. You may also want to make notes about which children understand that the rule must work for all the rows of their table.

Children who wrote complete responses on Day 1 can explain how they used the rule to find out how many chairs Luka will need for 18 desks.

Summarize Ask children to reflect on their work and revisions. Ask: *How did you improve your rule and table?* Answers vary. *How did you check your rule?* GMP8.1 Sample answer: I made sure it worked for all the rows.

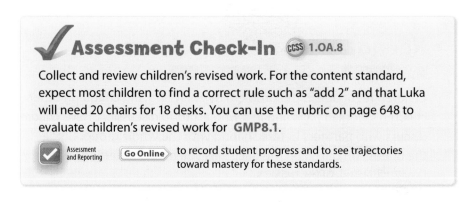

✓ **Assessment Check-In** CCSS 1.OA.8

Collect and review children's revised work. For the content standard, expect most children to find a correct rule such as "add 2" and that Luka will need 20 chairs for 18 desks. You can use the rubric on page 648 to evaluate children's revised work for GMP8.1.

[✓] Assessment and Reporting | Go Online to record student progress and to see trajectories toward mastery for these standards.

Go Online for optional generic rubrics in the *Assessment Handbook* that can be used to assess any additional GMPs addressed in the lesson.

Math Journal 2, p. 151

Math Boxes — Lesson 7-9
DATE

① Jake and Rosie are playing *Base-10 Exchange*.

Circle the blocks that need to be exchanged.

Jake   Rosie

② Write the time.

About __6__ o'clock

③ Use | and • to show 73.

Sample answer:

| | | | | | | •••

④ Lucy skips for 2 blocks, runs for 6 blocks, and walks for 5 blocks. How many blocks did she travel in all?

$2 + 6 + 5 = ?$

◯ 11 blocks
⬤ 13 blocks
◯ 31 blocks

⑤ A barn has 4 goats, 7 cows, and 4 pigs.

How many animals are in the barn?

$4 + 7 + 4 = $ __15__

Unit
animals

⑥ Add.

$\underline{7} = 4 + 3$

$4 + 5 = \underline{9}$

$\begin{array}{r} 5 \\ + 3 \\ \hline 8 \end{array}$

$\begin{array}{r} 4 \\ + 4 \\ \hline 8 \end{array}$

Unit

① 1.NBT.2, 1.NBT.2a, 1.NBT.2b ② 1.MD.3 ③ 1.NBT.2 ④ 1.OA.2, 1.OA.3, 1.OA.6 ⑤ 1.OA.2, 1.OA.3, 1.OA.6 ⑥ 1.OA.6, 1.OA.8 one hundred fifty-one 151

Sample Children's Work—Evaluated

See the sample in the margin. This work meets expectations for the content standard because this child found + 2 as the rule and 20 as the number of chairs that Luka will need. With revision, the work meets expectations for the mathematical practice because this child created a table with 4 rows that all support the rule of + 2. This child also wrote a justification for the rule in Problem 2 that refers to all four rows of the table. GMP8.1

Go Online > for other samples of evaluated children's work.

3 Practice 10–15 min

Go Online

ePresentations eToolkit Home Connections

▶ Math Boxes 7-9

Math Journal 2, p. 151

| WHOLE CLASS | SMALL GROUP | **PARTNER** | **INDEPENDENT** |

Mixed Practice Math Boxes 7-9 are paired with Math Boxes 7-11.

▶ Home Link 7-9

Math Masters, p. 211

Homework Children solve "What's My Rule?" problems with missing rules.

Sample child's work, "Meeting Expectations"

① Use pattern blocks to help you fill in the table.

Desks	Chairs
3	5
2	4
8	10
4	6

Example:

② Write a rule for your table.

Write how you found and checked your rule.

I know the Rule is Plus 2 becuse 3+②=5 2②=4 8+②=10 4②=6. So evry Mishin fact you add 2 in it.

Desks

Rule
+2

Chairs

③ How many chairs will Luka need for 18 desks?

_20_ chairs

Math Masters, p. 211

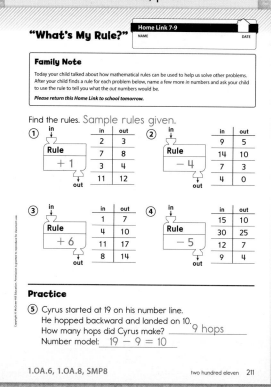

"What's My Rule?"   Home Link 7-9
NAME   DATE

Family Note

Today your child talked about how mathematical rules can be used to help us solve other problems. After your child finds a rule for each problem below, name a few more *in* numbers and ask your child to use the rule to tell you what the *out* numbers would be.

Please return this Home Link to school tomorrow.

Find the rules. Sample rules given.

①
Rule
+ 1

in	out
2	3
7	8
3	4
11	12

②
Rule
– 4

in	out
9	5
14	10
7	3
4	0

③
Rule
+ 6

in	out
1	7
4	10
11	17
8	14

④
Rule
– 5

in	out
15	10
30	25
12	7
9	4

Practice

⑤ Cyrus started at 19 on his number line. He hopped backward and landed on 10. How many hops did Cyrus make? __9 hops__
Number model: __19 – 9 = 10__

1.OA.6, 1.OA.8, SMP8   two hundred eleven 211

Addition Facts: "What's My Rule?"

Overview Children practice addition facts and find unknown numbers in number sentences in the "What's My Rule?" routine.

▶ **Before You Begin**
For Part 1, display the function machine depicted in the Math Message for children.

 Common Core State Standards

Focus Clusters
- Understand and apply properties of operations and the relationship between addition and subtraction.
- Add and subtract within 20.
- Work with addition and subtraction equations.
- Use place value understanding and properties of operations to add and subtract.

---

1 Warm Up   15–20 min

	Materials	
Mental Math and Fluency Children solve facts by making 10.	slate	1.OA.6
Daily Routines Children complete daily routines.	See pages 2–37.	See pages xiv–xvii.

---

2 Focus   35–40 min

Math Message Children determine the rule and the missing inputs for a function machine.		1.OA.6, 1.OA.8 SMP8
Finding Inputs for "What's My Rule?" Tables Children find rules, missing outputs, and missing inputs.	*Math Journal 2*, p. 152; *Math Masters*, p. TA37 (optional)	1.OA.6, 1.OA.7, 1.OA.8, 1.NBT.5 SMP1
✓ **Assessment Check-In** See page 656.	*Math Journal 2*, p. 152	1.OA.6, 1.OA.8
Introducing *Tric-Trac* **Game** Children practice addition facts.	*My Reference Book*, pp. 168–169 (optional); *Math Masters*, p. G54; per partnership: 2 dot dice, 20 pennies	1.OA.3, 1.OA.6 SMP1

1.OA.8 **Spiral Snapshot**

GMC Find the unknown in addition and subtraction equations.

7-1 through 7-4 Warm Up Focus Practice | **7-8** Focus Practice | **7-9** Warm Up Focus Practice | **7-10** Focus Practice | **8-1 through 8-3** Warm Up Practice | **8-11** Practice

Spiral Tracker   Go Online   to see how mastery develops for all standards within the grade.

---

3 Practice   10–15 min

Math Boxes 7-10 Children practice and maintain skills.	*Math Journal 2*, p. 153	See page 657.
Home Link 7-10 **Homework** Children practice the "What's My Rule?" routine.	*Math Masters*, pp. 213–214	1.OA.6, 1.OA.8, 1.NBT.2, 1.NBT.2c, 1.NBT.4, SMP8

---

connectED.mcgraw-hill.com

Plan your lessons online with these tools.

 ePresentations   Student Learning Center   Facts Workshop Game   eToolkit   Professional Development   Home Connections   Spiral Tracker   Assessment and Reporting   English Learners Support   Differentiation Support

Differentiation Options RtI

CCSS 1.OA.6, 1.OA.8, SMP1

Readiness 5–10 min

Solving Number Mysteries

| WHOLE CLASS |
| SMALL GROUP |
| PARTNER |
| INDEPENDENT |

To provide support for finding unknown addends, have children solve number mysteries. Tell children you have an unknown number. Give clues about your number, such as: *My number plus 3 equals 4* or *My number minus 2 equals 8.* Discuss how children figured out each unknown number. Have them check their answers by restating the clues with the unknown numbers. **GMP1.4**

CCSS 1.OA.4, 1.OA.8, 1.NBT.4, 1.NBT.5, SMP7

Enrichment 10–15 min

Filling in Frames and Rules

Math Masters, p. 212

| WHOLE CLASS |
| SMALL GROUP |
| PARTNER |
| INDEPENDENT |

To further children's experience with finding unknown numbers and rules, have them solve Frames-and-Arrows problems that extend beyond addition and subtraction facts. Children complete *Math Masters,* page 212. Discuss how filling in the first frames of a Frames-and-Arrows problem is similar to finding input numbers in a "What's My Rule?" table. **GMP7.2**

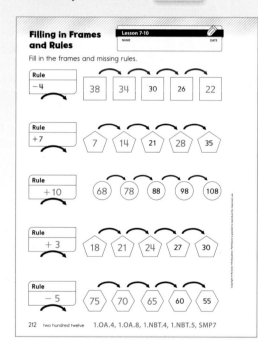

Filling in Frames and Rules — Lesson 7-10

Fill in the frames and missing rules.

Rule −4: 38 34 30 26 22

Rule +7: 7 14 21 28 35

Rule +10: 68 78 88 98 108

Rule +3: 18 21 24 27 30

Rule −5: 75 70 65 60 55

212 two hundred twelve 1.OA.4, 1.OA.8, 1.NBT.4, 1.NBT.5, SMP7

CCSS 1.OA.6, 1.NBT.3, SMP6

Extra Practice 10–15 min

Playing *Shaker Addition Top-It*

Math Masters, p. G50 (optional), p. G51; per group: two dot dice or two 10-sided dice; 20 counters

| WHOLE CLASS |
| SMALL GROUP |
| PARTNER |
| INDEPENDENT |

Children practice addition facts by playing *Shaker Addition Top-It.* **GMP6.4** For detailed instructions, see Lesson 7-4.

Observe

- Which addition facts do children know automatically?
- How do children compare sums?

Discuss

- *How did you add?*
- *What other strategy might be faster or easier?*

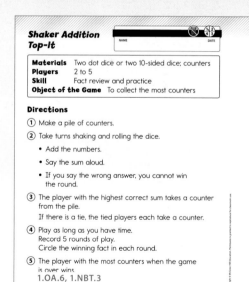

Shaker Addition Top-It

Materials	Two dot dice or two 10-sided dice; counters
Players	2 to 5
Skill	Fact review and practice
Object of the Game	To collect the most counters

Directions

1. Make a pile of counters.
2. Take turns shaking and rolling the dice.
   - Add the numbers.
   - Say the sum aloud.
   - If you say the wrong answer, you cannot win the round.
3. The player with the highest correct sum takes a counter from the pile.
   If there is a tie, the tied players each take a counter.
4. Play as long as you have time.
   Record 5 rounds of play.
   Circle the winning fact in each round.
5. The player with the most counters when the game is over wins.

1.OA.6, 1.NBT.3

English Language Learners Support

Beginning ELL Scaffold children's understandings of the terms used in solving "What's My Rule" problems. Rephrase statements as they work. For example, state and then rephrase the following: *figure out* as *find an answer, the missing numbers* as *the numbers that are not there,* and *unknown* as *number we don't know.* Encourage children to use these terms by completing sentence frames such as: "I have to find the answer or I have to figure out the answer."

 Go Online   ELL **English Learners Support**

1 Warm Up 15–20 min   Go Online ePresentations eToolkit

▶ Mental Math and Fluency

Have children solve each fact by making 10 and record sums on their slates. Then have them briefly explain how they used the strategy.

- ●○○ 9 + 2; 9 + 4; 9 + 7 11; 13; 16; Sample answer: I move 1 over to the 9 to make 10 and then add the rest.
- ●●○ 9 + 4; 8 + 3; 9 + 6 13; 11; 15; Sample answer: I move 1 or 2 over to the larger number to make 10 and then add the rest.
- ●●● 9 + 5; 8 + 4; 8 + 7 14; 12; 15; Sample answer: I move 1 or 2 over to the larger number to make 10 and then add the rest.

▶ Daily Routines

Have children complete the Daily Routines. For detailed instructions, see pages 2–37. For specifics on CCSS coverage, see pages xiv–xvii.

2 Focus 35–40 min   Go Online ePresentations eToolkit

▶ Math Message

Find and record the rule and the missing input numbers. GMP8.1

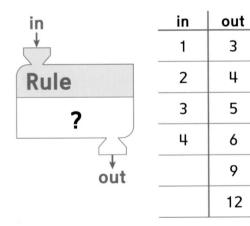

in	out
1	3
2	4
3	5
4	6
	9
	12

Math Journal 2, p. 152

More "What's My Rule?" Problems Lesson 7-10 DATE

❶ Find the rule. Sample rule given.

Rule	in	out
Add 4	5	9
	8	12
	10	14
out
Your Turn Answers vary.

❷ Fill in the blanks.

Rule	in	out
− 10	16	6
	22	12
	23	13
out
Your Turn Answers vary.

❸ What comes out?

Rule	in	out
+ 5	5	10
	6	11
	8	13
out
Your Turn Answers vary.

❹ What goes in?

Rule	in	out
− 2	7	5
	11	9
	5	3
out
Your Turn Answers vary.

❺ Make up your own.

Rule	in	out
out
Your Turn

❻ Make up your own.

Rule	in	out
out
Your Turn

Answers vary.

152 one hundred fifty-two 1.OA.6, 1.OA.7, 1.OA.8, 1.NBT.5

▶ Finding Inputs for "What's My Rule?" Tables

Math Journal 2, p. 152

WHOLE CLASS | SMALL GROUP | PARTNER | INDEPENDENT

Math Message Follow-Up Discuss children's answers to the Math Message. Children may phrase the rule in various ways, such as "Add 2," "Count up by 2," or "2 more."

Check that +2 works as the rule by writing number sentences for the function machine. **GMP1.4** Write:

$1 + \square = 3$

$2 + \square = 4$

$3 + \square = 5$

Then discuss determining whether placing 2 in the unknown box makes *all* of these number sentences true.

Write the rule in the function machine. Next have children share how they found the missing inputs. Illustrate the first missing input with a number sentence: $\square + 2 = 9$. Invite children to share their answers, and record the missing input in the box for the unknown. Repeat with $\square + 2 = 10$. Tell children that today they will learn more about finding missing input numbers.

Display the function machine with the rule "Add 1" shown in the margin. Point out that this table is different from the last one because the rule is known, but some of the inputs and outputs are missing. Write a number sentence for a row with a missing output, such as: $4 + 1 = \square$. Have children suggest a number to go in the box, and explain that finding the output is like finding the unknown number in a number sentence.

Academic Language Development Contrast the words *known* and *unknown*, explaining that *unknown* is the same as saying *not known*. Make connections to other words children might know that have the prefix *un-* to reinforce their understanding of *un-* meaning *not* or *no*.

Find the rest of the missing numbers. Ask: *What did you do to find the missing input numbers?* Sample answers: I found the number that comes before the output number. I subtracted 1 from the output number. Explain to children that they can check whether an input number is correct by applying the rule to the input number; the result should be the output number. They can also check by writing a number sentence and determining whether it is true. **GMP1.4**

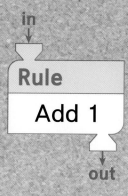

in	out
1	
4	
	7
	10
7	
	9

Common Misconception

Differentiate **IF** children apply rules down the columns rather than across the rows, **THEN** present "What's My Rule?" problems using *Math Masters*, page TA38. The arrows might remind children which direction to apply the rules.

Go Online Differentiation Support

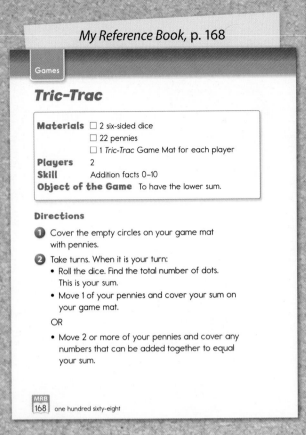

Games

Tric-Trac

Materials	☐ 2 six-sided dice
	☐ 22 pennies
	☐ 1 *Tric-Trac* Game Mat for each player
Players	2
Skill	Addition facts 0–10
Object of the Game	To have the lower sum.

Directions

❶ Cover the empty circles on your game mat with pennies.

❷ Take turns. When it is your turn:
  • Roll the dice. Find the total number of dots. This is your sum.
  • Move 1 of your pennies and cover your sum on your game mat.
  OR
  • Move 2 or more of your pennies and cover any numbers that can be added together to equal your sum.

MRB
168 one hundred sixty-eight

Tric-Trac
Game Mat

NAME _____ DATE _____

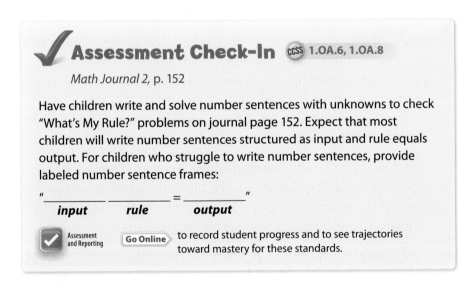

Display the following function machine with an "Add 3" rule.

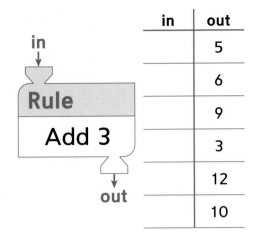

in	out
	5
	6
	9
	3
	12
	10

Have children find the missing numbers and describe their solution strategies. Sample answer: I subtracted 3 from the output number. Write number sentences for some of the inputs and outputs in the function machine. Use unknown boxes to represent missing inputs. Have children fill in the boxes to check their answers. **GMP1.4**

Practice finding inputs with several more machines in which all the inputs are missing. Then have children complete more "What's My Rule?" problems on journal page 152.

✓ **Assessment Check-In** ⒸⒸⓈⓈ **1.OA.6, 1.OA.8**

Math Journal 2, p. 152

Have children write and solve number sentences with unknowns to check "What's My Rule?" problems on journal page 152. Expect that most children will write number sentences structured as input and rule equals output. For children who struggle to write number sentences, provide labeled number sentence frames:

" _____ _____ = _____ "
  input   **rule**   **output**

☑ Assessment and Reporting   〈 Go Online 〉 to record student progress and to see trajectories toward mastery for these standards.

▶ # Introducing *Tric-Trac*

Math Masters, p. G54

| WHOLE CLASS | SMALL GROUP | **PARTNER** | INDEPENDENT |

Have children practice addition facts by playing *Tric-Trac.* Play a demonstration round with children. Have them share multiple ways to cover numbers for each sum. **GMP1.5**

Distribute the *Tric-Trac* game mat (*Math Masters,* page G54) to children. Directions can be found on *My Reference Book,* pages 168 and 169. Children play the game in partnerships.

Observe

- Which children cover numbers whose sum equals the sum of the dots?
- Which children use three addends?

Discuss

- *Is it better to cover sums with one penny or more than one penny? Why?*
- *Which numbers did you try to cover first? Why?*

| **Differentiate** | **Game Modifications** | Go Online | Differentiation Support |

Summarize Tell children that you rolled a 7 in *Tric-Trac.* Have them name all the possible ways you could cover a sum of 7 on the game mat. **GMP1.5**

③ # Practice 10–15 min

| Go Online | ePresentations | eToolkit | Home Connections |

▶ # Math Boxes 7-10

Math Journal 2, p. 153

| WHOLE CLASS | SMALL GROUP | **PARTNER** | **INDEPENDENT** |

Mixed Practice Math Boxes 7-10 are paired with Math Boxes 7-7.

▶ # Home Link 7-10

Math Masters, pp. 213–214

Homework Children practice addition and subtraction facts by solving "What's My Rule?" problems.

Math Journal 2, p. 153

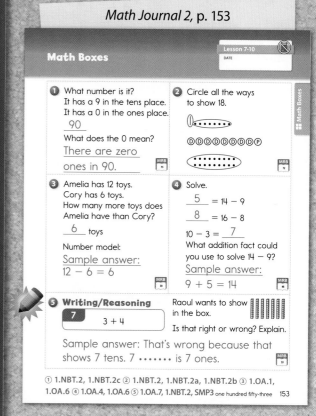

Math Masters, pp. 213–214

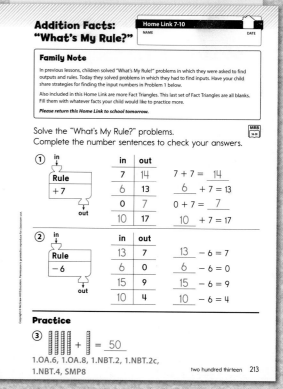

Digital Clocks

Overview Children tell time using digital and analog clocks.

▶ **Before You Begin**
Display a digital clock alongside your class demonstration clock. For Part 2, attach a minute hand to your demonstration clock or make a new demonstration clock from *Math Masters*, page 215. For the optional Extra Practice activity, prepare *Math Masters*, page TA39 with a variety of times in analog and digital notation before copying.

▶ **Vocabulary**
minute hand · clockwise · digital clock

 Common Core State Standards

Focus Cluster
Tell and write time.

---

1 Warm Up   15–20 min

Materials

Mental Math and Fluency Children find sums of three numbers.	slate	**1.OA.3, 1.OA.6**
Daily Routines Children complete daily routines.	See pages 2–37.	See pages xiv–xvii.

---

2 Focus   35–40 min

Math Message Children record things they think take one minute to do.		**1.MD.3**
Introducing the Minute Hand Children discuss the length of a minute. They identify and discuss the minute hand on an analog clock.	*Math Masters,* p. 215; analog clock; demonstration clock; scissors; brads	**1.MD.3** SMP1
Introducing Digital Time Children are introduced to digital time notation.	*Math Masters,* p. 216; analog clock; digital clock; slate	**1.MD.3** SMP2, SMP5
✓ **Assessment Check-In** See page 662.	toolkit clock	**1.MD.3, SMP5**
Introducing *Time Match* **Game** Children practice telling time.	*Math Masters,* pp. G55–G57	**1.MD.3** SMP2

CCSS 1.MD.3 **Spiral Snapshot**

GMC Tell and write time using digital clocks.

7-11
Focus Practice | 8-1 Warm Up | 8-3 Practice | 8-8 Focus Practice | 8-9 Practice | 9-1 Warm Up | 9-4 Practice | 9-9 Warm Up

Spiral Tracker   **Go Online**   to see how mastery develops for all standards within the grade.

---

3 Practice   10–15 min

Math Boxes 7-11 Children practice and maintain skills.	*Math Journal 2,* p. 154	See page 663.
Home Link 7-11 **Homework** Children practice telling and writing time.	*Math Masters,* p. 218	**1.OA.7, 1.NBT.3, 1.NBT.4, 1.MD.3** SMP5

---

connectED.mcgraw-hill.com

Plan your lessons online with these tools.

 ePresentations   Student Learning Center   Facts Workshop Game   eToolkit   Professional Development   Home Connections   Spiral Tracker   Assessment and Reporting   English Learners Support   Differentiation Support

Differentiation Options

 RtI

Readiness
10–15 min

Just Before and Just After

| WHOLE CLASS |
| SMALL GROUP |
| PARTNER |
| INDEPENDENT |

number cards 1–12,
hour-hand-only clocks

To help children gain more experience telling time with an hour-hand-only clock, have them draw number cards and practice displaying that hour on their hour-hand-only clocks. Emphasize phrases such as "about _____," "almost _____," "just before _____," and "a little after _____" when children discuss the time. GMP6.3

Enrichment
10–15 min

Calculating Elapsed Time

| WHOLE CLASS |
| SMALL GROUP |
| PARTNER |
| INDEPENDENT |

Activity Card 93;
Math Masters, p. 217

To apply children's understanding of the concept of time, have them use toolkit clocks to calculate the passage of time. GMP6.4

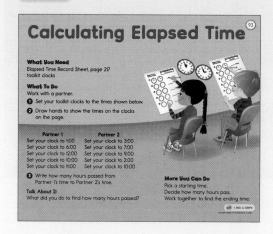

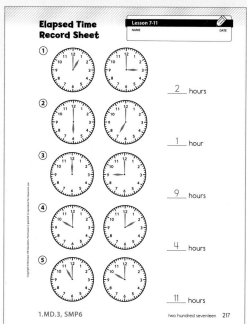

Extra Practice
10–15 min

Telling Time

| WHOLE CLASS |
| SMALL GROUP |
| PARTNER |
| INDEPENDENT |

Math Masters, p. TA39

Have children practice telling time on analog and digital clocks. GMP2.1

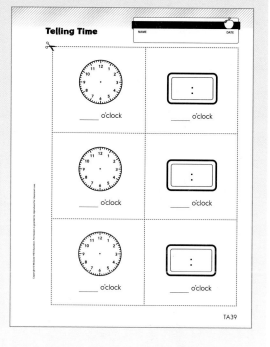

English Language Learners Support

Beginning ELL Children might associate the word *hands* with body parts. Review this term in the context of telling time, using a picture of a clock whose minute and hour hands have hands at the ends instead of arrowheads. Direct children to point to their hands and then to the minute and hour hands on the analog clock.

 Go Online   ELL English Learners Support

Standards and Goals for Mathematical Practice

SMP1 Make sense of your problems and persevere in solving them.
GMP1.4 Check whether your answer makes sense.

SMP2 Reason abstractly and quantitatively.
GMP2.3 Make connections between representations.

SMP5 Use appropriate tools strategically.
GMP5.2 Use tools effectively and make sense of your results.

One Hour	One Half Hour	One Second
• 2 TV episodes • math class		
	One Minute • lining up and walking to the water fountain • saying the Pledge of Allegiance 5 times.	

Time Museum
You may wish to use an index card for Second, an $8\frac{1}{2}$" by 11 piece of piece of paper for One Minute, a half-sheet of chart paper for One Half Hour, and a sheet of chart paper for One Hour.

1 Warm Up — 15–20 min

Go Online — ePresentations — eToolkit

▶ Mental Math and Fluency

Have children find and record the sums on their slates. Briefly discuss children's solution strategies. Emphasize strategies that add combinations of 10 and near doubles before adding the third number.

- ●○○ $2 + 3 + 5$ 10
- ●●○ $7 + 4 + 3$ 14
- ●●● $6 + 2 + 5$ 13

▶ Daily Routines

Have children complete the Daily Routines. For detailed instructions, see pages 2–37. For specifics on CCSS coverage, see pages xiv–xvii.

2 Focus — 35–40 min

Go Online — ePresentations — eToolkit

▶ Math Message

What can you do in one minute? Record your ideas.

▶ Introducing the Minute Hand

Math Masters, p. 215

WHOLE CLASS	SMALL GROUP	PARTNER	INDEPENDENT

Math Message Follow-up After children share their ideas about the length of a minute, ask them how they could check whether their lists make sense. **GMP1.4** If no one mentions it, tell children there are 60 seconds in a minute so they can count each second from 1 to 60 to estimate the length of one minute. Help children estimate one minute by counting and then clapping in unison: 1, (clap), 2, (clap), 3, (clap), and so on, to 60, (clap).

Ask children whether their ideas still make sense. Test a few by doing the activities while the class counts out 60 seconds. **GMP1.4** Add some of the ideas to the Time Museum started in Lesson 6-1. Write *One Minute* on a piece of paper smaller than what was used for the hour paper, and record things that take about a minute to complete. *For example:*

- Lining up and walking to the water fountain
- Writing the alphabet three times
- Saying the Pledge of Allegiance five times

Direct children's attention to the classroom analog clock. Ask them to recall what they learned about the hour hand. Sample answers: It's the shorter hand. It tells you about what hour it is. Remind children that the longer hand on the clock is called the **minute hand.**

Remind children that both hands of the clock always move from smaller to larger numbers, in a **clockwise** direction. Remind them that an hour hand takes an entire hour to get from one number to the next. When it points directly at the 3, the time is 3 o'clock. When it points directly to the 4, it is 4 o'clock. Explain that the minute hand needs to move all the way around the clock once in one hour. When the time is "o'clock," the minute hand points to the 12. When the minute hand gets back to the 12, an hour has passed. Gesture clockwise from 12 around to 12 to show children the minute hand's movement between "o'clock" times.

Tell children that it takes 1 minute for the minute hand to move from one mark on the clock face to the next. Move the minute hand on your demonstration clock slowly around the clock face, starting at 12 o'clock. Have children count by 1s as the minute hand passes each minute mark until the minute hand gets back to 12 and they say 60. Point out that the hour hand has moved from 12 to 1. Ask: *How many minutes are in 1 hour?* 60 minutes

Children might notice that it takes 5 minutes for the minute hand to move from one number on the clock face to the next. You may wish to repeat the motions and have children count by 5s to 60 as the minute hand passes each number.

Academic Language Development Reinforce children's understanding of the word *clockwise* using movement. For example, have a small group of children form a circle and direct them to move clockwise as the other children watch. Or have children move their pencils clockwise to draw circles on paper. Have children compare the direction of the movement to the direction of the hands on an analog clock.

Distribute *Math Masters*, page 215 and brads. Help children assemble new toolkit clocks. (Alternatively, you may wish to attach the minute hand from *Math Masters*, page 215 to the clocks they made in Lesson 6-1.) Have children practice reading times on the demonstration clock and showing times to the hour on their clocks. Emphasize that when the hour hand is in the o'clock position, the minute hand is pointing straight up.

Math Masters, p. 215

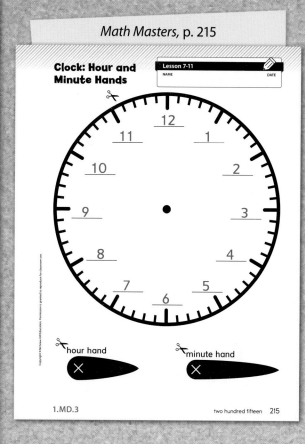

NOTE To make sturdier clocks, have children paste the clock face to a small paper plate before attaching the hands with the brad. Children may benefit from coloring each hour's section of the clock and writing the hour numbers in corresponding colors so clocks look similar to their clocks from Lesson 6-1.

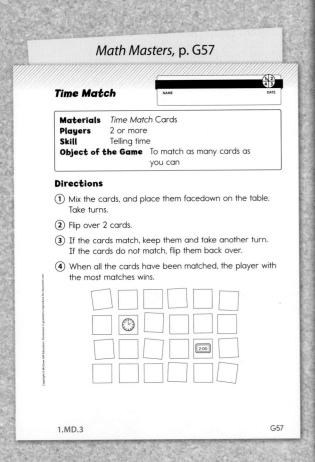

▶ Introducing Digital Time

Math Masters, p. 216

WHOLE CLASS | SMALL GROUP | PARTNER | INDEPENDENT

Show 5 o'clock on the demonstration clock. Ask whether anyone has a clock at home that shows time with numbers but no hands. Explain that this is called a **digital clock.** Display *Math Masters,* page 216, and share an example of a real digital clock.

Explain what the numbers and symbol on the digital clock mean. GMP5.2

- The numbers are separated by a colon.
- The number before the colon tells the hour.
- The number after the colon tells the minutes after the hour.
- *:00* means "0 minutes after the hour" or "o'clock."

Show 5 o'clock on an analog and a digital clock, and explain that they show the same time. GMP2.3 Call out times to the hour, and have children show them on their analog clocks and record the digital times on their slates. *For example:*

- Show 7 o'clock on the analog clock, and ask children to record the digital time on their slates. 7:00
- Show 9:00 on the digital clock, and ask children to show the time on the analog clock. Hour hand pointing to 9, minute hand pointing to 12

You may wish to have children do these activities in partnerships or over the next few weeks for additional practice. Encourage children to use their toolkit clocks to show the time of activities, such as recess, lunch, or music. Occasionally set an hour-hand-only clock to match the hour hand on the classroom clock and call on children to tell the time. Remind children to say "about" or "a little before" or "a little after" to indicate that their time measurements are not exact.

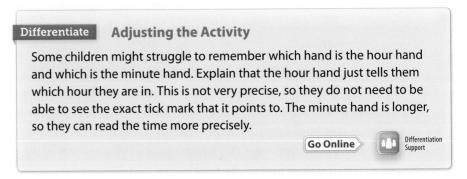

Differentiate **Adjusting the Activity**

Some children might struggle to remember which hand is the hour hand and which is the minute hand. Explain that the hour hand just tells them which hour they are in. This is not very precise, so they do not need to be able to see the exact tick mark that it points to. The minute hand is longer, so they can read the time more precisely.

Go Online ▶ | Differentiation Support

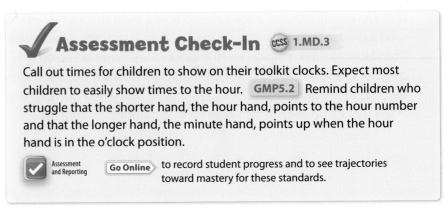

✓ Assessment Check-In CCSS 1.MD.3

Call out times for children to show on their toolkit clocks. Expect most children to easily show times to the hour. GMP5.2 Remind children who struggle that the shorter hand, the hour hand, points to the hour number and that the longer hand, the minute hand, points up when the hour hand is in the o'clock position.

✓ Assessment and Reporting | Go Online ▶ to record student progress and to see trajectories toward mastery for these standards.

▶ Introducing *Time Match*

Math Masters, pp. G55–G56

| WHOLE CLASS | SMALL GROUP | **PARTNER** | INDEPENDENT |

Demonstrate how to play *Time Match.* You may wish to distribute the game directions found on *Math Masters,* page G57.

Observe

- Which children correctly read the times to the hour on the analog clock?
- Which children correctly connect the two representations of the time? **GMP2.3**

Discuss

- *How do you know what time is shown on the clock?*
- *Which hand helps you tell the hour?*

| **Differentiate** | **Game Modifications** | **Go Online** | Differentiation Support |

Summarize Have children look at the classroom clock and name the hour. Keep a digital clock near the classroom analog clock to help children connect the two representations and learn to tell time using both. **GMP2.3**

3 Practice   10–15 min

Go Online   ePresentations   eToolkit   Home Connections

▶ Math Boxes 7-11

Math Journal 2, p. 154

| WHOLE CLASS | SMALL GROUP | **PARTNER** | **INDEPENDENT** |

Mixed Practice Math Boxes 7-11 are paired with Math Boxes 7-9.

▶ Home Link 7-11

Math Masters, p. 218

Homework Children practice telling time to the hour and writing it in digital notation.

Math Journal 2, p. 154

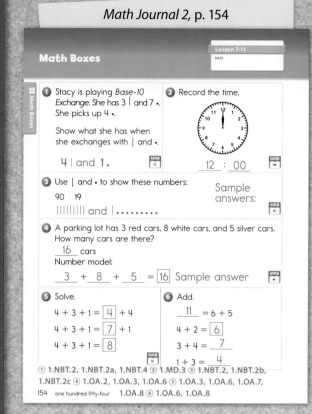

Math Boxes

Lesson 7-11
DATE

① Stacy is playing *Base-10 Exchange.* She has 3 | and 7 •. She picks up 4 •.

Show what she has when she exchanges with | and •.

____4____ | and ____1____ •.

② Record the time.

____12____ : ____00____

③ Use | and • to show these numbers:

90   19

||||||||| and | •••••••••

Sample answers:

④ A parking lot has 3 red cars, 8 white cars, and 5 silver cars. How many cars are there?

____16____ cars

Number model:

__3__ + __8__ + __5__ = 16 Sample answer

⑤ Solve.

$4 + 3 + 1 = \boxed{4} + 4$

$4 + 3 + 1 = \boxed{7} + 1$

$4 + 3 + 1 = \boxed{8}$

⑥ Add.

$\underline{11} = 6 + 5$

$4 + 2 = \boxed{6}$

$3 + 4 = \underline{7}$

$1 + 3 = \underline{4}$

① 1.NBT.2, 1.NBT.2a, 1.NBT.4 ② 1.MD.3 ③ 1.NBT.2, 1.NBT.2b, 1.NBT.2c ④ 1.OA.2, 1.OA.3, 1.OA.6 ⑤ 1.OA.3, 1.OA.6, 1.OA.7, 1.OA.8 ⑥ 1.OA.6, 1.OA.8

154   one hundred fifty-four

Math Masters, p. 218

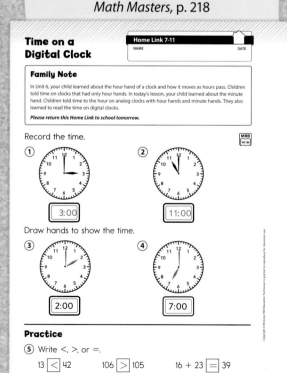

Time on a Digital Clock

Home Link 7-11
NAME   DATE

Family Note

In Unit 6, your child learned about the hour hand of a clock and how it moves as hours pass. Children told time on clocks that had only hour hands. In today's lesson, your child learned about the minute hand. Children told time to the hour on analog clocks with hour hands and minute hands. They also learned to read the time on digital clocks.

Please return this Home Link to school tomorrow.

Record the time.

① 3:00

② 11:00

Draw hands to show the time.

③ 2:00

④ 7:00

Practice

⑤ Write <, >, or =.

13 $\boxed{<}$ 42   106 $\boxed{>}$ 105   16 + 23 $\boxed{=}$ 39

218   two hundred eighteen   1.MD.3, 1.OA.7, 1.NBT.3, 1.NBT.4, SMP5

Unit 7 Progress Check

Overview **Day 1:** Administer the Unit Assessments.
Day 2: Administer the Open Response Assessment.

 Student Learning Center
Students may take assessments digitally.

 Assessment and Reporting
Record results and track progress toward mastery.

Day 1: Unit Assessments

1 Warm Up 5–10 min

Self Assessment
Children complete the Self Assessment.

Materials

Assessment Handbook, p. 45

2a Assess 35–50 min

Unit 7 Assessment
These items reflect mastery expectations to this point.

Assessment Handbook, pp. 46–48

Unit 7 Challenge (Optional)
Children may demonstrate progress beyond expectations.

Assessment Handbook, p. 49

CCSS Common Core State Standards	Goals for Mathematical Content (GMC)	Lessons	Self Assessment	Unit 7 Assessment	Unit 7 Challenge
1.OA.3	Apply properties of operations to add or subtract.*	7-1 to 7-3, 7-6, 7-10	1, 3	1, 2, 7	1, 2
1.OA.4	Understand subtraction as an unknown-addend problem.*	7-1 to 7-4, 7-6	1, 2	1–3	
1.OA.5	Relate counting to addition and subtraction.*	7-4	3	4	
1.OA.6	Add within 10 fluently.*	7-1 to 7-4, 7-6, 7-8 to 7-10	1	1, 2, 7	1, 2
	Subtract within 10 fluently.*	7-1 to 7-4, 7-6, 7-8 to 7-10	1–3	1, 2, 8	
	Add doubles automatically.*	7-1 to 7-4, 7-6, 7-8		3, 7	
	Subtract doubles.	7-1 to 7-4, 7-6, 7-8	2	3, 8	
	Add combinations of 10 automatically.*	7-1 to 7-4, 7-6, 7-8		1, 7	
	Subtract combinations of 10.	7-1 to 7-4, 7-6, 7-8	2	1, 8	
	Add and subtract within 20 using strategies.	7-1 to 7-4, 7-6, 7-8 to 7-10	1–3	3, 4	1, 2
1.OA.7	Understand the meaning of the equal sign.*	7-10		12	
	Determine whether equations involving addition or subtraction are true or false.*	7-10		12	
1.OA.8	Find the unknown in addition and subtraction equations.*	7-1 to 7-4, 7-8 to 7-10		2, 7, 8, 11	
1.NBT.5	Mentally find 10 more or 10 less than a 2-digit number.	7-8, 7-10		11	
1.MD.3	Tell and write time using analog clocks.	7-11	6	9, 10	
	Tell and write time using digital clocks.	7-11	6	9, 10	
1.G.1	Distinguish between defining and non-defining attributes.	7-5 to 7-7	4	5, 6	

*Instruction and most practice on this content is complete.

	Goals for Mathematical Practice (GMP)	Lessons	Self Assessment	Unit 7 Assessment	Unit 7 Challenge
SMP1	Check whether your answer makes sense. **GMP1.4**	7-9 to 7-11	5	12	
SMP3	Make mathematical conjectures and arguments. **GMP3.1**	7-1		4	
SMP5	Use tools effectively and make sense of your results. **GMP5.2**	7-11	6		
SMP6	Explain your mathematical thinking clearly and precisely. **GMP6.1**	7-3, 7-5, 7-7		3	
SMP7	Look for mathematical structures such as categories, patterns, and properties. **GMP7.1**	7-3 to 7-8		5, 6, 11	
	Use structures to solve problems and answer questions. **GMP7.2**	7-1 to 7-3, 7-6, 7-8		1, 2, 11	1, 2
SMP8	Create and justify rules, shortcuts, and generalizations. **GMP8.1**	7-1, 7-4, 7-8 to 7-10	4	11	

/// Spiral Tracker ⟨ **Go Online** ⟩ to see how mastery develops for all standards within the grade.

1 Warm Up   5–10 min

▶ Self Assessment

Assessment Handbook, p. 45

| WHOLE CLASS | SMALL GROUP | PARTNER | **INDEPENDENT** |

Children complete the Self Assessment to reflect on their progress in Unit 7.

Remind children of a time when they did each type of problem.

Item	Remind children that they ...
1	worked with fact families and fact triangles. (Lessons 7-1 and 7-2)
2	applied the think-addition strategy when playing *Salute!* (Lessons 7-3 and 7-6)
3	counted up to subtract. (Lesson 7-4)
4	sorted shapes by defining attributes. (Lessons 7-5 and 7-7)
5	wrote number sentences to make sure "What's My Rule?" answers made sense. (Lessons 7-8, 7-9, and 7-10)
6	showed times to the hour on toolkit clocks. (Lesson 7-11)

Assessment Handbook, p. 45

| NAME | DATE | Lesson 7-12 ✓ |

Unit 7 Self Assessment

Put a check in the box that tells how you do each skill.

Skills	I can do this by myself. I can explain how to do this.	I can do this by myself.	I can do this with help.
① Write fact families.			
② Think addition to subtract.			
③ Count up to subtract.			
④ Name a rule used to sort shapes.			
⑤ Check whether my answer makes sense.			
⑥ Tell time to the hour.			

Assessment Masters 45

2a Assess

35–50 min

Go Online

Assessment and Reporting

Differentiation Support

▶ Unit 7 Assessment

Assessment Handbook, pp. 46–48

| WHOLE CLASS | SMALL GROUP | PARTNER | **INDEPENDENT** |

Children complete the Unit 7 Assessment to demonstrate their progress on the Common Core State Standards covered in this unit.

Generic rubrics in the *Assessment Handbook* appendix can be used to evaluate children's progress on the Mathematical Practices.

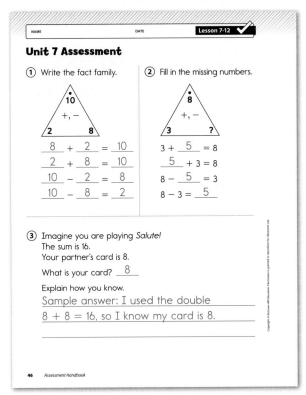

Assessment Handbook, p. 46

Differentiate | **Adjusting the Assessment**

Item(s)	Adjustments
1, 2	To scaffold items 1 and 2, provide blank parts-and-total diagrams or dominoes. To extend items 1 and 2, have children write fact families for other Fact Triangles. They may leave some numbers missing for a partner to solve.
3	To scaffold item 3, refer children to the Strategy Wall. You may wish to suggest that they think about addition doubles.
4	To scaffold item 4, provide a number line. Have children solve both by counting up and counting back. To extend item 4, have children list 5 subtraction facts they solve more easily by counting up and 5 they solve more easily by counting back.
5, 6	To scaffold items 5 and 6, provide pattern blocks or attribute blocks. To extend items 5 and 6, have children suggest some possible nondefining attributes of the shapes.
7, 8	To scaffold items 7 and 8, provide counters.
9, 10	To scaffold items 9 and 10, have children use their toolkit clocks. To extend items 9 and 10, have children tell what time it will be in a half hour, 1 hour, and 3 hours from each time.
11, 12	To scaffold items 11 and 12, have children use counters or a number grid. To extend item 11, have children complete additional rows for the table, continuing to follow the given rule.

Advice for Differentiation

All instruction and most practice is complete for the content that is marked with an asterisk (*) on page 664.

Use the online assessment and reporting tools to track children's performance. Differentiation materials are available online to help you assess children's needs.

> **NOTE** See the Unit Organizer on pages 588–589 or the online Spiral Tracker for details on Unit 7 focus topics and the spiral.

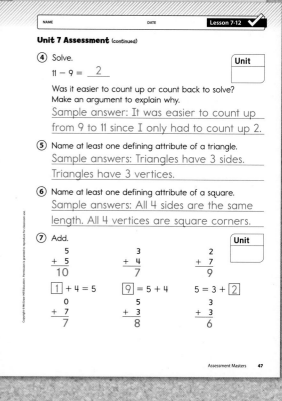

Assessment Handbook, p. 47

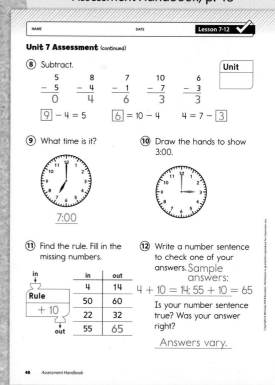

Assessment Handbook, p. 48

▶ Unit 7 Challenge (Optional)

Assessment Handbook, p. 49

| WHOLE CLASS | SMALL GROUP | PARTNER | **INDEPENDENT** |

Children can complete the Unit 7 Challenge after they complete the Unit 7 Assessment.

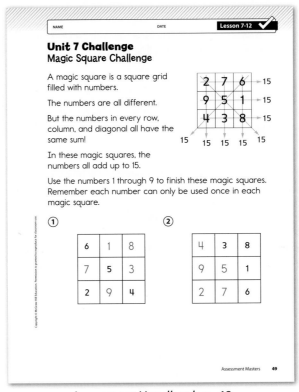

Assessment Handbook, p. 49

Day 2: Open Response Assessment

2b Assess   50–55 min

Materials

Solving the Open Response Problem
After a brief introduction, children solve a multistep comparison number story and explain how they found their answers.

Assessment Handbook, pp. 50–51; crayons or colored pencils; red, blue, and green counters (optional)

Discussing the Problem
After completing the problem, children share their answers and solution strategies.

Assessment Handbook, pp. 50–51

CCSS Common Core State Standards

	Goals for Mathematical Content (GMC)	Lessons
1.OA.1	Solve number stories by adding and subtracting.	7-3
1.OA.6	Add within 10 fluently.	7-1 to 7-4, 7-6, 7-8 to 7-10
	Subtract within 10 fluently.	7-1 to 7-4, 7-6, 7-8 to 7-10
	Add combinations of 10 automatically.	7-1 to 7-3, 7-6, 7-8, 7-10
	Subtract combinations of 10.	7-1 to 7-3, 7-6, 7-8, 7-10

	Goal for Mathematical Practice (GMP)	
SMP1	Make sense of your problem. **GMP1.1**	7-5

📊 **Spiral Tracker**   (Go Online) to see how mastery develops for all standards within the grade.

▶ **Evaluating Children's Responses**

Evaluate children's abilities to solve comparison number stories. Use the rubric below to evaluate their work based on **GMP1.1**.

Goal for Mathematical Practice **GMP1.1** Make sense of your problem.	Not Meeting Expectations	Partially Meeting Expectations	Meeting Expectations	Exceeding Expectations
	For Problem 1, does not draw, show, or write all three of the following: • 10 red crayons • fewer than 10 blue crayons • fewer than 10 green crayons	For Problem 1, draws, shows, or writes all three of the following: • 10 red crayons • fewer than 10 blue crayons • fewer than 10 green crayons	For Problem 1, draws, shows, or writes all three items under Partially Meeting Expectations, and provides evidence of correctly interpreting at least one of the following: • 6 more reds than blues • 3 more reds than greens	Meets expectations, and provides a complete explanation of a solution strategy in Problem 2.

3 Look Ahead   10–15 min

Materials

Math Boxes 7-12
Children preview skills and concepts for Unit 8.

Math Journal 2, p. 155

Home Link 7-12
Children take home the Family Letter that introduces Unit 8.

Math Masters, pp. 219–222

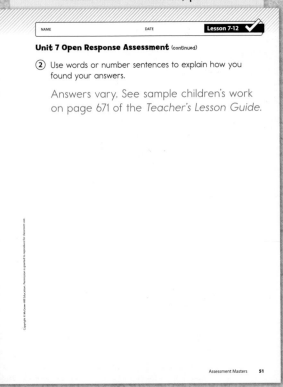

2b Assess

50–55 min

Go Online

Assessment and Reporting

▶ Solving the Open Response Problem

Assessment Handbook, pp. 50–51

| WHOLE CLASS | SMALL GROUP | **PARTNER** | **INDEPENDENT** |

This open response problem requires children to apply skills and concepts from Unit 7 to solve a multistep comparison number story. The focus of this task is **GMP1.1:** Make sense of your problem.

Distribute *Assessment Handbook*, pages 50–51. Read the directions aloud. Tell children that their job is to find the number of blue crayons and the number of green crayons that Rowan has and explain how they found their answers. Explain that they will have to think carefully about what the problem is asking them to do and they may have to try different strategies before they figure out how to solve it. GMP1.1

Make sure that crayons or colored pencils are available to children as they work on the problem. Some children may also benefit from using counters to model the situation.

Circulate and observe. You may wish to take notes about children's performance on the task. Expect many children to find the comparison context challenging. To help them get started, ask: *What do you know from the story? Does Rowan have more than 10 blue crayons or fewer than 10? How could you show 6 more red crayons than blue crayons?*

Differentiate   **Adjusting the Assessment**

For children who struggle to get started, model the following strategy for finding the number of blue crayons: *First, I will make a guess. I guess that Rowan has 5 blue crayons. Let's draw a picture of 10 reds and 5 blues. Does he have 6 more reds than blues? No, there are only 5 more reds than blues. So we can guess again. What other number of blues could we try?*

▶ Discussing the Problem

Assessment Handbook, pp. 50–51

| WHOLE CLASS | SMALL GROUP | PARTNER | INDEPENDENT |

After children complete the pages, invite individuals to share their answers. Encourage children who used effective solution strategies to share their strategies with the class.

Evaluating Children's Responses CCSS 1.OA.1, 1.OA.6

Collect children's work. For the content standards, expect many children to find the correct number of blue crayons (4), the correct number of green crayons (7), or both. However, some children may struggle to interpret and solve the problem. You can use the rubric on page 669 to evaluate children's work for **GMP1.1**.

See the sample in the margin. This work meets expectations for the content standards because this child found that Rowan has 4 blue crayons and 7 green crayons. The work exceeds expectations for the mathematical practice. This child's picture for Problem 1 shows 10 reds, fewer than 10 blues, and fewer than 10 greens. The reporting of correct answers via the picture and the numbers written on the answer blanks provides evidence of correct interpretations of "6 more reds than blues" and "3 more reds than greens." The words and number sentences written for Problem 2 provide a complete explanation of how subtraction was used to find the answers. **GMP1.1**

Go Online ✓ Assessment and Reporting

---

③ Look Ahead 10–15 min **Go Online** 🏠 Home Connections

▶ Math Boxes 7-12: Preview for Unit 8 ✐

Math Journal 2, p. 155

| WHOLE CLASS | SMALL GROUP | PARTNER | INDEPENDENT |

Mixed Practice Math Boxes 7-12 are paired with Math Boxes 7-8. These problems focus on skills and understandings that are prerequisite for Unit 8. You may want to use information from these Math Boxes to plan instruction and grouping in Unit 8.

▶ Home Link 7-12: Unit 8 Family Letter

Math Masters, pp. 219–222

Home Connection The Unit 8 Family Letter provides information and activities related to Unit 8 content.

Sample child's work, "Exceeding Expectations"

1. Rowan has 3 colors of crayons: red, blue, and green.
   He has 10 reds.
   He has 6 more reds than blues.
   He has 3 more reds than greens.
   How many blue crayons does he have?
   How many green?
   You can draw a picture to help.

   \|\|\|\|\|\|\|\|\|\|
   \|\|\|\|
   \|3\|\|\|\|

   __4__ blue crayons   __7__ green crayons

2. Use words or number sentences to explain how you found your answers.
   $10-6=4$ 4 blues
   $10-3=7$ 7 greens
   10 reds

NOTE Additional samples of evaluated children's work can be found in the *Assessment Handbook* appendix.

Math Journal 2, p. 155

Math Boxes
Preview for Unit 8   Lesson 7-12
                                        DATE

❶ How many corners?

   __5__ corners

❷ Draw a polygon with 3 sides.

   Shapes vary.

   What shape is it?
   Triangle

❸ Draw lines to match the shapes that look alike.

   Column A   Column B

❹ Write the numbers that are 10 more and 10 less than each number below. Use base-10 blocks if you need to.

   __4__ 14 __24__
   __48__ 58 __68__
   __80__ 90 __100__

❺ **Writing/Reasoning**

   How many of the smallest triangles are in this square?

   __4__ triangles

   How do you know if they are the same size?
   Sample answers: They look the same. Their sides are all the same because it's a square.

   ① 1.G.1 ② 1.G.1 ③ 1.G.1 ④ 1.NBT.2, 1.NBT.5
   ⑤ 1.G.1, 1.G.3, SMP7

   one hundred fifty-five 155

Unit 8 Organizer
Geometry

In this unit, children learn about attributes of 2- and 3-dimensional shapes, compose shapes and decompose composite shapes, and divide shapes into halves and fourths. Children also continue to practice telling and writing time, work with bar graphs, and use their understanding of place value and properties of operations to add and subtract larger numbers. Children's learning will focus on four clusters of the Common Core's content standards, as well as in-depth work on two of the Mathematical Practices.

CCSS Standards for Mathematical Content

Domain	Cluster
Number and Operations in Base Ten	Use place value understanding and properties of operations to add and subtract.
Measurement and Data	Tell and write time. Represent and interpret data.
Geometry	Reason with shapes and their attributes.

Because the standards within each domain can be broad, *Everyday Mathematics* has unpacked each standard into Goals for Mathematical Content **GMC**. For a complete list of Standards and Goals, see page EM1.

For an overview of the CCSS domains, standards, and mastery expectations in this unit, see the **Spiral Trace** on pages 678–679. See the **Mathematical Background** (pages 680–682) for a discussion of the following key topics:

- Attributes of Shapes and Composite Shapes
- Fractional Parts of Shapes
- Time and Data
- Adding and Subtracting 10

CCSS Standards for Mathematical Practice

SMP3 Construct viable arguments and critique the reasoning of others.

SMP6 Attend to precision.

For a discussion about how *Everyday Mathematics* develops these practices and a list of Goals for Mathematical Practice **GMP**, see page 683.

 **Virtual Learning Community** **Go Online** to **vlc.cemseprojects.org** to search for video clips on each practice.

©Echo/Cultura/Getty Images

▶ **Go Digital with these tools at connectED.mcgraw-hill.com**

ePresentations | Student Learning Center | Facts Workshop Game | eToolkit | Professional Development | Home Connections | Spiral Tracker | Assessment and Reporting | English Learners Support | Differentiation Support

Contents

*The standards listed here are addressed in the **Focus** of each lesson. For all the standards in a lesson, see the Lesson Opener.

Unit 8 Materials

VLC Virtual Learning Community

See how *Everyday Mathematics* teachers organize materials. Search "Classroom Tours" at **vlc.cemseprojects.org**.

Lesson	Math Masters	Activity Cards	Manipulative Kit	Other Materials
8-1	pp. 223–224; TA4; G58 (optional)	94	twist ties; straws	slate; demonstration clock; scissors; glue
8-2	pp. 225–228	95–96		slate; chart paper; per partnership: modeling clay or play dough, Pattern-Block Template; scissors; tape or glue; crayons or colored pencils
8-3	pp. 229–231; G55–G56; G57 (optional)	97–98		slate; prepared tally chart; paper circle, square, and rectangle; chart paper; scissors; glue or tape; Two Equal Shares Poster; Four Equal Shares Poster
8-4	pp. 232–234; TA3			Standards for Mathematical Practice Poster; per partnership: 1 prepared triangle; colored pencils; children's work from Day 1; chart paper (optional); *My Reference Book*
8-5	pp. 235–239; G59 (optional)	99	pattern blocks	slate; scissors; plastic bag or envelope; tape or glue; folder; Pattern-Block Template
8-6	pp. 240–241; TA4; TA40; G58 (optional)	100	base-10 blocks; 3-dimensional blocks; pattern blocks	slate; ball; can; box; everyday objects; paper bag, scissors, glue
8-7	pp. 235; 242; TA40; TA41; G37 (optional); G38	101–103	per small group: 20 or more pattern-block squares; 3-dimensional shape blocks; pattern blocks; per partnership: 4 each of number cards 0–9	slate; Pattern-Block Template; per small group: 15 quarter-sheets of paper; everyday objects; Number Cards; prepared index cards
8-8	pp. 216, 243–245; TA30; G48 (optional); G55–G56; G57 (optional); G60–G61	104	toolkit clock; per partnership: 2 dot dice, 20 dimes, 20 pennies; per partnership: 8 toolkit clocks	slate; demonstration clock; digital clock; *My Reference Book*; per partnership: 5 one-dollar bills
8-9	pp. 246–252; TA18; G55–G56; G57 (optional); G60–G61	105–106		crayons
8-10	pp. 253–254; G59 (optional)	107	base-10 blocks; pattern blocks	slate; prepared index cards; *My Reference Book*; Class Number Grid; number grid, stick-on notes; blindfold; folder
8-11	pp. 255–256; G34 (optional)	108	base-10 blocks; coin	slate; number grid
8-12	pp. 257–259; *Assessment Handbook*, pp. 52–63			

Literature Link **8-1** *Round Is a Mooncake: A Book of Shapes* (optional) **8-2** *Rabbit and Hare Divide an Apple* (optional); *The Little Mouse, the Red Ripe Strawberry, and the Big Hungry Bear* (optional) **8-3** *Picture Pie: A Circle Drawing Book* (optional) **8-9** *Lemonade for Sale* (optional)

Go Online for a complete literature list in Grade 1.

Problem Solving  Professional Development

Everyday Mathematics emphasizes equally all three of the Common Core's dimensions of **rigor:**

- conceptual understanding
- procedural skill and fluency
- applications

Math Messages, other daily work, Explorations, and Open Response tasks provide many opportunities for children to apply what they know to solve problems.

▶ Math Message

Math Messages require children to solve a problem they have not been shown how to solve. Math Messages provide almost daily opportunities for problem solving.

▶ Daily Work

Journal pages, Home Links, Writing/Reasoning prompts, and Differentiation Options often require children to solve problems in mathematical contexts and real-life situations. **Minute Math+** offers number stories and a variety other practice activities for transition times and for spare moments throughout the day. See Routine 6, pages 32–37.

▶ Explorations

In Exploration A of Lesson 8-7, children create composite shapes from squares and compose new shapes from those composite shapes. In Exploration B, they use 3-dimensional blocks and objects to build new shapes. And in Exploration C, children identify facts that can be solved using the near-doubles and making-10 strategies.

▶ Open Response and Reengagement

In Open Response Lesson 8-4, children model a real-world situation with a drawing and use their drawing to solve a problem. The reengagement discussion on Day 2 may focus on what types of drawings are more helpful in solving the problem or different ways the models can be interpreted. Using mathematical models to solve problems and answer questions is the focus Mathematical Practice for the lesson. As children gain experience with making mathematical models for real-world situations, they will develop an understanding of how mathematics can be used to help them solve problems in everyday life. GMP4.2

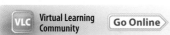 **Virtual Learning Community** **Go Online** to watch an Open Response and Reengagement lesson in action. Search "Open Response" at **vlc.cemseprojects.org**.

Look for GMP1.1–1.6 markers, which indicate opportunities for children to engage in SMP1: "Make sense of problems and persevere in solving them." Children also become better problem solvers as they engage in all of the CCSS Mathematical Practices. The yellow GMP markers throughout the lessons indicate places where you can emphasize the Mathematical Practices and develop children's problem-solving skills.

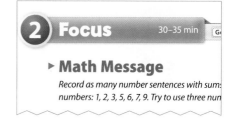

2 Focus 30–35 min G

▶ **Math Message**

Record as many number sentences with sums numbers: 1, 2, 3, 5, 6, 7, 9. Try to use three nun

5 Writing/Reasoning
How do you know what t

15 Sampl

Lesson **8-7** Explorations — Explori and Ad

Overview Child They identify fact

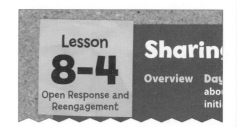

Lesson **8-4** Open Response and Reengagement — Sharin

Overview Day abo initi

Assessment and Differentiation

See pages xxii–xxv to learn about this comprehensive online system for recording, monitoring, and reporting children's progress using core program assessments.

 Virtual Learning Community | Go Online ▷ to **vlc.cemseprojects.org** for tools and ideas related to assessment and differentiation from *Everyday Mathematics* teachers.

✔ Ongoing Assessment

In addition to frequent informal opportunities for "kid watching," every lesson (except Explorations) offers an **Assessment Check-In** to gauge children's performance on one or more of the standards addressed in that lesson.

Lesson	Task Description	CCSS Common Core State Standards
8-1	Identify defining attributes of a triangle.	1.G.1, SMP6
8-2	Partition a rectangle into two equal shares.	1.G.3
8-3	Name one equal share of a whole.	1.G.3, SMP6
8-4	Use a model to compare halves and fourths.	1.G.3, SMP4
8-5	Create a composite shape.	1.G.2
8-6	Distinguish between defining and nondefining attributes of a cylinder.	1.G.1, SMP7
8-8	Read time to the half hour.	1.MD.3, SMP5
8-9	Represent data in a bar graph.	1.MD.4, SMP2
8-10	Find 10 more or 10 less than a number.	1.NBT.5
8-11	Solve 10 more and 10 less problems.	1.NBT.5

▶ Periodic Assessment

Unit 8 Progress Check This assessment focuses on the CCSS domains of *Number and Operations in Base Ten, Measurement and Data,* and *Geometry*. It also contains a Cumulative Assessment to help monitor children's learning and retention of content from the previous units.

NOTE Odd-numbered units include an **Open Response Assessment.** Even-numbered units include a **Cumulative Assessment.**

▶ Unit 8 Differentiation Activities

 Differentiation Support   English Learners Support

Differentiation Options Every regular lesson provides **Readiness, Enrichment, Extra Practice,** and **Beginning English Language Learner Support** activities that address the Focus standards for that lesson.

Activity Card 100

CCSS 1.OA.6, SMP2	CCSS 1.OA.6, SMP2	CCSS 1.OA.6, SMP6
Readiness 10–15 min	**Enrichment** 10–15 min	**Extra Practice** 10–15 min
Exploring Doubles WHOLE CLASS SMALL GROUP PARTNER	**Drawing Doubles Pictures** WHOLE CLASS SMALL GROUP PARTNER	**Playing *Roll and Record Doubles*** WHOLE CLASS SMALL GROUP PARTNER
number cards 1–10, counters,	Activity Card 76, crayons	

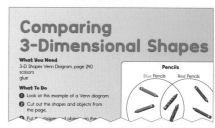

Activity Cards These activities, written to children, enable you to differentiate Part 2 of the lesson through small-group work.

English Language Learners Activities and point-of use support help children at different levels of English language proficiency succeed.

Differentiation Support online pages

Differentiation Support Two online pages for most lessons provide suggestions for game modifications, ways to scaffold lessons for children who need additional support, and language development suggestions for Beginning, Intermediate, and Advanced English language learners.

For **ongoing distributed practice,** see these activities:
- Mental Math and Fluency
- Differentiation Options: Extra Practice
- Part 3: Journal pages, Math Boxes, *Math Masters,* Home Links
- Print and online games

Ongoing Practice

 Differentiation Support

▶ Embedded Facts Practice

Basic Facts Practice can be found in every part of the lesson. Look for activities or games labeled with CCSS 1.OA.6; or go online to the Spiral Tracker and search using CCSS 1.OA.6.

▶ Games

Games in *Everyday Mathematics* are an essential tool for practicing skills and developing strategic thinking.

Lesson	Game	Skills and Concepts	CCSS **Common Core State Standards**
8-1 8-6	*I Spy*	Describing and identifying shapes	**1.G.1, SMP3, SMP6**
8-3 8-8 8-9	*Time Match*	Telling time	**1.MD.3, SMP2**
8-5 8-10	*Make My Design*	Creating composite shapes	**1.G.1, 1.G.2, SMP3, SMP6**
8-7	*Addition Top-It*	Solving addition facts and using relation symbols	**1.OA.6, 1.NBT.3, SMP7**
8-8	*Penny-Dime-Dollar Exchange*	Exchanging ones for tens and tens for hundreds	**1.NBT.2, 1.NBT.2a, SMP6**
8-10	*Before and After*	Finding numbers that are 1 less or 1 more than a given number	**1.NBT.1, SMP7**

VLC Virtual Learning Community [Go Online] to look for examples of *Everyday Mathematics* games at **vlc.cemseprojects.org**.

CCSS Spiral Trace: Skills, Concepts, and Applications

⭐ **Mastery Expectations** This Spiral Trace outlines instructional trajectories for key standards in Unit 8. For each standard, it highlights opportunities for Focus, Warm Up, and Practice activities, as well as formative and summative assessment. It describes the **degree of mastery**—as measured against the entire standard—expected at this point in the year.

Operations and Algebraic Thinking

1.OA.6   Add and subtract within 20, demonstrating fluency for addition and subtraction within 10. Use strategies such as counting on; making ten (e.g., $8 + 6 = 8 + 2 + 4 = 10 + 4 = 14$); decomposing a number leading to a ten (e.g., $13 - 4 = 13 - 3 - 1 = 10 - 1 = 9$); using the relationship between addition and subtraction (e.g., knowing that $8 + 4 = 12$, ones knows $12 - 8 = 4$); and creating equivalent but easier or known sums (e.g., adding $6 + 7$ by creating the known equivalent $6 + 6 + 1 = 12 + 1 = 13$).

⭐ By the end of Unit 8, expect children to **solve addition and subtraction facts within 10.**

Number and Operations in Base Ten

1.NBT.2   Understand that the two digits of a two-digit number represent amounts of tens and ones.

⭐ By the end of Unit 8, expect children to **apply place-value understanding to solve number-grid puzzles.**

1.NBT.5   Given a two-digit number, mentally find 10 more or 10 less than the number, without having to count; explain the reasoning used.

⭐ By the end of Unit 8, expect children to **mentally find 10 more or 10 less than a two-digit number.**

Measurement and Data

1.MD.3   Tell and write time in hours and half-hours using analog and digital clocks.

⭐ By the end of Unit 8, expect children to **tell time to the half-hour on digital and analog clocks.**

Go to **connectED.mcgraw-hill.com** for comprehensive trajectories that show how in-depth mastery develops across the grade.

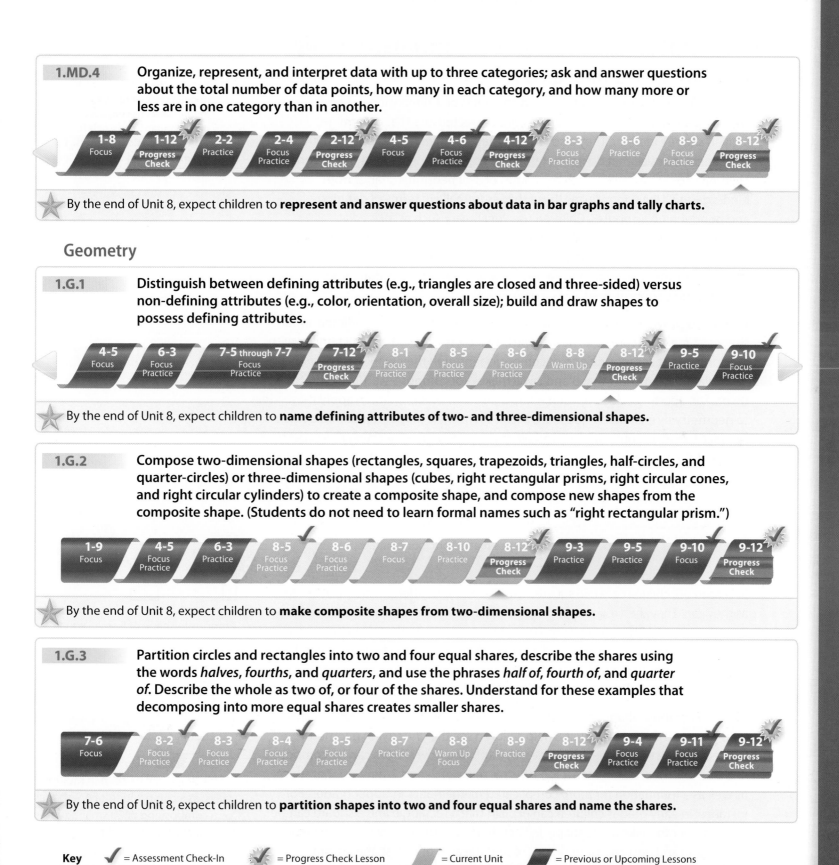

1.MD.4 Organize, represent, and interpret data with up to three categories; ask and answer questions about the total number of data points, how many in each category, and how many more or less are in one category than in another.

1-8 Focus | 1-12 Progress Check | 2-2 Practice | 2-4 Focus Practice | 2-12 Progress Check | 4-5 Focus | 4-6 Focus Practice | 4-12 Progress Check | 8-3 Focus Practice | 8-6 Practice | 8-9 Focus Practice | 8-12 Progress Check

⭐ By the end of Unit 8, expect children to **represent and answer questions about data in bar graphs and tally charts.**

Geometry

1.G.1 Distinguish between defining attributes (e.g., triangles are closed and three-sided) versus non-defining attributes (e.g., color, orientation, overall size); build and draw shapes to possess defining attributes.

4-5 Focus | 6-3 Focus Practice | 7-5 through 7-7 Focus Practice | 7-12 Progress Check | 8-1 Focus Practice | 8-5 Focus Practice | 8-6 Focus Practice | 8-8 Warm Up | 8-12 Progress Check | 9-5 Practice | 9-10 Focus Practice

⭐ By the end of Unit 8, expect children to **name defining attributes of two- and three-dimensional shapes.**

1.G.2 Compose two-dimensional shapes (rectangles, squares, trapezoids, triangles, half-circles, and quarter-circles) or three-dimensional shapes (cubes, right rectangular prisms, right circular cones, and right circular cylinders) to create a composite shape, and compose new shapes from the composite shape. (Students do not need to learn formal names such as "right rectangular prism.")

1-9 Focus | 4-5 Focus Practice | 6-3 Practice | 8-5 Focus Practice | 8-6 Focus Practice | 8-7 Focus | 8-10 Practice | 8-12 Progress Check | 9-3 Practice | 9-5 Practice | 9-10 Focus | 9-12 Progress Check

⭐ By the end of Unit 8, expect children to **make composite shapes from two-dimensional shapes.**

1.G.3 Partition circles and rectangles into two and four equal shares, describe the shares using the words *halves*, *fourths*, and *quarters*, and use the phrases *half of*, *fourth of*, and *quarter of*. Describe the whole as two of, or four of the shares. Understand for these examples that decomposing into more equal shares creates smaller shares.

7-6 Focus | 8-2 Focus Practice | 8-3 Focus Practice | 8-4 Focus Practice | 8-5 Focus Practice | 8-7 Practice | 8-8 Warm Up Focus | 8-9 Practice | 8-12 Progress Check | 9-4 Focus Practice | 9-11 Focus Practice | 9-12 Progress Check

⭐ By the end of Unit 8, expect children to **partition shapes into two and four equal shares and name the shares.**

Key ✓ = Assessment Check-In   = Progress Check Lesson   = Current Unit   = Previous or Upcoming Lessons

Mathematical Background: Content

 The discussion below highlights major content areas and Common Core State Standards addressed in Unit 8. See the online Spiral Tracker for complete information about learning trajectories for all standards.

▶ Attributes of Shapes and Composite Shapes
(Lessons 8-1, 8-5 through 8-7)

Children learned in Unit 7 that defining attributes of 2-dimensional shapes, such as the number of sides or vertices, are characteristics that determine a given shape's identity. They also learned that nondefining attributes are traits such as color or size that do not determine a shape's identity. **1.G.1** Although children should be familiar with the names of 3-dimensional shapes from Kindergarten, Unit 8 introduces these shapes formally in terms of defining and nondefining attributes. Children use vocabulary such as *face, edge,* and *vertex* to define 3-dimensional shapes. (For example: "All cubes have 6 square-shaped faces, 12 edges, and 8 vertices.") As they learn about attributes of shapes, children develop key geometric concepts that appear in increasingly sophisticated ways in later grades.

Unit 8 also provides opportunities for children to use both 2- and 3-dimensional shapes to make, or compose, new shapes. **1.G.2** Children are already familiar with the concepts of composing and decomposing through their work with numbers. For example, children know that 8 and 2 will make—or compose—a 10, and that 12 can be broken apart—or decomposed—into 3 and 9. In Unit 8, children apply these ideas to geometry by examining questions such as *What new shapes can be made when several triangles are put together?* and *What shape results from stacking two cubes?* In Lesson 8-5, children piece together four identical triangles to make new shapes.

Four composite shapes made from four triangles

In Lesson 8-7, children use five squares to make as many different composite shapes as they can. They then repeatedly trace one of the composite shapes to make a new composite shape. This activity emphasizes that a composite shape can be seen as a combination of smaller pieces and as a new single whole. As children manipulate and trace shapes to make new shapes, they also gain informal exposure to transformations, such as flips, rotations, and slides, which are studied in later grades.

Children compose shapes every time they create a pattern-block design or build with blocks. The lessons in Unit 8 help children see the underlying geometric concepts at the root of these activities. By exploring how shapes can be composed and decomposed, children develop a better sense of 2- and 3-dimensional space, as well as a foundation for more complicated mathematics in upper grades. For instance, in early grades children can explain that a rectangle and a square can be put together to make an "L" shape. In later grades, children apply this knowledge of composing and decomposing shapes to find the area of an L-shaped room.

 Standards and Goals for Mathematical Content

Because the standards within each domain can be broad, *Everyday Mathematics* has unpacked each standard into Goals for Mathematical Content GMC. For a complete list of Standards and Goals, see page EM1.

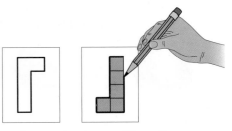

Composite shape of 5 small squares

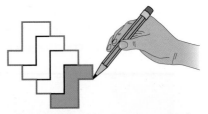

This shape can be seen in multiple ways: as a composition of 4 S-shapes, as a composition of 20 small squares, or as one 20-sided polygon.

Unit 8 Vocabulary

composite	half	quarter
edge	half-hour	surface
equal shares	half-past	vertex
face	mathematical model	whole
fourth	number-grid puzzle	

▶ Fractional Parts of Shapes

(Lessons 8-2 through 8-4)

Partitioning shapes, or dividing shapes into equal parts, is also an important topic in Unit 8. This work constitutes the foundation for the study of fractions throughout *Everyday Mathematics*.

Children who encounter formal standard notation for fractions without a firm conceptual understanding of fractional parts often struggle to understand fractions as quantities. They think, for example, of "$\frac{1}{2}$" as two separate numbers (1 and 2) and may develop other misconceptions as well. For these reasons, *First Grade Everyday Mathematics* does not introduce formal fractional notation or terms such as *numerator* or *denominator*.

Lessons 8-2 through 8-4 develop understanding of fraction concepts by emphasizing the idea of equal shares. Children partition objects into two and four equal shares and discuss how they know the shares are equal. Equal shares in these lessons are often the same shape because equality of the shares is easier for young children to verify if the shares are the same shape. However, although the children may not necessarily face the issue, it is important to note that equal shares need not be the same shape.

Two halves

As children work with equal shares, they discuss names for the equal partitions as well as for the recomposed whole. Children are encouraged to use a variety of ways to describe the partitions and wholes, such as "1 out of 2 parts," "quarter," "1 half," "2 halves," and "4 quarters." These terms not only help children distinguish between different kinds of partitions, but they also emphasize the relationship between equal shares and the whole, reinforcing the idea that putting together 2 halves (or 4 fourths) results in 1 whole.

The Open Response lesson in this unit helps children explore the relationship between the number of equal shares and the size of those shares. Children should conclude that if they divide a whole into 4 equal shares, then they will get smaller shares than if they divide the same whole into 2 equal shares. **1.G.3**

▶ Time and Data (Lessons 8-8 and 8-9)

The instructional sequence for time in *First Grade Everyday Mathematics* reflects the difficulties that children have in reading and placing the hour hand on an analog clock. When time is first introduced in Unit 6, children tell time to the nearest hour with hour-hand-only clocks. In Unit 7, they use standard analog and digital clocks to tell time to the nearest hour. In Unit 8, children begin to tell and write time to the nearest half-hour. **1.MD.3** This lesson also connects telling time to previous lessons about fractional parts. Children notice and discuss how two *half* hours make one *whole* hour.

In Lesson 8-9, children develop their skills for representing and interpreting data. **1.MD.4** Children conduct class surveys, organize the data into bar graphs, and then use the bar graphs to ask and answer questions about the data. The process of collecting data and constructing graphs helps children interpret and understand how data are represented in other graphs and charts. Independent practice using tally marks, labeling a graph, and creating bars that correspond to numbers of tallies also helps children develop their skills with representing data. Lesson 8-9 builds upon children's understanding of data concepts by reinforcing the *meaning* of a graph—and not just the mechanics of its construction. Question prompts are provided to help children ask and answer questions about the different types of information that their graphs can provide.

Toolkit clock

▶ Adding and Subtracting 10
(Lessons 8-10 and 8-11)

In previous units, children used number grids to count, to add and subtract, and to study place-value patterns. In Unit 8, they apply their knowledge of number grids to quickly and accurately add and subtract 10 from any 2-digit number. Children begin practicing this skill in Lesson 8-10 by solving number-grid puzzles. These puzzles are pieces of a number grid in which some, but not all, of the numbers are missing.

To solve these puzzles, children must understand number-grid patterns. For example, they must realize that adding or subtracting 10 (moving down or up a column) results in a change in the tens place while the number of ones stays the same. Lesson 8-10 also connects the number grid to children's experience with using base-10 blocks to add and subtract. Children discuss how moving within a row on a number grid is like adding and subtracting cubes (ones), and moving within a column is like adding and subtracting longs (tens). **1.NBT.4** In Lesson 8-11, children visualize number grids and base-10 blocks to practice *mentally* adding and subtracting 10 from two-digit numbers. **1.NBT.5**

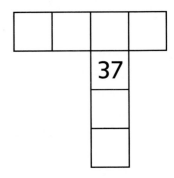

Number-grid puzzle

Mathematical Background: Practices

 *In Everyday Mathematics, children learn the **content** of mathematics as they engage in the **practices** of mathematics. As such, the Standards for Mathematical Practice are embedded in children's everyday work, including hands-on activities, problem-solving tasks, discussions, and written work. Read here to see how Mathematical Practices 3 and 6 are emphasized in this unit.*

▶ Standard for Mathematical Practice 3

Mathematical argument is central to Mathematical Practice 3. Children should be able to construct informal arguments—to tell *how* they know or *why* they think as they do—as well as listen to others' arguments, think about whether those arguments make sense, and ask questions. In Lesson 8-2, children construct and evaluate mathematical arguments by dividing a pancake into equal shares and showing how they know the shares are equal. They construct various informal arguments, with words as well as with objects. GMP3.1

Throughout Unit 8, children make sense of others' mathematical thinking. GMP3.2 For example, to play *I Spy* in Lesson 8-1, children interpret the Spy's clues to figure out which shape is being described. To play *Make My Design* in Lesson 8-5, children must make sense of others' thinking to recreate a design.

First Grade Everyday Mathematics fosters the development of Mathematical Practice 3 by providing numerous opportunities for class discussions. As children verbalize and demonstrate their ideas, they build on each other's discoveries, gain insights, and refine their understanding of mathematics.

▶ Standard for Mathematical Practice 6

Mathematical Practice 6 focuses on accuracy and efficiency when calculating and measuring, as well as the precise expression of ideas. Students are expected to use clear units, labels, and language. GMP6.1, GMP6.3, GMP6.4 To reinforce this practice, teachers can model the use of precise mathematical language, list important vocabulary, or refer to a word wall or chart during an explanation or discussion.

Children continue to use clear mathematical language throughout Unit 8. In Lesson 8-1, children use precise language about specific defining attributes to contrast various shapes. In Lesson 8-3, children label fractional parts with clear, descriptive names. And in Lesson 8-5, *Make My Design* challenges players to use accurate shape vocabulary. GMP6.3

Unit 8 also offers opportunities for children to clearly and precisely explain their thinking. GMP6.1 In Lesson 8-1, children explain why they think a particular shape is or is not a polygon. In Lesson 8-3, they explain how they created equal shares. And in Lesson 8-11, children explain their thought processes as they mentally add and subtract 10.

In addition to these examples, *First Grade Everyday Mathematics* offers consistent practice with Mathematical Practice 6 across all units. Like Mathematical Practice 3, this practice is reinforced every time children engage in mathematical discussions and are encouraged to express their ideas.

 Standards and Goals for Mathematical Practice

SMP3 **Construct viable arguments and critique the reasoning of others.**

GMP3.1 Make mathematical conjectures and arguments.

GMP3.2 Make sense of others' mathematical thinking.

SMP6 **Attend to precision.**

GMP6.1 Explain your mathematical thinking clearly and precisely.

GMP6.2 Use an appropriate level of precision for your problem.

GMP6.3 Use clear labels, units, and mathematical language.

GMP6.4 Think about accuracy and efficiency when you count, measure, and calculate.

Go Online to the *Implementation Guide* for more information about the Mathematical Practices.

For children's information on the Mathematical Practices, see *My Reference Book,* pages 1–22.

Building Shapes with Defining Attributes

Overview Children construct 2-dimensional shapes and identify defining and nondefining attributes of those shapes.

▶ **Before You Begin**

For Part 2, each child will need 15 twist ties and 3 each of 8-inch straws, 6-inch straws, and 4-inch straws. For the Assessment Check-In, make a demonstration triangle with sides longer than 8 inches. Pattern the triangle with stripes or polka dots. (Alternatively you may wish to use an attribute block.) Also, display the Defining/Nondefining Attributes Posters from Lesson 7-7. For the optional Extra Practice activity, obtain a copy of **Round Is a Mooncake: A Book of Shapes** by Roseanne Thong (Chronicle Books, 2000).

Common Core State Standards

Focus Cluster
Reason with shapes and their attributes.

① Warm Up 15–20 min

	Materials	
Mental Math and Fluency Children read an analog clock and record the time digitally.	demonstration clock, slate	**1.MD.3**
Daily Routines Children complete daily routines.	See pages 2–37.	See pages xiv–xvii.

② Focus 30–35 min

Math Message Children use descriptions of attributes to draw shapes.		**1.G.1**
Constructing Straw Polygons Children review defining attributes and create polygons.	*Math Journal 2*, pp. 156–157; (*See Before You Begin.*)	**1.G.1** SMP1, SMP6
✓ **Assessment Check-In** See page 687.	demonstration triangle or attribute block	**1.G.1** SMP6
Introducing *I Spy* **Game** Children practice describing and identifying shapes.	*Math Masters*, p. G58 (optional)	**1.G.1** SMP3, SMP6

CCSS 1.G.1 **Spiral Snapshot**

GMC Distinguish between defining and non-defining attributes.

| 4-5
Focus | 6-3
Focus
Practice | 7-5 through 7-7
Focus
Practice | 8-1
Focus | 8-5
Focus
Practice | 8-6
Focus
Practice | 9-5
Practice |

Spiral Tracker   **Go Online** to see how mastery develops for all standards within the grade.

③ Practice 15–20 min

Practicing Fact Families Children write fact families.	*Math Journal 2*, p. 158	**1.OA.3, 1.OA.4, 1.OA.6, 1.OA.8, SMP7**
Math Boxes 8-1 Children practice and maintain skills.	*Math Journal 2*, p. 159	See page 689.
Home Link 8-1 **Homework** Children play *I Spy* with someone at home.	*Math Masters*, p. 224	**1.OA.7, 1.NBT.3, 1.NBT.4, 1.G.1** SMP3, SMP6

connectED.mcgraw-hill.com ▶

Plan your lessons online with these tools.

ePresentations Student Learning Center Facts Workshop Game eToolkit Professional Development Home Connections Spiral Tracker Assessment and Reporting English Learners Support Differentiation Support

Differentiation Options RtI

 CCSS 1.G.1, SMP2

Readiness
5–10 min

WHOLE CLASS
SMALL GROUP
PARTNER
INDEPENDENT

Guessing the Shape

To provide experience identifying shapes, have partners take turns using their fingers to "draw" pattern-block shapes on each other's backs. Children guess the shape being drawn based on defining attributes (such as the number of sides). **GMP2.2**

CCSS 1.G.1, SMP7

Enrichment
15–20 min

WHOLE CLASS
SMALL GROUP
PARTNER
INDEPENDENT

Comparing 2-Dimensional Shapes

Activity Card 94;
Math Masters, p. 223; scissors; glue

To further explore attributes of shapes, have children compare shapes using a Venn diagram. **GMP7.1** You may wish to use the illustration from Step 1 on the Activity Card to explain or review how to complete a Venn diagram. After children cut out and glue the shapes and figures on *Math Masters,* page 223, encourage them to draw additional shapes in each section of the Venn diagram.

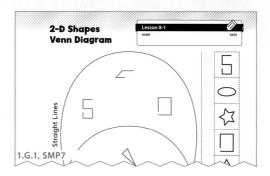

CCSS 1.G.1, SMP7

Extra Practice
10–15 min

WHOLE CLASS
SMALL GROUP
PARTNER
INDEPENDENT

Reading about Geometry

Math Masters, p. TA4;
Round Is a Mooncake: A Book of Shapes

Literature Link To provide practice with geometry skills, read ***Round Is a Mooncake: A Book of Shapes*** by Roseanne Thong (Chronicle Books, 2000). On an Exit Slip (*Math Masters,* page TA4), have children draw a shape with 4 sides and 4 square corners that they see in the classroom. Discuss the names of the shapes they drew. **GMP7.1**

English Language Learners Support

Beginning ELL To review defining attributes, have children play a guessing game with *yes* or *no* questions. One child secretly puts an attribute block inside a bag. The guesser asks questions, such as: "Does it have vertices? Does it have three sides? Is it a _____?" The first player may reach into the bag to touch the block while answering questions. Demonstrate one round with a child, with you as the guesser. Then reverse roles so children can practice using the terms to ask questions. You may wish to play as a class or form partnerships.

  Go Online **ELL** English Learners Support

Lesson 8-1   **685**

SMP1 **Make sense of problems and persevere in solving them.**
    GMP1.1 Make sense of your problem.

SMP3 **Construct viable arguments and critique the reasoning of others.**
    GMP3.2 Make sense of others' mathematical thinking.

SMP6 **Attend to precision.**
    GMP6.1 Explain your mathematical thinking clearly and precisely.

Math Journal 2, p. 156

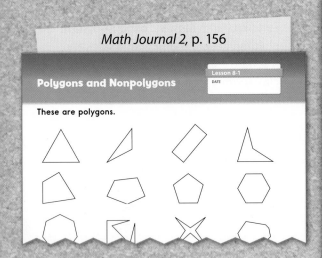

NOTE You may wish to enlarge journal page 157 to poster size for children to use as a reference or display the Two-Dimensional Shapes Poster.

Math Journal 2, p. 157

1 Warm Up   15–20 min   Go Online

ePresentations   eToolkit

▶ Mental Math and Fluency

Display the following times on the demonstration clock. Have children record the times in digital notation on their slates.

● ○ ○   3:00; 10:00

● ● ○   6:00; 12:00

● ● ●   4:00; 11:00

▶ Daily Routines

Have children complete the Daily Routines. For detailed instructions, see pages 2–37. For specifics on CCSS coverage, see pages xiv–xvii.

2 Focus   30–35 min   Go Online

ePresentations   eToolkit

▶ Math Message

Ben says, "My shape has exactly 3 sides and 3 corners." Draw Ben's shape.

Sasha says, "My shape is small and blue." Draw Sasha's shape.

Could Ben and Sasha have drawn the same shape? Explain why or why not. Record your answer.

▶ Constructing Straw Polygons

Math Journal 2, pp. 156–157

Math Message Follow-Up Have children display the shapes they drew in response to the Math Message. Ask: *What shape could Ben be describing?* Triangle *How can you be sure?* Sample answer: I am sure because the only shape that has exactly 3 sides and 3 corners is a triangle. Call attention to children's drawings for Sasha's description. Ask: *Do you know for sure what shape Sasha has? Why or why not?* Sample answer: No. The words *small* and *blue* could apply to many different shapes. **GMP1.1** *Could Ben and Sasha have drawn the same shape?* Sample answer: Yes. A triangle can be blue and small.

Remind children about the Defining/Nondefining Attributes Posters from Lesson 7-7. Ask: *What if Sasha said, "My shape is small, blue, and has 6 sides." Could she and Ben still have drawn the same shape?* No *What shape could Sasha be talking about?* Hexagon *Which words in Sasha's new description gave the most important clues that helped you know the shape?* The shape has 6 sides.

Tell children that today they will continue learning about attributes of shapes by making their own shapes and by playing a new game.

Have children examine journal page 156. Discuss why some shapes are labeled *polygons* and others are *not polygons*. Review that a polygon is a flat, closed shape with three or more straight sides that meet only at their vertices. (*See Lesson 7-7.*) Ask children to examine these shapes and those on both journal page 157 and the Two-Dimensional Shapes Poster. Have them trace the sides of the polygons with their fingers. Discuss the defining attributes of various shapes.

Tell children that they will be constructing polygons using straws and twist ties. Show children how to join straws to make polygons.

1. Insert one end of a twist tie into one end of a straw.

2. Insert the other end of the twist tie into another straw.

3. Push the straws together until they meet to form a corner, or vertex.

4. Connect other straws similarly. The straws form the polygon's sides.

Have children practice connecting straws and twist ties to build triangles. Remind them that triangles (and other polygons) are "closed"—their sides make fences with no openings. Figures that are not closed are "open"— the fence is open. Have children pull apart one corner of their triangles to make open figures, then close the triangles and set them aside.

Have partnerships use straws of various lengths to build other polygons, such as those shaped like pattern blocks. Explain that ideally each polygon should lay flat, as if it were made on a geoboard. Each partnership should end up with at least three different polygons.

 Assessment Check-In  **CCSS** 1.G.1

Ask children to show you the triangles they created earlier. Showing the demonstration triangle, ask: *How can my shape and your shape both be triangles even though they look very different?* Expect most children to respond that both are triangles because they have 3 sides and 3 vertices, or corners. Size and fill do not matter. **GMP6.1** Have children who have difficulty explaining defining attributes explain the difference between a triangle and a square so that a Kindergartner would understand. Point out any wording they use that calls attention to a defining attribute.

Assessment and Reporting **Go Online** to record student progress and to see trajectories toward mastery for these standards.

After partners have built at least three polygons, display the Defining and Nondefining Attributes of Rectangles Poster from Lesson 7-7. Review which defining attributes of rectangles also define squares to remind children that a square is a special kind of rectangle. Make a new poster with columns for Defining and Nondefining Attributes of Squares. Have children suggest attributes to fill in the columns.

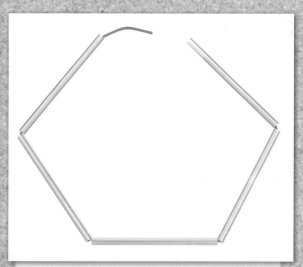

Making a polygon with straws and twist ties

I Spy

NAME   **DATE**

Materials	None
Players	2 to 4
Skill	Describing and identifying shapes
Object of the Game	To correctly guess the shape

Directions

① Play with a partner or in a small group.
Take turns being the Spy.

② When you are the Spy:

• Look for a shape in the room.
Don't tell anyone the name of the shape you spy.

• Say, "I spy with my little eye a shape with. . . ."
Then give clues that describe your shape.
Be sure to use defining attributes.

③ Other players guess which shape was spied.

④ Play again with another player as the Spy.

I spy with my little eye a shape with 4 sides that are all the same length.

G58   1.G.1

Fact Families with Fact Triangles and Dominoes

Lesson 8-1

DATE

Write the fact families.

1

11
+, −
6   5

$5 + 6 = 11$
$6 + 5 = 11$
$11 - 6 = 5$
$11 - 5 = 6$

2

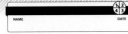

$8 + 6 = 14$
$6 + 8 = 14$
$14 - 8 = 6$
$14 - 6 = 8$

3 Write the missing number. Complete the fact family.

12
+, −
9   3

$3 + \underline{9} = 12$
$9 + 3 = 12$
$12 - 3 = 9$
$12 - 9 = 3$

4 How can $9 + 3 = 12$ help you solve $12 - 9$?

Sample answer: I know $9 + 3 = 12$, so $12 - 9$ must be 3 because they are in the same fact family.

158   one hundred fifty-eight   1.OA.3, 1.OA.4, 1.OA.6, 1.OA.8, SMP7

Have children compare the posters for rectangles and squares. Ask children who made squares to hold up their constructions. Ask: *Which attributes of squares make them special rectangles?* **GMP6.1** Allow all children opportunities to generate ideas, though some may have explained this in Lesson 7-7. Highlight that all four sides are the same length for all squares, but that is not a defining attribute of rectangles.

Professional Development Squares are often defined as special rectangles by their equal side lengths. However, other special attributes of squares could be used to define them. For example, squares have more degrees of rotational symmetry than nonsquare rectangles, and their diagonals form congruent right triangles. You may wish to demonstrate these attributes by showing quarter and half turns with the rectangle and square pattern blocks, and by drawing and comparing diagonals in a square and a nonsquare rectangle.

Ask a child who constructed a polygon with more than four sides to push in one of the corners to make a concave polygon.

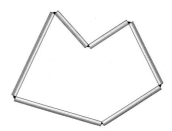

A concave polygon

Refer to the characteristics of a polygon that children reviewed at the beginning of the lesson. Ask: *Is this new shape still a polygon?* Yes *How do you know?* **GMP6.1** Sample answer: It is still closed and flat and has straight sides and corners.

Have children build two final shapes based on these attributes:

• Shape 1: *Make a shape that has 4 sides that are all the same length, and 4 corners that all look the same.*

• Shape 2: *Make a shape that is large and pointy.*

Compare children's shapes. Ask: *Did everyone make the same shape for Shape 1? Why or why not?* Yes. Any shape with 4 identical (square) corners and 4 same-length sides has to be a square.

Emphasize that the list of defining attributes made it clear that Shape 1 had to be a square. Ask: *For Shape 2, did everyone make the same shape? Why or why not?* No. Large and pointy are not defining attributes of any specific shape.

NOTE Since many children enjoy making these constructions, you may wish to make the materials available to children for several days.

▶ Introducing *I Spy*

| WHOLE CLASS | SMALL GROUP | PARTNER | INDEPENDENT |

Demonstrate a few rounds of *I Spy* with the whole class, focusing on 2-dimensional shapes and using words such as *corners, sides,* and *vertices* to describe the shapes. Then have children play in small groups or partnerships. Remind them to pay close attention to the clues so they can figure out what shape the child is describing. **GMP3.2** You may wish to distribute game directions found on *Math Masters,* page G58.

Observe
- Do children recognize shapes of everyday objects?
- Do children focus on the defining attributes of the shapes they are describing (using vocabulary such as *sides, corners,* and *vertices*)?

Discuss
- Point to an object in the room shaped like a square and another shaped like a nonsquare rectangle. Ask: *What is the same about the shapes? What is different?*
- *I spy with my little eye something blue. Is this enough for you to figure out what shape I am spying? Why or why not?* **GMP6.1**

Differentiate **Game Modifications** ⟨Go Online⟩ Differentiation Support

Summarize Have a few children share *I Spy* riddles with the whole class. Encourage children to use riddles focusing on defining attributes.

③ Practice 15–20 min ⟨Go Online⟩ ePresentations eToolkit Home Connections

▶ Practicing Fact Families
Math Journal 2, p. 158

| WHOLE CLASS | SMALL GROUP | PARTNER | INDEPENDENT |

Have children practice writing fact families. **GMP7.2**

▶ Math Boxes 8-1
Math Journal 2, p. 159

| WHOLE CLASS | SMALL GROUP | PARTNER | INDEPENDENT |

Mixed Practice Math Boxes 8-1 are paired with Math Boxes 8-3.

▶ Home Link 8-1
Math Masters, p. 224

Homework Children practice naming defining attributes by playing *I Spy* at home.

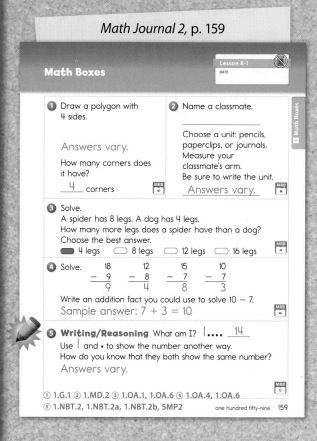

Math Journal 2, p. 159

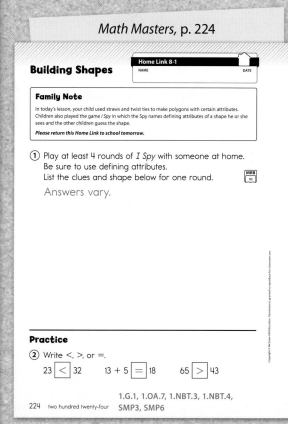

Math Masters, p. 224

Halves

Overview Children divide shapes into two equal shares and discuss how to name the shares.

▶ **Before You Begin**

For Part 2, you may wish to have children cut out the crackers on *Math Masters*, page 225; each child will need two crackers. Store remaining crackers for Lesson 8-3. For the optional Enrichment activity, get the book *Rabbit and Hare Divide an Apple* by Harriet Ziefert (Puffin, 1998). For the optional Extra Practice activity, read *The Little Mouse, the Red Ripe Strawberry, and the Big Hungry Bear* by Don and Audrey Wood (Child's Play International, 1984) to your class.

▶ **Vocabulary**

equal shares • half • whole

**Common Core
State Standards**

Focus Cluster
Reason with shapes and their attributes.

1 Warm Up 15–20 min

	Materials	
Mental Math and Fluency Children solve subtraction facts.	slate	1.OA.4, 1.OA.5, 1.OA.6
Daily Routines Children complete daily routines.	See pages 2–37.	See pages xiv–xvii.

2 Focus 35–40 min

Math Message Children find objects that are divided into equal parts.		1.G.3 SMP3
Partitioning Pancakes in Halves Children divide "pancakes" into two equal shares and discuss how they know the shares are equal.	*Math Journal 2*, p. 160; chart paper; per partnership: modeling clay or play dough, Pattern-Block Template	1.G.3 SMP2, SMP3
Partitioning Crackers in Halves Children divide "crackers" into halves in multiple ways.	*Math Journal 2*, p. 160; *Math Masters*, p. 225; scissors; tape or glue	1.G.3 SMP1
✓ **Assessment Check-In** See page 693.	*Math Masters*, p. 225	1.G.3
Naming Shares Children discuss how they could name shares.	*Math Journal 2*, p. 160	1.G.3, SMP3

CCSS 1.G.3   **Spiral Snapshot**

GMC Partition shapes into equal shares.

7-6 Focus	8-2 Focus Practice	8-3 through 8-5 Focus Practice	8-7 Practice	8-8 Warm Up Focus	9-4 Focus Practice	9-11 Focus Practice

III **Spiral Tracker**   **Go Online** to see how mastery develops for all standards within the grade.

3 Practice 10–15 min

Practicing "What's My Rule?" Children practice addition and subtraction.	*Math Journal 2*, p. 161	1.OA.8, 1.NBT.4, 1.NBT.5
Math Boxes 8-2 Children practice and maintain skills.	*Math Journal 2*, p. 162	See page 695.
Home Link 8-2 **Homework** Children divide squares into two equal shares.	*Math Masters*, p. 228	1.NBT.2, 1.G.3 SMP1

connectED.mcgraw-hill.com

Plan your lessons online with these tools.

ePresentations Student Learning Center Facts Workshop Game eToolkit Professional Development Home Connections Spiral Tracker Assessment and Reporting English Learners Support Differentiation Support

Differentiation Options

 RtI

CCSS 1.G.3, SMP1

Readiness   10–15 min

WHOLE CLASS
SMALL GROUP
PARTNER
INDEPENDENT

Determining Equal Shares

Math Masters, p. 226; scissors

To provide experience determining whether shares are equal, have children cut apart the cards on *Math Masters,* page 226. Ask them to sort the cards into two groups—shapes divided into two equal shares and shapes not divided into two equal shares. Have children share their sorts and discuss some of the ways they might check whether a shape has equal shares. GMP1.4 Allow them to fold the cards or cut the shapes apart and match the pieces as needed.

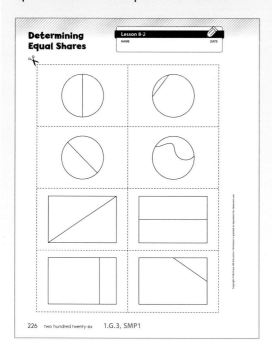

Determining Equal Shares
Lesson 8-2
NAME   DATE

226   two hundred twenty-six   1.G.3, SMP1

CCSS 1.G.3, SMP4

Enrichment   15–20 min

WHOLE CLASS
SMALL GROUP
PARTNER
INDEPENDENT

Dividing an Apple

Activity Card 95;
Math Masters, p. 227;
Rabbit and Hare Divide an Apple

Literature Link To extend children's understanding of dividing shapes in half, have them read **Rabbit and Hare Divide an Apple** by Harriet Ziefert (Puffin, 1998). Children show how Rabbit and Hare could have divided the two pieces of an apple to share equally. GMP4.1

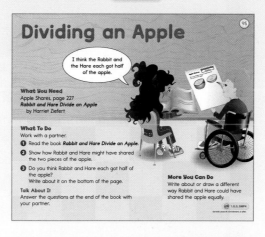

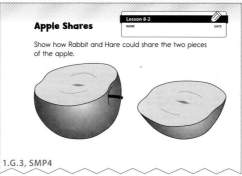

Apple Shares
Lesson 8-2
NAME   DATE

Show how Rabbit and Hare could share the two pieces of the apple.

1.G.3, SMP4

CCSS 1.G.3, SMP4

Extra Practice   15–20 min

WHOLE CLASS
SMALL GROUP
PARTNER
INDEPENDENT

Sharing Food

Activity Card 96;
The Little Mouse, the Red Ripe Strawberry, and the Big Hungry Bear; crayons or colored pencils

Literature Link To provide practice dividing shapes into two equal parts, read **The Little Mouse, the Red Ripe Strawberry, and the Big Hungry Bear** by Don and Audrey Wood (Child's Play International, 1984). Children draw pictures of food they could share and show how they would share the food equally with a friend. GMP4.1

English Language Learners Support

Beginning ELL To provide practice differentiating between the terms *equal* and *unequal,* have children show examples of equal-size pairs of objects and unequal-size pairs of objects. Use *not equal* interchangeably with *unequal* to help children understand and connect the terms.

Go Online   **ELL** English Learners Support

Standards and Goals for
Mathematical Practice

SMP1 **Make sense of problems and persevere in solving them.**
　GMP1.5 Solve problems in more than one way.

SMP2 **Reason abstractly and quantitatively.**
　GMP2.1 Create mathematical representations using numbers, words, pictures, symbols, gestures, tables, graphs, and concrete objects.

SMP3 **Construct viable arguments and critique the reasoning of others.**
　GMP3.1 Make mathematical conjectures and arguments.

NOTE Children will complete a single journal page across several activities in this lesson, so be sure that children complete only the part specified in each activity and do not work ahead.

Math Journal 2, p. 160

Partitioning Pancakes and Crackers in Halves

Lesson 8-2
DATE

① Show how you divided the pancake into 2 equal shares.
Answers vary.

② Show how to share 1 cracker between 2 people.
Answers vary.

③ Show another way to share 1 cracker between 2 people.
Answers vary.

④ Write a name for 1 of the equal parts of the cracker or pancake.
Sample answers: half; 1 half; one-half; 1 out of 2 equal shares

---

---

1 Warm Up　15–20 min
Go Online
ePresentations　eToolkit

▶ Mental Math and Fluency

Ask children to solve the first fact in each pair and share their strategies. Elicit counting up to subtract, and ask children to apply that strategy to solve the second fact. Children record answers on their slates.

●○○　$8 - 6$　2; $11 - 9$　2
●●○　$9 - 7$　2; $11 - 8$　3
●●●　$9 - 6$　3; $12 - 9$　3

▶ Daily Routines

Have children complete the Daily Routines. For detailed instructions, see pages 2–37. For specifics on CCSS coverage, see pages xiv–xvii.

2 Focus　35–40 min
Go Online
ePresentations　eToolkit

▶ Math Message

Find and draw an object that is divided into equal-size parts. Tell a partner what your object is and how many equal-size parts it has. How do you know it is divided into equal-size parts? **GMP3.1** *Record your answer.*

▶ Partitioning Pancakes in Halves

Math Journal 2, p. 160

| WHOLE CLASS | SMALL GROUP | PARTNER | INDEPENDENT |

Math Message Follow-Up Ask children to identify the objects they found and the number of equal-size parts, or **equal shares,** each has. Objects might include windows with glass panes, bookcases with shelves, a whiteboard with sections, or the floor with tiles. Have children share how they knew the objects were divided into equal shares. Point to an object that is divided into unequal-size parts, and ask children to explain whether it is divided into equal shares.

Distribute a ball of clay to each partnership. Tell partners to create a "pancake" by flattening the ball of clay with their hands. Then ask them to use the edge of a Pattern-Block Template to divide the pancake so that two people could share it equally. **GMP2.1**

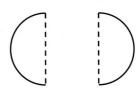

Some partnerships may divide the pancake into four pieces and make two shares with two pieces each. Acknowledge that this is a solution, but for the naming discussion later in the lesson, use only examples that show two pieces.

Ask partners to discuss how they can show that their shares are equal, and have several partnerships share their ideas. Sample answer: I put one piece on top of the other, and they are the same size. **GMP3.1** Show examples from children who created unequal shares, and discuss whether these are equal shares of the pancake.

Have children draw the two equal shares of the pancake on journal page 160. Label a poster Two Equal Shares and ask a child to draw the two equal shares on it. **GMP2.1**

▶ Partitioning Crackers in Halves

Math Journal 2, p. 160; *Math Masters*, p. 225

Have children cut out the crackers on *Math Masters,* page 225. Collect and store extra crackers for use in Lesson 8-3. Ask them to talk with their partners about how to share one cracker equally between two people. Have each child cut a cracker into two equal shares and attach the two parts to journal page 160. Then have them record a different way to share the second cracker equally. **GMP1.5** Encourage children to discuss how they know the two shares are equal as they work.

Ask volunteers to share how they divided the crackers and how they knew the two shares were equal. Record one or more ways children divided the crackers on the Two Equal Shares Poster.

✓ Assessment Check-In CCSS 1.G.3

Math Masters, p. 225

Observe as children cut apart crackers and determine whether the two shares are equal. Expect most children to cut the crackers either horizontally or vertically and overlay the pieces to determine that they are equal. Some children may have difficulty sharing the crackers more than one way. Encourage children who struggle to fold the crackers before they cut them to experiment with different ways of dividing them. Challenge children who excel to find a third way to share the cracker equally.

 Assessment and Reporting ⟨Go Online⟩ to record student progress and to see trajectories toward mastery for these standards.

NOTE The Two Equal Shares Poster is completed throughout the lesson.

Two Equal Shares

Name of one share

half

1 half

1 out of 2 parts

Name of all shares

whole

2 halves

2 out of 2 parts

Example of a completed
Two Equal Shares Poster

Math Masters, p. 225

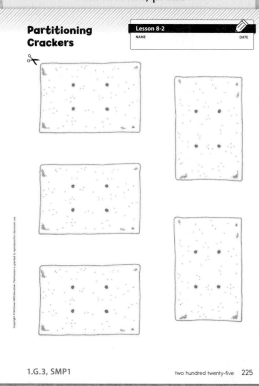

Partitioning Crackers   Lesson 8-2
NAME   DATE

1.G.3, SMP1   two hundred twenty-five   225

NOTE Although it may come up in discussion, avoid using formal fractional notation, such as $\frac{1}{2}$, to name one or all of the shares. The focus in Grade 1 is using words and pictures to describe equal shares. Fractional notation will be introduced in later grades.

► Naming Shares

Math Journal 2, p. 160

| **WHOLE CLASS** | SMALL GROUP | PARTNER | INDEPENDENT |

Display the Two Equal Shares Poster. Point to one of the pancake shares on the poster. Ask: *How could you describe the share that you got when you divided the pancake equally with a partner? What could you call this piece of the pancake?* Sample answers: half, 1 half, half of a pancake, half of a circle, 1 out of 2 equal parts Point to one of the cracker shares. *What could you call this piece of the cracker? How could we describe the share that you got when you divided a cracker equally with your partner?* Sample answers: half, 1 half, half of a cracker, a small rectangle, 1 out of 2 equal shares

Remind children that they recorded many names for a number using a name-collection box. Explain that they can also use many names to describe a share when there are two equal parts. Write "Name of one share" on the poster, and record relevant names, including **half.** Ask children to think of other situations in which they might describe a share using these terms. Sample answers: I ate half of a sandwich for lunch. I borrowed half of a sheet of paper from my friend. I ate 1 out of 2 shares of a granola bar.

Hold up two parts of a cracker from *Math Masters,* page 225. Tell children to imagine that they are having crackers for a snack. Their partners are not hungry, so they are going to eat their own shares and their partners' shares. Ask: *What could you call these two shares together?* Sample answers: whole, 2 halves, the whole cracker, 2 out of 2 equal shares Write "Name of all shares" on the poster, and record the names children generate, including **whole.**

Academic Language Development Children might not be familiar with the plural form of *half.* To provide oral practice, say and have children repeat *one-half, two-halves* as you talk about equal shares or parts. Show the singular and plural forms of the word in writing. Facilitate a discussion about the spellings and pronunciations of both words.

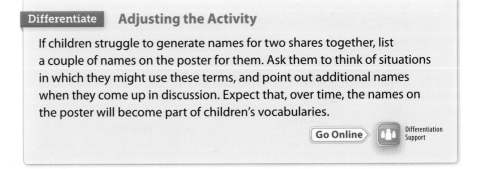

| **Differentiate** | **Adjusting the Activity** |

If children struggle to generate names for two shares together, list a couple of names on the poster for them. Ask them to think of situations in which they might use these terms, and point out additional names when they come up in discussion. Expect that, over time, the names on the poster will become part of children's vocabularies.

| Go Online ▷ | 📖 | Differentiation Support |

Summarize Show children a cracker that has been divided unequally. Ask: *Did I divide this cracker into halves? Why or why not?* | GMP3.1 | No; the two parts are not equal sizes.

Math Journal 2, p. 161

"What's My Rule?"

Lesson 8-2
DATE

Find the rule and the missing numbers. Use the number grid if you like.

① in

Rule +1

in	out
14	15
22	23
4	5
18	19

out

② in

Rule +8

in	out
11	19
24	32
10	18
29	37

out

③ in

Rule −10

in	out
13	3
51	41
18	8
29	19

out

④ in

Rule +10

in	out
23	33
15	25
7	17
37	47

out

⑤ How can you find the rule in Problem 4?
Sample answer: The ones digit stays the same, and the tens digit goes up by 1. So, the rule is +10.

1.OA.8, 1.NBT.4, 1.NBT.5 — one hundred sixty-one 161

3 Practice 10–15 min

Go Online

ePresentations   eToolkit   Home Connections

▶ Practicing "What's My Rule?"

Math Journal 2, p. 161

| WHOLE CLASS | SMALL GROUP | **PARTNER** | INDEPENDENT |

Have children practice addition and subtraction by solving "What's My Rule?" problems.

▶ Math Boxes 8-2

Math Journal 2, p. 162

| WHOLE CLASS | SMALL GROUP | **PARTNER** | INDEPENDENT |

Mixed Practice Math Boxes 8-2 are paired with Math Boxes 8-5.

▶ Home Link 8-2

Math Masters, p. 228

Homework Children practice finding different ways to divide a square into halves.

Math Journal 2, p. 162

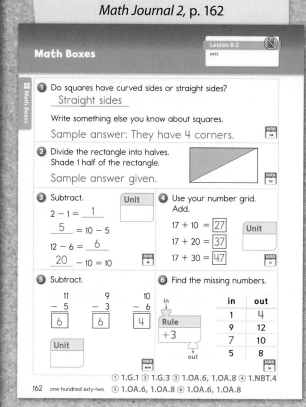

Math Journal 2, p. 162

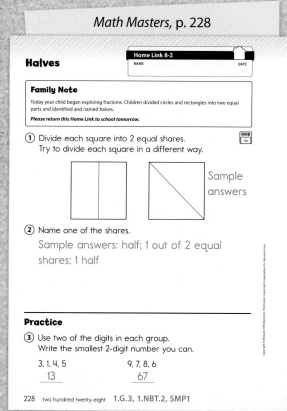

Math Masters, p. 228

Fourths

Overview Children divide shapes into four equal shares and name the shares. They compare the sizes of the shares to the number of equal shares.

▶ **Before You Begin**
For Part 2, display the Two Equal Shares Poster from Lesson 8-2. For the Math Message, display a small rectangle to represent a fruit bar and a blank tally chart. Title the tally chart "How Much Fruit Bar We Want," and label the choices: Whole Fruit Bar, Half of a Fruit Bar, and Don't Know. For the optional Readiness activity, prepare a paper circle, square, and rectangle for each child. For the optional Enrichment activity, you may wish to get a copy of *Picture Pie: A Circle Drawing Book* by Ed Emberly.

▶ **Vocabulary**
fourth • quarter

Common Core State Standards

Focus Clusters
• Represent and interpret data.
• Reason with shapes and their attributes.

1 Warm Up 15–20 min

	Materials	
Mental Math and Fluency Children solve subtraction facts.	slate	1.OA.6, 1.OA.8
Daily Routines Children complete daily routines.	See pages 2–37.	See pages xiv–xvii.

2 Focus 35–40 min

Math Message Children decide whether they would want half of a fruit bar or a whole fruit bar.		1.MD.4, 1.G.3
Discussing the Size of the Whole Children discuss the idea that the size of an equal share is dependent on the size of the whole.	Two Equal Shares Poster	1.G.3 SMP6
Partitioning into Halves and Fourths Children divide shapes into 2 and 4 equal shares and name the parts.	*Math Journal 2*, p. 163; leftover crackers from *Math Masters*, p. 225; slate; Two Equal Shares Poster; chart paper; scissors; glue or tape	1.G.3 SMP1, SMP6
✓ **Assessment Check-In** See page 700.	*Math Journal 2*, p. 163	1.G.3, SMP6
Comparing Sizes of Shares Children discuss the idea that increasing the number of shares results in smaller shares.	*Math Journal 2*, pp. 160 and 163; Two Equal Shares Poster; Four Equal Shares Poster	1.G.3 SMP6

CCSS 1.G.3 Spiral Snapshot

GMC Describe equal shares using fraction words.

8-2 Focus Practice	8-3 Focus Practice	8-4 Focus Practice	8-5 Focus	8-7 Practice	8-8 Warm Up Focus Practice	9-4 Focus Practice	9-11 Focus Practice

Spiral Tracker **Go Online** to see how mastery develops for all standards within the grade.

3 Practice 10–15 min

Playing *Time Match* **Game** Children practice telling time.	*Math Masters*, pp. G55–G56; p. G57 (optional)	1.MD.3 SMP2
Math Boxes 8-3 Children practice and maintain skills.	*Math Journal 2*, p. 164	See page 701.
Home Link 8-3 **Homework** Children divide squares into 4 equal parts.	*Math Masters*, p. 231	1.OA.6, 1.MD.4, 1.G.3 SMP1, SMP6

connectED.mcgraw-hill.com

Plan your lessons online with these tools.

 ePresentations   Student Learning Center   Facts Workshop Game   eToolkit   Professional Development   Home Connections   Spiral Tracker   Assessment and Reporting   English Learners Support   Differentiation Support

Differentiation Options RtI

Readiness
10–15 min

WHOLE CLASS
SMALL GROUP
PARTNER
INDEPENDENT

Folding Paper Pizzas

per child: 1 paper circle, square, and rectangle

To review dividing shapes into equal parts using a concrete model, have children fold paper shapes into two equal parts. Give each child a paper circle, square, and rectangle. Children imagine each shape is a pizza and work in small groups or partnerships to divide each shape into two equal parts. Children "top" their pizzas by drawing a different ingredient on each part; for example, a pizza might be half green pepper and half mushroom. Have children share their strategies for dividing the shapes into equal parts. Ask them to describe the shares using language from the Two Equal Shares Poster from Lesson 8-2. GMP6.3

Enrichment
15–20 min

WHOLE CLASS
SMALL GROUP
PARTNER
INDEPENDENT

Making Rectangle Designs

Activity Card 97;
Math Masters, p. 229;
scissors; crayons; glue or tape;
Picture Pie: A Circle Drawing Book

Literature Link To apply children's understanding of wholes, halves, and fourths, have them create and label new shapes or pictures using fractional pieces of a rectangle. GMP6.3 When children have finished, consider having them look at the pictures made from fractional parts of a circle in ***Picture Pie: A Circle Drawing Book*** by Ed Emberly (Little, Brown, 2006).

Extra Practice
15–20 min

WHOLE CLASS
SMALL GROUP
PARTNER
INDEPENDENT

Making an Equal Shares Book

Activity Card 98;
Math Masters, p. 230;
crayons or markers

To provide practice naming equal shares, have children make an equal shares book. GMP2.2

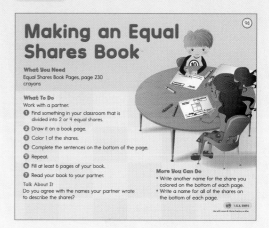

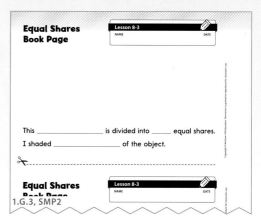

1.G.3, SMP2

English Language Learners Support

Beginning ELL Children may be familiar with the word *hole*. Introduce the homophone *whole*, and point out the spelling. Provide visual support for understanding the *whole* as all of something. Point to classroom objects, gesturing to indicate the *whole* object and then a *part* of the object. Follow up by showing children pictures of whole objects and equal shares of objects, such as apples and sandwiches. Use *yes* or *no* questions or Total Physical Response prompts with children to identify the part and the whole.

Go Online **ELL** English Learners Support

Standards and Goals for
Mathematical Practice

SMP1 **Make sense of problems and persevere in solving them.**
GMP1.5 Solve problems in more than one way.

SMP6 **Attend to precision.**
GMP6.1 Explain your mathematical thinking clearly and precisely.
GMP6.3 Use clear labels, units, and mathematical language.

Professional Development

The sample answers given in Lessons 8-2 and 8-3 reflect how children are likely to draw or check halves and quarters. Other valid answers have equal shares that are the same size, but are not congruent or drawn with straight lines. For example, the following rectangles are divided into 4 equal shares.

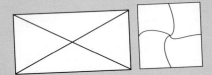

1 Warm Up 15–20 min | Go Online | ePresentations | eToolkit

▸ Mental Math and Fluency

Display the following subtraction equations. Have children show their answers on their slates.

- ●○○ $10 - 6 = \boxed{4}$; $8 - 4 = \boxed{4}$; $10 - 1 = \boxed{9}$
- ●●○ $10 - \boxed{5} = 5$; $14 - 7 = \boxed{7}$; $10 - \boxed{7} = 3$
- ●●● $3 = \boxed{6} - 3$; $12 - \boxed{6} = 6$; $\boxed{10} - 2 = 8$

▸ Daily Routines

Have children complete the Daily Routines. For detailed instructions, see pages 2–37. For specifics on CCSS coverage, see pages xiv–xvii.

2 Focus 35–40 min | Go Online | ePresentations | eToolkit

▸ Math Message

Look at the rectangle. It represents a fruit bar. If it were snack time, which would you want—a whole fruit bar or half of a fruit bar? Make a tally mark to show your vote.

▸ Discussing the Size of the Whole

| **WHOLE CLASS** | SMALL GROUP | PARTNER | INDEPENDENT |

Math Message Follow-Up Encourage children to explain their votes. Expect most children to choose a whole fruit bar, although some children may have valid reasons for choosing a half. Ask: *If you wanted a larger piece, would you want a whole or a half of a fruit bar? Why?* Sample answer: a whole fruit bar because if you split it in half, the parts are smaller During this discussion, have a child demonstrate dividing the fruit bar into two equal shares, and ask children how to name one part. **GMP6.3** Sample answers: half; 1 half; 1 out of 2 equal shares Refer to the Two Equal Shares Poster, if needed.

Next display a fruit bar three or four times the size of the first bar alongside it. Ask: *If you were hungry, would you want the whole small fruit bar or half of the large fruit bar?* Invite children to explain their ideas. Emphasize that the size of a half depends on the size of the whole. In this case, half of the large fruit bar is larger than the whole small fruit bar. Ask children to list some other items that they have divided in half to share. Point out that the sizes of the halves are dependent on the sizes of the whole items children share.

Tell children that today they will continue to talk about dividing things equally and naming the equal shares.

▶ Partitioning into Halves and Fourths

Math Journal 2, p. 163; Math Masters, p. 225

`WHOLE CLASS` `SMALL GROUP` `PARTNER` `INDEPENDENT`

Ask children to divide their slates into two equal shares and to write a correct name for one of the shares. Have them discuss with their partners how they know the parts they made are equal size. `GMP6.1`

Have children display their slates, and highlight some of the different divisions. Most children divide their slates vertically or horizontally, but look for other instances of equal divisions. Allow children who divided their slates unequally to make the shares equal.

Have children suggest names for the parts and whole of the slate. Refer to the Two Equal Shares Poster, and add any new names that arise.

- *What would you call one of the parts on your slate?* Sample answers: half; one half; 1 half; 1 out of 2 equal shares
- *How can you name both of the shares together?* Sample answers: two halves; 2 halves; 2 out of 2 equal shares, whole

Have children erase their slates and divide them into 4 equal shares. Ask them how they know the shares they made are equal. `GMP6.1` Have children display their slates, and highlight some of the different ways children divided the slates into 4 equal shares. Label a poster Four Equal Shares, and ask a child to draw an example on it. Display the Four Equal Shares Poster next to the Two Equal Shares Poster. Sample answers given.

- *How are the drawings on the two posters different?* One has shapes divided in 2 equal shares. One has shapes divided into 4 equal shares.
- *Look at your slates. How many equal shares are on your slates?* 4
- *What would you call one of the equal shares on your slate?* 1 out of 4 equal shares; 1 fourth; 1 quarter; one-fourth; one quarter
- *How can you name all the equal shares together?* 4 out of 4 parts; 4 fourths; 4 quarters; four-fourths; four quarters; the whole

Add names children generate for one equal part and all four equal parts to the Four Equal Shares Poster.

Academic Language Development Use the term *quarter* with children to describe one of four equal shares. **1.G.3** Though they have not been formally introduced to quarter hours or coins, they may be familiar with *quarter* from personal experiences. Help children understand how *quarter* in a fraction context (1 out of 4 equal parts) relates to a quarter coin (4 quarters in 1 dollar).

`Differentiate` **Adjusting the Activity**

Ask children to divide their slates into three equal parts. Compare children's solutions, and discuss whether the shares they made are equal or not.

`Go Online` Differentiation Support

NOTE During the lesson, you will be creating a Four Equal Shares Poster similar to the Two Equal Shares Poster your class created in Lesson 8-2.

Four Equal Shares

<u>Name of one share</u>

quarter

1 fourth

1 out of 4 equal shares

<u>Name of all shares</u>

whole

4 fourths

4 out of 4 equal shares

Completed Four Equal Shares Poster

Common Misconception

`Differentiate` Watch for children who name any equal share *half*, regardless of the number of equal parts that make up a whole. Encourage these children to count the number of equal parts and use terms that are more descriptive of the number of parts, such as *1 out of 4 equal shares*.

`Go Online` Differentiation Support

Math Journal 2, p. 163

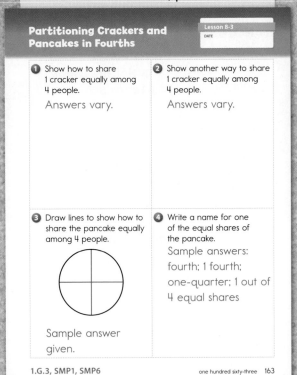

1. Show how to share 1 cracker equally among 4 people.

Answers vary.

2. Show another way to share 1 cracker equally among 4 people.

Answers vary.

3. Draw lines to show how to share the pancake equally among 4 people.

Sample answer given.

4. Write a name for one of the equal shares of the pancake.

Sample answers: fourth; 1 fourth; one-quarter; 1 out of 4 equal shares

1.G.3, SMP1, SMP6

one hundred sixty-three   163

Math Masters, p. G57

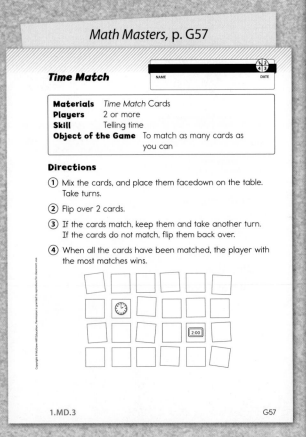

Time Match

NAME _____ DATE _____

Materials	Time Match Cards
Players	2 or more
Skill	Telling time
Object of the Game	To match as many cards as you can

Directions

1. Mix the cards, and place them facedown on the table. Take turns.

2. Flip over 2 cards.

3. If the cards match, keep them and take another turn. If the cards do not match, flip them back over.

4. When all the cards have been matched, the player with the most matches wins.

1.MD.3   G57

Distribute the extra crackers from Lesson 8-2, or have children cut out the crackers on *Math Masters,* page 225. Ask partners to discuss how they could share one cracker equally among four people. Have each child cut a cracker into four equal parts and attach the parts to journal page 163. Then have them record a different way to share a second cracker into four equal parts. **GMP1.5** Encourage children to discuss how they know the shares are equal as they then complete the rest of the journal page.

Ask several children to share how they named the equal shares of the pancake in Problem 4. On the poster, draw a pancake divided into 4 equal shares and add names for one equal share and all equal shares.

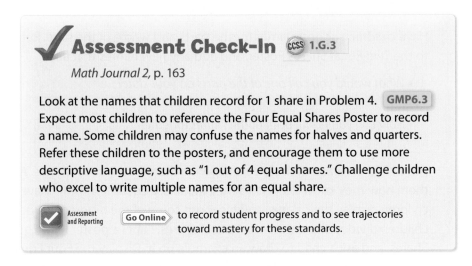

✓ **Assessment Check-In** CCSS **1.G.3**

Math Journal 2, p. 163

Look at the names that children record for 1 share in Problem 4. **GMP6.3** Expect most children to reference the Four Equal Shares Poster to record a name. Some children may confuse the names for halves and quarters. Refer these children to the posters, and encourage them to use more descriptive language, such as "1 out of 4 equal shares." Challenge children who excel to write multiple names for an equal share.

☑ Assessment and Reporting   Go Online to record student progress and to see trajectories toward mastery for these standards.

▶ Comparing Sizes of Shares

Math Journal 2, pp. 160 and 163

WHOLE CLASS   SMALL GROUP   **PARTNER**   INDEPENDENT

Pose this problem: *Imagine you are very hungry and have 1 cracker. Would you rather share the cracker with 1 friend (among 2 children) or with 3 friends (among 4 children)? Why?* Have partnerships discuss and share their thoughts with the class. **GMP6.1** Encourage children to refer to journal pages 160 and 163 or the class posters to compare the sizes of the shares of one cracker. Draw out the idea that the greater the number of equal shares of an object, the smaller the shares will be.

Extend the conversation by asking children to consider equal shares among 3 friends and 5 friends. Ask questions like: *If you were hungry, would you rather equally share a pancake among 2 friends or 5 friends?* 2 friends *Would the pancake pieces be smaller if you equally shared them among 3 friends or 4 friends?* 4 friends Refer to the two posters throughout this discussion. Emphasize that the whole must be the same when comparing the sizes of shares.

Summarize Ask children to explain how equally dividing a cracker into 4 shares is different from equally dividing a cracker into 2 shares. Sample answers: One-fourth is smaller than 1 half. They are named differently.

③ **Practice** 10–15 min

Go Online

ePresentations | eToolkit | Home Connections

▶ Playing *Time Match*

Math Masters, pp. G55–G56

| WHOLE CLASS | **SMALL GROUP** | **PARTNER** | INDEPENDENT |

Have children play *Time Match* to practice telling time to the hour using digital and analog clocks. For directions for playing the game, use *Math Masters*, page G57. For more information, see Lesson 7-11.

Observe
- Which children correctly read the times to the hour on the analog clock?
- Which children connect the two representations of the time? **GMP2.3**

Discuss
- *How do you know what time is shown on the clock?*
- *Which hand tells you the hour?*

▶ Math Boxes 8-3

Math Journal 2, p. 164

| WHOLE CLASS | SMALL GROUP | **PARTNER** | **INDEPENDENT** |

Mixed Practice Math Boxes 8-3 are paired with Math Boxes 8-1.

▶ Home Link 8-3

Math Masters, p. 231

Homework Children practice different ways of dividing a square into fourths.

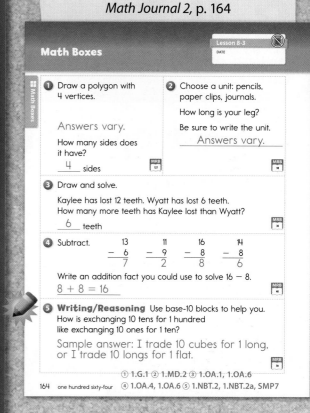

Math Journal 2, p. 164

Math Boxes

Lesson 8-3
DATE

① Draw a polygon with 4 vertices.

Answers vary.

How many sides does it have?

__4__ sides

② Choose a unit: pencils, paper clips, journals.

How long is your leg?

Be sure to write the unit.

___Answers vary.___

③ Draw and solve.

Kaylee has lost 12 teeth. Wyatt has lost 6 teeth. How many more teeth has Kaylee lost than Wyatt?

__6__ teeth

④ Subtract.

$$\begin{array}{cccc} 13 & 11 & 16 & 14 \\ -\ 6 & -\ 9 & -\ 8 & -\ 8 \\ \hline 7 & 2 & 8 & 6 \end{array}$$

Write an addition fact you could use to solve $16 - 8$.

$8 + 8 = 16$

⑤ **Writing/Reasoning** Use base-10 blocks to help you. How is exchanging 10 tens for 1 hundred like exchanging 10 ones for 1 ten?

Sample answer: I trade 10 cubes for 1 long, or I trade 10 longs for 1 flat.

164 one hundred sixty-four

① 1.G.1 ② 1.MD.2 ③ 1.OA.1, 1.OA.6
④ 1.OA.4, 1.OA.6 ⑤ 1.NBT.2, 1.NBT.2a, SMP7

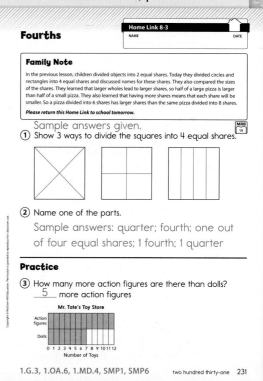

Math Masters, p. 231

Fourths

Home Link 8-3
NAME   DATE

Family Note

In the previous lesson, children divided objects into 2 equal shares. Today they divided circles and rectangles into 4 equal shares and discussed names for these shares. They also compared the sizes of the shares. They learned that larger wholes lead to larger shares, so half of a large pizza is larger than half of a small pizza. They also learned that having more shares means that each share will be smaller. So a pizza divided into 6 shares has larger shares than the same pizza divided into 8 shares.

Please return this Home Link to school tomorrow.

Sample answers given.

① Show 3 ways to divide the squares into 4 equal shares.

② Name one of the parts.

Sample answers: quarter; fourth; one out of four equal shares; 1 fourth; 1 quarter

Practice

③ How many more action figures are there than dolls?

__5__ more action figures

Mr. Tate's Toy Store

Action figures

Dolls

0 1 2 3 4 5 6 7 8 9 10 11 12
Number of Toys

1.G.3, 1.OA.6, 1.MD.4, SMP1, SMP6   two hundred thirty-one 231

Sharing Paper Squares

2-Day Lesson

Overview **Day 1:** Children use drawings to answer a question about sharing paper squares. **Day 2:** The class discusses some initial drawings and explanations, and children revise their work.

Day 1: Open Response

▶ **Before You Begin**
Cut out one paper triangle per partnership from *Math Masters,* page 232, or plan to have children cut them out. Have scissors available for children when they work on the Math Message. Display the Equal Shares Posters from Lessons 8-2 and 8-3. Solve the open response problem and consider ways children might draw a picture of the situation. If possible, schedule time to review children's work and plan for Day 2 of this lesson with your grade-level team.

▶ **Vocabulary**
mathematical model

CCSS **Common Core State Standards**

Focus Cluster
Reason with shapes and their attributes.

1 Warm Up 15–20 min

	Materials	
Mental Math and Fluency Children find 10 more and 10 less than numbers.		1.NBT.5
Daily Routines Children complete daily routines.	See pages 2–37.	See pages xiv–xvii.

2a Focus 45–55 min

Math Message Children divide a triangular piece of cheese into equal shares.	per partnership: 1 triangle cut from *Math Masters,* p. 232; scissors	1.G.3 SMP4
Modeling Equal Shares Children discuss how they divided the cheese into equal shares and what it means for shares to be equal.	*Math Journal 2,* p. 165; Standards for Mathematical Practice Poster	1.G.3 SMP4
Solving the Open Response Problem Children use a drawing to decide which is a larger share, a rectangle divided into 4 equal pieces or the same rectangle divided into 2 equal pieces.	*Math Masters,* p. 233; Equal Shares Posters	1.G.3 SMP1, SMP4

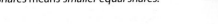

Getting Ready for Day 2 →

Review children's work and plan discussion for reengagement. *Math Masters,* p. TA3; children's work from Day 1

CCSS 1.G.3 **Spiral Snapshot**

GMC Understand that *more equal shares* means *smaller equal shares.*

| 8-3
Focus | 8-4
Focus
Practice | 8-7
Practice | 8-9
Practice | 9-4
Focus
Practice | 9-7
Practice | 9-9
Practice | 9-11
Focus
Practice |

 Spiral Tracker **Go Online** to see how mastery develops for all standards within the grade.

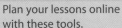
 ePresentations
 Student Learning Center
 Facts Workshop Game
 eToolkit
 Professional Development
 Home Connections
 Spiral Tracker
 Assessment and Reporting
 English Learners Support
 Differentiation Support

1 Warm Up 15–20 min

Go Online
ePresentations eToolkit

▶ Mental Math and Fluency

Have children solve.

● ○ ○ Find 10 more than 70. 80 Find 10 less than 70. 60

● ● ○ Find 10 less than 15. 5 Find 10 more than 15. 25

● ● ● Find 10 more than 28. 38 Find 10 less than 28. 18

▶ Daily Routines

Have children complete the Daily Routines. For detailed instructions, see pages 2–37. For specifics on CCSS coverage, see pages xiv–xvii.

2a Focus 45–55 min

Go Online
ePresentations eToolkit

▶ Math Message

Math Masters, p. 232

Pretend a triangle is a piece of cheese. GMP4.1 *Work with a partner to divide the cheese into two equal shares. Talk about how you know you each have an equal share.* GMP4.2

English Language Learners English language learners may need to review different ways to refer to halves and quarters, for example, half, one-quarter, one out of four. Use total physical response activities with models or pictures, such as: *Pick up one out of four pieces. Point to one-half of the rectangle.*

CCSS
Standards and Goals for
Mathematical Practice

SMP1 Make sense of problems and persevere in solving them.
  GMP1.1 Make sense of your problem.

SMP4 Model with mathematics.
  GMP4.1 Model real-world situations using graphs, drawings, tables, symbols, numbers, diagrams, and other representations.

  GMP4.2 Use mathematical models to solve problems and answer questions.

Professional Development

The focus for this lesson is GMP4.2. A *mathematical model* is a representation of a real-world situation using mathematical objects, such as numbers, shapes, and graphs. For example, the number sentences children write for number stories are mathematical models because they use numbers to represent what is happening in real-world stories. In this lesson, children use a paper triangle as a mathematical model for a piece of cheese and drawings as mathematical models of paper squares.

Go Online for information about SMP4 in the *Implementation Guide*.

Math Masters, p. 232

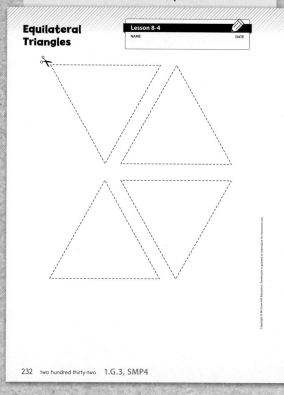

Equilateral Triangles   Lesson 8-4
NAME   DATE

232 two hundred thirty-two   1.G.3, SMP4

Sharing Cheese

Lesson 8-4

DATE

① Show how you shared your piece of cheese into two equal shares. **Sample answer:**

② Is this piece of cheese divided into two equal shares? __No__

Explain how you know.
Sample answers: The top part is little and the bottom part is big. If I folded them on the line, they wouldn't fit exactly on top of each other.

1.G.3, SMP4   one hundred sixty-five   165

▶ Modeling Equal Shares

Math Journal 2, p. 165

| **WHOLE CLASS** | SMALL GROUP | PARTNER | INDEPENDENT |

Math Message Follow-Up After children have shared the piece of cheese and talked to their partners about how they know the shares are equal, ask them to share their ideas. Ask:

- *How did you show your shares?* **GMP4.1** Sample answers: We drew a line down the middle. We cut the piece of cheese in half. We folded the triangle.
- *What are names for one share?* Sample answers: half, 1 half, 1 out of 2 parts
- *How do you know you have equal shares or halves?* **GMP4.2** Sample answers: We cut the triangle and put one share on top of the other; if they match, then they are the same size. We drew a line down the middle. If you fold it on the line, you can see that the two parts are the same size.

> **Differentiate** | **Adjusting the Activity**
>
> For children who have trouble explaining their thinking, ask them to show you how they know the shares are equal. Remind them that they can draw, cut, or fold the triangle to help them with their explanations.

Have children complete journal page 165 by drawing a line to show their equal shares in Problem 1 and explaining why the shares in Problem 2 are not equal. Ask: *How can you tell the shares in Problem 2 are not equal?* **GMP4.2** Sample answers: There is more room in the bottom part than in the top part. If you fold on the line, you can see that the shares are not the same size.

Tell children that the paper triangle and the drawings on the journal page are mathematical models. Explain that a **mathematical model** is something mathematicians use to represent, or show, something in the real world that they do not actually have in front of them. Ask: *Did we have pieces of cheese when we solved the Math Message problem?* No *What did we use as mathematical models?* Paper triangles, drawings of triangles

Refer to **GMP4.2** on the Standards for Mathematical Practice Poster. Point out that children used a mathematical model to solve a problem. Ask: *How did our mathematical models help us figure out if we had equal shares?* Sample answer: We could fold the paper to check that our shares matched up.

Tell children that they will use a drawing to model another problem situation and use their model to help solve a problem. **GMP4.2**

Academic Language Development Two important terms used in this lesson—*model* and *share*—are words that are used routinely as nouns and verbs. This might confuse some children. Prior to the lesson and as needed, highlight the connection between the noun and verb usages of each of these terms. Use pictures, diagrams, and role-plays to connect different everyday uses of the terms with the ways they will be used in this lesson.

▶ Solving the Open Response Problem

Math Masters, p. 233

| WHOLE CLASS | SMALL GROUP | **PARTNER** | INDEPENDENT |

Distribute *Math Masters*, page 233 to each child. Tell children the following story:

> *In an art class, children sit at tables for a project. Two girls sit at one table and four boys sit at another table. Each table gets one large square of construction paper. The two girls share their paper square evenly, and the four boys share their paper square evenly. Who gets a larger share of paper, one girl or one boy?*

Explain to children that their job is to draw a picture of the situation and use their picture to figure out who will get a larger share of paper, one boy or one girl. They should be ready to explain their answers. Encourage children to refer to the Equal Shares Posters from Lessons 8-2 and 8-3. Have children work with a partner to solve the problem, but encourage them to make their drawings and write their answers individually.

Circulate and reread the problem to children as needed. Remind children that they need to show how they can share the paper squares evenly, tell who will get a larger share of paper, and explain their reasoning. As you observe children, ask them to explain their pictures. Ask: *Where is the girls' paper in your drawing? Where is the boys' paper?* GMP4.1 *How do you know who will get a larger share?* GMP4.2 *What does it mean to have a larger share of paper?* GMP1.1

Differentiate **Adjusting the Activity**

To make this activity more concrete for children, give them two squares of paper and ask them to think about how they can share the paper equally by drawing lines. Have children fold one square into 2 equal shares and the other into 4 equal shares and compare one share for a girl and one share for a boy. Then have them draw a picture of their paper squares. Have children who struggle to explain their thinking in writing dictate their answers to you or another child to write.

Summarize Ask: *How did making a drawing help you find out who had a larger share of paper?* GMP4.2 Answers vary.

Collect children's work so that you can evaluate it and prepare for Day 2.

Math Masters, p. 233

Sharing Paper Squares

Lesson 8-4

NAME   DATE

Two girls share one paper square.

Four boys share the same-size square of paper.

Tell who will get a larger share of paper, one girl or one boy.

Make a drawing of the problem. Explain your answer.

Answers vary. See sample children's work on page 711 of the Teacher's Lesson Guide.

1.G.3, SMP1, SMP4   two hundred thirty-three   233

Lesson 8-4   **705**

Math Masters, p. TA3

Planning a Follow-Up Discussion

Review children's work. Use the Reengagement Planning Form (*Math Masters,* page TA3) and the rubric on page 708 to plan ways to help children meet expectations on both the content and the practice standards. Look for common misconceptions about the relationship between the size of the shares and the number of shares, as well as interesting ways children modeled the problem in their drawings.

Reengagement Planning Form

Common Core State Standard (CCSS):

1.G.3 Partition circles and rectangles into two and four equal shares. . . . Understand for these examples that decomposing into more equal shares creates smaller shares.

Goal for Mathematical Practice (GMP): *GMP4.2 Use mathematical models to solve problems and answer questions.*

Differentiate   **Common Misconception**

Children sometimes do not understand that decomposing shapes into a larger number of equal shares creates smaller shares. In this problem, for example, children may say that the boys get more because there are more boys (4 pieces is more than 2 pieces). As you review children's work, plan to address this misconception through the reengagement discussion. Children may more easily make sense of the problem if you relate it to sharing food: *Would you get more pizza if you share a pizza between 2 people or if you share the same size pizza among 4 people?* Sample answer: You get more pizza when only 2 people share.

Organize the discussion of children's work in one of the ways below or in another way you choose. If children's work is unclear or if you prefer to show work anonymously, rewrite the work for display.

Go Online for sample children's work that you can use in your discussion.

1. Compare two different drawings of the situation. One drawing might show the two pieces of paper, as in Child A's work, and the other might show the number of children in the problem, as in Child B's work. Ask:
   - *How are these drawings different?* **GMP4.1** Sample answer: One drawing shows the paper and the shares, but doesn't show all the boys and girls. The other one shows the boys and girls.
   - *Besides drawing a picture, what else were we supposed to do?* **GMP1.1** Sample answer: Tell if a boy or a girl gets a larger share of paper.

- *Which picture do you think will be more helpful when you're trying to figure out who got a larger share of paper?* GMP4.2 Sample answers: The drawing of the paper shows how big the pieces are. The drawing of the children shows how many children there are, but I'm not sure how much paper they get.
- *How could these children make their drawings and answers better?* GMP4.1, GMP4.2 Sample answer: They could make sure the drawings show who got larger shares of paper and make sure they say who got more.

2. Show work that claims that one girl will get a larger share of paper, as in Child A's response, and work that claims that one boy will get a larger share of paper, as in Child C's response. Ask:
- *What is the same about these responses? What is different?* GMP4.1, GMP4.2 Sample answer: Both of these children drew pictures of the paper. The pictures look different because one drew whole paper squares and one showed the paper already cut apart. One child said a girl would get more paper because a girl gets a whole half. The other child said a boy would get more paper because the boys have more.
- *What do you think each child meant?* Sample answer: The first child meant that when you cut a paper in two pieces, that's a half; but when you cut it in more pieces, that's not a whole half. The other child meant that there are 4 boys but only 2 girls, so there are more boys.
- *Who do you agree with?* Sample answers: I agree the boys got more because there are more pieces of paper. I agree the girls got more because the pieces are bigger.
- *What does it mean for one child to have a larger share of paper?* GMP1.1 Sample answer: It means the pieces are bigger.
- *Who has a larger share of paper, a boy or a girl? How do you know?* GMP4.1 Sample answers: I can see that the girls' pieces are bigger in the picture. It's easier to see when the paper is still together, not cut apart. The boys have to share the same-size paper with 4 people instead of 2, so one boy will get a smaller piece than one girl.

3. Display two pieces of work that show the paper divided in different ways. For example, one child might have used a diagonal line to divide a square into 2 equal shares and another child might have used a vertical line. Ask:
- *Do these drawings both show equal shares? How can you tell?* GMP4.2 Sample answer: Yes. The pieces are the same size in each drawing.

4. Display a piece of work that shows the paper divided into unequal shares. Ask:
- *What do you notice about this drawing?* GMP4.2 Sample answers: The pieces aren't equal. They are not equal shares.
- *How could this child make the drawing better?* GMP4.1 Sample answer: This child could make sure all the shares are the same size.

Planning for Revisions

Have copies of *Math Masters,* page 233 or extra paper available for children to use in revisions. You might want to ask children to use colored pencils so you can see what they revised.

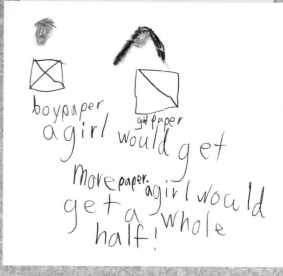

Sample child's work, Child A

Sample child's work, Child B

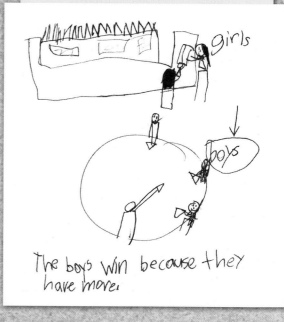

Sample child's work, Child C

Sharing Paper Squares

Overview **Day 2:** The class discusses some initial drawings and explanations, and children revise their work.

Day 2: Reengagement

▶ **Before You Begin**
Have extra copies available of *Math Masters,* page 233 for children to revise their work.

Focus Cluster
Reason with shapes and their attributes.

2b Focus   50–55 min

	Materials	
Setting Expectations Children review the open response problem and discuss what a good mathematical model might look like. They review how to respectfully discuss others' work.	*My Reference Book,* pp. 12–13; Guidelines for Discussions Poster	1.G.3 SMP4
Reengaging in the Problem Children discuss various drawings and how drawings help answer the question.	selected samples of children's work	1.G.3 SMP1, SMP4
Revising Work Children revise their work from Day 1.	*Math Masters,* p. 233 (optional); children's work from Day 1; colored pencils (optional)	1.G.3 SMP4
✓ **Assessment Check-In** See page 710 and the rubric below.		1.G.3 SMP4

Goal for Mathematical Practice **GMP4.2** Use mathematical models to solve problems and answer questions.	Not Meeting Expectations	Partially Meeting Expectations	Meeting Expectations	Exceeding Expectations
	Does not make a drawing of the two divided paper squares that clearly shows the relative size of the shares, and does not answer the question.	Makes a drawing of the two divided paper squares that clearly shows the relative size of the shares, but does not answer the question. **OR** Answers the question but does not make a drawing that supports the answer.	Makes a drawing of the two divided paper squares that clearly shows the relative size of the shares and provides an answer to the question that is supported by the drawing.	Meets expectations and explains that a larger number of children will lead to smaller shares (or a smaller number of children will lead to larger shares).

3 Practice   10–15 min

Math Boxes 8-4 Children practice and maintain skills.	*Math Journal 2,* p. 166	See page 710.
Home Link 8-4 **Homework** Children divide circles into equal shares and name the shares.	*Math Masters,* p. 234	1.OA.6, 1.OA.8, 1.G.3 SMP4

2b Focus
50–55 min | Go Online | ePresentations | eToolkit

NOTE These Day 2 activities will ideally take place within a few days of Day 1. Prior to beginning Day 2, see Planning a Follow-Up Discussion from Day 1.

▶ Setting Expectations

My Reference Book, pp. 12–13

WHOLE CLASS | **SMALL GROUP** | PARTNER | INDEPENDENT

Remind children that on Day 1, they talked about how to use mathematical models. Read *My Reference Book,* pages 12–13 as a class. Ask: *What mathematical models did Rosa use?* A tally chart and a number model *What else can be a mathematical model?* Sample answers: a drawing, a paper triangle

Review the open response problem from Day 1. Remind children that their job was to draw a picture of the art class situation and use their picture to decide who got a larger share of paper, one boy or one girl. Ask: *What kind of mathematical model did you use?* GMP4.1 A drawing *What would a good response include?* GMP4.1, GMP4.2 Sample answers: It would have a picture of how the boys and girls shared the paper. It would say who got more paper and how I know.

After this brief discussion, tell children that they are going to look at other children's work and see if they understand their pictures and agree with their answers. Remind children that it is all right to agree or disagree with others as long as they are respectful. Refer to your discussion guidelines and encourage children to use these sentence frames:

- I agree because _____.
- I disagree because _____.

Standards for Mathematical Practice

Make Models to Solve Problems

Rosa and Li want to know how many pets live in their neighborhood.

Rosa gathers data with a tally chart. She makes a tally mark for each pet in the neighborhood.

Pets	Tallies
Dogs	~~HHT~~
Cats	~~HHT~~ //
Rabbits	//

MRB 12 twelve

My Reference Book, p. 12

My Reference Book, p. 13

Standards for Mathematical Practice

I count the tally marks for each type of pet. I use those numbers to write an addition model.

Rosa

5 dogs + 7 cats + 2 rabbits = ? pets

5 + 7 + 2 = 14 pets

Use mathematical models to solve problems and answer questions.
Rosa uses tallies to keep track of the different pets. She uses a number model to find how many pets there are in all.

thirteen MRB 13

▶ Reengaging in the Problem

| **WHOLE CLASS** | SMALL GROUP | **PARTNER** | INDEPENDENT |

Children reengage in the problem by analyzing and critiquing other children's work in pairs and in a whole-group discussion. Have children discuss with partners before sharing with the whole group. Guide this discussion based on the decisions you made in Getting Ready for Day 2. **GMP1.1, GMP4.1, GMP4.2**

▶ Revising Work

| WHOLE CLASS | SMALL GROUP | **PARTNER** | **INDEPENDENT** |

Pass back children's work from Day 1. Before children revise anything, ask them to examine their drawings and explanations and decide how to improve them. Ask the following questions one at a time. Have partners discuss their responses and then give a thumbs-up or thumbs-down based on their own work.

- *Did you draw a picture of the problem?* **GMP4.1**
- *Can you tell from your picture who got a larger share of paper?* **GMP4.2**
- *Did you clearly say who you think got a larger share and why?* **GMP4.2**

Tell children that they now have a chance to revise their work. Those who drew clear pictures on Day 1 can add to their explanations. Tell children to add to their earlier work using colored pencils or to use another sheet of paper, instead of erasing their original work.

Summarize Ask: *How can drawing a picture help you solve problems?* **GMP4.2** Answers vary.

✓ **Assessment Check-In** CCSS 1.G.3

Collect and review children's revised work. Expect children to improve their drawings and explanations based on the class discussion. For the content standard, expect most children to correctly divide the squares of paper into equal shares and state that one girl will get a larger share of paper (or one boy will get a smaller share of paper). You can use the rubric on page 708 to evaluate children's revised work for **GMP4.2**.

 Assessment and Reporting ⟨Go Online⟩ to record student progress and to see trajectories toward mastery for these standards.

⟨Go Online⟩ for optional generic rubrics in the *Assessment Handbook* that can be used to assess any additional GMPs addressed in the lesson.

Math Journal 2, p. 166

Math Boxes

Lesson 8-4
DATE

① Is this a triangle? Explain why or why not.

Sample answer: No, because it has too many sides.

② What time does the clock show?

Choose the best answer.
○ 12:15   ○ 12:03
● 3:00   ○ 3:12

③ Circle the ones digit in each number.
9⃝0   9⃝

④ Write the fact family.

Total
9

Part	Part
9	0

$9 + 0 = 9$
$0 + 9 = 9$
$9 - 0 = 9$
$9 - 9 = 0$

⑤ **Writing/Reasoning**
How can knowing 7 + 7 help you solve 8 + 6?

Sample answer: 8 is 1 more than 7 and 6 is 1 less than 7. 1 more and 1 less even out, so 8 + 6 = 7 + 7.

166 one hundred sixty-six

① 1.G.1 ② 1.MD.3 ③ 1.NBT.2
④ 1.OA.3, 1.OA.6 ⑤ 1.OA.6

Sample Children's Work—Evaluated

See the sample in the margin. This work meets expectations for the content standard because the child divided the two squares into equal shares and stated that each girl will get a larger share (expressed as the girls having "more"). With revision, the work meets expectations for the mathematical practice because the drawing shows both paper squares and clearly shows that the girls' shares are larger. The answer is supported by the drawing.

Go Online ▷ for other samples of evaluated children's work.

Sample child's work, "Meeting Expectations"

3 Practice   10–15 min   Go Online

ePresentations   eToolkit   Home Connections

▶ Math Boxes 8-4 ✎

Math Journal 2, p. 166

| WHOLE CLASS | SMALL GROUP | **PARTNER** | **INDEPENDENT** |

Mixed Practice   Math Boxes 8-4 are paired with Math Boxes 8-6.

▶ Home Link 8-4

Math Masters, p. 234

Homework   Children divide circles into equal shares and name the shares.

Math Masters, p. 234

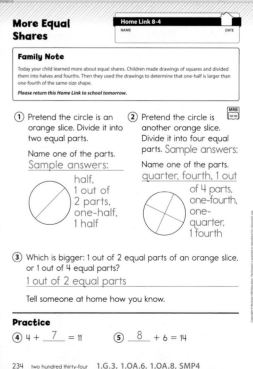

More Equal Shares   Home Link 8-4

NAME   DATE

Family Note

Today your child learned more about equal shares. Children made drawings of squares and divided them into halves and fourths. Then they used the drawings to determine that one-half is larger than one-fourth of the same-size shape.

Please return this Home Link to school tomorrow.

① Pretend the circle is an orange slice. Divide it into two equal parts.

Name one of the parts.
Sample answers:
half,
1 out of
2 parts,
one-half,
1 half

② Pretend the circle is another orange slice. Divide it into four equal parts. Sample answers:

Name one of the parts.
quarter, fourth, 1 out
of 4 parts,
one-fourth,
one-quarter,
1 fourth

③ Which is bigger: 1 out of 2 equal parts of an orange slice, or 1 out of 4 equal parts?
1 out of 2 equal parts

Tell someone at home how you know.

Practice

④ 4 + _7_ = 11   ⑤ _8_ + 6 = 14

234   two hundred thirty-four   1.G.3, 1.OA.6, 1.OA.8, SMP4

Combining 2-Dimensional Shapes

Overview Children combine 2-dimensional shapes to create composite shapes. They compose new shapes from the composite shapes.

▶ **Before You Begin**
For the Math Message, provide each child with one set (one row) of circles and circle pieces from *Math Masters,* page 235. For the first Part 2 activity, have children cut out triangles from *Math Masters,* page 236 before the lesson. (If possible, use colored paper.)

▶ **Vocabulary**
composite

Common Core State Standards

Focus Cluster
Reason with shapes and their attributes.

① Warm Up 15–20 min	**Materials**	
Mental Math and Fluency Children mentally find 10 more or 10 less than a number.	slate	**1.NBT.5**
Daily Routines Children complete daily routines.	See pages 2–37.	See pages xiv–xvii.

② Focus 35–40 min		
Math Message Children make a shape using whole and fractions of circles.	*Math Masters,* p. 235; scissors; plastic bag or envelope	**1.G.2, 1.G.3**
Composing New Shapes from Circles and Triangles Children use triangles, circles, and fractions of circles to create composite shapes.	*Math Journal 2,* pp. 167–168; *Math Masters,* pp. 235–236; tape or glue	**1.G.1, 1.G.2, 1.G.3** SMP1
Composing New Shapes from Composite Shapes Children use composite shapes to create new shapes.	*Math Masters,* pp. 235–236; blank paper; tape; glue	**1.G.2** SMP6
✓ **Assessment Check-In** See page 716.		**1.G.2**
Introducing *Make My Design* **Game** Children create and describe composite shapes.	*Math Masters,* p. G59 (optional); pattern blocks; folder	**1.G.1, 1.G.2** SMP3, SMP6

CCSS 1.G.2 Spiral Snapshot

GMC Build composite shapes.

| 4-5 Focus Practice | 6-3 Practice | 8-5 Focus Practice | 8-6 Focus Practice | 8-7 Focus | 8-10 Practice | 9-5 Practice | 9-10 Focus |

Spiral Tracker   Go Online   to see how mastery develops for all standards within the grade.

③ Practice 10–15 min		
Math Boxes 8-5 Children practice and maintain skills.	*Math Journal 2,* p. 169	See page 717.
Home Link 8-5 **Homework** Children use pattern blocks to fill a triangle.	*Math Masters,* p. 239	**1.G.1, 1.G.2** SMP1

connectED.mheducation.com

Plan your lessons online with these tools.

ePresentations   Student Learning Center   Facts Workshop Game   eToolkit   Professional Development   Home Connections   Spiral Tracker   Assessment and Reporting   English Learners Support   Differentiation Support

Differentiation Options RtI

 CCSS 1.G.1, SMP7

Readiness   10–15 min

Drawing Shapes with Defining Attributes

WHOLE CLASS
SMALL GROUP
PARTNER
INDEPENDENT

Pattern-Block Template

To support children's ability to describe shapes, provide them with shape descriptions and have them use their Pattern-Block Templates to draw shapes to match: GMP7.2

- *a shape with 6 sides* Large or small hexagon
- *3 different shapes with exactly 4 vertices* Any 3 of the following: large or small rhombus, trapezoid, square, rectangle, parallelogram
- *a shape with a corner that looks like the corner of your math journal; this is called a square corner* Square or rectangle
- *a shape with no square corners* Any shape except a square or rectangle
- *a shape with no straight sides* Any circle
- *a shape with 3 sides and 3 corners* Large or small triangle

CCSS 1.G.2, SMP1

Enrichment   15–20 min

Shape Challenges with Triangles

WHOLE CLASS
SMALL GROUP
PARTNER
INDEPENDENT

Activity Card 99;
paper triangles cut from
Math Masters, p. 236; crayons

To further explore making composite shapes, children challenge their partners to solve Shape Challenges they create using the triangles cut out for Part 2 of the lesson. GMP1.3

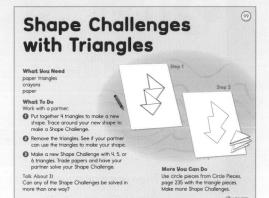

CCSS 1.G.2, SMP6

Extra Practice   10–15 min

Composing New Shapes from Pattern Blocks

WHOLE CLASS
SMALL GROUP
PARTNER
INDEPENDENT

Math Masters, pp. 237–238;
3 each of pattern blocks:
triangle, square, trapezoid,
fat rhombus, hexagon

Have children practice making composite shapes by tracing around pattern blocks to complete *Math Masters,* pages 237 and 238. Encourage them to use geometry terms (*polygon, trapezoid, rhombus, side, vertex,* and so on) as they discuss their new shapes. GMP6.3

Challenge children to use different pattern blocks to create the shapes they made for Problems 3 and 4.

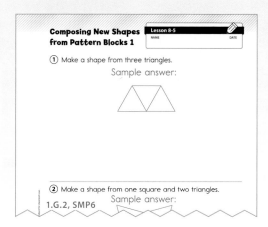

English Language Learners Support

Beginning ELL Before you formally introduce the game *Make My Design,* think aloud to model making and describing a design using a few pattern blocks at a time. Then provide children with practice describing their own designs using only three pattern blocks and sitting face-to-face without the divider folder between them. Use verbal clues and demonstration questions, if necessary. For example: *Is your triangle above the square? Is your triangle below the square?* Once children are comfortable, have them practice using the divider folders and adding more blocks in their designs.

Go Online   ELL English Learners Support

SMP1 **Make sense of problems and persevere in solving them.**
    GMP1.3 Keep trying when your problem is hard.

SMP3 **Construct viable arguments and critique the reasoning of others.**
    GMP3.2 Make sense of others' mathematical thinking.

SMP6 **Attend to precision.**
    GMP6.3 Use clear units, labels, and mathematical language.

1 Warm Up · 15–20 min · Go Online · ePresentations · eToolkit

▶ Mental Math and Fluency

Have children solve and record their answers on their slates. Briefly discuss how children found their answers, highlighting mental strategies.

● ○ ○ 10 more than 40. 50 10 less than 40. 30

● ● ○ 10 less than 53. 43 10 more than 53. 63

● ● ● 10 more than 90. 100 10 less than 10. 0

▶ Daily Routines

Have children complete the Daily Routines. For detailed instructions, see pages 2–37. For specifics on CCSS coverage, see pages xiv–xvii.

2 Focus · 35–40 min · Go Online · ePresentations · eToolkit

▶ Math Message

Math Masters, p. 235

Carefully cut out three circles.
Then cut along all the dotted lines.
Record some names for the pieces.
Then arrange your circle pieces on paper to make a new shape.
Trace your new shape.

▶ Composing New Shapes from Circles and Triangles

Math Journal 2, pp. 167–168; *Math Masters*, pp. 235–236

| WHOLE CLASS | SMALL GROUP | PARTNER | INDEPENDENT |

Math Message Follow-Up Hold up a set of circles cut from *Math Masters*, page 235, and ask children to suggest names for the shapes that represent partial circles. If no one suggests these terms, introduce *half circle* and *quarter circle*. Connect this vocabulary to previous lessons by emphasizing that the half circle is *one-half* or *1 out of 2 equal shares* of the whole circle and that the quarter circle is *one-fourth* or *1 out of 4 equal shares* of the whole circle. Have children show the shapes they traced for the Math Message. Point out that they created many different shapes with the same circle pieces.

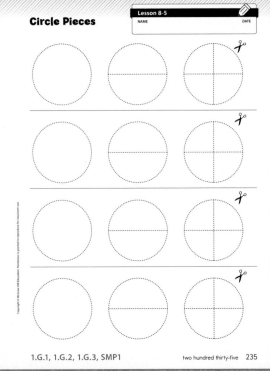

Math Masters, p. 235

1.G.1, 1.G.2, 1.G.3, SMP1   two hundred thirty-five 235

Pose the following tasks to challenge children to make specific shapes with their half-circle and quarter-circle pieces:

- *Use exactly 4 pieces to make a whole circle.*
  Children arrange 4 quarter circles to make 1 whole circle.
- *Use exactly 2 pieces to make a whole circle.*
  Children arrange 2 half circles to make 1 whole circle.
- *Use exactly 3 pieces to make a whole circle.*
  Children arrange 1 half circle and 2 quarter circles to make 1 whole circle.
- *Use exactly 2 pieces to make a half circle.*
  Children arrange 2 quarter circles to make a half circle.

Have children set the circle pieces aside. Tell them that today they will have many opportunities to put shapes together to make new shapes.

Have each child take out four paper triangles cut from *Math Masters,* page 236. Allow children time to explore and arrange the triangles any way they choose.

Ask children to identify the shapes on journal pages 167 and 168. Sample answers: rectangle; triangle; trapezoid; square or rhombus **Challenge** children to make each of the shapes using exactly four triangles. When they are sure that they have solved each challenge and that all four pieces fit, they should tape or glue the four triangles in place. Point out that the activity is challenging and that children should keep trying even if they do not solve the challenges quickly. GMP1.3

Differentiate **Adjusting the Activity**

Scaffold composing triangles to make new shapes by pairing children who struggle with partners who are able to make the shapes. Their partners can display completed new shapes as models, if needed, for the other children to replicate.

Go Online ▷ | Differentiation Support

Allow children to work for 5–10 minutes. Choose a few children's shapes to display. Do not expect children to solve all of the Shape Challenges during this lesson. You may wish to encourage children to continue working on their shapes throughout the week. GMP1.3

Math Journal 2, p. 167

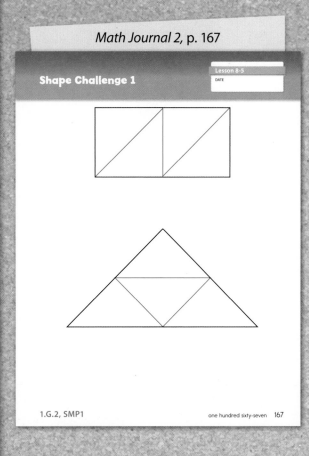

Shape Challenge 1 Lesson 8-5 DATE

1.G.2, SMP1   one hundred sixty-seven   167

Math Journal 2, p. 168

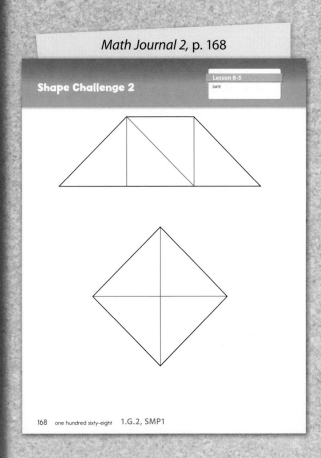

Shape Challenge 2 Lesson 8-5 DATE

168   one hundred sixty-eight   1.G.2, SMP1

triangle and half-circle

triangle and quarter-circle

half-circle and quarter-circle

Sample composite shapes

Math Masters, p. G59

Make My Design

	NAME	DATE

Materials	Pattern blocks, folder
Players	2
Skill	Create composite shapes
Object of the Game	To create the same shape as your partner

Directions

① Player 1 chooses 6 blocks.
   Player 2 chooses the same 6 blocks.

② Players sit face-to-face with a folder between them. They stand the folder to hide the designs.

③ Player 1 creates a shape with the blocks.

④ Using only words, Player 1 tells Player 2 how to "Make My Design." Player 2 can ask questions.

⑤ Players remove the folder and look at the two designs. Players discuss how closely the designs match.

⑥ Players trade roles and play again.

1.G.1, 1.G.2   G59

► # Composing New Shapes from Composite Shapes

Math Masters, pp. 235–236

WHOLE CLASS | SMALL GROUP | PARTNER | **INDEPENDENT**

Have each child gather 1 triangle, 1 half-circle, and 1 quarter-circle. Tell children to put together *any two* of these pieces to create a new shape. They should use a piece of tape to secure the two pieces side-by-side so they become one unit. (*See margin.*)

Have children display their new shapes. Explain to children that when they put shapes together to make new shapes, they are making **composite** shapes. You may also wish to use the verb form of *composite,* which is *compose,* with children. Provide paper, and have them trace their new shape multiple times to create another larger shape.

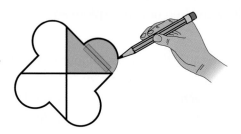

Making a new composite shape
from another composite shape

Have children share how they used two shapes to make new shapes and then used copies of their new shapes to create other new shapes. As they discuss their shapes, encourage them to describe the positions of the shapes and to use correct shape names.   **GMP6.3**   For example, "My first shape was a triangle with a vertex pointing up. Then I flipped it so the vertex pointed down. I turned my new shape on its other side and then traced around it again." You may wish to have children color their new shapes and display them in the classroom.

✓ **Assessment Check-In**   **CCSS** 1.G.2

Ask children to use two or more triangles to create a square or a rectangle. Observe as children work. Expect most children to create a composite shape. Provide hints to children who struggle to make a specific shape. For example, ask: *What happens if you turn the triangles before you put them together?* To challenge children, ask them to use more than 4 triangles to make a square or a rectangle.

 Assessment and Reporting   **Go Online** to record student progress and to see trajectories toward mastery for these standards.

▶ Introducing *Make My Design*

| WHOLE CLASS | SMALL GROUP | **PARTNER** | INDEPENDENT |

In this game, children explain how to make composite geometric shapes. Their partners interpret these descriptions and then create the same shapes. GMP3.2 You may wish to have children refer to *Math Masters*, page G59 as you demonstrate how to play the game.

Observe

- Do children describe the composite shape before they give details about individual blocks?
- Do children use shape vocabulary such as *side* and *vertex* to describe their designs? GMP6.3

Discuss

- *What was the most difficult part about describing your design?*
- *What kinds of questions did you ask to help you make the design your partner described?*

| Differentiate | **Game Modifications** | Go Online | Differentiation Support |

Summarize Remind children that they made new shapes from circle pieces, triangles, and pattern blocks. They also combined their new shapes to make other different shapes. Have children tell their partners which activity they liked best and why.

③ Practice 10–15 min

Go Online | ePresentations | eToolkit | Home Connections

▶ Math Boxes 8-5

Math Journal 2, p. 169

| WHOLE CLASS | SMALL GROUP | **PARTNER** | **INDEPENDENT** |

Mixed Practice Math Boxes 8-5 are paired with Math Boxes 8-2.

▶ Home Link 8-5

Math Masters, p. 239

Homework Children practice making composite shapes.

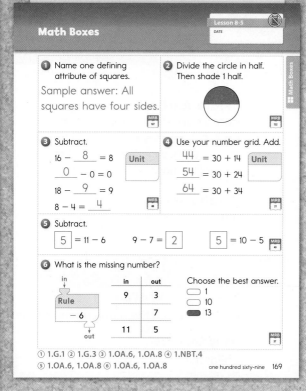

Math Journal 2, p. 169

Math Boxes Lesson 8-5 DATE

① Name one defining attribute of squares.
Sample answer: All squares have four sides.

② Divide the circle in half. Then shade 1 half.

③ Subtract.
$16 - \underline{8} = 8$ Unit
$\underline{0} - 0 = 0$
$18 - \underline{9} = 9$
$8 - 4 = \underline{4}$

④ Use your number grid. Add.
$\underline{44} = 30 + 14$ Unit
$\underline{54} = 30 + 24$
$\underline{64} = 30 + 34$

⑤ Subtract.
$\boxed{5} = 11 - 6$ $9 - 7 = \boxed{2}$ $\boxed{5} = 10 - 5$

⑥ What is the missing number?
in ↓
Rule
$- 6$
↓ out

in	out
9	3
	7
11	5

Choose the best answer.
○ 1
○ 10
⬤ 13

① 1.G.1 ② 1.G.3 ③ 1.OA.6, 1.OA.8 ④ 1.NBT.4
⑤ 1.OA.6, 1.OA.8 ⑥ 1.OA.6, 1.OA.8   one hundred sixty-nine 169

Math Masters, p. 239

Composing Shapes Home Link 8-5 NAME DATE

Family Note
Today your child used shapes such as triangles, squares, trapezoids, half circles, and quarter circles to make new shapes and designs.
Please return this Home Link to school tomorrow.

① Cut out the four shapes from the side of the page.
Use two or more shapes to fill one triangle. Trace around the pieces to show how they fit together.
Then use different shapes to fill the other triangle.   Sample answers:

Practice

② Find 3 shapes that have 4 straight sides in your home. Draw them on the back of this page.
Answers vary.

1.G.2, 1.G.1, SMP1   two hundred thirty-nine 239

3-Dimensional Shapes

Overview Children identify defining attributes of 3-dimensional shapes. They combine 3-dimensional shapes to create composite shapes.

▶ **Before You Begin**
Place a ball (sphere), a can (cylinder), and a box (rectangular prism) near the Math Message. For Part 2, you may wish to have children cut out the 3-Dimensional Shape Cards from *Math Masters*, page TA40 prior to the lesson. These cards will also be used in Lesson 8-7. Display some of the objects children were asked to bring to school in the Home Link from Lesson 8-1.

▶ **Vocabulary**
surface · face · edge · vertex

Common Core State Standards

Focus Cluster
Reason with shapes and their attributes.

1 Warm Up   15–20 min

	Materials	
Mental Math and Fluency Children record the number represented by base-10 blocks.	base-10 blocks, slate	1.NBT.2, 1.NBT.2a, 1.NBT.2b, 1.NBT.2c
Daily Routines Children complete daily routines.	See pages 2–37.	See pages xiv–xvii.

2 Focus   30–35 min

Math Message Children compare 3-dimensional objects.	ball, can, box	1.G.1
Defining Attributes of 3-Dimensional Shapes Children distinguish defining and nondefining attributes of 3-dimensional shapes.	*Math Journal 2*, p. 170; 3-dimensional shape blocks; everyday objects	1.G.1 SMP2
Describing 3-Dimensional Shapes Children identify 3-dimensional shapes based on attributes.	*Math Journal 2*, p. 170; 3-dimensional shape blocks; everyday objects	1.G.1 SMP2, SMP7
✓ **Assessment Check-In** See page 722.	Exit Slip (*Math Masters*, p. TA4); shape blocks including cylinder	1.G.1 SMP7
Combining 3-Dimensional Shapes Children create composite shapes.	*Math Masters*, p. TA40; 3-dimensional shape blocks; everyday objects	1.G.1, 1.G.2 SMP2

CCSS 1.G.1   **Spiral Snapshot**

GMC Distinguish between defining and non-defining attributes.

◁ | 6-3 Focus Practice | 7-5 through 7-7 Focus Practice | 8-1 Focus Practice | 8-5 Focus Practice | 8-6 Focus Practice | 9-5 Practice | 9-10 Focus Practice | ▷

 Spiral Tracker   **Go Online** to see how mastery develops for all standards within the grade.

3 Practice   15–20 min

Making a Shapes Bar Graph Children create and graph composite shapes.	*Math Journal 2*, p. 171; triangle, square, and trapezoid pattern blocks	1.OA.3, 1.OA.6, 1.MD.4, 1.G.2 SMP4
Math Boxes 8-6 Children practice and maintain skills.	*Math Journal 2*, p. 172	See page 723.
Home Link 8-6 **Homework** Children identify attributes of a cube.	*Math Masters*, p. 241	1.G.1 SMP7

 ePresentations   Student Learning Center   Facts Workshop Game   eToolkit   Professional Development   Home Connections   Spiral Tracker   Assessment and Reporting   English Learners Support   Differentiation Support

Differentiation Options

1.G.1, SMP2, SMP7   1.G.1, SMP7   1.G.1, SMP6

Readiness   10–15 min

| WHOLE CLASS |
| SMALL GROUP |
| PARTNER |
| INDEPENDENT |

Identifying Shapes Using Touch

Math Journal 2, p. 170; 3-dimensional shapes; paper bag

To explore geometric attributes, have children describe 3-dimensional shapes. Each group takes a paper bag containing a 3-dimensional shape. Without looking into the bag, each child takes a turn reaching into the bag and feeling and describing the object. After all children have described the object, they remove it from the bag and compare it to the shapes pictured on journal page 170. **GMP7.1**

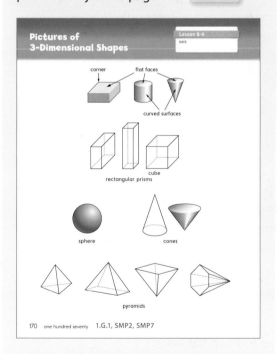

Enrichment   15–20 min

| WHOLE CLASS |
| SMALL GROUP |
| PARTNER |
| INDEPENDENT |

Comparing 3-Dimensional Shapes

Activity Card 100; *Math Masters*, p. 240; scissors; glue

To further explore attributes of 3-dimensional shapes, have children compare 3-dimensional shapes and objects using a Venn diagram. **GMP7.1** You may wish to use the illustration in Step 1 of the Activity Card to explain or review how to complete a Venn diagram.

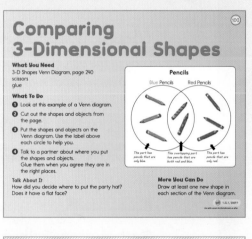

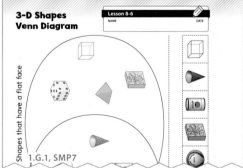

Extra Practice   10–15 min

| WHOLE CLASS |
| SMALL GROUP |
| PARTNER |
| INDEPENDENT |

Playing *I Spy*

Math Masters, p. G58 (optional)

To provide experience describing and identifying shapes, have children play *I Spy* with 2- and 3-dimensional shapes. For detailed instructions, see Lesson 8-1.

Observe

- Which children describe attributes of shapes correctly? **GMP6.1**

Discuss

- *Which words might help you describe a 3-dimensional shape?*

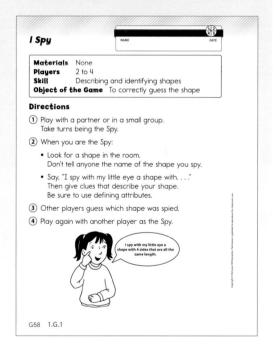

English Language Learners Support

Beginning ELL Help children associate descriptive terms with attributes by showing them examples of shapes with a given attribute and then asking children to identify non-examples. For example, show children a sphere and say: *This is round.* Then show them a few objects, all round except for one, and ask children to point to the one that is not round. Repeat with other attributes.

Go Online   **ELL** English Learners Support

Standards and Goals for
Mathematical Practice

SMP2 **Reason abstractly and quantitatively.**

GMP2.2 Make sense of the representations you and others use.

GMP2.3 Make connections between representations.

SMP7 **Look for and make use of structure.**

GMP7.1 Look for mathematical structures such as categories, patterns, and properties.

---

1 Warm Up 15–20 min

ePresentations   eToolkit

▶ Mental Math and Fluency

Display the following base-10 blocks. Have children write the numbers represented on their slates.

●○○ 13 cubes 13; 1 long, 6 cubes 16; 20 cubes 20

●●○ 21 cubes 21; 4 longs 40; 2 longs 11 cubes 31

●●● 3 longs 14 cubes 44; 5 longs, 10 cubes 60; 7 longs, 21 cubes 91

▶ Daily Routines

Have children complete the Daily Routines. For detailed instructions, see pages 2–37. For specifics on CCSS coverage, see pages xiv–xvii.

---

2 Focus 30–35 min

ePresentations   eToolkit

▶ Math Message

Look at the ball, the can, and the box. What do these objects have in common? What makes them different? Record your ideas.

▶ Defining Attributes of 3-Dimensional Shapes

Math Journal 2, p. 170

| WHOLE CLASS | SMALL GROUP | PARTNER | INDEPENDENT |

Math Message Follow-Up Discuss the similarities and differences among the objects. Remind children that they learned about the attributes of flat shapes in Unit 7. Tell children that today they will learn about shapes that they can put their hands around.

Ask children how they would describe the ball, the can, and the box. Expect answers such as the following:

• The ball is round.

• The can has a curved side and two flat sides.

• The flat sides of the can are circles.

• All the sides of the box are flat.

Begin using formal names for these shapes such as *sphere, cylinder,* and *rectangular prism,* but do not expect children to consistently use them. Gather examples of each shape. Invite children to describe the differences among the spheres. Sample answers: The marble is filled, but the ping pong ball is empty. Some are heavy, and some are light. Ask questions such as: *This sphere is blue, but are all spheres blue?* No *Do all spheres roll?* Yes Repeat with the cylinder and the rectangular prism.

Remind children that a defining attribute of a shape is an attribute that is always true for that shape. Explain that the outside or "skin" of any of these 3-dimensional shapes is called its **surface.** Describing the similarities of the surfaces helps define the shapes.

Invite children to describe the similarities among the surfaces of the rectangular prisms. Sample answers: They have flat sides. All the sides are rectangles. They have 8 corners. They have no holes. All their edges touch. Ask someone to point to the flat sides of the rectangular prism. Tell children that these sides are called **faces.** Point to an **edge,** and explain that using *face* instead of *side* helps people know whether you are talking about a face or an edge. Point to a corner of the rectangular prism. Explain that, on a 3-dimensional shape, a **vertex** is a point where at least three edges meet. A cylinder does not have any vertices because its flat faces do not meet. Ask: *How many vertices does the rectangular prism have?* Point to each vertex while children count aloud. 8 vertices

Have children look at journal page 170. Compare these representations to everyday objects and 3-dimensional shape blocks. Discuss how 3-dimensional shapes can be represented with drawings, pictures, and real objects. Point out that the dotted lines in drawings show the edges of the shapes that you would need X-ray vision to see. GMP2.2, GMP2.3

▶ **Describing 3-Dimensional Shapes**

Math Journal 2, p. 170

| WHOLE CLASS | SMALL GROUP | PARTNER | INDEPENDENT |

Divide children into 6 groups, one group for each of the following shapes: cube, non-cube rectangular prism, sphere, cylinder, pyramid, and cone. Provide each group with blocks or everyday objects of their shape. After children examine the 3-dimensional shapes, encourage them to match their objects to representations on journal page 170. GMP2.2, GMP2.3

Have children stand. Explain that you will give clues about one of the shapes. If the clue fits a group's shape, the group remains standing. If the clue does not fit, the group sits down. Give the following clues: GMP7.1

- *There are objects in real life that look like it.* All remain standing. (Briefly discuss how this is nondefining: It could describe any of the 3-dimensional shapes, and does not always describe any one shape.)
- *It has at least 1 face.* Sphere group sits.
- *It has curved parts that let it roll.* Cube, rectangular prism, and pyramid groups sit.
- *It has a point.* Cylinder group sits. The remaining shape is the cone.

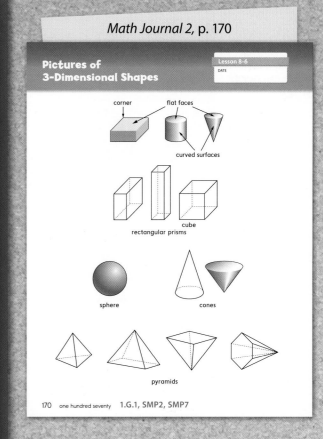

Math Journal 2, p. 170

Pictures of 3-Dimensional Shapes

Lesson 8-6
DATE

corner   flat faces

curved surfaces

rectangular prisms   cube

sphere   cones

pyramids

170   one hundred seventy   1.G.1, SMP2, SMP7

Adjusting the Activity

Differentiate Allow children who need more tactile exploration with the 3-dimensional shapes to hold and explore the blocks and objects as they practice describing the shapes using formal and informal descriptions. When children use informal descriptions, reinforce the formal vocabulary for the shapes and attributes. For example, if a child uses the word *skin* or *outside,* consider asking them to repeat the word *surface* as they touch the outside of the shape.

Go Online Differentiation Support

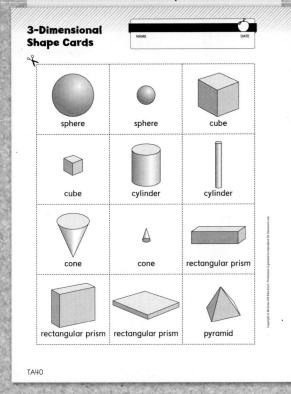

3-Dimensional Shape Cards

NAME · DATE

sphere	sphere	cube
cube	cylinder	cylinder
cone	cone	rectangular prism
rectangular prism	rectangular prism	pyramid

TA40

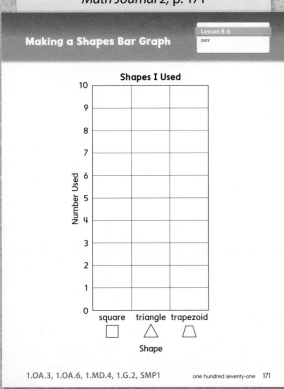

Making a Shapes Bar Graph

Lesson 8-6

DATE

Shapes I Used

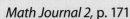

Number Used (0–10)

square   triangle   trapezoid

Shape

1.OA.3, 1.OA.6, 1.MD.4, 1.G.2, SMP1   one hundred seventy-one   171

Repeat this process. Include clues about nondefining attributes (for example, *It is small, made of cardboard, blue,* or *its pointed end is facing down*) to prompt further discussion about defining and nondefining attributes. Emphasize that defining attributes of a particular shape are always true for that shape by asking questions such as: *Does a cube always have to rest on a face?* No *Does a cube always have to have 6 faces?* Yes Discuss whether clues about nondefining or defining attributes—vertices, edges, and the shape and number of faces—are more useful for identifying the correct 3-dimensional shape.

✓ **Assessment Check-In** **GCSS** 1.G.1

Math Masters, p. TA4

Display a variety of 3-dimensional blocks, including a cylinder. Of the following clues, ask children to identify the two that are defining attributes of a cylinder on an Exit Slip (*Math Masters,* page TA4). **GMP7.1**

1. *It can roll.*

2. *It is wooden (or plastic, or whatever material your blocks are).*

3. *It has two faces.*

Expect most children to choose clues 1 and 3, but some children might struggle to understand why they should eliminate clue 2. Suggest that those who struggle reword the clues to include the shape name and the word *always. For example:* A cylinder can always roll. A cylinder is always wooden. A cylinder always has two faces. To challenge children, have them complete the Enrichment activity.

☑ Assessment and Reporting   **Go Online** to record student progress and to see trajectories toward mastery for these standards.

▶ **Combining 3-Dimensional Shapes**

Math Masters, p. TA40

WHOLE CLASS | **SMALL GROUP** | **PARTNER** | INDEPENDENT

Tell children that they can combine 3-dimensional shapes to make larger—and often different—shapes. Hold up two cylinders of the same size. Ask: *If I combined these two cylinders, flat face to flat face, what would my new shape be?* A longer cylinder

Distribute *Math Masters,* page TA40, and have children cut out the 3-Dimensional Shape Cards. (They will use the cards in Lesson 8-7 as well, so decide how you will store them.) Place the 3-dimensional shape blocks and everyday objects in the front of the room. Tell children to display the cards showing shapes that, when combined, would look like a new crayon. Cylinder and cone Then ask volunteers to compose the shape with the blocks or everyday objects. **GMP2.2, GMP2.3**

Go Online for more examples of combining 3-dimensional shapes.

Emphasize again that different 3-dimensional shapes can be combined to make new shapes. Have children suggest more examples.

Have two children make and display composite shapes of their choice. Ask the rest of the class to imagine combining the two composite shapes in various ways and to describe what the resulting shapes might look like. Discuss how composites of 3-dimensional shapes can be combined with other composites to make something new, demonstrating with blocks and objects. If enough blocks or objects are available, partnerships can create composite shapes and then combine them with another partnerships' structures to make new composite shapes.

Summarize Have partnerships discuss the 3-dimensional shapes that make up their desks or tables. Share responses as time permits.

3 Practice 15–20 min

Go Online

ePresentations   eToolkit   Home Connections

▶ Making a Shapes Bar Graph

Math Journal 2, p. 171

Have children practice building composite shapes and representing data about the shapes in bar graphs. Have them create designs with no more than 10 each of square, triangle, and trapezoid pattern blocks. GMP4.1 Ask them to record how many of each shape they used in the bar graph on journal page 171.

Then have them work in partnerships to ask questions about the graphs. Model a few sample questions such as: *How many more squares than trapezoids did you use? How many blocks did you use all together?* Encourage children to explain their answers. Listen for children who group two numbers to make adding three numbers easier or who count up to find differences. Have them share their strategies with the whole class.

▶ Math Boxes 8-6

Math Journal 2, p. 172

WHOLE CLASS | SMALL GROUP | **PARTNER** | INDEPENDENT

Mixed Practice Math Boxes 8-6 are paired with Math Boxes 8-4.

▶ Home Link 8-6

Math Masters, p. 241

Homework Children practice identifying defining attributes of a cube.

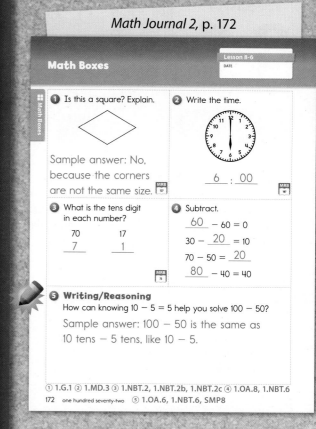

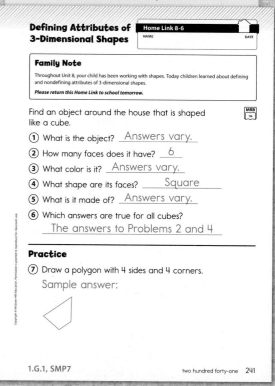

Exploring Composition of Shapes and Addition Fact Strategies

Overview Children create composite shapes from 2- and 3-dimensional shapes. They identify facts that can be solved with different fact strategies.

▶ **Before You Begin**

For Exploration A, you may wish to use colored paper for the quarter sheets. For Exploration B, have children gather the 3-Dimensional Shape Cards cut out from *Math Masters,* page TA40 in Lesson 8-6. For the optional Readiness activity, write the words *sphere, rectangular prism, cube, pyramid, cylinder,* and *cone* on index cards.

Common Core State Standards

Focus Clusters
- Add and subtract within 20.
- Reason with shapes and their attributes.

	Materials	
① Warm Up 15–20 min		
Mental Math and Fluency Children solve number stories involving 2-digit numbers.	number grid, slate	1.OA.1, 1.NBT.4, 1.NBT.6
Daily Routines Children complete daily routines.	See pages 2–37.	See pages xiv–xvii.
② Focus 35–40 min		
Math Message Children determine which pattern blocks they could use to compose a hexagon.	Pattern-Block Template	1.G.2
Making a Hexagon Children share different ways to compose a hexagon with other shapes.	Pattern-Block Template	1.G.2
Exploration A: Making New Shapes with 5 Squares Children create composite shapes from squares. Then they compose new shapes from the composite shapes.	per small group: 20 or more pattern-block squares, 15 quarter-sheets of paper, 4 full-size sheets of blank paper	1.G.2 SMP1
Exploration B: Building with 3-Dimensional Shapes Children use 3-dimensional blocks and objects to build new shapes.	Activity Card 101; *Math Journal 2,* p. 173; *Math Masters,* p. TA40 (3-Dimensional Shape Cards); 3-dimensional shape blocks; everyday objects, crayons or markers	1.G.2 SMP2
Exploration C: Sorting by Strategies Children identify facts that can be solved using the near-doubles and making-10 strategies.	Activity Card 102; *Math Journal 2,* p. 174; Fact Triangles	1.OA.6 SMP7
③ Practice 10–15 min		
Drawing and Naming Equal Shares Children practice partitioning shapes into equal parts and naming the parts.	*Math Journal 2,* p. 175	1.G.3 SMP6
Math Boxes 8-7 Children practice and maintain skills.	*Math Journal 2,* p. 176	See page 729.
Home Link 8-7 **Homework** Children practice using both near-doubles and making-10 strategies and comparing the two.	*Math Masters,* p. 242	1.OA.1, 1.OA.6, 1.OA.8 SMP7

Differentiation Options

RtI

CCSS 1.G.1, SMP2

Readiness   10–15 min

Describing and Naming 3-Dimensional Shapes

WHOLE CLASS
SMALL GROUP
PARTNER
INDEPENDENT

3-dimensional shape blocks; everyday objects that represent 3-dimensional shapes; index cards labeled: *sphere, rectangular prism, cube, pyramid, cylinder, cone*

To provide experience categorizing and naming 3-dimensional shapes, have children name shapes and describe them based on defining attributes. Children connect shape representations and 3-dimensional blocks, everyday objects you collected, and labeled index cards. **GMP2.3**

Have one child select an object or shape block—for example, a poster tube. The next child chooses the index card labeled *cylinder* as the group says "cylinder" aloud. The following child chooses another object or shape block that is a cylinder (if one is available). The next two children each describe a defining attribute of the cylinder. It has two flat faces that are circles. It rolls. Continue with other 3-dimensional shapes.

CCSS 1.G.2, SMP7

Enrichment   15–20 min

My New Shape

WHOLE CLASS
SMALL GROUP
PARTNER
INDEPENDENT

Activity Card 103; *Math Masters,* p. 235 (circle pieces) and p. TA41; pattern blocks

To further explore creating composite shapes with 2-dimensional shapes, children trace pattern blocks and circle pieces (prepared in Lesson 8-5) to create new composite shapes for their partner to fill. **GMP7.2**

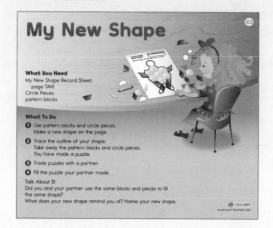

My New Shape Record Sheet

Make a new shape. Trace the outline of your new shape.

CCSS 1.OA.6, 1.NBT.3, SMP7

Extra Practice   10–15 min

Playing *Addition Top-It*

WHOLE CLASS
SMALL GROUP
PARTNER
INDEPENDENT

Math Masters, p. G37 (optional), p. G38; per partnership: 4 each of number cards 0–9

To practice addition facts using relation symbols, have children play *Addition Top-It.* For detailed instructions, see Lesson 5-5.

Observe

- Which strategies work well for the facts children do not know?

Discuss

- *How can you use doubles or combinations of 10 to help you solve facts you do not know?* **GMP7.2**

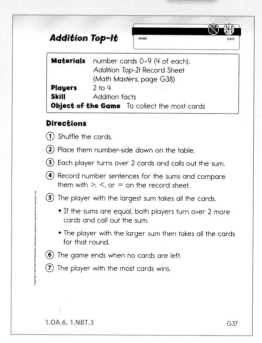

Addition Top-It

Materials	number cards 0–9 (4 of each), *Addition Top-It* Record Sheet (*Math Masters,* page G38)
Players	2 to 4
Skill	Addition facts
Object of the Game	To collect the most cards

Directions

① Shuffle the cards.
② Place them number-side down on the table.
③ Each player turns over 2 cards and calls out the sum.
④ Record number sentences for the sums and compare them with >, <, or = on the record sheet.
⑤ The player with the largest sum takes all the cards.
  - If the sums are equal, both players turn over 2 more cards and call out the sum.
  - The player with the larger sum then takes all the cards for that round.
⑥ The game ends when no cards are left.
⑦ The player with the most cards wins.

1.OA.6, 1.NBT.3   G37

English Language Learners Support

Beginning ELL Provide practice associating shapes with the written terms *sphere, rectangular prism, cube, pyramid, cylinder,* and *cone* by having children match the written words with pictures of the shapes.

Go Online **ELL** English Learners Support

Activity Card 101

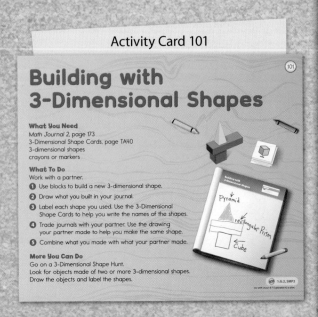

**Building with
3-Dimensional Shapes**

What You Need
Math Journal 2, page 173
3-Dimensional Shape Cards, page TA40
3-dimensional shapes
crayons or markers

What To Do
Work with a partner.
1 Use blocks to build a new 3-dimensional shape.
2 Draw what you built in your journal.
3 Label each shape you used. Use the 3-Dimensional Shape Cards to help you write the names of the shapes.
4 Trade journals with your partner. Use the drawing your partner made to help you make the same shape.
5 Combine what you made with what your partner made.

More You Can Do
Go on a 3-Dimensional Shape Hunt.
Look for objects made of two or more 3-dimensional shapes.
Draw the objects and label the shapes.

1 Warm Up   15–20 min   Go Online
ePresentations   eToolkit

Mental Math and Fluency

Encourage children to use number grids to solve the problems and record answers on their slates. Briefly discuss children's solution strategies.

◉○○ The peach tree has 17 blossoms on it. The apple tree has 8 blossoms. How many blossoms are there in all? 25 blossoms

◉◉○ Hannah had 84 paper clips in her bag. She dropped 30 of them. How many paper clips are left? 54 paper clips

◉◉◉ There were 21 black cars in the parking lot. There were also 15 red cars. How many black and red cars were there in all? 36 cars

▶ **Daily Routines**

Have children complete the Daily Routines. For detailed instructions, see pages 2–37. For specifics on CCSS coverage, see pages xiv–xvii.

2 Focus   35–40 min   Go Online

ePresentations   eToolkit

▶ **Math Message**

Dave needs a hexagon to finish a pattern-block puzzle.
He has no hexagon blocks left.
What other blocks can Dave use to make a hexagon?
Use your Pattern-Block Template to draw the blocks Dave could use.

▶ **Making a Hexagon**

| **WHOLE CLASS** | SMALL GROUP | PARTNER | INDEPENDENT |

Math Message Follow-Up Have children share their responses to the Math Message. Expect them to suggest that Dave could make a hexagon with 2 trapezoid blocks, 6 triangle blocks, 3 rhombus blocks, or some combination of trapezoid, triangle, and rhombus blocks. Display blocks to represent children's responses.

Tell children that in two of today's Explorations, they will learn more about how shapes combine to make new shapes. In the third Exploration, they will sort Fact Triangles.

▶ Exploration A: Making New Shapes with 5 Squares

| WHOLE CLASS | **SMALL GROUP** | PARTNER | **INDEPENDENT** |

Before beginning Exploration A, remind children about the new shapes they made with triangles in Lesson 8-5. Today they will work in small groups to find how many different shapes they can make from 5 squares.

Explain that children will make different shapes using exactly 5 square pattern blocks arranged in different ways. For each shape, they should make sure that the sides of the squares touch and line up evenly.

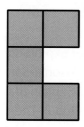   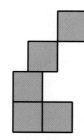

This is correct. Each square touches another square along a side. The sides line up evenly.

This is incorrect. Some squares touch only at corners. Not all sides line up evenly.

Have a child demonstrate how to make and trace the composite shape on a quarter sheet of paper. Tell children that they should make as many different shapes as they can, tracing each new shape on a new quarter sheet. Then display the arrangements below to illustrate how one shape can appear to be different, but is really the same shape turned a different direction or flipped over. Remind children to check that each shape is different by turning and flipping the shape to compare it with those they have already made.

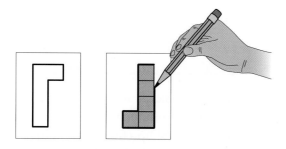

Differentiate **Adjusting the Activity**

Watch for children who need support tracing around their composite shapes. The illustrations do not have to be perfect, but accurate enough to determine that each new shape is different from the others.

Go Online 👥 Differentiation Support

Math Journal 2, p. 173

Building with 3-Dimensional Shapes Record Sheet

Lesson 8-7
DATE

Draw what you made.
Label the 3-dimensional shapes you used.

1.G.2, SMP2

one hundred seventy-three   173

Activity Card 102

Sorting by Strategies (102)

> I can solve these using the near-doubles strategy. I can solve the facts in this other pile using the making-10 strategy.

What You Need
Math Journal 2, page 174
Fact Triangles

What To Do
1. Look through your Fact Triangles. Figure out which addition facts they show.
2. Sort them by strategies you can use to solve the facts. Make 3 piles:
   • near doubles
   • making 10
   • other
3. Use your piles to complete the journal page.

Talk About It
Share the facts you have in your piles. See if there are any you are missing. With your group, make a list of facts that could fit in either pile.

More You Can Do
Think about the subtraction facts your Fact Triangles show. Sort them by the subtraction strategies on the Strategy Wall.

1.OA.6, SMP7

Math Journal 2, p. 174

Sorting by Strategies

Lesson 8-7
DATE

Record four facts that you could solve using near doubles.

Record four facts that you could solve by making 10.

Near-doubles strategy	Making-10 strategy
Sample answers:	Sample answers:
4 + 5; 6 + 5; 7 + 6;	8 + 7; 8 + 6; 8 + 5;
8 + 9; 7 + 8	9 + 6; 9 + 7; 9 + 5;
	9 + 4; 7 + 5

List a fact that you could solve using *either* strategy:

Sample answers: 8 + 7; 9 + 8; 7 + 6

Explain how you could solve it using the near-doubles strategy.

Sample answer: I can solve 7 + 8 because I know
7 + 7 = 14, so I add one more to get 8 + 7 = 15.

Explain how you could solve it using the making-10 strategy.

Sample answer: I can solve 7 + 8 because I can
take 2 from the 7 and add that to the 8 to make 10.
I have 5 more left, so 10 + 5 = 15. 15 is the answer.

174 one hundred seventy-four 1.OA.6, SMP7

Math Journal 2, p. 175

Drawing and Naming Equal Shares

Lesson 8-7
DATE

① Divide the circle into 2 equal shares.

Sample answer given.

② Write a name for one share. half-circle, 1 half
③ Write a name for both shares. whole, 2 out of 2 equal shares
④ Divide the circle into 4 equal shares.

Sample answer given.

⑤ Write a name for one share. quarter, 1 fourth
⑥ Which is bigger, 1 half of the circle or 1 fourth of the same circle? Why?
1 half is bigger because there are fewer
equal shares.

Try This

⑦ Which is smaller, 1 half of a rectangle or 1 quarter of the same rectangle? Why?
Sample answer: 1 quarter of a rectangle
because 1 out of 4 equal shares would be
smaller than 1 out of 2 equal shares.

1.G.3, SMP6   one hundred seventy-five 175

Point out that this is a challenging task and that they should try to find more shapes even after they think they've found them all. **GMP1.3**

NOTE There are 12 different shapes that can be made with 5 squares. This set of 12 configurations is often referred to as pentominoes. Do not expect children to find all 12 shapes. For your reference, the 12 configurations are shown below:

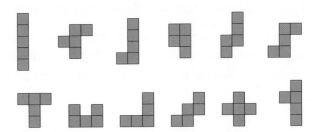

About halfway through the time allotted for Exploration A, remind children that they made bigger shapes in Lesson 8-5 by tracing combinations of triangles and circle pieces. Explain that they should choose one of the 5-square shapes and cut it out. Then they should use it to build another composite shape on a separate sheet of paper. You may wish to demonstrate with the following example.

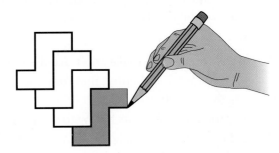

Repeating a composite shape to build another composite shape

▶ Exploration B: Building with 3-Dimensional Shapes

Activity Card 101; *Math Journal 2*, p. 173; *Math Masters*, p. TA40

| WHOLE CLASS | SMALL GROUP | PARTNER | INDEPENDENT |

Children build new 3-dimensional shapes from blocks and everyday objects. Children then represent their new shapes on journal page 173, labeling each 3-dimensional shape they used so their partners can make the same shape. Explain that they should not worry about accurately representing 3-dimensionality in their drawings and that their drawings will be 2-dimensional, or flat, even though their block structures are 3-dimensional. **GMP2.3** Keep copies of *Math Masters*, page TA40 (3-Dimensional Shape Cards) on hand to help children remember shape names.

After children have drawn their shapes, encourage them to combine one of their 3-dimensional constructions with their partner's to see what new shape can be formed.

▶ Exploration C: Sorting by Strategies

Activity Card 102; *Math Journal 2,* p. 174

| WHOLE CLASS | **SMALL GROUP** | PARTNER | INDEPENDENT |

Children sort their Fact Triangles according to two addition strategies introduced in Unit 6—the near-doubles and making-10 strategies. Children use their Fact Triangles to identify facts they could solve by using one or both of those strategies. **GMP7.2** They record examples on journal page 174.

Summarize Call attention to some of the composite shape drawings from Exploration A and the drawings of 3-dimensional creations from Exploration B. Emphasize that our world is full of shapes that are combined in many different ways. If classroom space permits, you might display children's 5-square shapes as well as their labeled drawings.

3 Practice 10–15 min

Go Online · ePresentations · eToolkit · Home Connections

▶ Drawing and Naming Equal Shares

Math Journal 2, p. 175

| WHOLE CLASS | SMALL GROUP | **PARTNER** | **INDEPENDENT** |

Have children practice partitioning shapes into equal shares and naming the parts. **GMP6.3**

▶ Math Boxes 8-7

Math Journal 2, p. 176

| WHOLE CLASS | SMALL GROUP | **PARTNER** | **INDEPENDENT** |

Mixed Practice Math Boxes 8-7 are paired with Math Boxes 8-10.

▶ Home Link 8-7

Math Masters, p. 242

Homework Children practice using near-doubles and making-10 strategies and comparing the two.

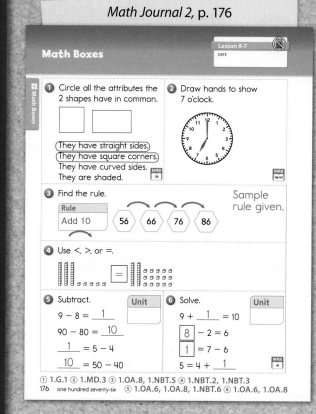

Math Journal 2, p. 176

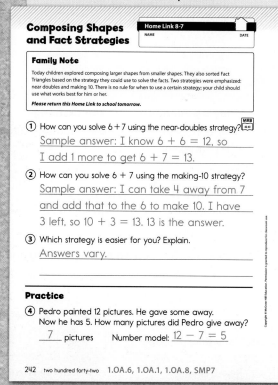

Math Masters, p. 242

Time to the Half Hour

Overview Children are introduced to half hours. They tell and write time to the half hour.

▶ **Before You Begin**
For the optional Extra Practice activity, prepare additional *Time Match* Cards from *Math Masters*, pages G60–G61.

▶ **Vocabulary**
half hour • half-past

 **Common Core State Standards**

Focus Clusters
• Tell and write time.
• Reason with shapes and their attributes.

1 Warm Up   15–20 min

	Materials	
Mental Math and Fluency Children draw shapes, partition them into equal shares, and name the parts.	slate	1.G.1, 1.G.3
Daily Routines Children complete daily routines.	See pages 2–37.	See pages xiv–xvii.

2 Focus   30–35 min

Math Message Children shade half of a clock face.		1.MD.3, 1.G.3 SMP2
Introducing Time to the Half Hour Children are introduced to the time half-past an hour.	Time Museum, demonstration clock, toolkit clocks	1.MD.3, 1.G.3 SMP5, SMP6
✓ **Assessment Check-In** See page 733.	toolkit clocks	1.MD.3, SMP5
Reading Digital Clocks to the Half Hour Children practice reading and telling time to the half hour.	*Math Masters*, p. 216; demonstration clock; digital clock	1.MD.3 SMP2, SMP5

CCSS 1.MD.3 **Spiral Snapshot**

GMC Tell and write time with analog clocks.

| 7-11
Focus
Practice | 8-1
Warm Up | 8-3
Practice | 8-8
Focus
Practice | 8-9
Practice | 9-1
Warm Up | 9-4
Practice | 9-9
Warm Up
Practice |

 Spiral Tracker   Go Online   to see how mastery develops for all standards within the grade.

3 Practice   15–20 min

Playing *Penny-Dime-Dollar Exchange* **Game** Children practice exchanging pennies for a dime and dimes for a dollar.	*Math Masters*, p. TA30, p. G48 (optional); per partnership: 2 dot dice, 5 one-dollar bills, 20 dimes, and 20 pennies	1.NBT.2, 1.NBT.2a SMP6
Math Boxes 8-8: Preview for Unit 9 Children practice prerequisite skills for Unit 9.	*Math Journal 2*, p. 177	See page 735.
Home Link 8-8 **Homework** Children practice telling and showing time.	*Math Masters*, pp. 244–245	1.OA.1, 1.OA.3, 1.OA.6, 1.MD.3, 1.G.1 SMP5

connectED.mheducation.com

Plan your lessons online
with these tools.

 ePresentations   Student Learning Center   Facts Workshop Game   eToolkit   Professional Development   Home Connections   Spiral Tracker   Assessment and Reporting   English Learners Support   Differentiation Support

Differentiation Options   RtI

Readiness
10–15 min

WHOLE CLASS
SMALL GROUP
PARTNER
INDEPENDENT

Hickory Dickory Dock with Clocks

toolkit clocks

To provide experience telling time to the hour, lead children in singing the familiar nursery rhyme *Hickory Dickory Dock*. For each verse, let a child call out the time for the others to show on their toolkit clocks. GMP2.1

> ### *Hickory Dickory Dock*
>
> *Hickory dickory dock,*
>
> *The mouse ran up the clock;*
>
> *The clock struck _____,*
>
> *The mouse ran down;*
>
> *Hickory dickory dock.*

Enrichment
10–15 min

WHOLE CLASS
SMALL GROUP
PARTNER
INDEPENDENT

Ordering Clocks by Times

Activity Card 104;
Math Masters, p. 243;
per partnership: 8 toolkit clocks

To further explore telling time, have children order clocks that show a variety of times. GMP2.2

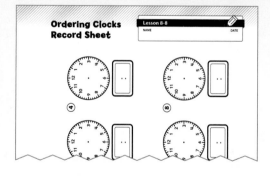

Extra Practice
10–15 min

WHOLE CLASS
SMALL GROUP
PARTNER
INDEPENDENT

Playing *Time Match*

Math Masters, pp. G55–G56 and G60–G61; p. G57 (optional)

To provide practice telling time to the hour and half hour, have children play *Time Match* with additional half-hour cards. For detailed instructions, see Lesson 7-11.

Observe
- Which children tell time on both analog and digital clocks correctly?

Discuss
- When a digital clock displays 4:00, where does the hour hand point on an analog clock? GMP2.3
- When the minute hand on an analog clock points to 6, what numbers are displayed on a digital clock? GMP2.3

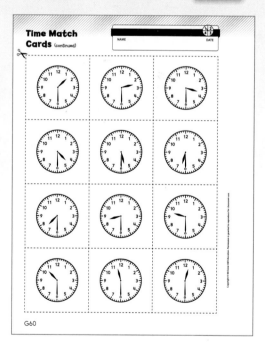

English Language Learners Support

Beginning ELL Think aloud and demonstrate to introduce the phrase *to set a clock*. Demonstrate the process of *setting* a clock, as you say: *I will set the clock to _____*, so children can see that it means you are adjusting the time on the clock. Direct children with Total Physical Response prompts to set clocks to specified times.

Go Online   **ELL** English Learners Support

SMP2 Reason abstractly and quantitatively.

GMP2.1 Create mathematical representations using numbers, words, pictures, symbols, gestures, tables, graphs, and concrete objects.

SMP5 Use appropriate tools strategically.

GMP5.2 Use tools effectively and make sense of your results.

SMP6 Attend to precision.

GMP6.1 Explain your mathematical thinking clearly and precisely.

1 Warm Up 15–20 min Go Online ePresentations eToolkit

▶ Mental Math and Fluency

Have children draw and partition shapes on their slates.

- ◐○○ Draw a circle. Divide it into 2 equal shares. What would you call one of the parts? Sample answers: half; 1 out of 2 parts; one-half
- ◐◐○ Draw a rectangle. Divide it into 2 equal shares. What would you call both of the parts? Sample answers: whole; 2 halves; 2 out of 2 parts
- ◐◐◐ Draw a rectangle. Divide it into 4 equal shares. What would you call one of the parts? Sample answers: quarter; 1 fourth; 1 out of 4 parts

▶ Daily Routines

Have children complete the Daily Routines. For detailed instructions, see pages 2–37. For specifics on CCSS coverage, see pages xiv–xvii.

2 Focus 30–35 min Go Online ePresentations eToolkit

▶ Math Message

Draw the clock face, and shade half of the clock. How much time has passed if the minute hand begins at 12 and goes through all of the shaded part?
GMP2.1

▶ Introducing Time to the Half Hour

| WHOLE CLASS | SMALL GROUP | PARTNER | INDEPENDENT |

Math Message Follow-Up Children share their drawings. Compare the different representations and discuss how both pieces of the clock must be equal to be divided in half. (*See margin.*) Ask: *What would you name the part of the clock that you shaded?* Sample answer: one half of the clock Tell children that today they will use what they know about halves to learn more about telling time.

Review how far the minute and hour hands move in an hour. Set the demonstration clock to 4 o'clock. Ask: *If I move the minute hand halfway around the clock, in which direction will it be pointing?* Straight down Ask children to watch the hands carefully as you move the minute hand so that it is pointing straight down. Ask: *How far around the clock did the minute hand travel?* Halfway around the clock *How far did the hour hand travel?* Halfway from one number to the next number

A clock face with no hands, shaded to show a half hour

Explain that it is now a **half hour** past 4 o'clock. The minute hand traveled halfway around the clock, and the hour hand now points halfway between the 4 and the 5, so the time shown is between 4 o'clock and 5 o'clock. The time is read as **half-past** 4 o'clock, or half-past 4. GMP5.2

Connect half hours to the fraction work started in Lesson 8-2. Ask:

• *How is the half hour similar to other halves you have seen?* Sample answers: It's just like half a circle. It's the same shape as a pancake cut in half.

• *How many half hours are in one hour? How do you know?* GMP6.1 Sample answers: two, because there are 2 halves in a whole; two, because if you move the minute hand one more half hour, it is back to where it started

Have children read the times as "half-past _____ o'clock" or "half an hour past _____ o'clock" as you display times on the demonstration clock. Then have children model times to the half hour on their toolkit clocks. GMP5.2 Emphasize placing the hour hand halfway between the hour numbers and the minute hand halfway around the clock.

Guide children to think of things that take about a half hour to do from personal experience. Sample answers: lunch time; one television episode; art class Add to the Time Museum by labeling a piece of paper "One Half Hour" and by recording some of the children's personal references. The paper should be smaller than the hour paper and larger than the minute paper, but it does not need to be proportional.

Have children practice telling time to the hour and half hour by showing each other times to read from their toolkit clocks or by setting their toolkit clocks to various times. Have partners take turns setting their clocks and telling the time to the half hour. GMP5.2 After children have had some time to practice, ask:

• *When it is half-past an hour, where does the minute hand point?* Straight down

• *When it is half-past an hour, where does the hour hand point?* Sample answer: halfway between two numbers

 Assessment Check-In CCSS 1.MD.3

Observe as partnerships read the times to the half hour. As this is the first introduction to half hours, do not expect all children to succeed. Expect most children to identify a given time correctly as half past an hour. Remind children who struggle that the position of the hour hand shows that the time is between two hours. It is not the next hour until the hour hand moves all the way to the next number on the clock. GMP5.2

Assessment and Reporting | Go Online | to record student progress and to see trajectories toward mastery for these standards.

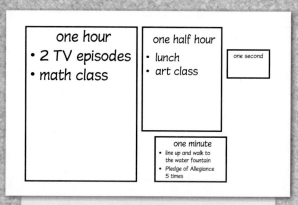

Time Museum

Math Masters, p. 216

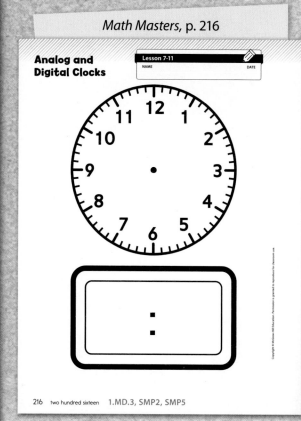

216 two hundred sixteen   1.MD.3, SMP2, SMP5

Math Masters, p. G48

Penny-Dime-Dollar Exchange

NAME | DATE

Materials	1 dollar, 20 dimes, 20 pennies
	1 Place-Value Mat with Money per player
	(Math Masters, page TA30)
	2 dot dice
	1 sheet of paper labeled "Bank"
Players	2
Skill	Making place-value exchanges
Object of the Game	To exchange for a dollar

Directions

1. Place all of the money in the bank.
2. Players take turns. When it is your turn:
   - Roll the dice.
   - Take that number of cents from the bank. Place the coins on the mat.
   - If you have 10 or more pennies in the Pennies column, exchange 10 pennies for 1 dime.
   - If you have 10 or more dimes in the Dimes column, exchange 10 dimes for 1 dollar.
3. The winner is the first player to make the exchange for a dollar.

G48   1.NBT.2, 1.NBT.2a

Math Journal 2, p. 177

Math Boxes Preview for Unit 9

Lesson 8-8
DATE

1. Draw a line to match each face to the correct picture of the 3-dimensional shape.

   Column A   Column B

2. Is the blue part of each circle a half? What are other names for the blue parts?

   Yes. 1 of 2 equal shares; 1 half; both together are whole.

3. Circle the tens digit in each number.

   36   120   59   66

4. Write <, >, or =.

   DDD < DDDD
   PPPP   PPPP

5. (base-ten blocks)

   + (base-ten blocks)
   = 60

6. True or False?

   4 + 1 = 10 − 5 true
   12 − 6 = 11 − 7 false
   4 − 0 = 14 − 4 false
   9 + 9 = 20 − 2 true

① 1.G.1 ② 1.G.3 ③ 1.NBT.2 ④ 1.NBT.2, 1.NBT.3
⑤ 1.NBT.2, 1.NBT.2a, 1.NBT.4 ⑥ 1.OA.6, 1.OA.7 one hundred seventy-seven **177**

▶ Reading Digital Clocks to the Half Hour

Math Masters, p. 216

WHOLE CLASS | SMALL GROUP | PARTNER | INDEPENDENT

Remind children that there are 60 minutes in an hour. Ask them to think about how many minutes are in a half hour. 30 minutes If needed, count the minutes while moving the minute hand one half hour on the demonstration clock.

Differentiate   **Adjusting the Activity**

Help children connect 30 minutes to 1 half of an hour. Remind them that 2 halves are 2 equal shares that make up the whole. Ask: *If 30 minutes is a half hour, then how many minutes are in a whole hour?* 30 + 30 = 60; 60 minutes Display a clock shaded as for the Math Message, and write the number sentence below it. **GMP2.3**

Go Online Differentiation Support

30 + 30 = 60

Remind children that digital clocks display the time using numbers instead of hands. Display and distribute *Math Masters*, page 216. Draw hands to show half-past 2 o'clock. Ask whether anyone knows how to record this time on the digital clock. 2:30 Remind children that the number before the colon tells what hour it is and that the number after the colon tells how many minutes it is past the hour. **GMP2.3** Demonstrate how to record 30 after the colon to show 30 minutes after the hour. Explain that we read this time as *two thirty*. **GMP5.2**

Continue to model different times on the demonstration clock, and have children record the time in digital notation. Alternate this with providing digital times and having children draw hands on the clock.

Summarize Ask children to talk in partnerships about how a circle divided in half is like half an hour.

③ **Practice** 15–20 min

Go Online

ePresentations eToolkit Home Connections

▶ Playing *Penny-Dime-Dollar Exchange*

Math Masters, p. TA30

| WHOLE CLASS | SMALL GROUP | **PARTNER** | INDEPENDENT |

Have children practice place-value concepts in the context of money by playing *Penny-Dime-Dollar Exchange.* For detailed instructions, see Lesson 6-11.

Observe

- Which children make correct exchanges when they have 10 pennies or 10 dimes? **GMP6.4**

Discuss

- *If you have 9 pennies and 9 dimes and you roll a 1, what exchanges do you make?*

▶ Math Boxes 8-8: Preview for Unit 9

Math Journal 2, p. 177

| WHOLE CLASS | SMALL GROUP | **PARTNER** | **INDEPENDENT** |

Mixed Practice Math Boxes 8-8 are paired with Math Boxes 8-12. These problems focus on skills and understandings that are prerequisites for Unit 9. You may want to use information from these Math Boxes to plan instruction and grouping in Unit 9.

▶ Home Link 8-8

Math Masters, pp. 244–245

Homework Children practice telling time to the hour and half hour.

Math Masters, p. 244

Telling Time to the Half Hour   Home Link 8-8

NAME   DATE

Family Note

Today your child began telling time to the nearest half hour on analog and digital clocks. Work together to complete these pages. Tell your child at which times, on the hour or half hour, he or she wakes up and goes to bed on school days. Have your child practice telling the time at home when it is close to the hour or half hour.

Please return these Home Link pages to school tomorrow.

Record the time.

①   ②

__5__ o'clock   half-past __7__ o'clock

③   ④

| 2:30 | | 9:30 |

1.MD.3, 1.OA.1, 1.OA.3, 1.OA.6, 1.G.1, SMP5

244   two hundred forty-four

Math Masters, p. 245

Telling Time to the Half Hour (cont.)   Home Link 8-8

NAME   DATE

Draw the hour hand and the minute hand to show the time. Then write the time.

⑤ This is about the time I wake up on a school day.

    Sample answer:

| 6:30 |

⑥ This is about the time I go to bed on a school night.

    Sample answer:

| 8:00 |

Practice

⑦ Alex has 3 baseballs, 4 marbles, and 5 pencils. How many spheres does Alex have?

__7__ spheres

Number model: __7 = 3 + 4__

Sample number model given.

1.MD.3, 1.OA.1, 1.OA.3, 1.OA.6, 1.G.1, SMP5   two hundred forty-five   245

Review: Data

Overview Children create bar graphs. They ask and answer questions about data shown in bar graphs.

▶ **Before You Begin**
Decide how you will display the tally chart shown on page 738 for the Math Message. For Part 2, prepare Survey Forms (*Math Masters*, page 247) according to the margin note on page 739. You may wish to read *Lemonade for Sale* by Studart J. Murphy (HarperCollins, 1997) as it relates to the content of this lesson, bar graphs.

**Common Core
State Standards**

Focus Clusters
- Understand and apply properties of operations and the relationship between addition and subtraction.
- Add and subtract within 20.
- Represent and interpret data.

1 Warm Up 15–20 min

Materials

Mental Math and Fluency Children determine whether equations are true or false.		1.OA.6, 1.OA.7
Daily Routines Children complete daily routines.	See pages 2–37.	See pages xiv–xvii.

2 Focus 35–40 min

Math Message Children answer a question about data in a tally chart.		1.OA.6, 1.MD.4 SMP1
Reviewing Tally Charts and Bar Graphs Children use a tally chart to create a bar graph. They learn about asking good questions about data.	*Math Masters*, pp. TA18 and 246	1.OA.3, 1.OA.6, 1.MD.4 SMP2, SMP4
Making Survey Bar Graphs Children complete a survey and use the class results to create a bar graph. They ask and answer questions about the data.	*Math Masters*, pp. TA18 and 246–247	1.OA.6, 1.MD.4 SMP2, SMP4
✓ **Assessment Check-In** See page 740.	*Math Masters*, p. TA18	1.MD.4, SMP2

CCSS 1.MD.4 **Spiral Snapshot**

GMC Answer questions about data.

1-8 Focus	2-2 Practice	2-4 Focus Practice	4-5 Focus	4-6 Focus Practice	8-3 Focus Practice	8-6 Practice	8-9 Focus Practice

Spiral Tracker   Go Online   to see how mastery develops for all standards within the grade.

3 Practice 10–15 min

Playing *Time Match* Game Children practice telling time.	*Math Masters*, pp. G55–G56 and G60–G61; p. G57 (optional)	1.MD.3 SMP2
Math Boxes 8-9 Children practice and maintain skills.	*Math Journal 2*, p. 178	See page 741.
Home Link 8-9 Homework Children create a bar graph and answer questions about the data it shows.	*Math Masters*, p. 252	1.OA.6, 1.MD.4, 1.G.3 SMP4

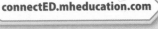
connectED.mheducation.com

Plan your lessons online
with these tools.

 ePresentations   Student Learning Center   Facts Workshop Game   eToolkit   Professional Development   Home Connections   Spiral Tracker   Assessment and Reporting   ELL English Learners Support   Differentiation Support

CCSS 1.OA.3, 1.OA.6, 1.MD.4, SMP2

Readiness   10–15 min

Completing a Bar Graph

| WHOLE CLASS |
| SMALL GROUP |
| PARTNER |
| INDEPENDENT |

Math Masters, p. 248

To provide additional exposure to creating and interpreting bar graphs, have children complete *Math Masters*, page 248. Discuss whether the bar graph matches the tally chart. Then have children complete the bar graph so that it represents the data in the tally chart. GMP2.3 Have them use their completed bar graphs to answer questions such as:

- *How many children like* Top-It *best?* 6
- *How many children voted?* 20 *How do you know?* Sample answer: I added 6 + 8 + 6.
- *How many more children like* Roll and Total *than* Subtraction Bingo? 2 *How can you tell by looking at the graph?* Sample answer: I can see that the column for *Roll and Total* is 2 votes taller than for *Subtraction Bingo*.

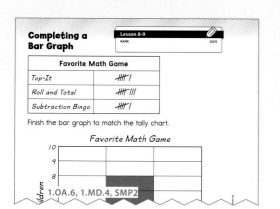

Completing a Bar Graph

Lesson 8-9
NAME   DATE

Favorite Math Game	
Top-It	⊦⊦⊦ I
Roll and Total	⊦⊦⊦ III
Subtraction Bingo	⊦⊦⊦ I

Finish the bar graph to match the tally chart.

Favorite Math Game

1.OA.6, 1.MD.4, SMP2

CCSS 1.MD.4, SMP4, SMP6

Enrichment   15–20 min

Data with 4 Categories

| WHOLE CLASS |
| SMALL GROUP |
| PARTNER |
| INDEPENDENT |

Activity Card 105;
Math Masters, pp. 249–250;
crayons

To apply children's understanding of data sets, have them make a prediction and then collect and analyze data for a question they ask their classmates. GMP4.2 They make and label bar graphs to display their data. GMP6.3

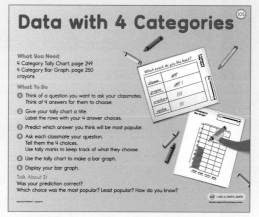

Data with 4 Categories 105

What You Need
4 Category Tally Chart, page 249
4 Category Bar Graph, page 250
crayons

What To Do
1. Think of a question you want to ask your classmates. Think of 4 answers for them to choose.
2. Give your tally chart a title. Label the rows with your 4 answer choices.
3. Predict which answer you think will be most popular.
4. Ask each classmate your question. Tell them the 4 choices. Use tally marks to keep track of what they choose.
5. Use the tally chart to make a bar graph.
6. Display your bar graph.

Talk About It
Was your prediction correct?
Which choice was the most popular? Least popular? How do you know?

4 Category Bar Graph

Lesson 8-9
NAME   DATE

10
9
8
7
6
5

1.MD.4, SMP4, SMP6

CCSS 1.OA.6, 1.MD.4, SMP4

Extra Practice   10–15 min

Asking and Answering Questions about Data

| WHOLE CLASS |
| SMALL GROUP |
| PARTNER |
| INDEPENDENT |

Activity Card 106;
Math Masters, pp. 246 and 251

To provide practice interpreting survey results, children ask and answer questions about data displayed in a tally chart and a bar graph. GMP4.2 Encourage children to share solution strategies by asking a follow-up question like *How do you know?*

Asking and Answering Questions about Data 106

What You Need
Question Cards, page 246
Tally Chart and Bar Graph, page 251
scissors

How many cheese sandwiches all together?

What To Do
Work with a partner.
1. Cut out one set of question cards. Mix the cards and place them facedown.
2. Look carefully at the tally chart and bar graph.
3. Draw question cards to help you ask questions. Take turns asking and answering questions.

Talk About It
After each question is answered, ask: "How do you know?"

More You Can Do
Think of questions about the lunch food that you cannot answer with this graph. What would you do to find those answers?

1.OA.4, 1.MD.4, SMP4

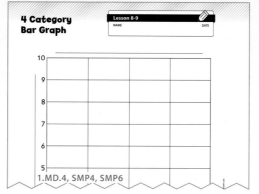

Question Cards

Lesson 8-9
NAME   DATE

Ask a question about:

How many ____ all together?

Ask a question about:

How many more ____ than ____?

Ask a question about:

How many fewer ____ than ____?

Ask a question about:

How many in one column (or row)?

English Language Learners Support

Beginning ELL Move your hand upward to show that when you ask a question, the pitch of your voice rises toward the end of the sentence. Then move your hand downward to show that your voice lowers toward the end of the sentence when you answer. Write short questions and answers, slanting the last words up and down, respectively. Then read them aloud with children as you gesture for them to raise or lower their pitch on the last word.

Go Online   ELL   English Learners Support

**Standards and Goals for
Mathematical Practice**

**SMP1 Make sense of problems and persevere
in solving them.**
GMP1.1 Make sense of your problem.

SMP2 Reason abstractly and quantitatively.
GMP2.3 Make connections between
representations.

SMP4 Model with mathematics.
GMP4.2 Use mathematical models to solve
problems and answer questions.

Literature Link You may wish to
read *Lemonade for Sale* by Stuart
J. Murphy (HarperCollins, 1997) as it
relates to the content of this lesson,
bar graphs.

Math Masters, p. TA18

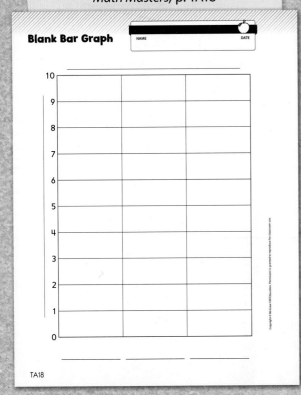

Blank Bar Graph

TA18

---

1 Warm Up 15–20 min Go Online ePresentations eToolkit

▶ Mental Math and Fluency

Display the following equations. Have children show thumbs up or down
to indicate whether each equation is true or false. Discuss ways to make
the false equations true.

●○○ $6 = 5 + 1$ True $6 = 9 - 2$ False $6 = 6$ True

●●○ $4 + 1 = 5 + 1$ False $3 + 2 = 2 + 3$ True $8 + 2 = 6 + 4$ True

●●● $5 + 2 = 7 - 2$ False $3 + 3 = 9 - 3$ True $4 - 1 = 3 - 1$ False

▶ Daily Routines

Have children complete the Daily Routines. For detailed instructions,
see pages 2–37. For specifics on CCSS coverage, see pages xiv–xvii.

---

2 Focus 35–40 min Go Online ePresentations eToolkit

▶ Math Message

*Mr. Chan's class took a survey to figure out which pet is the favorite.
They made a tally chart, but the tallies for Turtle got erased.*

Favorite Class Pet		
Fish	ЖЖ I	
Rabbit	ЖЖ IIII	
Turtle		

If 19 children voted, how many voted for Turtle?

*What do you need to do to solve this problem? Record your answer,
and explain how you found it.* GMP1.1

▶ Reviewing Tally Charts and Bar Graphs

Math Masters, pp. TA18 and 246

| WHOLE CLASS | SMALL GROUP | PARTNER | INDEPENDENT |

Math Message Follow-Up Have children share their responses to the Math Message. 4 tallies should be in the Turtle row. Share how children knew what to do. Call attention to subtraction and missing-addend strategies.

Remind children of the Super Powers bar graph they created from the tally chart in Unit 4. Tell them that today they will continue to learn about bar graphs and how to ask and answer questions about bar graphs.

Display *Math Masters,* page TA18. Have children use the information from the tally chart in the Math Message to suggest labels for each part of the bar graph and demonstrate coloring the bars. GMP2.3 Ask: *How can you tell from the bar graph that 19 children answered the question?* Sample answer: Look at the number each bar reaches on the side of the graph, and add the numbers together.

Remind children that bar graphs are useful tools for organizing information because they can help answer questions. GMP4.2 Display question cards (*Math Masters,* page 246). Tell children that asking good questions is part of being a good mathematician.

Have children cut out the question cards. With a volunteer, model how to use the cards to generate questions about the Favorite Class Pet bar graph. Select a card, and read it aloud. *For example:*

Ask a question about: How many more _____ than _____?

Then think aloud as you formulate a question: *I'm supposed to ask a question about how many more of one thing compared to another. I can see that there are more votes in the Rabbit column than in the Fish column, so I'll ask about that. How many more children chose Rabbit than chose Fish?* Have a volunteer answer the question; then switch roles. GMP4.2 These cards will be used in the next activity and in the optional Extra Practice activity.

Academic Language Development Role playing and thinking aloud to model formulating questions may help children formulate their own.

▶ Making Survey Bar Graphs

Math Masters, pp. TA18 and 246–247

| WHOLE CLASS | SMALL GROUP | PARTNER | INDEPENDENT |

Tell children that they will now take a survey, tally the results, create bar graphs, and ask and answer questions. Distribute one prepared survey form (*Math Masters,* page 247) to each child. (*See margin note.*)

Math Masters, p. 246

Question Cards
Lesson 8-9
NAME   DATE

Ask a question about:
How many _____ all together?

Ask a question about:
How many more _____ than _____?

Ask a question about:
How many fewer _____ than _____?

Ask a question about:
How many in one column (or row)?

Ask a question about:
How many in two columns (or rows) combined?

WILD CARD
Ask any question!

246 two hundred forty-six   1.OA.6, 1.MD.4, SMP4

NOTE To prepare the survey form (*Math Masters,* page 247), write a question at the top and include three response options. *For example:*

Which math activity did you like best?

☐ Building with 3-dimensional shapes

☐ Making new shapes with 3 triangles

☐ Playing *Make My Design*

Other survey topics include children's favorite lunch food, read-aloud book, juice, recess activity, or math game.

Math Masters, p. 247

Making Survey Bar Graphs
Lesson 8-9
NAME   DATE

Mark ONE box.

☐ _____

☐ _____

☐ _____

Mark ONE box.

1.OA.6, 1.MD.4   _____

Adjusting the Activity

Differentiate If children would benefit from more practice with tally charts, have each child create a tally chart (or create a class tally chart) as you read each survey result aloud.

Go Online | Differentiation Support

Have each child complete a form. Collect the surveys or have children place them in a ballot box. Take a few minutes to have children predict the results and discuss their reasoning. Sample answer: I predict that Playing *Make My Design* was the most popular math activity because I saw lots of kids play it again during free-choice time. Select a few children to read each survey response aloud, one by one, as they sort them into piles that correspond to the responses. Then have children count the number of forms in each pile, displaying each category and total.

Distribute a blank bar graph (*Math Masters,* page TA18) to each child. Have children fill in the title and the labels and then complete their bar graphs independently, using the information displayed. **GMP2.3**

> **NOTE** If one of the categories receives more than 10 votes, children can extend the graph by taping an additional blank bar graph to the top of their original graph. Remind them to continue the numbering along the side and to add a title.

When children finish their bar graphs, have them work in partnerships to ask and answer questions about the graph. **GMP4.2** Encourage children to use the question cards they cut out in the previous activity to help them ask a variety of questions. Remind those who are answering questions to explain how they arrived at their answers.

Summarize Discuss how children's predictions of the survey results compare to the actual results.

✔ Assessment Check-In CCSS 1.MD.4

Math Masters, p. TA18

Observe as children create their bar graphs. Expect most children to correctly represent the survey results in bar graph form. **GMP2.3** Have children who struggle arrange the survey results on the floor in categories, stacking the papers end-to-end to make a large, real-object graph. To challenge children, have them predict the results of the same survey if it were given to the whole school.

Assessment and Reporting | Go Online | to record student progress and to see trajectories toward mastery for these standards.

Math Masters, p. G57

Time Match

NAME   DATE

Materials	*Time Match* Cards
Players	2 or more
Skill	Telling time
Object of the Game	To match as many cards as you can

Directions

1. Mix the cards, and place them facedown on the table. Take turns.
2. Flip over 2 cards.
3. If the cards match, keep them and take another turn. If the cards do not match, flip them back over.
4. When all the cards have been matched, the player with the most matches wins.

1.MD.3   G57

3 Practice   10–15 min

Go Online · ePresentations · eToolkit · Home Connections

▶ Playing *Time Match*

Math Masters, pp. G55–G56 and G60–G61

| WHOLE CLASS | SMALL GROUP | **PARTNER** | INDEPENDENT |

Children practice telling time to the hour and to the half hour. For directions for playing *Time Match,* use *Math Masters,* page G57. For more information, see Lesson 7-11. Pages G60–G61 provide additional half-hour cards.

Observe

- Which children correctly connect the analog and digital representations of the time? **GMP2.3**
- Which children use "half-past" or "thirty" to describe the time to the half hour?

Discuss

- *How do you know if the clock shows an o'clock time or a half-past time?*
- *What do the numbers after the colon on a digital clock mean?*

▶ Math Boxes 8-9

Math Journal 2, p. 178

| WHOLE CLASS | SMALL GROUP | **PARTNER** | **INDEPENDENT** |

Mixed Practice Math Boxes 8-9 are paired with Math Boxes 8-11.

▶ Home Link 8-9

Math Masters, p. 252

Homework Children practice displaying data on bar graphs and answering questions about the data.

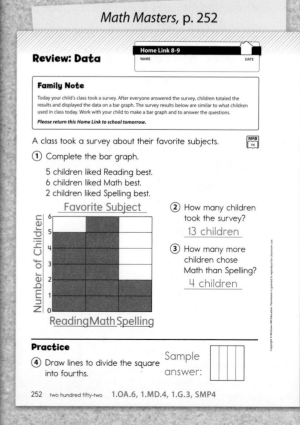

Math Journal 2, p. 178

Math Masters, p. 252

Number-Grid Puzzles

Overview Children review place-value patterns in the number grid and use them to solve number-grid puzzles.

▶ **Before You Begin**
Display the piece of the number grid shown in the Math Message. For the optional Extra Practice activity, create four sets of numbered index cards 50–99 for each partnership.

▶ **Vocabulary**
number-grid puzzle

Common Core State Standards

Focus Clusters
- Understand place value.
- Use place value understanding and properties of operations to add and subtract.

1 Warm Up   15–20 min

	Materials	
Mental Math and Fluency Children write and compare numbers represented by base-10 blocks.	slate, base-10 blocks	1.NBT.2, 1.NBT.2a, 1.NBT.2b, 1.NBT.2c, 1.NBT.3
Daily Routines Children complete daily routines.	See pages 2–37.	See pages xiv–xvii.

2 Focus   35–40 min

Math Message Children fill in missing numbers on a number grid.		1.NBT.2, 1.NBT.5 SMP7
Reviewing Place Value on the Number Grid Children review place-value patterns shown on the number grid and model the patterns with base-10 blocks.	Class Number Grid, base-10 blocks	1.NBT.2, 1.NBT.5 SMP1, SMP7
Introducing Number-Grid Puzzles Children use patterns in the number grid to help them identify hidden numbers on a number grid.	*Math Journal 2*, pp. 179–180; Class Number Grid; stick-on notes	1.NBT.2, 1.NBT.5 SMP7
✓ **Assessment Check-In** See page 746.	*Math Journal 2*, p. 180; Class Number Grid and base-10 blocks (optional)	1.NBT.5

CCSS 1.NBT.5 Spiral Snapshot

GMC Mentally find 10 more or 10 less than a 2-digit number.

8-2 Practice	8-4 Warm Up	8-5 Warm Up	8-10 Focus Practice	8-11 Focus Practice	9-3 Warm Up Focus Practice	9-8 Warm Up Practice	9-9 Focus Practice

Spiral Tracker   **Go Online** to see how mastery develops for all standards within the grade.

3 Practice   10–15 min

Playing *Make My Design* **Game** Children create and describe composite shapes.	*Math Masters*, p. G59 (optional); pattern blocks; folder	1.G.1, 1.G.2 SMP3, SMP6
Math Boxes 8-10 Children practice and maintain skills.	*Math Journal 2*, p. 181	See page 747.
Home Link 8-10 **Homework** Children solve number-grid puzzles.	*Math Masters*, p. 254	1.NBT.2, 1.NBT.5, 1.NBT.6 SMP7

connectED.mheducation.com

Plan your lessons online with these tools.

 ePresentations   Student Learning Center   Facts Workshop Game   eToolkit   Professional Development   Home Connections   Spiral Tracker   Assessment and Reporting   English Learners Support   Differentiation Support

Differentiation Options

RtI

CCSS 1.NBT.2, 1.NBT.5, SMP7

Readiness
10–15 min

WHOLE CLASS
SMALL GROUP
PARTNER
INDEPENDENT

Pinning the Number on the Number Grid

Class Number Grid, stick-on notes, blindfold

Have children explore place-value patterns on the Class Number Grid. The patterns they should discover include the following:

- The numbers get larger as you move down or to the right.
- The tens digit increases as you move down.
- The ones digit increases as you move to the right.
- The ones digit stays the same down a column.
- The numbers directly to the right and left of a number are 1 more and 1 less than the number.
- The tens digit stays the same across a row until the last number, the decade number.

Point out that the numbers directly above and below a number are 10 less and 10 more than the number, respectively. Blindfold one of the children. Say a number between 1 and 100. The blindfolded child tries to place a stick-on note on the number. Other children may give hints in terms of patterns that were described. For example: "You have to move 10 more." GMP7.2

CCSS 1.NBT.2, 1.NBT.5, SMP7

Enrichment
10–15 min

WHOLE CLASS
SMALL GROUP
PARTNER
INDEPENDENT

Solving Number Codes

Activity Card 107;
Math Masters, p. 253;
number grid

To apply children's understanding of place-value patterns, have them create and solve number codes that use arrows to represent changes in tens and ones. GMP7.2

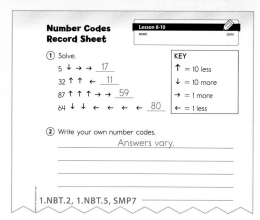

CCSS 1.NBT.1, SMP7

Extra Practice
10–15 min

WHOLE CLASS
SMALL GROUP
PARTNER
INDEPENDENT

Playing *Before and After*

My Reference Book, pp. 140–141; per partnership: 4 each of index cards numbered 50–99

To provide practice finding numbers that are 1 less or 1 more than a given number, have children play *Before and After*. For detailed instructions, see Lesson 5-6.

Observe

- Which children quickly recognize numbers just before and just after a given number?
- Which children rely on the number line or number grid?

Discuss

- *How do you know whether you are holding a number card that is just before or just after the numbers on the table?* GMP7.2

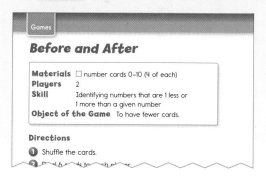

English Language Learners Support

Beginning ELL Scaffold the use of the term *missing* in number-grid puzzles with simple puzzles that have a few pieces missing, and make statements, such as: *There is a piece missing.* Ask children to point to where other pieces are missing. Ask *yes* or *no* questions as you point to other holes, such as: *Is something missing from here?* Have children fill the holes.

Go Online ⟩ ELL English Learners Support

1 Warm Up   15–20 min

ePresentations   eToolkit

▶ Mental Math and Fluency

Display base-10 blocks for two numbers. Children write the numbers on their slates and compare them using <, >, or =.

●○○ 1 long, 8 cubes	4 longs	18 < 40
●●○ 6 longs, 18 cubes	7 longs, 8 cubes	78 = 78
●●● 5 longs, 10 cubes	1 long, 5 cubes	60 > 15

▶ Daily Routines

Have children complete the Daily Routines. For detailed instructions, see pages 2–37. For specifics on CCSS coverage, see pages xiv–xvii.

2 Focus   35–40 min

ePresentations   eToolkit

▶ Math Message

Draw this piece of a number grid. Fill in the missing numbers, and record how you found them. GMP7.2

Reviewing Place Value on the Number Grid

WHOLE CLASS	SMALL GROUP	PARTNER	INDEPENDENT

Math Message Follow-Up Discuss children's answers to the Math Message. Have children explain how they found the missing numbers. 38 on the top and 58 on the bottom They may mention that they used the Class Number Grid or counted up and back by 10s in their heads.

Tell children that they will further explore number grid patterns and use them to solve puzzles in today's lesson.

Distribute base-10 blocks, and have children use them to check their answers by modeling 48. GMP1.4 Ask: *How can you show 10 less than 48?* Take away one long from 48 to get 38. *How can you show 10 more than 48?* Add one long to 48 to get 58. Tell children that today they are going to find more missing numbers on the number grid. Review place-value patterns on the number grid. GMP7.2 Have children name patterns they see, making sure they include the following:

• Except for the last number in the row, all numbers in a row have the same tens digit. When you move within in a row, you count up or down by 1s.
• Numbers in a column have the same ones digit. When you move within a column, you count up or down by 10s.

Model some number-grid patterns using base-10 blocks. Have children show 55 with base-10 blocks. Ask: *How could you show the number that is 1 more than 55?* Add 1 cube. *1 less than 55?* Remove 1 cube. Point out that moving within a row on a number grid is like adding or subtracting cubes. Have children name the number of tens and ones in numbers within the same row, emphasizing that the number of tens is the same until the last number in the row, which is the decade number.

Then ask: *How could you show the number that is 10 more than 55?* Add 1 long. *10 less than 55?* Remove 1 long. Explain that moving within a column on the number grid is like adding or subtracting longs. Have children name the number of tens and ones in numbers within the same column, emphasizing that the number of ones is always the same.

Remind children that every number has two "ten friends," numbers that are 10 more and 10 less than a given number. Point to 67 on the number grid and ask children to find the ten friends for 67. 57, 77 Discuss how they can use number-grid patterns to find the ten friends. GMP7.2

▶ Introducing Number-Grid Puzzles

Math Journal 2, pp. 179–180

| WHOLE CLASS | SMALL GROUP | PARTNER | INDEPENDENT |

Tell children that they are going to solve some **number-grid puzzles** in which they use number-grid patterns to find missing numbers. GMP7.2 Have children close their eyes while you cover 33, 42, 44, and 53 on the Class Number Grid with stick-on notes. After children open their eyes, direct them to look at the covered numbers.

21	22	23	24	25
31	32		34	35
41		43		45
51	52		54	55
61	62	63	64	65

NOTE If children are interested in the top and bottom rows of the grid, you may wish to discuss the "ten friends" that do not appear on the number grid. Encourage children to apply the same patterns mentally to find those ten friends.

Math Journal 2, p. 179

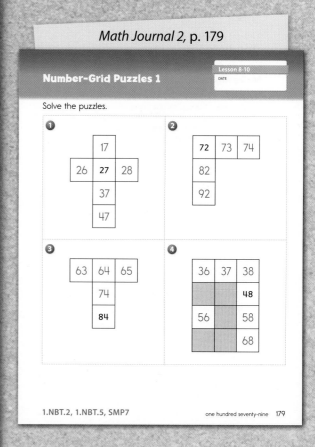

Lesson 8-10

Number-Grid Puzzles 1

DATE

Solve the puzzles.

❶

	17	
26	27	28
	37	
	47	

❷

72	73	74
82		
92		

❸

63	64	65
	74	
	84	

❹

36	37	38
		48
56		58
		68

1.NBT.2, 1.NBT.5, SMP7   one hundred seventy-nine   179

Number-Grid Puzzles 2

Lesson 8-10
DATE

Fill in the missing numbers.

									0
	2	3	4	5		7		9	10
						17			20
21	22	23				27			30
	32					37			40
	42		44						
	52		54						60
61			64		66	67	68		70
	72		74				78		80
		83	84	85	86		88		90
91									100
101	102	103	104						110

180 one hundred eighty 1.NBT.2, 1.NBT.5, SMP7

Make My Design

NAME

DATE

Materials	Pattern blocks, folder
Players	2
Skill	Create composite shapes
Object of the Game	To create the same shape as your partner

Directions

① Player 1 chooses 6 blocks.
   Player 2 chooses the same 6 blocks.

② Players sit face-to-face with a folder between them.
   They stand the folder to hide the designs.

③ Player 1 creates a shape with the blocks.

④ Using only words, Player 1 tells Player 2 how to "Make My Design." Player 2 can ask questions.

⑤ Players remove the folder and look at the two designs. Players discuss how closely the designs match.

⑥ Players trade roles and play again.

1.G.1, 1.G.2   G59

Ask: *How could you find the missing numbers?* Sample answers: Find 1 more than 32, 41, 43, and 52. Find 10 more than 32, 23, 43, and 34. Then have children try just using the 43 to find all the missing numbers. *How do you find the number to the left of 43?* Sample answers: Count back 1. Subtract 1 from 43. *How do you find the number directly above 43?* Sample answers: Count back 10. Find 10 less than 43. Continue until children have discovered all the missing numbers.

Display and complete journal page 179 as a class. Explain that these number-grid puzzles show only part of the number grid to make the problems more challenging. Children should rely on what they know about number-grid patterns to solve them.

Differentiate Adjusting the Activity

When children begin solving the number-grid puzzles, some may need to see the pieces from the first journal page (*Math Journal 2*, p. 179) within the context of a complete number grid to visualize the place-value patterns. For these children, copy a complete number grid, and outline the part for the puzzle in black marker. When children become more proficient at solving the puzzles, have them return to using only the puzzle pieces.

Go Online   Differentiation Support

Then have children complete the next journal page independently or in partnerships. Discuss the patterns children use to solve the problems.
GMP7.2

Academic Language Development Children may benefit from using sentence frames to describe what is missing in the number grid. Write examples, such as: "I know it is 57 because the numbers get larger by 1 when I go to the right. I know it is 36 because numbers get smaller by 10 when I move up a column."

✓ **Assessment Check-In** CCSS 1.NBT.5

Math Journal 2, p. 180

Observe as children find numbers 10 more and 10 less than a given number on journal page 180. Expect children to be able to complete the page, using the number grid as needed. For children who struggle, provide base-10 blocks to model adding and subtracting 10. Encourage children who exceed expectations to do the Enrichment activity without using a number grid.

 Assessment and Reporting   Go Online to record student progress and to see trajectories toward mastery for these standards.

Summarize Have children close their eyes. Tell them to think of the number 22. Ask: *What numbers are 1 more and 1 less than 22?* 23, 21 Then ask: *What numbers are 10 more and 10 less than 22?* 32, 12 Encourage them to think of the number grid or base-10 blocks in their heads to help them find 10 more and 10 less. GMP7.2 Discuss whether it was easier to find 1 more and 1 less than 22, or 10 more and 10 less than 22 in their heads. Tell children to keep practicing until it is just as easy to find 10 more and 10 less than a number as it is to find 1 more and 1 less than a number.

3 Practice 10–15 min

Go Online ePresentations eToolkit Home Connections

▶ Playing *Make My Design*

| WHOLE CLASS | SMALL GROUP | **PARTNER** | INDEPENDENT |

Have children practice describing composite shapes and spatial relationships by playing *Make My Design*. For directions for playing the game, use *Math Masters,* page G59. For more information, see Lesson 8-5.

Observe

- Which children create accurate composite shapes based on their partners' instructions? GMP3.2
- Which children accurately describe positions of pattern blocks relative to each other? GMP6.3

Discuss

- *Which part of your design was most difficult to describe?*
- *Try a round using 6 pattern blocks of the same shape. Was it easier or harder to describe the big shape? Why?*

▶ Math Boxes 8-10

Math Journal 2, p. 181

| WHOLE CLASS | SMALL GROUP | **PARTNER** | **INDEPENDENT** |

Mixed Practice Math Boxes 8-10 are paired with Math Boxes 8-7.

▶ Home Link 8-10

Math Masters, p. 254

Homework Children show someone at home how to fill in pieces of the number grid.

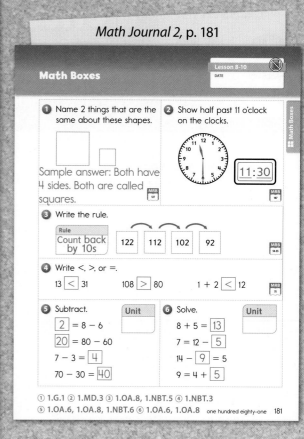

Math Journal 2, p. 181

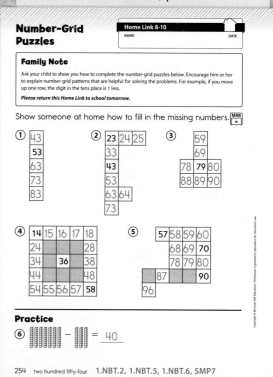

Math Masters, p. 254

Lesson 8-11

Mentally Finding 10 More and 10 Less

Overview Children use place value to mentally add 10 to and subtract 10 from a given number.

▶ **Before You Begin**

For Part 2, prepare a set of cards from *Math Masters,* page 255 for each partnership.

 Common Core State Standards

Focus Clusters

- Represent and solve problems involving addition and subtraction.
- Understand place value.
- Use place value understanding and properties of operations to add and subtract.

1 Warm Up 15–20 min	**Materials**	
Mental Math and Fluency Children write numbers and identify the tens digit.	slate	1.NBT.2
Daily Routines Children complete daily routines.	See pages 2–37.	See pages xiv–xvii.

2 Focus 35–40 min		
Math Message Children mentally find 10 more than 25.		1.OA.1, 1.NBT.5 SMP1
Adding and Subtracting 10 Mentally Children apply strategies for adding or subtracting 10.		1.NBT.5 SMP7
Being a Digit Detective Children discuss patterns as they add and subtract 1 and 10.	*Math Masters,* p. 255; base-10 blocks	1.NBT.2, 1.NBT.5 SMP7
Solving 10 More, 10 Less Problems Mentally Children mentally find numbers that are 10 more or 10 less.	*Math Journal 2,* p. 182	1.NBT.4, 1.NBT.5, 1.NBT.6 SMP6
✓ **Assessment Check-In** See page 752.	*Math Journal 2,* p. 182; base-10 blocks or number grid (optional)	1.NBT.5

CCSS 1.NBT.5 **Spiral Snapshot**

GMC Mentally find 10 more or 10 less than a 2-digit number.

| 8-4
Warm Up | 8-5
Warm Up | 8-10
Focus
Practice | 8-11
Focus
Practice | 9-3
Warm Up
Focus
Practice | 9-8
Warm Up
Practice | 9-9
Focus
Practice | 9-11
Warm Up
Practice |

Spiral Tracker **Go Online** to see how mastery develops for all standards within the grade.

3 Practice 10–15 min		
Applying and Finding Rules Children practice "What's My Rule?" and Frames and Arrows.	*Math Journal 2,* p. 183	1.OA.8, 1.NBT.5 SMP8
Math Boxes 8-11 Children practice and maintain skills.	*Math Journal 2,* p. 184	See page 753.
Home Link 8-11 **Homework** Children mentally find 10 more and 10 less.	*Math Masters,* p. 256	1.OA.2, 1.OA.3, 1.OA.6, 1.NBT.5, SMP7

 connectED.mheducation.com

Plan your lessons online with these tools.

ePresentations Student Learning Center Facts Workshop Game eToolkit Professional Development Home Connections Spiral Tracker Assessment and Reporting English Learners Support Differentiation Support

CCSS 1.NBT.2, 1.NBT.5, SMP2

Readiness
5–10 min

WHOLE CLASS
SMALL GROUP
PARTNER
INDEPENDENT

Adding and Subtracting Longs

base-10 blocks

To explore digit patterns when adding and subtracting 10 from a number, have children model 2-digit numbers using base-10 blocks. Then ask them to add or remove a long to find 10 more or 10 less than the number. Discuss the new number represented by the blocks. Ask them to describe patterns they notice in finding 10 more and 10 less than a number using the blocks. Sample answer: I add a long to find 10 more and subtract a long to find 10 less. The cubes stay the same. GMP2.2 Tell children that when they are finding 10 more or 10 less than a number, they can mentally add or subtract a long.

CCSS 1.NBT.2, 1.NBT.5, SMP7

Enrichment
10–15 min

WHOLE CLASS
SMALL GROUP
PARTNER
INDEPENDENT

Adding and Subtracting 9s

Activity Card 108,
number grid, coin

To further explore using tens and place-value patterns to add and subtract numbers, have children solve + 9 and − 9 problems. GMP7.2

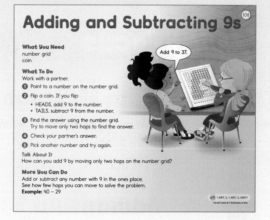

CCSS 1.NBT.5, SMP6

Extra Practice
10–15 min

WHOLE CLASS
SMALL GROUP
PARTNER
INDEPENDENT

Playing *What's Your Way?*

Math Masters, p. G34 (optional);
per partnership: 1 coin,
number grid

For practice finding 10 more and 10 less than a 2-digit number, have children play *What's Your Way?* For detailed instructions, see Lesson 4-11.

Observe
- Which children can explain how they found their answers? GMP6.1
- Which children find 10 more or 10 less mentally?

Discuss
- *How can you find 10 more or 10 less in your head?*

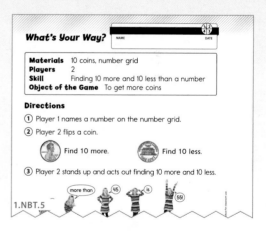

English Language Learners Support

Beginning ELL Think aloud to demonstrate the meaning of the phrase *find the answer mentally*. Use actions to help illustrate the meaning of this phrase, such as closing your eyes and pointing to your head.

Go Online > **ELL** English Learners Support

1 Warm Up 15–20 min [Go Online] ePresentations eToolkit

▶ Mental Math and Fluency

Dictate each number. Have children write it on their slates and circle the digit in the tens place.

●○○ 12 ①2; 18 ①8; 30 ③0
●●○ 70 ⑦0; 28 ②8; 13 ①3
●●● 190 1⑨0; 262 2⑥2; 514 5①4

▶ Daily Routines

Have children complete the Daily Routines. For detailed instructions, see pages 2–37. For specifics on CCSS coverage, see pages xiv–xvii.

2 Focus 35–40 min [Go Online] ePresentations eToolkit

▶ Math Message

Toya put 25 cents in the vending machine for a box of raisins. It says she needs to put in 10 more cents. How much do the raisins cost? Find the answer in your head, and record how you know it. GMP1.2

▶ Adding and Subtracting 10 Mentally

| WHOLE CLASS | SMALL GROUP | PARTNER | INDEPENDENT |

Math Message Follow-Up Discuss how children found their answers to the Math Message. 35 cents Emphasize strategies, including visualizing a number grid and moving down a row from 25 to 35, thinking about adding a long to base-10 blocks representing 25, and adding 1 to the tens digit of 25. Ask children whether this problem reminds them of anything they did in the previous lesson. Finding 10 more than a number

Remind children of previous times they have found 10 more and 10 less than a number, including using the number grid. Tell children that today they will add and subtract 10 from numbers using only their brains. Encourage them to use visualizing strategies from previous lessons to find 10 more and 10 less, until it becomes easy for them to find 10 more and 10 less mentally. GMP7.2

Academic Language Development Explain that there are different ways to describe the same action. For example, finding 10 more than a number is the same as adding 10 to the number or counting up 10; and finding 10 less than a number is the same as subtracting 10 from the number or counting back 10. Provide sentence frames for children to use when describing their strategies. *For example:* "I added 10 to _____. I counted up 10 from _____."

▶ # Being a Digit Detective

Math Masters, p. 255

| WHOLE CLASS | SMALL GROUP | PARTNER | INDEPENDENT |

Have children think about the number 63. Ask: *What is the tens digit in 63?* 6 *What is the ones digit?* 3 *If you add 10 to 63, what are the tens and ones digits in the new number?* 7 is the tens digit; 3 is the ones digit. *What do you notice about the tens and ones digits when you add 10 to a number?* The tens digit goes up by 1. The ones digit stays the same.

Explain that thinking about digits can make adding and subtracting 10s easier. If you add 10, the tens digit increases by 1; if you subtract 10, the tens digit decreases by 1. The ones digit stays the same.

Have children explore this relationship by acting as digit detectives. Provide each partnership with a set of Digit Detective Cards from *Math Masters*, page 255. Partners place the cards facedown in a pile. One child says a 2-digit number and identifies the tens and ones digits in the number. The partner turns over a card, adds or subtracts according to the card, and says the new number and the tens and ones digits in it. Partners compare the tens and ones in each number before putting the card back in the pile. Partners switch roles and repeat the activity.

Discuss patterns children see in the digits. **GMP7.2** Emphasize that the ones digit stays the same in a number when children add or subtract 10. Point out that the tens digit *usually* stays the same when children add or subtract 1 to or from a number. Ask children to give examples in which both digits change when they add or subtract 1 to or from a number. Sample answers: when I subtract 1 from 60; when I add 1 to 29 Model these situations with base-10 blocks, and explain that these digits change when children make a new ten by adding 1 and exchanging their 10 ones for 1 ten or when they exchange a ten for ones so they can subtract 1.

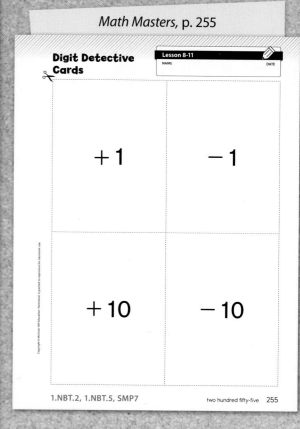

Math Masters, p. 255

Digit Detective Cards

Lesson 8-11
NAME   DATE

$+1$ -1

$+10$ -10

1.NBT.2, 1.NBT.5, SMP7   two hundred fifty-five   255

Adjusting the Activity

Differentiate If children struggle to answer the digit detective questions correctly, begin with just $+1$ and -1 cards. Continue to remind children to think about the number grid as they do the activity. After practicing with $+1$ and -1, change to just using the $+10$ and -10 cards. Have children play with all the cards when they are ready.

 Go Online Differentiation Support

Math Journal 2, p. 182

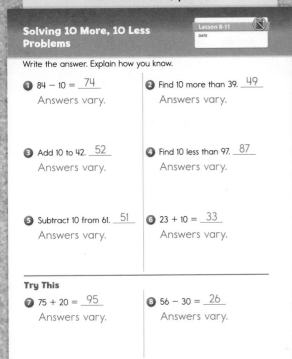

Solving 10 More, 10 Less Problems

Lesson 8-11
DATE

Write the answer. Explain how you know.

1 84 − 10 = __74__
Answers vary.

2 Find 10 more than 39. __49__
Answers vary.

3 Add 10 to 42. __52__
Answers vary.

4 Find 10 less than 97. __87__
Answers vary.

5 Subtract 10 from 61. __51__
Answers vary.

6 23 + 10 = __33__
Answers vary.

Try This

7 75 + 20 = __95__
Answers vary.

8 56 − 30 = __26__
Answers vary.

182   one hundred eighty-two   1.NBT.4, 1.NBT.5, 1.NBT.6, SMP6

Math Journal 2, p. 183

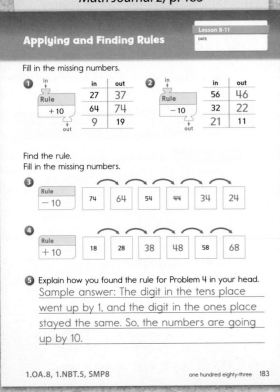

Applying and Finding Rules

Lesson 8-11
DATE

Fill in the missing numbers.

1

in	out
27	37
64	74
9	19

Rule +10

2

in	out
56	46
32	22
21	11

Rule −10

Find the rule.
Fill in the missing numbers.

3 Rule −10

| 74 | 64 | 54 | 44 | 34 | 24 |

4 Rule +10

| 18 | 28 | 38 | 48 | 58 | 68 |

5 Explain how you found the rule for Problem 4 in your head.
Sample answer: The digit in the tens place
went up by 1, and the digit in the ones place
stayed the same. So, the numbers are going
up by 10.

1.OA.8, 1.NBT.5, SMP8   one hundred eighty-three   183

▶ Solving 10 More, 10 Less Problems Mentally

Math Journal 2, p. 182

| WHOLE CLASS | SMALL GROUP | **PARTNER** | **INDEPENDENT** |

Have children solve the problems on journal page 182. Explain that children should complete the problems mentally, without a number grid or other tool. As time permits, discuss the strategies children used when solving 10 more and 10 less problems. **GMP6.1**

✔ **Assessment Check-In** (CCSS) **1.NBT.5**

Math Journal 2, p. 182

Observe as children work on journal page 182. Expect children to complete Problems 1 through 6 correctly. Make sure children who struggle understand that *find 10 more, add 10,* and *+ 10* mean the same thing, as do *find 10 less, subtract 10,* and *− 10.* For children who incorrectly answer any of the problems, have them use base-10 blocks or a number grid to help them solve one of those problems. Explain that they can think about base-10 blocks or the number grid when they add or subtract 10 mentally. Encourage children who exceed expectations to do the optional Enrichment activity.

✔ Assessment and Reporting   (Go Online)   to record student progress and to see trajectories toward mastery for these standards.

Summarize Have children close their eyes and think about the number 32. Have them add 1 to 32. Then have them add 10 to 32. Discuss whether it was easier to add 1 or 10 to 32. Tell children to keep practicing until it is just as easy to add 10 to a number as it is to add 1 to a number.

3 Practice

10–15 min   Go Online

ePresentations   eToolkit   Home Connections

▶ Applying and Finding Rules

Math Journal 2, p. 183

| WHOLE CLASS | SMALL GROUP | **PARTNER** | **INDEPENDENT** |

Have children practice solving "What's My Rule?" and Frames-and-Arrows problems and explain how to figure out a rule. **GMP8.1**

▶ Math Boxes 8–11

Math Journal 2, p. 184

| WHOLE CLASS | SMALL GROUP | **PARTNER** | **INDEPENDENT** |

Mixed Practice Math Boxes 8-11 are paired with Math Boxes 8-9.

▶ Home Link 8–11

Math Masters, p. 256

Homework Children practice adding and subtracting 10 with someone at home.

Math Journal 2, p. 184

Math Boxes

Lesson 8-11
DATE

① Which is bigger: 1 half or 1 quarter of this circle?
   1 half

② Find the rule. Choose the best answer.

in	out
9	2
14	7
15	8

Rule

◯ + 7
⬛ − 7
◯ + 5
◯ + 1

③ At recess, 7 children played tag, 13 children jumped rope, and 9 children played basketball. Complete the tally chart.

Recess Activities

Tag	ᚏᚏ //
Jump Rope	ᚏᚏ ᚏᚏ ////
Basketball	ᚏᚏ ////

④ Books Read in a Week

Write a question you could answer with this graph.
Sample answer: How many books did Sam read?

⑤ **Writing/Reasoning** Explain why this is NOT a triangle.
   Sample answer: It has 3 sides, but they don't all touch.
   Draw lines to make it a triangle.

① 1.G.3 ② 1.OA.6, 1.OA.8 ③ 1.MD.4
184 one hundred eighty-four ④ 1.MD.4 ⑤ 1.G.1, SMP3

Math Masters, p. 256

Mentally Finding 10 More and 10 Less

Home Link 8-11
NAME   DATE

Family Note

In earlier lessons, children used classroom tools, such as a number grid, base-10 blocks, and dimes, to help them add and subtract 10 from a given number. Today they made calculations mentally, using only their brains. Work with your child on adding and subtracting 10 mentally. Have your child do the routine below several times.

Please return this Home Link to school tomorrow.

Ask someone at home to say any number between 10 and 99. Record the number and fill in the blanks after it two times below. Answers vary.

① Number: _____

There are _____ tens and _____ ones in _____.

10 more than _____ is _____. 10 less than _____ is _____.

② Number: _____

There are _____ tens and _____ ones in _____.

10 more than _____ is _____. 10 less than _____ is _____.

Practice

③ Elaine's farm has 4 cows, 2 goats, and 8 chickens. How many animals are there all together?

   __14__ animals   __4__ + __2__ + __8__ = __14__

256 two hundred fifty-six 1.NBT.5, 1.OA.2, 1.OA.3, 1.OA.6, SMP7

Unit 8 Progress Check ✔️

Overview **Day 1:** Administer the Unit Assessments.
Day 2: Administer the Cumulative Assessment.

2-Day Lesson

 Student Learning Center
Students may take
assessments digitally.

 Assessment and Reporting
Record results and track
progress toward mastery.

Day 1: Unit Assessments

▶ **Before You Begin**
For the Unit 8 Challenge, make enough copies of *Assessment Handbook*, p. 58 on construction paper so
that each child has one set of pieces.

① Warm Up   5–10 min

Materials

Self Assessment
Children complete the Self Assessment.

Assessment Handbook, p. 52

②a Assess   35–50 min

Unit 8 Assessment
These items reflect mastery expectations to this point.

Assessment Handbook, pp. 53–57

Unit 8 Challenge (Optional)
Children may demonstrate progress beyond expectations.

Assessment Handbook, pp. 58–60

CCSS Common Core State Standards	Goals for Mathematical Content (GMC)	Lessons	Self Assessment	Unit 8 Assessment	Unit 8 Challenge
1.NBT.2	Understand place value.*	8-10, 8-11		12, 13	
1.NBT.5	Mentally find 10 more or 10 less than a 2-digit number.*	8-10, 8-11	6	12, 13, 16, 17	
1.MD.3	Tell and write time using analog clocks.*	8-8	4	14, 15	
	Tell and write time using digital clocks.	8-8	4	14, 15	
1.MD.4	Organize and represent data.*	8-3, 8-9		10	
	Ask questions about data.*	8-9	5	11	
	Answer questions about data.*	8-9	5	11	
1.G.1	Distinguish between defining and non-defining attributes.	8-1, 8-6		1, 6–9	
	Build and draw shapes to possess defining attributes.*	8-1, 8-5	1	1	
1.G.2	Build composite shapes.	8-5 to 8-7		6, 7	1–5
1.G.3	Partition shapes into equal shares.	8-2 to 8-4, 8-8	2	2, 3	
	Describe equal shares using fraction words.	8-2 to 8-5, 8-8		4, 5	

*Instruction and most practice on this content is complete.

	Goals for Mathematical Practice (GMP)	Lessons	Self Assessment	Unit 8 Assessment	Unit 8 Challenge
SMP1	Keep trying when your problem is hard. GMP1.3	8-5, 8-7			1–5
SMP3	Make mathematical conjectures and arguments. GMP3.1	8-2	3		
	Make sense of others' mathematical thinking. GMP3.2	8-1, 8-5		8, 9	
SMP4	Model real-world situations using graphs, drawings, tables, symbols, numbers, diagrams, and other representations. GMP4.1	8-4		10	
	Use mathematical models to solve problems and answer questions. GMP4.2	8-4, 8-9	5	11	
SMP7	Use structures to solve problems and answer questions. GMP7.2	8-7, 8-10, 8-11		8, 9, 16, 17	

/// Spiral Tracker   **Go Online** to see how mastery develops for all standards within the grade.

1 Warm Up 5–10 min

▶ Self Assessment

Assessment Handbook, p. 52

| WHOLE CLASS | SMALL GROUP | PARTNER | **INDEPENDENT** |

Children complete the Self Assessment to reflect on their progress in Unit 8.

Remind children of a time when they did each type of problem.

Item	Remind children that they . . .
1	built shapes using straws and twist-ties. (Lesson 8-1)
2	shared crackers among two and four people. (Lessons 8-2 and 8-3)
3	explained how they knew shares were equal. (Lessons 8-2 and 8-3)
4	told the time shown on their partner's clock. (Lesson 8-8)
5	answered questions about the Favorite Class Pet bar graph. (Lesson 8-9)
6	solved + 10 and – 10 problems. (Lesson 8-11)

Assessment Handbook, p. 52

NAME   DATE   Lesson 8-12 ✓

Unit 8 Self Assessment

Put a check in the box that tells how you do each skill.

Skills	I can do this by myself. I can explain how to do this.	I can do this by myself.	I can do this with help.
① Build shapes with defining attributes.			
② Divide shapes into 2 or 4 equal parts.			
③ Make mathematical arguments.			
④ Tell time to the half hour.			
⑤ Ask and answer questions about bar graphs.			
⑥ Mentally find 10 more and 10 less.			

52   *Assessment Handbook*

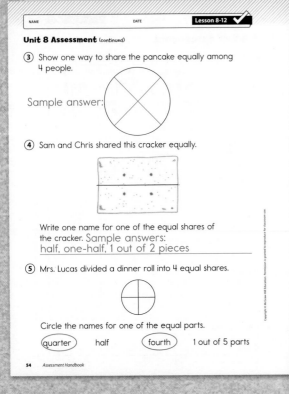

Unit 8 Assessment

Lesson 8-12

① Draw a shape that has 4 sides and that has 4 corners that all look the same.

Sample answers:

Make your shape blue with yellow spots.

Name one defining attribute of your shape.
Sample answers: It has 4 sides. It has 4 corners.

Name one nondefining attribute of your shape.
Sample answers: It is blue. It has yellow spots.

What is your shape called?
Sample answers: rectangle, square

② Show one way you could share the piece of paper equally with one friend.

Sample answer:

Assessment Masters **53**

▶ # Unit 8 Assessment

Assessment Handbook, pp. 53–57

| WHOLE CLASS | SMALL GROUP | PARTNER | **INDEPENDENT** |

Children complete the Unit 8 Assessment to demonstrate their progress on the Common Core State Standards covered in this unit.

Generic rubrics in the *Assessment Handbook* appendix can be used to evaluate children's progress on the Mathematical Practices.

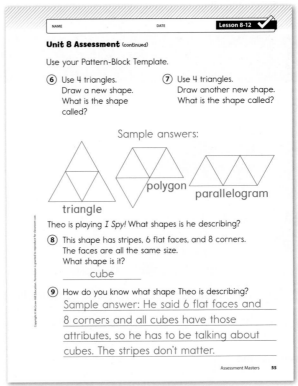

Unit 8 Assessment (continued) Lesson 8-12

Use your Pattern-Block Template.

⑥ Use 4 triangles. Draw a new shape. What is the shape called?

⑦ Use 4 triangles. Draw another new shape. What is the shape called?

Sample answers:

triangle   polygon   parallelogram

Theo is playing *I Spy!* What shapes is he describing?

⑧ This shape has stripes, 6 flat faces, and 8 corners. The faces are all the same size. What shape is it?
cube

⑨ How do you know what shape Theo is describing?
Sample answer: He said 6 flat faces and 8 corners and all cubes have those attributes, so he has to be talking about cubes. The stripes don't matter.

Assessment Masters **55**

Assessment Handbook, p. 55

Unit 8 Assessment (continued) Lesson 8-12

③ Show one way to share the pancake equally among 4 people.

Sample answer:

④ Sam and Chris shared this cracker equally.

Write one name for one of the equal shares of the cracker. Sample answers:
half, one-half, 1 out of 2 pieces

⑤ Mrs. Lucas divided a dinner roll into 4 equal shares.

Circle the names for one of the equal parts.

(quarter)   half   (fourth)   1 out of 5 parts

54 Assessment Handbook

Differentiate **Adjusting the Assessment**

Item(s)	Adjustments
1	To scaffold item 1, provide children with a Pattern-Block Template. To extend item 1, have children list as many defining and nondefining attributes of their shape as they can.
2, 3	To scaffold items 2 and 3, provide paper rectangles and circles.
4, 5	To extend items 4 and 5, have children list the names for 2 equal shares of the cracker and 4 equal shares of the dinner roll.
6, 7	To scaffold items 6 and 7, provide children with pattern blocks. To extend items 6 and 7, have children draw more shapes using 4 triangles.
8, 9	To scaffold items 8 and 9, provide 3-dimensional solids.
10	To scaffold item 10, have children represent each tally mark with a counter and then transfer counters to the bar graph.
11	To extend item 11, have children answer the questions their partners wrote about the bar graph.
12, 13	To scaffold items 12 and 13, have children represent numbers with base-10 blocks. To extend items 12 and 13, have children extend the number grids at least one cell in every direction.
14, 15	To scaffold items 14 and 15, have children use their toolkit clocks. To extend items 14 and 15, have children tell what time it will be in a half hour, 1 hour, and 3 hours from each time.
16, 17	To scaffold items 16 and 17, have children use a number grid.

Advice for Differentiation

All instruction and most practice is complete for the content that is marked with an asterisk (*) on page 754.

Use the online assessment and reporting tools to track children's performance. Differentiation materials are available online to help you address children's needs.

> **NOTE** See the Unit Organizer on pages 678–679 or the online Spiral Tracker for details on Unit 8 focus topics and the spiral.

Assessment Handbook, p. 56

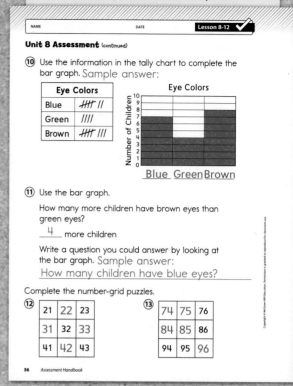

Assessment Handbook, p. 57

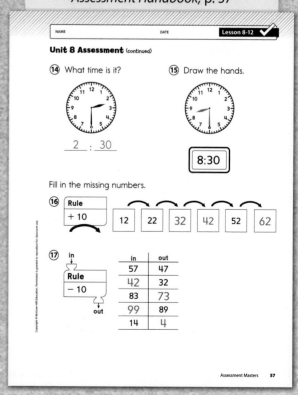

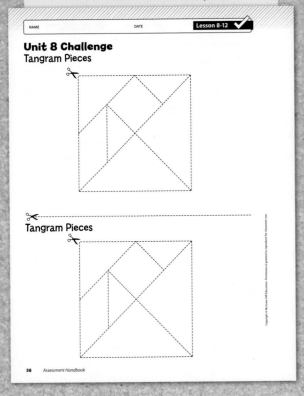

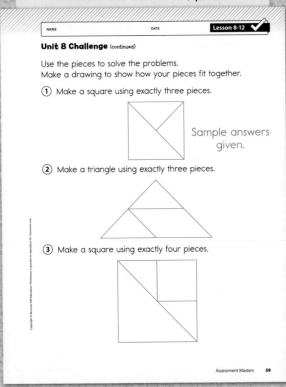

▶ Unit 8 Challenge (Optional)

Assessment Handbook, pp. 58–60

| WHOLE CLASS | SMALL GROUP | PARTNER | **INDEPENDENT** |

Children can complete the Unit 8 Challenge after they complete the Unit 8 Assessment.

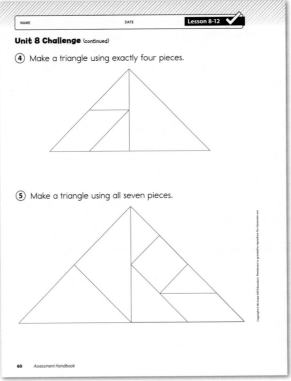

Assessment Handbook, p. 60

Unit 8 Progress Check ✔

Day 2: Cumulative Assessment

2b Assess   30–45 min

Cumulative Assessment

These items reflect mastery expectations to this point.

Materials

Assessment Handbook, pp. 61–63

CCSS Common Core State Standards

Common Core State Standards	Goals for Mathematical Content (GMC)	Cumulative Assessment
1.OA.1	Solve number stories by adding and subtracting.*	1, 2
	Model parts-and-total, change, and comparison situations.*	1, 2
1.OA.2	Model and solve number stories involving the addition of 3 addends.*	8
1.OA.3	Apply properties of operations to add or subtract.*	5, 8
1.OA.4	Understand subtraction as an unknown-addend problem.*	3
1.OA.5	Relate counting to addition and subtraction.*	4
1.OA.6	Add within 10 fluently.*	2, 3, 5
	Subtract within 10 fluently.*	2, 3, 6
	Add doubles automatically.*	5
	Subtract doubles.	6
	Add combinations of 10 automatically.*	3, 5
	Subtract combinations of 10.	3, 6
	Add and subtract within 20 using strategies.	1, 4, 8
1.OA.8	Find the unknown in addition and subtraction equations.*	5, 6
1.NBT.4	Understand adding 2-digit numbers and 1-digit numbers.*	7

	Goals for Mathematical Practice (GMP)	
SMP1	Reflect on your thinking as you solve your problem. **GMP1.2**	8
	Compare the strategies you and others use. **GMP1.6**	4
SMP4	Model real-world situations using graphs, drawings, tables, symbols, numbers, diagrams, and other representations. **GMP4.1**	1, 2, 8
SMP5	Choose appropriate tools. **GMP5.1**	7
	Use tools effectively and make sense of your results. **GMP5.2**	7
SMP6	Explain your mathematical thinking clearly and precisely. **GMP6.1**	3, 7, 8
	Use clear labels, units, and mathematical language. **GMP6.3**	8

*Instruction and most practice on this content is complete.

 Spiral Tracker   Go Online ▷   to see how mastery develops for all standards within the grade.

3 Look Ahead   10–15 min

Math Boxes 8-12

Children preview skills and concepts for Unit 9.

Materials

Math Journal 2, p. 185

Home Link 8-12

Children take home the Family Letter that introduces Unit 9.

Math Masters, pp. 257–259

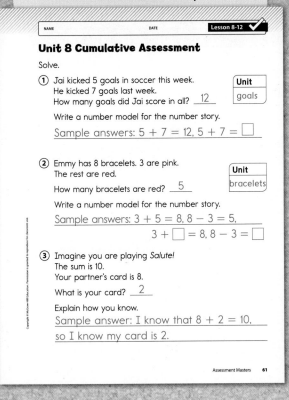

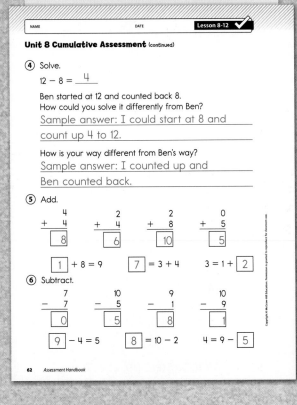

2b Assess 30–45 min

Go Online

Assessment and Reporting   Differentiation Support

▶ Cumulative Assessment

Assessment Handbook, pp. 61–63

| WHOLE CLASS | SMALL GROUP | PARTNER | **INDEPENDENT** |

Children complete the Cumulative Assessment. The problems in the Cumulative Assessment address content from Units 1–7.

Monitor children's progress on the Common Core State Standards using the online assessment and reporting tools.

Generic rubrics in the *Assessment Handbook* appendix can be used to evaluate children's progress on the Mathematical Practices.

Differentiate   Adjusting the Assessment

Item(s)	Adjustments
1, 2	To scaffold items 1 and 2, suggest that children solve using counters, a number line, or situation diagrams. To extend items 1 and 2, have children write number stories for classmates to solve.
3	To scaffold item 3, provide counters. To extend item 3, have children play *Salute!*
4	To scaffold item 4, suggest children use a number grid. To extend item 4, have children list four subtraction problems they might solve by counting back and four subtraction problems they might solve by counting up.
5, 6	To scaffold items 5 and 6, provide counters.
7	To scaffold item 7, determine what tool children should use and provide it to help children find the sum of 48 + 8. To extend item 7, have children use their tools to find the sum of 48 + 38.
8	To scaffold item 8, provide counters. To extend item 8, have children tell other number stories with three addends for classmates to solve.

Advice for Differentiation

All instruction and most practice is complete for the content that is marked with an asterisk (*) on page 759.

Use the online assessment and reporting tools to track children's performance. Differentiation materials are available online to help you address children's needs.

3 Look Ahead 10–15 min

Go Online

Home Connections

▶ Math Boxes 8-12: Preview for Unit 9

Math Journal 2, p. 185

| WHOLE CLASS | SMALL GROUP | **PARTNER** | **INDEPENDENT** |

Mixed Practice Math Boxes 8-12 are paired with Math Boxes 8-8. These problems focus on skills and understandings that are prerequisite for Unit 9. You may want to use information from these Math Boxes to plan instruction and grouping in Unit 9.

▶ Home Link 8-12: Unit 9 Family Letter

Math Masters, pp. 257–259

Home Connection The Unit 9 Family Letter provides information and activities related to Unit 9 content.

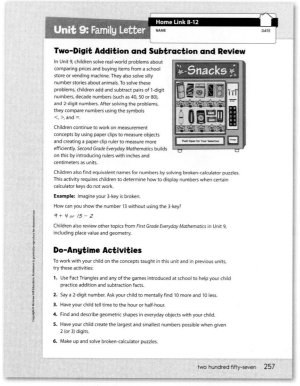

Math Masters, pp. 257–259

Assessment Handbook, p. 63

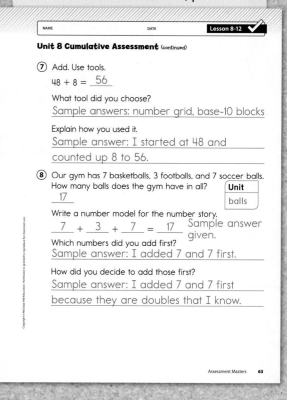

Math Journal 2, p. 185

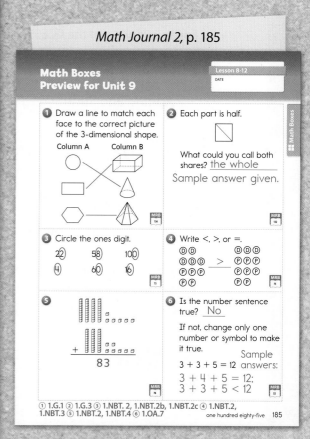

Unit 9 Organizer
Two-Digit Addition and Subtraction and Review

In this unit, children focus on adding and subtracting with 2-digit numbers. They also review other topics. Children's learning will focus on five clusters of the Common Core's content standards, as well as in-depth work on two of the Mathematical Practices.

CCSS Standards for Mathematical Content

Domain	Cluster
Operations and Algebraic Thinking	Represent and solve problems involving addition and subtraction.
Number and Operations in Base Ten	Understand place value. Use place value understanding and properties of operations to add and subtract.
Measurement and Data	Measure lengths indirectly and by iterating length units.
Geometry	Reason with shapes and their attributes.

Because the standards within each domain can be broad, *Everyday Mathematics* has unpacked each standard into Goals for Mathematical Content GMC . For a complete list of Standards and Goals, see page EM1.

For an overview of the CCSS domains, standards, and mastery expectations in this unit, see the **Spiral Trace** on pages 768–769. See the **Mathematical Background** (pages 770–771) for a discussion of the following key topics:

- Adding and Subtracting within 100
- Review and Assessment

CCSS Standards for Mathematical Practice

SMP1	Make sense of problems and persevere in solving them.
SMP5	Use appropriate tools strategically.

For a discussion about how *Everyday Mathematics* develops these practices and a list of Goals for Mathematical Practice GMP , see pages 772–773.

 Virtual Learning Community Go Online to **vlc.cemseprojects.org** to search for video clips on each practice.

IS2011-09/Alamy

Go Digital with these tools at connectED.mcgraw-hill.com

ePresentations | Student Learning Center | Facts Workshop Game | eToolkit | Professional Development | Home Connections | Spiral Tracker | Assessment and Reporting | English Learners Support | Differentiation Support

Contents

*The standards listed here are addressed in the **Focus** of each lesson. For all the standards in a lesson, see the Lesson Opener.

Unit 9 Materials

Virtual Learning Community See how *Everyday Mathematics* teachers organize materials. Search "Classroom Tours" at **vlc.cemseprojects.org.**

Lesson	Math Masters	Activity Cards	Manipulative Kit	Other Materials
9-1	pp. 260–263; G49 (optional)	109–110		slate; demonstration clock; 10 paper clips; glue; 1" by 15" strip of construction paper; 1" by 18" strip of paper; scissors; Fact Triangles; calculator
9-2	pp. 264–265; G46 (optional)	111	base-10 blocks; dimes (optional); counters	Animal Cards
9-3	pp. 266–267; TA3			Standards for Mathematical Practice Poster; *My Reference Book;* colored pencils (optional); Class Number Grid (optional)
9-4	p. 268; G54	112–113	per partnership: 2 dot dice, 20 pennies	slate; calculator; markers or crayons; colored pencils; three large name-collection boxes; *My Reference Book*
9-5	pp. 269–270; TA42; G42 (optional); G43–G45	114	base-10 blocks; pennies; dimes (optional); two sets of number cards 0–9; counters	*My Reference Book* (optional); slate; glue; scissors; Number Cards; number grid
9-6	pp. 271–273; TA42; G42 (optional); G43–G45	115	Quick Look Cards 103, 107, 115; base-10 blocks; pennies; dimes (optional); counters	*My Reference Book* (optional); slate; crayons
9-7	p. 274; G46 (optional); G54	116	base-10 blocks (optional); per partnership: 2 dot dice, 20 pennies; two sets of number cards 0–9	slate; Number Cards; number grid; Animal Cards; *My Reference Book*
9-8	pp. 275–277	117	base-10 blocks; number cards 1–9; coins; counters (optional)	slate; scissors; Number Cards; number grid; number lines
9-9	pp. 278; TA43; G35 (optional)	118	base-10 blocks; per partnership: 4 each of number cards 0–9	slate; per partnership: Number Cards; Class Number Grid; demonstration clock; stick-on notes
9-10	pp. 279–283; TA40 (optional); G55–G56; G57–G59 (optional); G60–G61	119	3-dimensional shape blocks; about 30 four-inch straws and about 15 twist ties; pattern blocks	slate; demonstration clock; plastic egg; everyday objects; tape; scissors; folder
9-11	pp. 284–289; G49 (optional)	120	counters	slate; scissors; tape or glue; markers or crayons; calculator; Fact Triangles
9-12	pp. 290–293; TA43; *Assessment Handbook* pp. 64–72			

Go Online to download all Quick Look Cards.

Problem Solving Professional Development

Everyday Mathematics emphasizes equally all three of the Common Core's dimensions of **rigor**: conceptual understanding, procedural skill and fluency, and applications.

Math Messages, other daily work, Explorations, and Open Response tasks provide many opportunities for children to apply what they know to solve problems.

► Math Message

Math Messages require children to solve a problem they have not been shown how to solve. Math Messages provide almost daily opportunities for problem solving.

► Daily Work

Journal pages, Home Links, Writing/Reasoning prompts, and Differentiation Options often require children to solve problems in mathematical contexts and real-life situations. **Minute Math+** offers number stories and a variety other practice activities for transition times and spare moments throughout the day. See Routine 6, pages 32–37.

► Explorations

In Exploration A of Lesson 9-4, children solve broken-calculator puzzles. In Exploration B, they divide a piece of paper into equal parts and use it to plan a garden. And, in Exploration C, children conduct a final review of their Facts Inventory Record.

► Open Response and Reengagement

In Lesson 9-3, children solve a problem about shopping at the school store and explain their solution strategies. The reengagement discussion on Day 2 may focus on characteristics of a clear explanation or how clear explanations can help children find and correct mistakes in their work. Explaining mathematical thinking clearly and precisely is the focus Mathematical Practice for the lesson. Writing clear and precise explanations can help children organize their thoughts, confirm that their answers are correct, and gain confidence in their problem-solving abilities. GMP6.1

 **Virtual Learning Community** Go Online to watch an Open Response and Reengagement lesson in action. Search "Open Response" at **vlc.cemseprojects.org**.

► Open Response Assessment

In Lesson 9-12, children identify and correct a mistake in a number-grid puzzle and explain how they corrected it. Using mathematical structures to solve problems is the focus Mathematical Practice for this assessment. GMP7.2 As children note and apply mathematical structures such as place-value patterns, they learn to apply familiar problem-solving techniques to new problems.

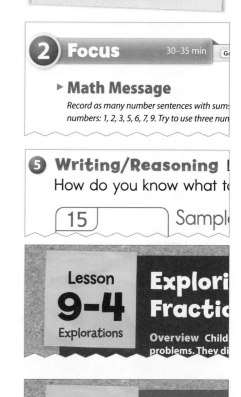

Look for GMP1.1–1.6 markers, which indicate opportunities for children to engage in **SMP1**: "Make sense of problems and persevere in solving them." Children also become better problem solvers as they engage in all of the CCSS Mathematical Practices. The yellow GMP markers throughout the lessons indicate places where you can emphasize the Mathematical Practices and develop children's problem-solving skills.

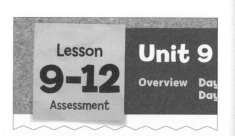

Assessment and Differentiation

See pages xxii–xxv to learn about this comprehensive online system for recording, monitoring, and reporting children's progress using core program assessments.

VLC Virtual Learning Community | **Go Online** to **vlc.cemseprojects.org** for tools and ideas related to assessment and differentiation from *Everyday Mathematics* teachers.

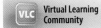

Ongoing Assessment

In addition to frequent informal opportunities for "kid watching," every lesson (except Explorations) offers an **Assessment Check-In** to gauge children's performance on one or more of the standards addressed in that lesson.

Lesson	Task Description	**CCSS** Common Core State Standards
9-1	Measure length accurately with nonstandard units.	1.MD.2, SMP5, SMP6
9-2	Solve comparison number stories using a number grid or base-10 blocks.	1.OA.1, 1.NBT.6
9-3	Add 2-digit numbers and explain strategies used.	1.NBT.4, SMP6
9-5	Use strategies to add 2-digit numbers.	1.NBT.4, SMP6
9-6	Solve a 2-digit subtraction number story and explain strategies used.	1.NBT.6, SMP6
9-7	Use tools and strategies to solve 2-digit number stories and record number models.	1.NBT.4
9-8	Record comparison number models using < and >.	1.NBT.3
9-9	Identify the number of tens and ones in a 2-digit number and the value of each digit.	1.NBT.2
9-10	Compose and name 3-dimensional shapes.	1.G.1, 1.G.2, SMP6
9-11	Record a name for all the parts of a shape divided into equal shares.	1.G.3, SMP6

▶ Periodic Assessment

Unit 9 Progress Check This assessment focuses on the CCSS domains of *Operations and Algebraic Thinking, Number and Operations in Base Ten, Measurement and Data,* and *Geometry.* It also contains an Open Response Assessment to assess children's understanding of using structures to solve problems. **GMP7.2**

End-of-Year Assessment This assessment provides a final opportunity to check children's understanding of the concepts and skills presented in *First Grade Everyday Mathematics.* See the *Assessment Handbook.*

NOTE Odd-numbered units include an **Open Response Assessment**. Even-numbered units include a **Cumulative Assessment.**

Unit 9 Differentiation Activities

 Differentiation Support   ELL English Learners Support

Differentiation Options Every regular lesson provides **Readiness**, **Enrichment**, **Extra Practice**, and **Beginning English Language Learner Support** activities that address the Focus standards for that lesson.

CCSS 1.OA.6, SMP2	CCSS 1.OA.6, SMP2	CCSS 1.OA.6, SMP6
Readiness 10–15 min	**Enrichment** 10–15 min	**Extra Practice** 10–15 min
WHOLE CLASS / SMALL GROUP / PARTNER	WHOLE CLASS / SMALL GROUP / PARTNER	WHOLE CLASS / SMALL GROUP / PARTNER
Exploring Doubles	**Drawing Doubles Pictures**	**Playing *Roll and Record Doubles***
number cards 1–10, counters,	Activity Card 76, crayons	

Activity Cards These activities, written to the children, enable you to differentiate Part 2 of the lesson through small-group work.

English Language Learners Activities and point-of-use support help children at different levels of English language proficiency succeed.

Differentiation Support Two online pages for most lessons provide suggestions for game modifications, ways to scaffold lessons for children who need additional support, and language development suggestions for Beginning, Intermediate, and Advanced English language learners.

Activity Card 111

Spending Money at the School Store

Differentiation Support online pages

For **ongoing distributed practice**, see these activities:
- Mental Math and Fluency
- Differentiation Options: Extra Practice
- Part 3: Journal pages, Math Boxes, *Math Masters*, Home Links
- Print and online games

Ongoing Practice

 Differentiation Support

▶ Embedded Facts Practice

Basic Facts Practice can be found in every part of the lesson. Look for activities or games labeled with CCSS 1.OA.6; or go online to the Spiral Tracker and search using CCSS 1.OA.6.

▶ Games

Games in *Everyday Mathematics* are an essential tool for practicing skills and developing strategic thinking.

Lesson	Game	Skills and Concepts	CCSS Common Core State Standards
9-1 9-11	*Beat the Calculator*	Using mental addition	1.OA.6, SMP5, SMP6
9-2 9-7	*Animal Weight Top-It*	Adding 2-digit numbers	1.NBT.3, 1.NBT.4, SMP1, SMP6
9-4 9-7	*Tric-Trac*	Adding within 10	1.OA.3, 1.OA.6, SMP1
9-5 9-6	*Stop and Go*	Adding and subtracting 2-digit numbers	1.NBT.4, 1.NBT.6, SMP5, SMP6
9-8	*Top-It* with School Store Cards	Comparing numbers using addition	1.NBT.3, 1.NBT.4, SMP3
9-9	*The Digit Game*	Comparing 2-digit numbers based on place value	1.NBT.2, 1.NBT.3, SMP3
9-10	*Time Match*	Telling time	1.MD.3, SMP2
9-10	*Make My Design*	Creating composite shapes	1.G.2, SMP 6
9-10	*I Spy*	Describing and identifying shapes	1.G.1, SMP3

ⒸⒸⓈⓈ Spiral Trace: Skills, Concepts, and Applications

⭐ **Mastery Expectations** This Spiral Trace outlines instructional trajectories for key standards in Unit 9. For each standard, it highlights opportunities for Focus instruction, Warm Up and Practice activities, as well as formative and summative assessment. It describes the **degree of Tmastery**— as measured against the entire standard—expected at this point in the year.

Operations and Algebraic Thinking

1.OA.7 Understand the meaning of the equal sign, and determine if equations involving addition and subtraction are true or false. *For example, which of the following equations are true and which are false?* $6 = 6, 7 = 8 - 1, 5 + 2 = 2 + 5, 4 + 1 = 5 + 2.$

| 5-10 Warm Up Practice | 5-13 Progress Check | 6-3 Warm Up Focus | 6-9 Focus Practice | 6-12 Progress Check | 7-2 Practice | 7-10 Focus Practice | 7-12 Progress Check | 8-9 Warm Up | 9-4 Focus Practice | 9-6 Practice | 9-8 Focus |

⭐ By the end of Unit 9, expect children to **determine whether equations involving addition and subtraction are true or false.**

Number and Operations in Base Ten

1.NBT.2 Understand that the two digits of a two-digit number represent amounts of tens and ones.

| 7-7 Practice | 8-6 Warm Up Practice | 8-8 Practice | 8-10 Warm Up Focus Practice | 8-11 Warm Up Focus | 8-12 Progress Check | 9-2 Warm Up Practice | 9-3 Warm Up | 9-5 Warm Up Practice | 9-8 Practice | 9-9 Focus Practice | 9-12 Progress Check |

⭐ By the end of Unit 9, expect children to **identify the number of tens and ones in a two-digit number and the value of the digit in each place.**

1.NBT.3 Compare two two-digit numbers based on meanings of the tens and ones digits, recording the results of comparisons with the symbols $>$, $=$, and $<$.

| 5-9 Focus Practice | 5-11 Practice | 5-13 Progress Check | 6-2 Warm Up Practice | 6-8 Focus Practice | 6-11 Warm Up Focus Practice | 6-12 Progress Check | 7-4 Practice | 8-10 Warm Up Practice | 9-5 Warm Up | 9-8 Focus Practice | 9-12 Progress Check |

⭐ By the end of Unit 9, expect children to **use place-value understanding to record comparisons of two-digit numbers using relation symbols.**

1.NBT.4 Add within 100, including adding a two-digit number and a one-digit number, and adding a two-digit number and a multiple of 10, using concrete models or drawings and strategies based on place value, properties of operations, and/or the relationship between addition and subtraction; relate the strategy to a written method and explain the reasoning used. Understand that in adding two-digit numbers, one adds tens and tens, ones and ones; and sometimes it is necessary to compose a ten.

| 7-1 Warm Up Practice | 7-5 Practice | 7-9 Warm Up | 8-2 Practice | 8-7 Warm Up | 8-11 Focus | 8-12 Progress Check | 9-1 through 9-8 Focus Practice | 9-9 Practice | 9-11 Warm Up Practice | 9-12 Progress Check |

⭐ By the end of Unit 9, expect children to **add within 100 and explain their strategies.**

Spiral Tracker

Go to **connectED.mcgraw-hill.com** for comprehensive trajectories that show how in-depth mastery develops across the grade.

1.NBT.6 Subtract multiples of 10 in the range 10–90 from multiples of 10 in the range 10–90 (positive or zero differences), using concrete models or drawings and strategies based on place value, properties of operations, and/or the relationship between addition and subtraction; relate the strategy to a written method and explain the reasoning used.

| 6-7 Focus | 6-8 Warm Up | 7-1 Practice | 7-5 Practice | 8-7 Warm Up Practice | 8-11 Focus | 9-2 Focus | 9-5 Focus | 9-6 Focus | 9-9 Practice | 9-11 Warm Up Practice | 9-12 Progress Check |

⭐ By the end of Unit 9, expect children to **subtract multiples of 10 from multiples of 10 within 100 and explain their strategies.**

Measurement and Data

1.MD.2 Express the length of an object as a whole number of length units, by laying multiple copies of a shorter object (the length unit) end to end; understand that the length measurement of an object is the number of same-size length units that span it with no gaps or overlaps. *Limit to contexts where the object being measured is spanned by a whole number of length units with no gaps or overlaps.*

| 4-2 through 4-4 Focus Practice | 4-5 Practice | 4-9 Practice | 4-11 Practice | 4-12 Progress Check | 5-7 Focus Practice | 5-8 Focus Practice | 5-13 Progress Check | 6-12 Progress Check | 9-1 Focus Practice | 9-12 Progress Check |

⭐ By the end of Unit 9, expect children to **measure the length of an object with same-size units.**

Geometry

1.G.2 Compose two-dimensional shapes (rectangles, squares, trapezoids, triangles, half-circles, and quarter-circles) or three-dimensional shapes (cubes, right rectangular prisms, right circular cones, and right circular cylinders) to create a composite shape, and compose new shapes from the composite shape. (Students do not need to learn formal names such as "right rectangular prism.")

| 1-9 Focus | 4-5 Focus Practice | 6-3 Practice | 8-5 Focus Practice | 8-6 Focus Practice | 8-7 Focus | 8-10 Practice | 8-12 Progress Check | 9-3 Practice | 9-5 Practice | 9-10 Focus | 9-12 Progress Check |

⭐ By the end of Unit 9, expect children to **construct composite shapes from two- and three-dimensional shapes.**

1.G.3 Partition circles and rectangles into two and four equal shares, describe the shares using the words *halves*, *fourths*, and *quarters*, and use the phrases *half of*, *fourth of*, and *quarter of*. Describe the whole as two of, or four of the shares. Understand for these examples that decomposing into more equal shares creates smaller shares.

| 7-6 Focus | 8-2 Focus Practice | 8-3 Focus Practice | 8-4 Focus Practice | 8-5 Focus Practice | 8-7 Practice | 8-8 WarmUp Focus | 8-9 Practice | 8-12 Progress Check | 9-4 Focus Practice | 9-11 Focus Practice | 9-12 Progress Check |

⭐ By the end of Unit 9, expect children to **partition shapes into two or four equal shares, describe the shares, and understand that making more equal shares results in smaller shares.**

Key ✓ = Assessment Check-In ✓ = Progress Check Lesson ▱ = Current Unit ▰ = Previous or Upcoming Lessons

Mathematical Background: Content

 The discussion below highlights major content areas and Common Core State Standards addressed in Unit 9. See the online Spiral Tracker for complete information about learning trajectories for all standards.

▶ Adding and Subtracting within 100
(Lessons 9-2, 9-3, 9-5 through 9-9)

Throughout the year, children have done considerable work with addition and subtraction. They have developed strategies for adding and subtracting within 20 and worked toward fluency with basic addition and subtraction facts. They have used place-value understanding and manipulatives such as base-10 blocks to add ones to ones and tens to tens, and to add or subtract 10 from any 2-digit number. In Unit 9, children bring together all these skills to add within 100 and to subtract multiples of 10 from larger multiples of 10 within 100. **1.NBT.4, 1.NBT.6** They develop and practice these skills as they create and solve problems about school supplies in Lessons 9-2, 9-3, and 9-8, vending machines in Lessons 9-5 and 9-6, and silly animal stories in Lesson 9-7.

Because children have been encouraged to use tools and strategies throughout the year, the leap to working with larger numbers is a natural one; children realize that they can apply familiar tools and strategies to new situations. For example, children have already had many experiences using base-10 blocks to exchange 10 cubes (ones) for 1 long (ten) in problems such as 25 + 7. Now, in Unit 9, when they encounter problems with larger numbers (for example, calculating the combined height of a 48-inch cheetah and a 36-inch octopus), they can apply the same tool and strategy: represent both numbers with base-10 blocks, combine the blocks, and make exchanges to represent the total.

Similarly, children already know that they can count up or back on a number grid to solve addition or subtraction problems like 15 + 3 or 23 − 4. In Lesson 9-9, children use the same tool (the number grid) and the same strategy (counting up or back) to calculate with larger numbers (for example, 28 + 60, or 62 − 50)—but this time they skip up or down entire rows to add or subtract 10s. With enough practice, most children will make the leap to mentally adding and subtracting 10s by the end of Unit 9.

(t)Ed-Imaging; (b)Mazer Creative Services

 Standards and Goals for Mathematical Content

Because the standards within each domain can be broad, *Everyday Mathematics* has unpacked each standard into Goals for Mathematical Content GMC. For a complete list of Standards and Goals, see page EM1.

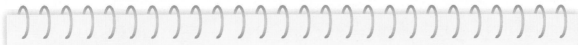

Unit 9 Vocabulary

ruler

▶ Review and Assessment

(Lessons 9-1, 9-8, 9-10 through 9-12)

The authors of *Everyday Mathematics* believe that children must be exposed to concepts and skills many times and in many ways before they attain mastery and command of those concepts and skills. With each new exposure, most children need reminders of what they have already been exposed to, even for concepts and skills previously learned. Also, most teachers find that children learn at different rates; only rarely are children all at the same level at the same time for any given skill or concept.

The review lessons in Unit 9 play an important role in this spiral format, offering a final exposure to key concepts developed throughout first grade. Some review lessons provide additional practice in new contexts. For example, in Lesson 9-11, children review fraction concepts by creating equal shares of granola bars and watermelon slices. Other review lessons reintroduce a familiar context with a new twist. For example, in Lesson 9-8, children revisit the School Store mini-posters to review the concept of equivalence with more sophisticated problems. In Lesson 9-9, children review place value with familiar number-grid puzzles, but this time, they create their own puzzles. And in Lesson 9-10, children work with familiar 3-dimensional shapes, but are challenged to construct their own 3-dimensional models with cut-out templates. And, in some review lessons, children get a preview of more sophisticated concepts that they will encounter in the next grade. For example, in Lesson 9-1, after reviewing first grade measurement concepts, children discover the usefulness of "collecting units" as they make their own rulers with nonstandard units.

Assessment, like review, is revisited regularly throughout the year, with a culminating wrap-up at the end of the year. While *Everyday Mathematics* is carefully designed to promote mastery of the *Common Core State Standards for Mathematics,* its ongoing and regular assessments allow teachers and children to focus on *progress* as much as mastery. In particular, the end-of-year monitoring gives both you and the second-grade teachers a sense of each child's progress and the general level at which the second-grade classes might be able to start after suitable review. End-of-year assessment also gives each child an awareness of what he or she has accomplished over the course of the school year.

Mathematical Background: Practices

 In Everyday Mathematics, *children learn the* **content** *of mathematics as they engage in the* **practices** *of mathematics. As such, the Standards for Mathematical Practice are embedded in children's everyday work, including hands-on activities, problem-solving tasks, discussions, and written work. Read here to see how Mathematical Practices 1 and 5 are emphasized in this unit.*

▶ Standard for Mathematical Practice 1

When children begin Kindergarten and first grade, many have a limited notion that mathematics is something they memorize or simply "know" in a rote manner. For example, children might proudly state, "I know what 5 plus 5 is!"— even when they have little understanding of the process of addition. Now, by the end of the year, through a regular focus on Mathematical Practice 1, most children view mathematics in a more complex way. They "do" mathematics— and doing mathematics involves solving problems and discussing how they solved them.

In Unit 9, children have many opportunities to engage with Mathematical Practice 1 as they make sense of and solve problems. In Lesson 9-2, children are encouraged to estimate answers before solving problems and to use the estimates as landmarks to reflect on their thinking and to check the reasonableness of their solutions. GMP1.2 In Lessons 9-4 and 9-8, children are asked explicitly to check whether their answers make sense. GMP1.4 In Lesson 9-4, children are encouraged to check their answers to broken-calculator problems by trying them out on a calculator. And, in Lesson 9-8, children check their answers to name-collection box problems by using the expressions to write equations.

Children are also encouraged to solve problems in more than one way. GMP1.5 In Lesson 9-5, children are challenged to solve vending-machine problems using a variety of strategies, and in Lesson 9-11, children find different ways to divide shapes into 4 equal shares. In Lessons 9-2 and 9-7, children share their strategies and discuss the similarities and differences among their problem-solving methods. GMP1.6

As the school year comes to a close, children should be reminded often about the progress they've made in tackling difficult problems. They should see themselves as problem-solvers, ready to take on new challenges in second grade.

Standards and Goals for Mathematical Practice

SMP1 **Make sense of problems and persevere in solving them.**

GMP1.1 Make sense of your problem.

GMP1.2 Reflect on your thinking as you solve your problem.

GMP1.3 Keep trying when your problem is hard.

GMP1.4 Check whether your answer makes sense.

GMP1.5 Solve problems in more than one way.

GMP1.6 Compare the strategies you and others use.

SMP5 **Use appropriate tools strategically.**

GMP5.1 Choose appropriate tools.

GMP5.2 Use tools effectively and make sense of your results.

Go Online to the *Implementation Guide* for more information about the Mathematical Practices.

For children's information on the Mathematical Practices, see *My Reference Book,* pages 1–22.

▶ Standard for Mathematical Practice 5

As children in first grade develop their skills as problem-solvers, they learn to appreciate the value of tools to help them solve mathematical problems. Children have used a variety of tools throughout first grade—counters, number lines, number grids, base-10 blocks, calculators, and other tools. In Unit 9, children focus on choosing appropriate tools and using those tools effectively. GMP5.1, GMP5.2 In Lesson 9-1, as children review measurement techniques, they discuss how stringing paper clip units together to make a chain can create a more efficient tool for measuring length. In Lessons 9-6, 9-7, and 9-8, children discuss and compare the tools they use to solve problems. For example, when children are asked to describe ways to solve $46 + 38$, they may mention counting on fingers, using counters, counting up on a number grid, or modeling with base-10 blocks. Each tool will result in the same answer, but some tools are more efficient than others. Throughout Unit 9, discussions focus on which tool makes the most sense for a given problem.

As children develop as mathematicians, their use of tools will reflect their growing understanding of mathematical concepts. With practice, many tools used in first grade will eventually be replaced by mental math and other shortcuts, while other more sophisticated tools, such as rulers, protractors, and geometry software, will be introduced in later grades. By emphasizing the use of tools as an important mathematical practice in first grade, young children develop habits and attitudes about problem-solving processes that will serve them well into the upper grades.

Review: Measurement

Overview Children create rulers using paper clips as units of length.
They use the rulers to measure objects.

▶ **Before You Begin**
For Part 2, each child will need a 1" by 15" piece of construction paper to make a paper-clip ruler.
Children take these rulers home but will use them again in the Unit 9 Assessment, so you may wish to
make extras. Prepare one demonstration ruler by gluing or taping 10 paper clips end-to-end to a strip
of paper. For the optional Readiness and Enrichment activities, prepare 1" by 18" strips of paper.

▶ **Vocabulary**
ruler

 **Common Core
State Standards**

Focus Clusters
• Use place value understanding
and properties of operations to
add and subtract.
• Measure lengths indirectly and
by iterating length units.

1 Warm Up 15–20 min

Materials

Mental Math and Fluency Children read an analog clock and record the time digitally.	demonstration clock, slate	1.MD.3
Daily Routines Children complete daily routines.	See pages 2–37.	See pages xiv–xvii.

2 Focus 30–35 min

Math Message Children use paper clips to measure the height of their desks.	per child: 10 paper clips	1.MD.2 SMP6
Making Rulers Children iterate units to make a ruler.	*Math Masters*, p. 260; glue; per child: 1" by 15" strip of construction paper, scissors	1.MD.2 SMP5, SMP6
Measuring with Rulers Children discuss how to use a ruler to measure. They practice measuring objects with it.	*Math Journal 2*, p. 186; paper-clip ruler; slate; per child: 10 paper clips	1.NBT.4, 1.MD.1, 1.MD.2 SMP5, SMP6
✓ **Assessment Check-In** See page 780.	*Math Journal 2*, p. 186; paper clips; paper-clip ruler	1.MD.2 SMP5, SMP6

CCSS 1.MD.2 **Spiral Snapshot**

GMC Express length as a whole
number of units.

4-2 Focus Practice	4-3 Focus Practice	4-4 Focus Practice	4-5 Practice	4-9 Practice	5-7 Focus Practice	5-8 Focus Practice	9-1 Focus Practice

▷

III Spiral Tracker   Go Online to see how mastery develops for all standards within the grade.

3 Practice 15–20 min

Playing *Beat the Calculator* **Game** Children practice addition facts.	*Math Masters*, p G49 (optional); Fact Triangles; calculator	1.OA.3, 1.OA.6 SMP6
Math Boxes 9-1 Children practice and maintain skills.	*Math Journal 2*, p. 187	See page 781.
Home Link 9-1 **Homework** Children find lengths with paper-clip rulers.	*Math Masters*, p. 263	1.NBT.3, 1.MD.2 SMP5, SMP6

connectED.mcgraw-hill.com ▷

Plan your lessons online
with these tools.

 ePresentations   Student Learning Center   Facts Workshop Game   eToolkit   Professional Development   Home Connections   Spiral Tracker   Assessment and Reporting   English Learners Support   Differentiation Support

Differentiation Options

RtI

Readiness   5–10 min

WHOLE CLASS
SMALL GROUP
PARTNER
INDEPENDENT

Measuring Body Parts

1 paper clip, slate,
1" by 18" strips of paper

To provide additional exposure to measuring with individual units, have children measure around different body parts. Demonstrate measuring a child's head. Wrap one of the strips of paper you prepared around the child's head. Mark the paper to show the distance, lay the paper flat, and measure accurately by iterating one paper clip. GMP6.4 Have children find the distance in paper-clip units around their wrists, heads, ankles, and necks, and have them record the results on their slates. Discuss the results, asking questions such as: *Who has the smallest wrist? The biggest head? Which ankle size is most common?*

Enrichment   20–25 min

WHOLE CLASS
SMALL GROUP
PARTNER
INDEPENDENT

Making a Digit Ruler

Activity Card 109;
Math Masters, p. 261;
1" by 18" strips of paper

To further explore length measurement, children make their own rulers with units that are one index finger wide. They measure items and record the results on *Math Masters*, page 261. GMP5.2

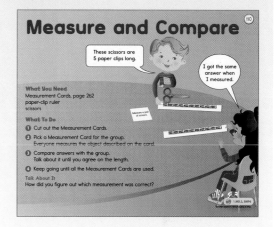

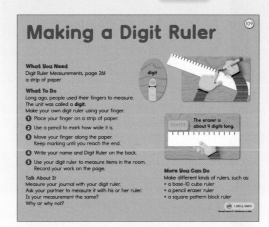

Extra Practice   10–15 min

WHOLE CLASS
SMALL GROUP
PARTNER
INDEPENDENT

Measure and Compare

Activity Card 110;
Math Masters, p. 262;
paper-clip rulers (made in Part 2 of the lesson); scissors

To provide practice measuring length accurately, have children use their paper-clip rulers to measure classroom objects. GMP6.4

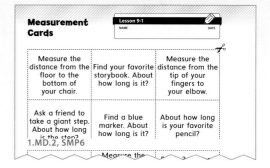

English Language Learners Support

Beginning ELL Use visuals and actions to help children understand that the terms *tall* and *high* refer to height. Model using questions and statements such as: *About how many paper clips tall are you? What a tall skyscraper! That is a very tall tree! About how many paper clips high is your desk? About how many paper clips high is this door? What a high mountain!* Provide question and answer sentence frames for oral practice, such as: *Is that a tall _____? That's a tall _____! Is the top of that _____ high? Yes, that _____ is high!*

  English Learners Support

Standards and Goals for
Mathematical Practice

SMP5 Use appropriate tools strategically.
    GMP5.1 Choose appropriate tools.
    GMP5.2 Use tools effectively and make sense of your results.
SMP6 Attend to precision.
    GMP6.4 Think about accuracy and efficiency when you count, measure, and calculate.

1 Warm Up   15–20 min   Go Online

ePresentations   eToolkit

▶ Mental Math and Fluency

Display the following times on the demonstration clock. Have children record the times in digital notation on their slates.

● ○ ○   3:00, 7:00, 11:00

● ● ○   12:00, 2:00, 10:30

● ● ●   4:30, 5:30, 7:30

▶ Daily Routines

Have children complete the Daily Routines. For detailed instructions, see pages 2–37. For specifics on CCSS coverage, see pages xiv–xvii.

2 Focus   30–35 min   Go Online

ePresentations   eToolkit

▶ Math Message

Take 10 paper clips. Use them to measure the height of your desk (or table). Record the total height.

What was difficult about measuring height accurately using paper clips? **GMP6.4** *Record what you think.*

▶ Making Rulers

Math Masters p. 260

| WHOLE CLASS | SMALL GROUP | PARTNER | INDEPENDENT |

Math Message Follow-Up Have children share their responses to the Math Message. Highlight responses about the difficulties of placing individual paper clips end-to-end to measure vertically, making a straight path with the paper clips, or keeping track of the total. Emphasize the idea of linking paper clips together to make a single chain. If no one tried this, demonstrate it.

Math Masters, p. 260

Paper Clips for Ruler

Lesson 9-1

NAME   DATE

260   two hundred sixty   1.NBT.4, 1.MD.2, SMP5

You may wish to use the rhyme introduced in Lesson 4-2 to review good measurement techniques and to help children explain why measuring accurately with individual units can sometimes be difficult:

> To find the length of your desk or your friend,
> use the same unit from **end to end.**
> Measurement will be a snap
> if you **never gap or overlap.**

Guide the discussion to conclude that connecting small units into a larger collection of units makes it easier to measure quickly and accurately. Emphasize the importance of choosing or creating an appropriate tool when measuring. GMP5.1, GMP6.4

Tell children that today they will make a special collection of units called a **ruler.** Children may be familiar with rulers and the terms *inch* and *foot.* Explain that a ruler can be made with any unit. In second grade, they will use inches and feet as units. But today they will make rulers using a unit they have used often in first grade—the paper clip.

Demonstrate that even a chain of 10 paper clips is a difficult tool for measuring; the paper clips overlap when the chain gets tangled or is not pulled taut. Remedy this by gluing or taping 10 paper clips end-to-end on a strip of paper. Tell children that they will create their rulers using drawings of paper clips. Display *Math Masters,* page 260.

Demonstrate making a ruler by cutting out the 10 paper clips and pasting them along the edge of the paper. (Avoid starting at the very end of the strip.) Position the pieces incorrectly, and ask children for help. They should tell you to place the units in a straight line along (or parallel to) the edge of the paper without gaps or overlaps. Point out that you are not trying to measure the paper strip, so you do not need to span the whole width. You are making a ruler to use for measuring other items accurately.

Paper-clip ruler

Distribute a strip of 1" by 15" paper and *Math Masters,* page 260 to each child. To make it easier to see the extent of each unit, you may wish to have children color the paper-clip segments in contrasting colors before cutting them out. Have children write their names and "Paper-Clip Ruler" on one side and then glue the paper-clip pieces on the other to make their rulers.

NOTE Some children may make the first unit and the end of the paper strip align. This is fine, but do not insist that children use the end of the paper strip as the starting point. Many rulers are not made this way. Furthermore, when the first unit does not coincide with the end, children must correctly line up the units with the object being measured.

▶ Measuring with Rulers

Math Journal 2, p. 186; *Math Masters*, p. 260

| WHOLE CLASS | SMALL GROUP | PARTNER | INDEPENDENT |

Have children measure a few small classroom items. Demonstrate aligning objects at different starting points on the ruler. Discuss which alignments allow you to most easily count the number of paper-clip units to determine the length of the item. GMP5.2, GMP6.4

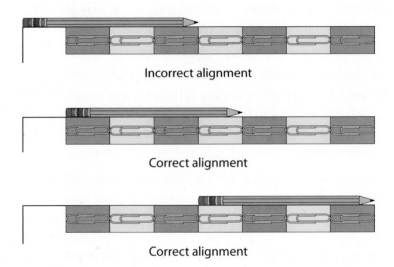

Incorrect alignment

Correct alignment

Correct alignment

Have children use their rulers to measure markers and then have them record their answers on their slates. Ask: *About how long is your marker? How do you know?* Sample answer: I counted the number of units from one end of the marker to the other end.

Math Journal 2, p. 186

Measuring with Rulers

Lesson 9-1
DATE

1. How wide is your *Math Journal*? Measure by lining up **paper clips.**
Sample answers: 6, 7

My journal is about _____ paper-clips wide.

Measure again using your **paper-clip ruler.**
Did you get the same answer? Explain why or why not.
_____ Answers vary. _____
_____

Measure more objects.
Use only your paper-clip ruler. Answers vary.

2. I measured _____.
It is about _____ paper-clip units long.

3. I measured _____.
It is about _____ paper-clip units long.

4. I measured _____.
It is about _____ paper-clip units long.

5. I measured _____.
It is about _____ paper-clip units long.

186   one hundred eighty-six   1.NBT.4, 1.MD.2, SMP5, SMP6

If the length of an item does not coincide with an exact number of paper-clip units, some children may use the term *one-half* to more precisely describe the length. Call attention to this connection with previous fraction lessons, but do not insist on this level of precision. Encourage children to describe their measures to an appropriate level of precision using phrases such as *about _____ paper clips, a little less than _____ paper clips, almost _____ paper clips, close to _____ paper clips,* or *between _____ and _____ paper clips.* For example, a marker may be described as about 6 paper clips long, or a little less than 6 paper clips long, or between 5 and 6 paper clips long.

Discuss how writing numbers on to the ruler would make it easier to count the number of paper-clip lengths. GMP5.2 Have children write numbers in the center of each unit, rather than between two units.

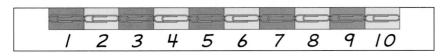

Ruler with numbered units

Have children measure longer objects to practice moving the ruler along bigger spans. Remind them how they measured in pencil lengths in Lesson 4-3. Have children explain their strategies for determining the total number of paper clips when the ruler is moved several times. For example: *Maria said that the bulletin board is 3 rulers plus 7 more paper clips wide. How many paper clips are in 1 ruler?* 10 *How do you find the number of paper clips in 3 whole rulers?* Sample answer: Count by 10s: 10, 20, 30. *What should you do to find the total width of the bulletin board?* Add 7 paper clips to 30; 37 paper clips in all.

Ask children to draw unit boxes labeled *paper clips* in one corner of their slates. Using their paper-clip rulers, have partnerships remeasure the heights of their desks or tables and record their answers, including any calculations, on their slates. Observe as children measure, looking for those who carefully reposition the ruler to span the entire height. Be sure children hold the paper taut so their units remain the same length with no overlaps. Also look for children who count by 10s to find the total number of paper-clip lengths. GMP5.2, GMP6.4 Have several children demonstrate and share their measurement and addition strategies.

Have children complete journal page 186. Children select classroom items to measure, or you may wish to display a list and have them choose from the list. *Suggestions:* a favorite book, a blue crayon, the smallest pencil you can find, a plant in the classroom, your friend's arm, the reading table, a bookcase, the distance you can walk in 2 giant steps, your shoe, the teacher's desk, the class calendar. You may wish to suggest that children select at least two items that are longer than the ruler. After children have completed the page, ask them to explain how they could use their paper-clip ruler to tell which of two objects is longer without directly comparing the objects.

Beat the Calculator

Materials	Fact Triangles
	1 calculator
Players	1 "Caller," 1 "Calculator," and 1 "Brain"
Skill	Mental addition
Object of the Game	To add numbers faster than a player using a calculator

Directions

① Mix the Fact Triangles. Place them facedown on the table.

② The Caller:
- takes a Fact Triangle from the pile.
- covers the sum.
- says the fact *without* the sum.

For example, "3 + 4 is equal to what?"

③ The Calculator solves the problem *with* a calculator.
The Brain solves it *without* a calculator.
The Caller decides who found the sum first.

④ Play about 10 rounds.
Then trade roles.

Another Way to Play

The Caller covers other numbers on the Fact Triangles. Players find the covered number. They can subtract to find the missing addend.

1.OA.3, 1.OA.4, 1.OA.6   G49

✓ **Assessment Check-In** CCSS 1.MD.2

Math Journal 2, p. 186

Observe as children complete Problem 1 on journal page 186. Expect most children to accurately place paper clips along the entire length of the object with no gaps or overlaps and to correctly express the length in paper clips. GMP6.4 Some children may use the ruler, but do not expect all children to do so successfully, as this is their first formal exposure to rulers. GMP5.2 Have children who struggle to accurately measure objects longer than their rulers think about how many paper clips each ruler has. Remind them that they wrote paper clips in the unit box so their answers should be the total number of paper clips. To challenge children, encourage them to complete the Enrichment activity.

 Assessment and Reporting | Go Online ▸ to record student progress and to see trajectories toward mastery for these standards.

Summarize Have children use their rulers to measure from the bases of their hands to the tips of their longest fingers. Have children stand up as you call out various measurements: *Stand up if your hand is about 3 paper clips long. Stand up if your hand is about halfway between 3 and 4 paper clips long. Stand up if . . .* Continue until everyone is standing.

3 Practice 15–20 min | Go Online ▸ | ePresentations eToolkit Home Connections

▸ **Playing *Beat the Calculator***

| WHOLE CLASS | SMALL GROUP | PARTNER | INDEPENDENT |

To practice addition facts, have children play *Beat the Calculator*. For directions for playing the game, use *Math Masters*, page G49. For more information, see Lesson 7-2.

Observe
- Which children solve facts quicker mentally than with a calculator? GMP6.4
- Which facts are more challenging?

Discuss
- *Why is it important to be able to beat the calculator?*

▶ # Math Boxes 9-1

Math Journal 2, p. 187

| WHOLE CLASS | SMALL GROUP | **PARTNER** | **INDEPENDENT** |

Mixed Practice Math Boxes 9-1 are paired with Math Boxes 9-3.

▶ # Home Link 9-1

Math Masters, p. 263

Homework Children use their paper-clip rulers to measure items at home. Remind children to bring the rulers back to class to use again.

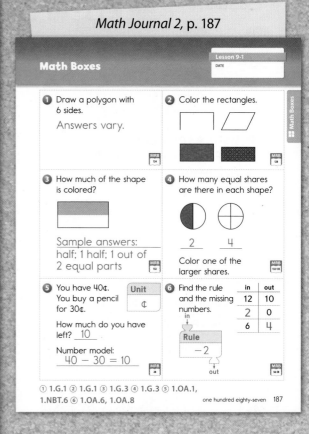

Math Journal 2, p. 187

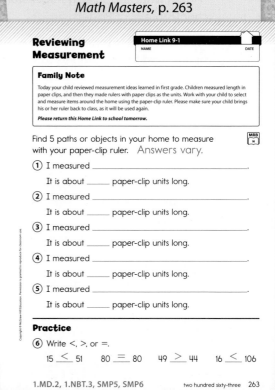

Math Masters, p. 263

Two-Digit Number Stories

Overview Children tell, model, and solve number stories with two and three addends.

▶ **Before You Begin**
For Part 2 and the optional Readiness activity, display parts-and-total, comparison, change-to-more, and change-to-less diagrams.

 Common Core State Standards

Focus Clusters
- Represent and solve problems involving addition and subtraction.
- Use place value understanding and properties of operations to add and subtract.

1 Warm Up 15–20 min

Materials

Mental Math and Fluency Children write the number represented by base-10 blocks.	base-10 blocks, slate	1.NBT.2, 1.NBT.2a, 1.NBT.2c
Daily Routines Children complete daily routines.	See pages 2–37.	See pages xiv–xvii.

2 Focus 35–40 min

Math Message Children estimate the solution to and solve a number story.		1.OA.1, 1.NBT.4 SMP1
Solving School Store Number Stories Children tell and solve number stories using the School Store Mini-Posters.	*Math Journal 2,* pp. 188–189 and inside back cover; base-10 blocks	1.OA.1, 1.NBT.4, 1.NBT.6 SMP1, SMP4
✓ **Assessment Check-In** See page 786.	*Math Journal 2,* inside back cover; base-10 blocks; dimes (optional)	1.OA.1, 1.NBT.6
Recording School Store Number Stories Children write and solve number stories using the School Store Mini-Posters.	*Math Journal 2,* pp. 188–190	1.OA.1, 1.NBT.4, 1.NBT.6 SMP4
Solving Number Stories with Three Addends Children add three numbers to solve number stories.	*Math Journal 2,* pp. 188–189	1.OA.2, 1.NBT.4 SMP1, SMP4

CCSS 1.NBT.6 **Spiral Snapshot**

GMC Subtract multiples of 10 from multiples of 10.

7-5 Practice	8-7 Warm Up Practice	8-11 Focus	9-2 Focus	9-5 Focus	9-6 Focus	9-9 Practice	9-11 Warm Up Practice

 Spiral Tracker   **Go Online** ▷ to see how mastery develops for all standards within the grade.

3 Practice 10–15 min

Math Boxes 9-2 Children practice and maintain skills.	*Math Journal 2,* p. 191	See page 787.
Home Link 9-2 **Homework** Children write and solve a number story using a money context.	*Math Masters,* p. 265	1.OA.1, 1.NBT.4, 1.MD.3 SMP4

 connectED.mcgraw-hill.com ▷

Plan your lessons online with these tools.

 ePresentations   Student Learning Center   Facts Workshop Game   eToolkit   Professional Development   Home Connections   Spiral Tracker   Assessment and Reporting   English Learners Support   Differentiation Support

Differentiation Options RtI

 CCSS 1.OA.1, 1.NBT.4, 1.NBT.6, SMP4

Readiness   10–15 min

Matching Number Stories to Number Sentences

| WHOLE CLASS |
| SMALL GROUP |
| PARTNER |
| INDEPENDENT |

Math Journal 2, inside back cover; *Math Masters,* p. 264; counters

To provide additional exposure to modeling number stories with situation diagrams and number models, work with children to complete *Math Masters,* page 264. Read each number story. Have children decide which number sentence best represents the number story. Use situation diagrams as needed. GMP4.1 Be sure to acknowledge that many number stories can be represented by several different situation diagrams and number models, and accept all reasonable suggestions. Then have children solve the number stories using counters, number grids, or number lines. GMP4.2

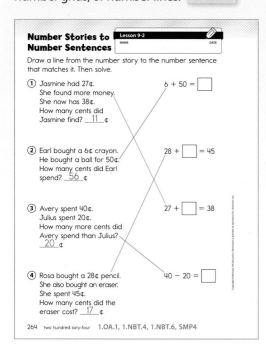

Number Stories to Number Sentences
Lesson 9-2
NAME   DATE

Draw a line from the number story to the number sentence that matches it. Then solve.

① Jasmine had 27¢. She found more money. She now has 38¢. How many cents did Jasmine find? __11__ ¢   6 + 50 = ☐

② Earl bought a 6¢ crayon. He bought a ball for 50¢. How many cents did Earl spend? __56__ ¢   28 + ☐ = 45

③ Avery spent 40¢. Julius spent 20¢. How many more cents did Avery spend than Julius? __20__ ¢   27 + ☐ = 38

④ Rosa bought a 28¢ pencil. She also bought an eraser. She spent 45¢. How many cents did the eraser cost? __17__ ¢   40 − 20 = ☐

264   two hundred sixty-four   1.OA.1, 1.NBT.4, 1.NBT.6, SMP4

CCSS 1.NBT.4, SMP1

Enrichment   15–20 min

Spending Money at the School Store

| WHOLE CLASS |
| SMALL GROUP |
| **PARTNER** |
| **INDEPENDENT** |

Activity Card 111; *Math Journal 2,* pp. 188–189

To extend children's understandings of adding 2-digit numbers, have them determine how to spend $3.00 at the school store. GMP1.1

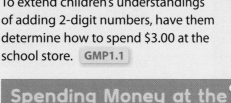

Spending Money at the School Store

I only spent 136¢. That's the same as $1.36. I still need to buy more.

What You Need
Math Journal 2, pages 188 and 189
paper

What To Do
① Pretend that you have $3.00 to spend at the school store.
② List the items you would buy and their prices.
③ Spend as much money as you can. Do not spend more than $3.00.
④ Check your shopping list by adding the prices. What is the sum?
⑤ Record the sum and tell how you added.

Talk About It
Compare sums with a partner. Who spent closer to $3.00? How do you know?

More You Can Do
Imagine you have $5.00. Spend as much money as you can. Do not spend more than $5.00.

CCSS 1.NBT.3, 1.NBT.4, SMP1

Extra Practice   10–15 min

Playing *Animal Weight Top-It*

| WHOLE CLASS |
| SMALL GROUP |
| **PARTNER** |
| INDEPENDENT |

Math Masters, p. G46 (optional); Animal Cards; base-10 blocks

To provide practice with addition of 2-digit numbers, have children play *Top-It* using Animal Cards. You may wish to distribute the directions on *Math Masters,* page G46. Encourage children to use base-10 blocks or number grids to find the total weights of the animals.

Observe
• What strategies do children use to find the total weights?

Discuss
• *How can you check whether your answer makes sense?* GMP1.4

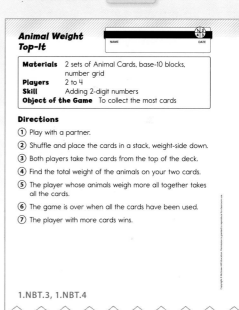

Animal Weight Top-It

Materials	2 sets of Animal Cards, base-10 blocks, number grid
Players	2 to 4
Skill	Adding 2-digit numbers
Object of the Game	To collect the most cards

Directions
① Play with a partner.
② Shuffle and place the cards in a stack, weight-side down.
③ Both players take two cards from the top of the deck.
④ Find the total weight of the animals on your two cards.
⑤ The player whose animals weigh more all together takes all the cards.
⑥ The game is over when all the cards have been used.
⑦ The player with more cards wins.

1.NBT.3, 1.NBT.4

English Language Learners Support

Beginning ELL Provide opportunities for children to practice creating number stories using the School Store Mini-Posters in the lesson. Display the posters and, if possible, actual objects representing the pictured items. Use Total Physical Response prompts to model, and have children practice naming the items for sale. Allow them to use pictures and real objects to tell number stories for given number models.

Standards and Goals for Mathematical Practice

SMP1 Make sense of problems and persevere in solving them.

GMP1.2 Reflect on your thinking as you solve your problem.

GMP1.6 Compare the strategies you and others use.

SMP4 Model with mathematics.

GMP4.1 Model real-world situations using graphs, drawings, tables, symbols, numbers, diagrams, and other representations.

---

① Warm Up   15–20 min

Go Online   ePresentations   eToolkit

▶ Mental Math and Fluency

Display combinations of base-10 blocks. Have children record the number represented by the blocks on their slates.

◉○○ 4 longs, 3 ones 43   5 longs, 7 ones 57   3 longs, 9 ones 39

◉◉○ 7 longs 70   5 longs, 16 ones 66   7 longs, 11 ones 81

◉◉◉ 1 flat, 5 longs, 5 ones 155   1 flat, 7 longs, 17 ones 187
      1 flat, 11 longs, 12 ones 222

▶ Daily Routines

Have children complete the Daily Routines. For detailed instructions, see pages 2–37. For specifics on CCSS coverage, see pages xiv–xvii.

---

② Focus   35–40 min

Go Online   ePresentations   eToolkit

▶ Math Message

Marcos wants to buy a colored pencil for 29¢ and a crayon for 6¢. Estimate whether he will pay more or less than 30¢. How do you know? Record your ideas and solve. GMP1.2

▶ Solving School Store Number Stories

Math Journal 2, pp. 188–189 and inside back cover

| WHOLE CLASS | SMALL GROUP | PARTNER | INDEPENDENT |

Math Message Follow-Up Discuss children's estimates, how they knew that the answer would be more than 30, and how estimating helped them reflect on their solutions. Sample answer: I knew the answer would be more than 30 because I was adding more than 1 to 29. Since 35 is more than 29, I knew I was right. Explain that estimating can be used to determine whether an answer is reasonable. Children can also tell whether they picked a good strategy if their answer is similar to their estimate. GMP1.2

Math Journal 2, p. 188

School Store Mini-Poster 1

Lesson 9-2
DATE

crayon
6¢

16 Crayons

ball
50¢

box of crayons
80¢

paper clip
2¢

pencil
28¢

rubber bands
8¢

eraser
17¢

188 one hundred eighty-eight   1.OA.1, 1.NBT.4, 1.NBT.6

Have children share strategies for finding the exact answer to the Math Message. Emphasize similarities and differences between their methods. **GMP1.6** Write a number model for the story. Ask: *Does this number story fit any of the situation diagrams?* Sample answer: parts-and-total Keep in mind that more than one model may be appropriate. Tell children that today they may use the situation diagrams you displayed to help them make up and solve different types of number stories.

Tell different types of stories using items from the School Store Mini-Posters on journal pages 188 and 189. Have children share multiple solution strategies and write number models for each story. **GMP4.1** You may also wish to discuss how children knew they were on the right track in solving the problem, based on estimating or determining a reasonable answer in another way. **GMP1.2**

Examples:

Parts-and-total story: *Omar bought an eraser (17¢) and stickers (25¢). How many cents did Omar spend?* 42¢

Possible strategies:

- Combining place-value groups and making a 10: I showed 17 and 25 with base-10 blocks. I had 3 longs and 12 cubes, so I exchanged 10 cubes for a long. Then I had 4 longs and 2 cubes, or 42¢.
- Counting on from the larger number: I started at 25 on the number grid. I moved down one row on the grid to 35. Then I counted 7 more spaces to get 42¢.

Possible number models:

$17 + 25 = 42; 25 + 17 = 42$

Change-to-more story: *Diego has 20 cents. His mom gives him 60 more cents. How many cents does Diego have now?* 80¢ *What one item could he buy using all of his money?* Box of crayons

Possible strategies:

- Counting up: I started at 60 on the number grid and counted up 20 by moving down 2 rows on the number grid. I landed at 80, so Diego has 80¢.
- Fact extension: I know that $2 + 6 = 8$, so 2 tens and 6 tens would be 8 tens, or 80¢.

Possible number models:

$60 + 20 = 80; 20 + 60 = 80$

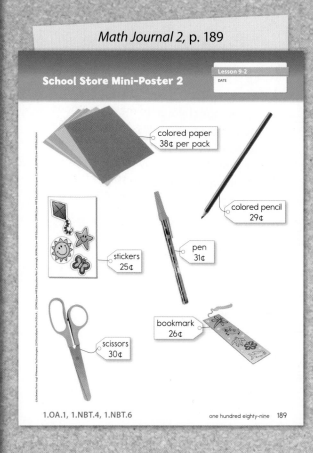

Math Journal 2, p. 189

School Store Mini-Poster 2 Lesson 9-2 DATE

colored paper 38¢ per pack

colored pencil 29¢

stickers 25¢

pen 31¢

scissors 30¢

bookmark 26¢

1.OA.1, 1.NBT.4, 1.NBT.6   one hundred eighty-nine   189

Comparison story: *Destiny is trying to decide whether to buy the ball (50¢) or the scissors (30¢). How much more does the ball cost than the scissors?* 20¢

Possible strategies:

- Counting back: I started at 50 on the number grid and counted back to 30 by moving up two rows on the number grid; that made 20¢.
- Counting up: I started at 30 and counted up to 50 by moving down two rows on the number grid; that made 20¢.

Possible number models:

- $50 - 20 = 30$
- $30 + 20 = 50$

After telling a variety of stories, have children make up number stories using items from the School Store Mini-Posters. Remind them that the situation diagrams are displayed to help them create different types of stories.

✓ **Assessment Check-In** CCSS 1.OA.1, 1.NBT.6

Observe as children solve comparison number stories involving finding differences between multiples of 10. Expect children to solve these problems accurately using the number grid or base-10 blocks. Encourage children who struggle with this to count up by 10s on the number grid. You may also model these problems using base-10 blocks or dimes.

Assessment and Reporting | Go Online | to record student progress and to see trajectories toward mastery for these standards.

▶ Recording School Store Number Stories

Math Journal 2, pp. 188–190

| WHOLE CLASS | SMALL GROUP | **PARTNER** | **INDEPENDENT** |

Have children record number stories on journal page 190 using items from the School Store Mini-Posters. Then they should write number models for their stories. GMP4.1 Encourage children who struggle to create stories to consult the displayed situation diagrams. Have children solve one of their partners' number stories.

Differentiate **Adjusting the Activity**

Children may need to role play or to use real objects to tell number stories to their partners. Help children use precise language by providing sentence frames, such as: "First I bought _____ for _____ cents. Then I bought _____ for _____ cents. I paid a total of _____ cents."

Go Online | Differentiation Support

Math Journal 2, p. 190

Recording Number Stories Lesson 9-2
DATE

Sample Story
I bought a bookmark and an eraser. I paid 43¢ in all.

Number Model: 26¢ + 17¢ = 43¢ Answers vary.

❶ Story 1

Number model: _____

❷ Story 2

Number model: _____

190 one hundred ninety 1.OA.1, 1.NBT.4, 1.NBT.6, SMP4

► Solving Number Stories with Three Addends

Math Journal 2, pp. 188–189

| WHOLE CLASS | SMALL GROUP | PARTNER | INDEPENDENT |

Tell children a few school store number stories in which they must add the prices of three items. Begin with 1-digit prices and then move to 2-digit prices. For each problem, discuss why children added certain numbers first to help them compare strategies. **GMP1.6** Write a number model for each story you tell. **GMP4.1**

Example:

Keisha bought rubber bands (8¢), a crayon (6¢), and a paper clip (2¢). How many cents did Keisha spend in all? 16¢

Possible strategies:

• Grouping addends—Making 10: I added 8 and 2 to make 10. Then I added 6 to 10 to get 16¢. 8 + 2 + 6 = 16

• Grouping addends—Using doubles: I added 6 and 2 to make 8. Then I solved the doubles fact 8 + 8 to get 16¢. 8 + 2 + 6 = 16

Summarize Have children make up number stories with three addends for their partners to solve. Discuss a few of those stories as a whole class.

③ Practice   10–15 min

Go Online   ePresentations   eToolkit   Home Connections

► Math Boxes 9-2

Math Journal 2, p. 191

| WHOLE CLASS | SMALL GROUP | **PARTNER** | **INDEPENDENT** |

Mixed Practice Math Boxes 9-2 are paired with Math Boxes 9-5.

► Home Link 9-2

Math Masters, p. 265

Homework Children make up number stories and write number models to go with their stories.

Math Journal 2, p. 191

Math Boxes

Lesson 9-2
DATE

① If a shape is a rectangle, what must be true? Circle ALL of the answers.
  A It is a polygon.
  B It is blue.
  C It has 4 sides.
  D It has 5 vertices (corners).
  MRB 126

② How many vertices (corners) does a rectangular prism have?
  __8__ vertices
  MRB 126

③ Divide the circle into 2 equal shares. Write one name for a share.
  Sample answers given. 1 half, 1 out of 2 equal pieces
  MRB 152

④ Write the numbers.
  __75__   __42__
  MRB 73

⑤ **Writing/Reasoning** Use base-10 blocks to add.
  43 + 49 = __92__
  Why might you add the tens first? Sample answer: Because I know 4 + 4 is 8. Then I add the ones and exchange once so it's 9 longs.
  MRB 91

① 1.G.1 ② 1.G.1 ③ 1.G.3 ④ 1.NBT.2, 1.NBT.2a
⑤ 1.NBT.2, 1.NBT.2a, 1.NBT.4, SMP1   one hundred ninety-one   191

Math Masters, p. 265

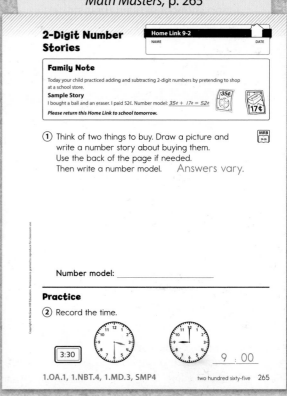

2-Digit Number Stories

Home Link 9-2
NAME   DATE

Family Note

Today your child practiced adding and subtracting 2-digit numbers by pretending to shop at a school store.
Sample Story
I bought a ball and an eraser. I paid 52¢. Number model: *35¢ + 17¢ = 52¢*
Please return this Home Link to school tomorrow.

① Think of two things to buy. Draw a picture and write a number story about buying them. Use the back of the page if needed. Then write a number model.   Answers vary.
  MRB 24-26

Number model: _____

Practice

② Record the time.

  3:30   9 : 00

1.OA.1, 1.NBT.4, 1.MD.3, SMP4   two hundred sixty-five   265

Shopping at the School Store

2-Day Lesson

Overview **Day 1:** Children find the total cost of three items and explain their strategies. **Day 2:** Children discuss explanations and revise their work.

Day 1: Open Response

▶ **Before You Begin**
Solve the open response problem and think about different strategies children might use. If possible, schedule time to review children's work and plan for Day 2 of this lesson with your grade-level team.

Common Core State Standards

Focus Cluster
Use place value understanding and properties of operations to add and subtract.

① Warm Up   15–20 min

	Materials	
Mental Math and Fluency Children mentally find 10 more and 10 less than given numbers and identify tens and ones digits.	slate	**1.NBT.2, 1.NBT.5**
Daily Routines Children complete daily routines.	See pages 2–37.	See pages xiv–xvii.

②a Focus   45–55 min

Math Message Children find the cost of buying two toys.	*Math Journal 2*, p. 192	**1.NBT.4** **SMP6**
Sharing Strategies Children share their strategies for solving the Math Message problem.	*Math Journal 2*, p. 192; Class Number Grid (optional); chart paper (optional)	**1.NBT.4** **SMP6**
Solving the Open Response Problem Children choose three items to buy, find the total cost, and show or explain their strategies.	*Math Journal 2*, p. 189; *Math Masters*, p. 266; number grid (*Math Journal 2*, inside back cover; optional)	**1.NBT.4** **SMP1, SMP6**

Getting Ready for Day 2 →

Review children's work and plan discussion for reengagement.   *Math Masters*, p. TA3;
children's work from Day 1

CCSS 1.NBT.4 Spiral Snapshot

GMC Understand adding 2-digit numbers.

7-1 Warm Up Practice	7-9 Warm Up	8-7 Warm Up	9-2 Focus Practice	9-3 Focus Practice	9-4 through 9-8 Focus Practice	9-9 Practice

 Spiral Tracker   Go Online to see how mastery develops for all standards within the grade.

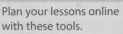
connectED.mcgraw-hill.com

Plan your lessons online
with these tools.

ePresentations Student Learning Center Facts Workshop Game eToolkit Professional Development Home Connections Spiral Tracker Assessment and Reporting English Learners Support Differentiation Support

788   Unit 9 | Two-Digit Addition and Subtraction and Review

1 Warm Up 15–20 min Go Online ePresentations eToolkit

▶ Mental Math and Fluency

Have children write the following numbers on their slates and circle the specified digit.

◉○○ 10 more than 13. Circle the tens digit. (2)3
   10 less than 28. Circle the tens digit. (1)8
   10 more than 20. Circle the tens digit. (3)0

◉◉○ 10 more than 30. Circle the tens digit. (4)0
   10 less than 48. Circle the tens digit. (3)8
   10 more than 105. Circle the hundreds digit. (1)15

◉◉◉ 10 more than 90. Circle the tens digit. 1(0)0
   10 less than 117. Circle the tens digit. 10(7)
   10 more than 190. Circle the hundreds digit. (2)00

▶ Daily Routines

Have children complete the Daily Routines. For detailed instructions, see pages 2–37. For specifics on CCSS coverage, see pages xiv–xvii.

2a Focus 45–55 min Go Online ePresentations eToolkit

▶ Math Message

Math Journal 2, p. 192

Complete journal page 192. Tell your partner how you solved the problem.
GMP6.1

▶ Sharing Strategies

Math Journal 2, p. 192

WHOLE CLASS	SMALL GROUP	PARTNER	INDEPENDENT

Math Message Follow-Up Have children share their solutions. Record strategies on the board or chart paper. Encourage children to describe exactly what they did and clearly explain all the steps. **GMP6.1** Be sure to discuss both correct and incorrect strategies. Help children use precise language. For example, if they are explaining how to add the tens in 17 + 20, remind them to say "1 ten + 2 tens" or "10 + 20." If no one suggests it, model counting up by 10s to find the total. Explain that the cost of the second toy is 20 cents, or 2 tens. *To find the total, start at 17 cents, then count up 1 ten (27 cents), then another ten (37 cents). I paid 37 cents in all.*

Differentiate **Adjusting the Activity**

If children struggle to add 10 mentally, encourage a volunteer to demonstrate how to count up by 10s on the Class Number Grid.

Professional Development

The focus of this lesson is **GMP6.1**. An explanation is *clear* if it is easily understood by the reader or listener. For example, a clear explanation will include all steps of a solution strategy. An explanation is *precise* if it uses specific mathematical language. For example, if a child is adding 21 and 32 and says, "I added 2 tens and 3 tens to get 5 tens," that would be a more precise explanation of this step than saying, "I added 2 and 3 to get 5."

Go Online for information about **SMP6** in the *Implementation Guide*.

Math Journal 2, p. 192

Buying Toys Lesson 9-3 DATE

You bought a toy that cost 17 cents.
You bought another toy that cost 20 cents.
How much did you pay in all?

I paid 37 cents

Tell your partner how you solved the problem.

192 one hundred ninety-two 1.NBT.4, SMP6

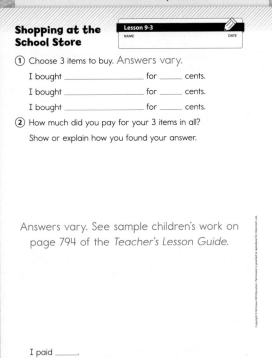

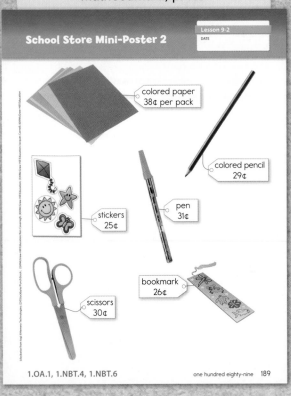

Point out that when children shared their strategies, it was helpful for them to be clear and detailed about what they did because this helped other children understand their strategies. Tell children that today they are going to pretend to go shopping. They should explain as thoroughly as they can how they found the amount they paid. GMP6.1

▶ Solving the Open Response Problem

Math Journal 2, p. 189; Math Masters, p. 266

WHOLE CLASS | SMALL GROUP | **PARTNER** | INDEPENDENT

Distribute *Math Masters,* page 266, and have children turn to *Math Journal 2,* page 189 (School Store Mini Poster 2). Tell children that they will work in pairs. Each child should choose three items from the poster and find the total amount needed to pay for the items, recording his or her work and answer on the *Math Masters* page. Then partners take turns explaining to each other how they solved the problem. Encourage partners to ask questions if they do not understand an explanation. After explaining their strategies to each other, children can use words or pictures to improve their explanations in Problem 2. Talking with their partners will help children reflect on their own thinking and write clearer explanations. GMP1.2, GMP6.1

Tell children that their explanations in Problem 2 are an important part of the task. They need to show or tell how they found the total.

Academic Language Development Some children may need support in developing their explanations for how they solved the problem. Sentence frames may help children get started, such as: "First I _____. Then I _____. I paid _____ cents." Alternatively, encourage children to use a number model to show how they solved the problem.

Children can use the number grid on the inside back cover of *Math Journal 2* to help them solve the problem. Circulate and observe children's strategies for finding the total. Note those who are satisfied with short statements, such as "I used my number grid." Ask these children to show or tell how they used the number grid. Remind children to use clear language when they refer to tens and ones and to label their answer with cents (¢). Also encourage them to make sure they wrote all the steps of their solution strategy in their explanation. Ask: *How can you use pictures, words, or number models to help make your explanation clearer?* GMP6.1

Differentiate | **Adjusting the Activity**

For children who finish the task before others, ask them to try to represent and explain their strategy in a different way. For example, you might ask a child who wrote an explanation in words to draw a picture to represent the strategy. Alternatively, ask children to solve the problem in a different way and to show their explanation for the second strategy on the back of their paper or on a new sheet.

Summarize Ask: *How did you find the amount you paid in all?* Answers vary. Collect children's work so that you can evaluate it and prepare for Day 2.

Getting Ready for Day 2

Math Masters, p. TA3

Planning a Follow-Up Discussion

Review children's work. Use the Reengagement Planning Form (*Math Masters,* page TA3) and the rubric on page 792 to plan ways to help children meet expectations on both the content and practice standards. Look for common misconceptions or errors in the addition as well as interesting ways children found the total, effectively or ineffectively.

Organize the discussion of children's work in one of the ways below or in another way you choose. If children's work is unclear or if you prefer to show work anonymously, rewrite the work for display.

Go Online for sample children's work that you can use in your discussion.

1. Display work that shows an incorrect total cost, as in Child A's work. Have partnerships use the explanation to help identify the child's error. Ask:
   - *Talk to your partner about this child's answer. What do you notice?* Sample answer: This child counted up. It says the child started at 29 and counted up 30 and landed on 89. But I would land on 59 if I counted up 30 from 29: 39, 49, 59.
   - *If this was how you added, what would you do next?* **GMP1.2** Sample answer: I would start at 59 and count up 38 more because that's the other price: 69, 79, 89, 90, 91, 92, 93, 94, 95, 96, 97. The total is 97 cents.
   - *This child said the total was 89 cents. What do you think the mistake was?* **GMP6.4** Sample answer: It is hard to tell because the child only talked about adding on 30¢ and not 38¢.
   - *Do you think it is important to explain how you solve a problem? Why or why not?* **GMP6.1, GMP6.4** Sample answer: Yes. If you don't get the right answer, your explanation can help your teacher or friend see what you did wrong.

2. Show work with an incomplete explanation, and have partners try to use the explanation to do the same thing the child did. For example, show a paper where a child skipped a step in the explanation, as in Child A's work, or show a paper where a child mentioned a tool but did not explain how the tool was used, as in Child B's work. Ask:
   - *What does this explanation tell us?* Sample answers: Child A counted up to add. Child B used a number grid.
   - *Work with a partner to try to use this child's strategy. What do you notice?* Sample answers: Child A: I can start at 29 and count up 30, but I don't land on 89. I'm not sure what this child did next. Child B: I can use my number grid, but I am not sure I made the same jumps because this child doesn't say how he or she jumped.
   - *How could this child make the answer to Problem 2 better?* **GMP6.1** Sample answers: Child A could make sure to write all the steps. Child B could tell how he or she used the number grid and write the answer with cents (¢).

Planning for Revisions

Have copies of *Math Masters,* page 266 or extra paper available for children to use in revisions. You might want to ask children to use colored pencils so you can see what they revised.

Sample child's work, Child A

① Choose 3 items to buy.

I bought ____a Pencil____ for _29¢_ cents.

I bought ____Some Paper____ for _38¢_ cents.

I bought ____Scissors____ for _30¢_ cents.

② How much did you pay for your 3 items in all? Show or explain how you found your answer.

I STartd at 29
then I conted
up to 30 I 1ad on 89
then it was 89cents.

I paid _89¢_.

Sample child's work, Child B

① Choose 3 items to buy.

I bought ____pen____ for _31_ cents.

I bought ____stickers____ for _25_ cents.

I bought ____Bookmark____ for _26_ cents.

② How much did you pay for your 3 items in all? Show or explain how you found your answer.

Me and my parcher
Yos are number grid to
make 31+25+26=82

I paid _82_.

Shopping at the School Store

Day 2: Reengagement

▶ **Before You Begin**
Have extra copies available of *Math Masters,* page 266 for children to revise their work.

Common Core State Standards

Focus Cluster
Use place value understanding and properties of operations to add and subtract.

2b Focus   50–55 min

	Materials	
Setting Expectations Children review the open response problem and discuss what a clear explanation might include. They review how to respectfully discuss others' work.	*My Reference Book,* p. 18; Standards for Mathematical Practice Poster	1.NBT.4 SMP6
Reengaging in the Problem Children discuss various solutions to the open response problem.	selected samples of children's work	1.NBT.4 SMP1, SMP6
Revising Work Children work to make their explanations clearer and more complete.	*Math Masters,* p. 266 (optional); children's work from Day 1; colored pencils (optional)	1.NBT.4 SMP1, SMP6
✓ **Assessment Check-In** See page 794 and the rubric below.		1.NBT.4 SMP6

Goal for Mathematical Practice **GMP6.1** Explain your mathematical thinking clearly and precisely.	**Not Meeting Expectations**	**Partially Meeting Expectations**	**Meeting Expectations**	**Exceeding Expectations**
	Does not explain a strategy for finding the total amount paid.	Explains a strategy for finding the total amount paid, but the explanation skips steps, identifies a tool without describing how the tool was used, or otherwise lacks detail.	Completely explains all steps of a strategy for finding the total amount paid. If any tools are mentioned, the explanation describes how the tools were used.	Meets expectations and describes the strategy using two different representations (e.g., pictures and words) or describes more than one strategy.

3 Practice   10–15 min

Math Boxes 9-3 Children practice and maintain skills.	*Math Journal 2,* p. 193; Pattern-Block Template	See page 795.
Home Link 9-3 **Homework** Children find the total amount they would pay for three school store items.	*Math Masters,* p. 267	1.OA.2, 1.OA.3, 1.OA.6, 1.OA.8, 1.NBT.4 SMP6

2b Focus · 50–55 min · Go Online · ePresentations · eToolkit

▶ Setting Expectations

My Reference Book, p. 18

| **WHOLE CLASS** | SMALL GROUP | PARTNER | INDEPENDENT |

Review the open response problem from Day 1. Remind children that their task was to record the prices of three items they bought from the school store and find out how much they would pay for all three items. Finally, they needed to explain how they figured out their answer. Refer to **GMP6.1** on the Standards for Mathematical Practice Poster. Ask: *What do you think a clear explanation should include?* **GMP6.1** Sample answers: It should say all the steps I used to find the total. It should say if I used a number grid or other tools and how I used them.

Have children turn to *My Reference Book*, page 18. Read the page together. Ask: *What details did Dell add to his explanation?* Sample answer: He talked about all his steps instead of just saying he added. He said, "20 is 2 tens." Tell children they are going to look at other children's work and think about what kinds of details could be added to make the explanations clearer. Remind children that it is all right to feel confused by an explanation and ask questions because this will help others improve their work, but it is important to be respectful. Encourage children to use these sentence frames:

- I don't understand _____.
- I wonder why _____.

▶ Reengaging in the Problem

| **WHOLE CLASS** | SMALL GROUP | **PARTNER** | INDEPENDENT |

Children reengage in the problem by analyzing and critiquing other children's work in pairs and in a whole-group discussion. Have children discuss with partners before sharing with the whole group. Guide the discussion based on the decisions you made in Getting Ready for Day 2. **GMP1.2, GMP6.1, GMP6.4**

NOTE These Day 2 activities will ideally take place within a few days of Day 1. Prior to beginning Day 2, see Planning a Follow-Up Discussion from Day 1.

My Reference Book, p. 18

Standards for Mathematical Practice

Be Careful and Accurate

14 stickers + 20 stickers = ? stickers

Dell told his teacher that he added 20 to 14 and got 34. Dell's teacher asks him to explain his thinking more clearly.

20 is 2 tens. I put 14 in my head, and then counted on 2 tens: 24, 34. That makes 34 stickers in all.

Dell

Explain your mathematical thinking clearly and precisely.
Dell uses details to tell how he added 14 and 20.

Use clear labels, units, and mathematical language.
Dell uses clear math language when he says, "20 is 2 tens." His answer uses the unit "stickers."

 18 eighteen

▶ Revising Work

WHOLE CLASS | SMALL GROUP | **PARTNER** | INDEPENDENT

Pass back children's work from Day 1. Children should confirm that their total amount in Problem 2 is correct, possibly using strategies that surfaced in the class discussion. Before children revise anything, ask them to read their explanations and decide how to improve them. Ask the following questions one at a time. Have partners discuss their responses and give a thumbs-up or thumbs-down based on their own work.

- *Did you find the correct total for the 3 items?* GMP6.4
- *Did you use clear language to describe all the steps you took when you added the prices?* GMP1.2, GMP6.1
- *Did you describe how you used tools to help you find the total?* GMP6.1

Tell children they now have a chance to revise their explanations. If children found the correct total and wrote a complete explanation on Day 1, they can use a different representation to describe their strategies (for example, pictures instead of words) or describe a different strategy to find the total. Remind children that the responses presented during the reengagement discussion are not the only correct responses. Tell children to add to their earlier work using colored pencils or to use another sheet of paper, instead of erasing their original work.

Summarize Ask children to reflect on their work and revisions. Ask: *What did you do to improve your explanation?* Answers vary.

✓ **Assessment Check-In** ⓒⓒⓢ 1.NBT.4

Collect and review children's revised work. Expect children to improve their explanations based on the class discussion. For the content standard, expect most children to find a correct total. You can use the rubric on page 792 to evaluate children's revised work for GMP6.1.

 | Assessment and Reporting | Go Online to record student progress and to see trajectories toward mastery for these standards.

Go Online for optional generic rubrics in the *Assessment Handbook* that can be used to assess any additional GMPs addressed in the lesson.

Sample Children's Work—Evaluated

See the sample in the margin. This work meets expectations for the content standard because the child found the correct total cost for the three items listed in Problem 1. With revision, this work meets expectations for the mathematical practice because the explanation covers the three steps of adding the tens, adding the ones, and then adding the tens and ones together. While the last step is described with only a number sentence, all three steps are included in the explanation. GMP6.1

Go Online for other samples of evaluated children's work.

Sample child's work, "Meeting Expectations"

① Choose 3 items to buy.

I bought colored paper for 38 cents.

I bought stickers for 25 cents.

I bought colored pencil for 29 cents.

② How much did you pay for your 3 items in all? Show or explain how you found your answer.

$$70 + 22 = 92$$

I added my tens first, then I added the ones.

$$20 + 30 + 20 = 70.$$

$$5 + 8 + 9 = 22.$$

I paid 92 .

(3) Practice 10–15 min

Go Online

ePresentations eToolkit Home Connections

▶ Math Boxes 9-3

Math Journal 2, p. 193

| WHOLE CLASS | SMALL GROUP | **PARTNER** | **INDEPENDENT** |

Mixed Practice Math Boxes 9-3 are paired with Math Boxes 9-1.

▶ Home Link 9-3

Math Masters, p. 267

Homework Children find the total amount they would pay for three items at the school store and explain their strategies to someone at home.

Math Journal 2, p. 193

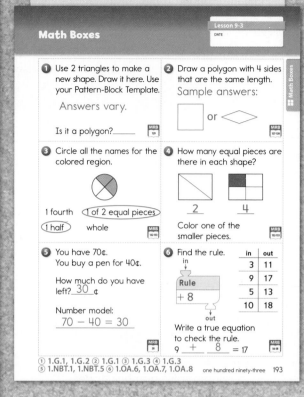

Math Boxes

Lesson 9-3

DATE

❶ Use 2 triangles to make a new shape. Draw it here. Use your Pattern-Block Template.

Answers vary.

Is it a polygon?_____

❷ Draw a polygon with 4 sides that are the same length.

Sample answers:

☐ or ◇

❸ Circle all the names for the colored region.

1 fourth   (1 of 2 equal pieces)
(1 half)   whole

❹ How many equal pieces are there in each shape?

___2___   ___4___

Color one of the smaller pieces.

❺ You have 70¢.
You buy a pen for 40¢.

How much do you have left? _30_ ¢

Number model:
70 − 40 = 30

❻ Find the rule.

in	out
3	11
9	17
5	13
10	18

Rule
+ 8

Write a true equation to check the rule.
9 _+_ _8_ = 17

① 1.G.1, 1.G.2 ② 1.G.1 ③ 1.G.3 ④ 1.G.3
⑤ 1.NBT.1, 1.NBT.5 ⑥ 1.OA.6, 1.OA.7, 1.OA.8   one hundred ninety-three   193

Math Masters, p. 267

Shopping for School Supplies

Home Link 9-3

NAME   DATE

Family Note

Today your child practiced explaining solution strategies clearly. Clear explanations make sense to the listener and include all of the steps used to solve the problem. For example, "I started at 21 and counted up," would not be a clear explanation for how a child added 21 and 7. An example of a clear explanation would be, "I started at 21. Then I counted up 7. I ended at 28. So, 21 + 7 = 28."

After your child solves the problem below, ask him or her to explain the strategy used. Ask questions to encourage your child to explain the strategy clearly.

Please return this Home Link to school tomorrow.

① You bought these items at the school store. How much did you pay in all?

crayon
6 ¢

scissors
30 ¢

stickers
25 ¢

I paid __61__ cents.

Explain to someone at home how you solved the problem.

Practice

② Solve.

20 = 8 + __7__ + 5   17 = 6 + __4__ + 7

1.OA.2, 1.OA.3, 1.NBT.4, 1.OA.6,
1.OA.8, SMP6   two hundred sixty-seven   267

Exploring Broken Calculators, Fractions, and Facts

Overview Children generate equivalent names for numbers in broken-calculator problems. They divide a rectangle into equal parts. They conduct a final Facts Inventory.

▶ **Before You Begin**
For the optional Readiness activity, display three large name-collection boxes with numbers selected to match children's skill level in each tag.

Common Core State Standards

Focus Clusters
- Add and subtract within 20.
- Work with addition and subtraction equations.
- Use place value understanding and properties of operations to add and subtract.
- Reason with shapes and their attributes.

1 Warm Up   15–20 min

	Materials	
Mental Math and Fluency Children solve number stories.	slate	1.OA.2, 1.OA.6
Daily Routines Children complete daily routines.	See pages 2-37.	See pages xiv–xvii.

2 Focus   30–35 min

Math Message Children record ways to get the number 12 on a calculator.	calculator	1.OA.6, 1.OA.7
Introducing Broken-Calculator Puzzles Children discuss equivalent names for 12.	calculator	1.OA.6, 1.OA.7 SMP6
Exploration A: Solving Broken-Calculator Puzzles Children imagine that their calculators are broken and explain how to display numbers without pressing the broken key.	*Math Journal 2*, p. 194; calculator	1.OA.6, 1.OA.7, 1.NBT.4 SMP1
Exploration B: Planning Gardens Children divide a piece of paper into equal parts and use it to plan a garden.	Activity Card 112, markers or crayons	1.G.3 SMP6
Exploration C: Final Facts Inventory Record Children conduct a final self assessment using their Facts Inventory Record.	*Math Journal 2*, pp. 217–223; colored pencils	1.OA.3, 1.OA.6 SMP6

3 Practice   15–20 min

Telling Time Children practice telling time to the nearest half-hour.	*Math Journal 2*, p. 195	1.MD.3 SMP5, SMP6
Math Boxes 9-4 Children practice and maintain skills.	*Math Journal 2*, p. 196	See page 801.
Home Link 9-4 **Homework** Children solve broken-calculator puzzles.	*Math Masters*, p. 268	1.OA.6, 1.OA.7, 1.NBT.2, 1.NBT.2a, 1.NBT.4 SMP4

Differentiation Options

 RtI

CCSS 1.OA.6, 1.OA.7, SMP1

Readiness
10–15 min

Reviewing Name-Collection Boxes

	WHOLE CLASS
	SMALL GROUP
	PARTNER
	INDEPENDENT

three large name-collection boxes

To provide additional exposure to generating equivalent names for numbers, review name-collection boxes. Then divide children into three groups and have them complete an activity with name-collection boxes. Each group begins at one of the three prepared, large name-collection boxes. They have two minutes to record as many names as they can in the name-collection box. After two minutes, they move to the next paper, review the names listed in the name-collection box, and continue adding more names. Encourage children to keep trying to write different names, as it will get more challenging to find new names as the activity progresses. **GMP1.3** Repeat until all groups have seen all the name-collection boxes. Collect the papers, and discuss the different names that children have listed.

CCSS 1.OA.6, 1.OA.7, 1.NBT.4, SMP2

Enrichment
15–20 min

More Broken-Calculator Problems

	WHOLE CLASS
	SMALL GROUP
	PARTNER
	INDEPENDENT

Activity Card 113, calculator

To extend children's work with equivalency, have them complete challenging broken-calculator problems. **GMP2.1**

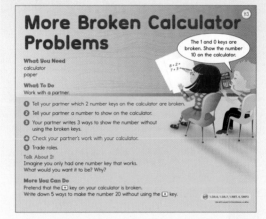

CCSS 1.OA.3, 1.OA.6, SMP1

Extra Practice
10–15 min

Playing *Tric-Trac*

	WHOLE CLASS
	SMALL GROUP
	PARTNER
	INDEPENDENT

My Reference Book, pp. 168–169; *Math Masters*, p. G54; per partnership: 2 dot dice, 20 pennies

To provide practice with addition sums to 10, have children play *Tric-Trac*. Encourage children to use two or three addends to cover their sum. **GMP1.5** For detailed instructions, see *My Reference Book*, pages 168–169.

Observe
- Which children use three addends to cover sums?

Discuss
- *How did you use three numbers to cover your sum?*

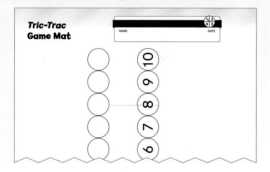

Tric-Trac Game Mat

English Language Learners Support

Beginning ELL Introduce children to *broken* by showing examples and non-examples. Show pictures or examples of common objects that are not broken. Point to each, and say: *This is not broken. It works.* Place them in one area. Say: *These are not broken.* Then show pictures or examples of the same objects broken; as you point to each, say: *This does not work. It's broken.* Place those in another area, and say: *These are all broken.* Have children sort pictures or objects. Each time, ask: *Is that broken? Is it working?*

Go Online **ELL** English Learners Support

① Warm Up 15–20 min Go Online
ePresentations eToolkit

▶ Mental Math and Fluency

Pose number stories with three addends. Have children record their answers on their slates. Briefly discuss solutions and strategies. Highlight finding and adding the combination of 10 first.

◕○○ A fruit salad has 5 strawberries, 7 grapes, and 5 blueberries. How many pieces of fruit are there in all? 17 pieces of fruit

◕◕○ Simone built a tower with 6 orange blocks, 4 purple blocks, and 7 green blocks. How many blocks did she use in all? 17 blocks

◕◕◕ I have 10 blue paper clips, 2 red paper clips, and 8 yellow paper clips. How many paper clips do I have in all? 20 paper clips

▶ Daily Routines

Have children complete the Daily Routines. For detailed instructions, see pages 2–37. For specifics on CCSS coverage, see pages xiv–xvii.

② Focus 30–35 min Go Online
ePresentations eToolkit

▶ Math Message

Use your calculator. Record three ways to get the number 12. Try using more than one step.

▶ Introducing Broken-Calculator Puzzles

WHOLE CLASS | SMALL GROUP | PARTNER | INDEPENDENT

Math Message Follow-Up Have children use both ⊕ *and* ⊖ to write ways to get to the number 12. Discuss answers as a class. Sample answers: $12; 1 + 11 =; 2 + 2 + 2 + 2 + 2 + 2 =; 13 - 1 =; 7 + 7 - 2 =; 9 - 6 + 9 =$ Demonstrate how to clearly label the steps and keys while recording the names. **GMP6.3** Tell children that in today's lesson, they will be using calculators to find names for numbers.

▶ Exploration A: Solving Broken-Calculator Puzzles

Math Journal 2, p. 194

| WHOLE CLASS | SMALL GROUP | **PARTNER** | **INDEPENDENT** |

Have children imagine that the 4-key on the calculator is broken—when they press 4, nothing happens. Ask them to try to make 14 appear on the calculator display without pressing the broken key. Sample answers: $5 + 9 =; 12 - 6 + 8 =; 2 + 2 + 2 + 2 + 2 + 2 + 2 =$ Have children share their strategies, and encourage them to use their calculators to check whether they are correct. GMP1.4

Pose other broken-calculator puzzles until children understand them. Then have them complete journal page 194. Encourage children to use their calculators to check their work. GMP1.4

Differentiate | **Adjusting the Activity**

If children have difficulty getting started, give them the keystrokes $12 + 2$ and have them verify that this works. Ask them to think about ways they can break apart the numbers 12 and 2 to help them generate additional solutions.

Go Online ▶ | Differentiation Support

▶ Exploration B: Planning Gardens

Activity Card 112

| WHOLE CLASS | SMALL GROUP | **PARTNER** | INDEPENDENT |

Partnerships plan gardens by dividing paper into equal parts, drawing what they would plant in each part, and describing each part using math language. GMP6.3 One partner divides a garden into 2 equal parts, and the other partner divides a garden into 4 equal parts. Then partners share papers and compare the sizes of the parts of their gardens.

Math Journal 2, p. 194

Broken-Calculator Puzzles

Lesson 9-4
DATE

Use + and − to solve the broken-calculator puzzles.
Use a calculator to check your answers.

❶ Imagine your 1-key is broken.
Write at least 3 ways to show 11 without using the 1-key.
Sample answers: $9 + 2 =$; $4 + 4 + 3 =$; $20 - 9 =$

❷ Imagine your 2-key is broken.
Write at least 3 ways to show 20 without using the 2-key.
Sample answers: $10 + 10 =$; $15 + 5 =$; $30 - 10 =$

❸ Imagine your 4-key is broken.
Write at least 3 ways to show 34 without using the 4-key.
Sample answers: $35 - 1 =$; $29 + 5 =$; $30 + 1 + 1 + 1 + 1 =$

❹ Write your own puzzle. Give it to your partner to solve.
Answers vary.

194   one hundred ninety-four   1.OA.6, 1.OA.7, 1.NBT.4, SMP1

Activity Card 112

Planning Gardens (112)

What You Need
paper
markers or crayons

What To Do
Work with a partner.
❶ One partner divides a paper into 2 equal parts.
The other partner divides another paper into 4 equal parts.
❷ Imagine your paper is a plan for a garden. You plant a different kind of plant in each part. Each part should have the same number of plants.
❸ Draw what you would plant.
❹ On the back, describe your garden using math language.

1 out of 2 equal shares of my garden has carrots.

Talk About It
Compare your garden to your partner's garden.
Whose garden has larger parts? Why?
How many different kinds of plants are growing in each garden?
Who has more kinds of plants? Why?

More You Can Do
Trade roles with your partner.
Divide your paper in a different way.
Plan a new garden.

1.G.3, SMP6

Math Journal 2, p. 217–223

My Facts Inventory Record, Part 1

DATE

Addition Fact	Know It	Don't Know It	How I Can Figure It Out...
0 + 1			
7 + 2			
0 + 3			
3 + 2			
0 + 5			
10 + 2			
3 + 1			
2 + 2			
6 + 0			
1 + 1			

1.OA.3, 1.OA.6   two hundred seventeen 217

▶ Exploration C: Final Facts Inventory Record

Math Journal 2, pp. 217– 223

| WHOLE CLASS | SMALL GROUP | PARTNER | **INDEPENDENT** |

Children conduct a final self assessment using their Facts Inventory Record on journal pages 217–223. Starting at the beginning of their records, children reexamine the facts that they did not know before and mark updates with colored pencils. Expect children to know some facts that they did not know before. Encourage them to use more sophisticated and efficient strategies when applicable (for example, substituting near doubles or making 10 for counting on) for remaining facts that they do not know. **GMP6.4** Many children will know their facts from Parts 1 and 2 of the Facts Inventory Record, but do not expect them to be automatic with many facts from Parts 3 and 4.

Summarize Have children share their favorite strategies for solving unknown facts in Exploration C.

3 Practice 15–20 min   Go Online

ePresentations   eToolkit   Home Connections

▶ Telling Time

Math Journal 2, p. 195

| WHOLE CLASS | SMALL GROUP | **PARTNER** | INDEPENDENT |

Children practice telling time to the nearest half hour. **GMP5.2, GMP6.2**

Math Journal 2, p. 195

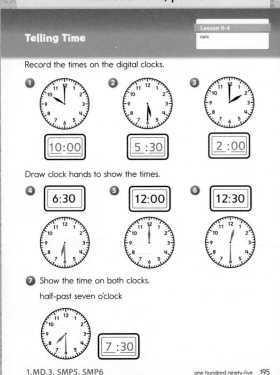

Telling Time

Lesson 9-4

DATE

Record the times on the digital clocks.

1. `10:00`
2. `5 :30`
3. `2 :00`

Draw clock hands to show the times.

4. `6:30`
5. `12:00`
6. `12:30`

7. Show the time on both clocks.

half-past seven o'clock

`7 :30`

1.MD.3, SMP5, SMP6   one hundred ninety-five 195

▶ # Math Boxes 9-4

Math Journal 2, p. 196

| WHOLE CLASS | SMALL GROUP | **PARTNER** | **INDEPENDENT** |

Mixed Practice Math Boxes 9-4 are paired with Math Boxes 9-6.

▶ # Home Link 9-4

Math Masters, p. 268

Homework Children practice solving broken-calculator problems.

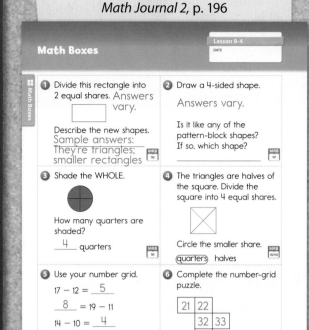

Math Journal 2, p. 196

Math Boxes

Lesson 9-4
DATE

① Divide this rectangle into 2 equal shares. Answers vary.

Describe the new shapes.
Sample answers:
They're triangles;
smaller rectangles

② Draw a 4-sided shape.
Answers vary.

Is it like any of the pattern-block shapes? If so, which shape?
_____

③ Shade the WHOLE.

How many quarters are shaded?
__4__ quarters

④ The triangles are halves of the square. Divide the square into 4 equal shares.

Circle the smaller share.
(quarters) halves

⑤ Use your number grid.
$17 - 12 = $ __5__
__8__ $ = 19 - 11$
$14 - 10 = $ __4__

What addition fact can you use to solve $14 - 10$?
Sample answer:
$10 + 4 = 14$

⑥ Complete the number-grid puzzle.

21	22	
	32	33
	42	
51		

196 one hundred ninety-six

① 1.G.1, 1.G.3 ② 1.G.1 ③ 1.G.3 ④ 1.G.3
⑤ 1.OA.4 ⑥ 1.NBT.2, 1.NBT.5

Math Masters, p. 268

Broken-Calculator Puzzles

Home Link 9-4
NAME   DATE

Family Note

In one of today's Exploration activities, your child solved broken-calculator puzzles. In these puzzles children came up with ways to make their calculators display a number, even though an important key was broken. They also divided a rectangle into equal shares, and they conducted a final Facts Inventory about which addition and subtraction facts they know.

Please return this Home Link to school tomorrow.

Use + and − to solve the broken-calculator puzzles.
Use a calculator to check your answers. Sample answers given.

① Imagine your 3-key is broken.
Write at least three ways to show 30.

$20 + 10 = $; $10 + 10 + 10 = $; $29 + 1 = $

② Imagine your 5-key is broken.
Write at least three ways to show 15.
Use subtraction in one of the ways.

$8 + 7 = $; $14 + 1 = $; $16 - 1 = $

③ Imagine your 1-key is broken.
Write at least three ways to show 18.

$9 + 9 = $; $20 - 2 = $; $8 + 8 + 2 = $

Practice

④ Jake has | | | | | | | | | and . .

Show one exchange he could make.

Sample answer: | | | | | | | | | . . . .

268 two hundred sixty-eight

1.OA.6, 1.OA.7, 1.NBT.4, 1.NBT.2, 1.NBT.2a, SMP1, SMP4

Vending Machine Addition and Subtraction

Overview Children apply a variety of strategies to add and subtract 2-digit numbers.

▶ **Before You Begin**
For Part 3, cut six rectangles per child from *Math Masters,* page 269.

 Common Core State Standards

Focus Clusters
- Represent and solve problems involving addition and subtraction.
- Use place value understanding and properties of operations to add and subtract.

	Materials	
① Warm Up 15–20 min		
Mental Math and Fluency Children write the largest possible numbers with given digits.	slate	**1.NBT.2, 1.NBT.3**
Daily Routines Children complete daily routines.	See pages 2–37.	See pages xiv–xvii.

② Focus 30–35 min		
Math Message Children find the sum of 55 and 35 to solve a number story.		**1.OA.1, 1.NBT.4** SMP1
Adding 2-Digit Vending Machine Prices Children apply a variety of strategies to add prices of items from a vending machine.	*Math Journal 2,* p. 197, inside back cover; *Math Masters,* p. TA42; base-10 blocks, dimes, pennies (optional)	**1.OA.1, 1.OA.2, 1.NBT.4** SMP1
✓ **Assessment Check-In** See page 806.	*Math Journal 2,* inside back cover and base-10 blocks (optional)	**1.NBT.4** SMP6
Finding Differences between 2-Digit Numbers Children apply a variety of strategies to find the difference between prices of items from a vending machine.	*Math Journal 2,* p. 197, inside back cover; *Math Masters,* p. TA42; base-10 blocks, pennies, dimes (optional)	**1.OA.1, 1.NBT.6** SMP6

CCSS 1.NBT.4 **Spiral Snapshot**

GMC Understand adding 2-digit numbers.

| 7-9
Warm Up | 8-7
Warm Up | 9-2 through 9-4
Focus
Practice | 9-5
Focus
Practice | 9-6 through 9-8
Focus
Practice | 9-9
Practice |

 Spiral Tracker   **Go Online** ⟩ to see how mastery develops for all standards within the grade.

③ Practice 15–20 min		
Turning Rectangles into Other Shapes Children rearrange composite shapes to create new ones.	*Math Journal 2,* p. 198; *Math Masters,* p. 269; glue; scissors	**1.G.1, 1.G.2**
Math Boxes 9-5 Children practice and maintain skills.	*Math Journal 2,* p. 199	See page 807.
Home Link 9-5 **Homework** Children add 2-digit numbers in context.	*Math Masters,* p. 270	**1.NBT.4, 1.G.2**

connectED.mcgraw-hill.com

Plan your lessons online with these tools.

 ePresentations   Student Learning Center   Facts Workshop Game   eToolkit   Professional Development   Home Connections   Spiral Tracker   Assessment and Reporting   English Learners Support   Differentiation Support

Differentiation Options RtI

Readiness   10–15 min

| WHOLE CLASS |
| SMALL GROUP |
| PARTNER |
| INDEPENDENT |

Practicing 2-Digit Addition and Subtraction

two sets of number cards 0–9, base-10 blocks

To explore adding and subtracting 2-digit numbers, each child turns over two number cards to create a 2-digit number and then represents that number with base-10 blocks. Partnerships add the 2-digit numbers by combining the base-10 blocks and then record a number sentence.

After a few rounds of adding, partners practice subtracting multiples of 10 from other multiples of 10. Lay out the two 0 cards. Each partner turns over one number card 1–9 and places it to the left of a 0 card. They represent their numbers with longs, subtract the smaller number from the larger number, and record the number sentence. **GMP2.1**

Enrichment   15–20 min

| WHOLE CLASS |
| SMALL GROUP |
| PARTNER |
| INDEPENDENT |

Making Change

Activity Card 114; *Math Journal*, p. 197

To extend children's understanding of adding and subtracting 2-digit numbers, have them make sense of problems involving making change. **GMP1.1** Children find the total costs of vending machine items and then calculate the change for the purchases that were made with $1.00.

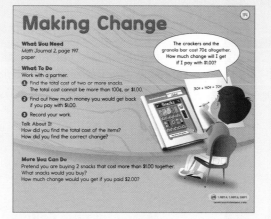

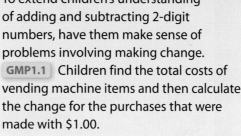

Extra Practice   10–15 min

| WHOLE CLASS |
| SMALL GROUP |
| PARTNER |
| INDEPENDENT |

Playing *Stop and Go*

Math Masters, p. G42 (optional); pp. G43, G44, and G45; *My Reference Book*, pages 164–165 (optional); number grid; counters; base-10 blocks

To provide practice adding and subtracting 2-digit numbers, have children play *Stop and Go*. For detailed instructions, see *My Reference Book*, pages 164–165 or use *Math Masters*, page G42.

Observe
- Which children add and subtract accurately? **GMP6.4**
- What strategies and tools do children use to add and subtract? **GMP5.1**

Discuss
- *What strategies and tools do you and your partner use to add and subtract?*

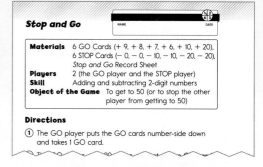

English Language Learners Support

Beginning ELL Introduce children to *vending machines* by pointing out and naming different parts of the machine using a photo of a real machine or the Vending Machine Poster. Expand vocabularies of children who may not be familiar with the food items shown on the poster by thinking aloud to name and describe the different items and bringing in real samples of unfamiliar items, if possible.

 Go Online **ELL** English Learners Support

Standards and Goals for
Mathematical Practice

SMP1 **Make sense of problems and persevere in solving them.**

    GMP1.5 Solve problems in more than one way.

SMP6 **Attend to precision.**

    GMP6.1 Explain your mathematical thinking clearly and precisely.

    GMP6.4 Think about accuracy and efficiency when you count, measure, and calculate.

Math Journal 2, p. 197

1 Warm Up   15–20 min   

ePresentations   eToolkit

▶ Mental Math and Fluency

Display two or three digits. Have children record on their slates the largest number they can make with the digits. *Suggestions:*

⬤◯◯ 5, 2 52; 3, 7 73; 1, 6 61; 4, 3 43

⬤⬤◯ 7, 2 72; 4, 9 94; 5, 8 85; 1, 9 91

⬤⬤⬤ 9, 1, 2 921; 4, 7, 7 774; 9, 8, 1 981; 5, 5, 6 655

▶ Daily Routines

Have children complete the Daily Routines. For detailed instructions, see pages 2–37. For specifics on CCSS coverage, see pages xiv–xvii.

2 Focus   30–35 min   

ePresentations   eToolkit

▶ Math Message

Hannah did 55 jumping jacks at recess. She did 35 jumping jacks after school. How many jumping jacks did Hannah do all together? How do you know? Try to solve in two different ways, and record your strategies and answer. **GMP1.5**

▶ Adding 2-Digit Vending Machine Prices

Math Journal 2, p. 197 and inside back cover; *Math Masters*, p. TA42

| **WHOLE CLASS** | SMALL GROUP | PARTNER | INDEPENDENT |

Math Message Follow-Up Have children share their strategies and solutions for adding 55 and 35. 90 Possible place-value strategies include using a concrete model such as base-10 blocks to compose a 10 or counting up on the number grid or number line. Children may also mentally count up from the larger number. Explain that today children will use many strategies to add 2-digit numbers when they buy snacks from a vending machine. **GMP1.5**

Display the Vending Machine Poster (*Math Masters*, page TA42), and have children turn to the poster on journal page 197. Point out the prices for the items on the poster.

Tell number stories that involve purchasing two or more items from the Vending Machine Poster, such as: *I want to buy a pack of cheese sticks (25¢) and a bag of pretzels (65¢). How much money do I need?* 90¢ Highlight a variety of tools and strategies, including those presented in the following stories. Encourage children to write number models for each problem. Keep in mind that more than one model may be appropriate. You may also wish to display situation diagrams.

The following problems list sample strategies that demonstrate place-value understanding.

Noah buys raisins (45¢), and Leo buys mints (35¢). How much did they spend all together? 80¢

Possible strategies:

- Combining place-value groups: I showed each price using dimes and pennies. I added the dimes and got 70¢; then I added 10 pennies and got the sum of 80¢.
- Composing a 10: I used base-10 blocks to show each price. I exchanged 10 cubes for a long, so I had 4 longs, 3 longs, and 1 long. Then I added the longs.
- Counting up from the larger addend: I counted up on the number grid. I started at 45, moved down 3 rows, and then right 5 spaces.

Possible number models:

- $40 + 5 + 30 + 5 = 70 + 10 = 80$
- $40 + 30 + 10 = 80$
- $45 + 30 + 5 = 80$

Stan bought crackers (30¢) and sunflower seeds (60¢). How much did he spend? 90¢

Possible strategy:

- Using fact extensions: I used dimes to represent the numbers and saw that I was just adding 3 dimes and 6 dimes. I know $3 + 6 = 9$, so $30 + 60 = 90$. So, 9 dimes is 90¢.

Possible number model:

- $30 + 60 = 90$

The following problems list sample strategies that demonstrate understanding of properties of operations.

Sonia wants to buy a pack of cheese sticks (25¢) and a bottle of water (75¢). How much will she have to pay? 100¢, or $1.00

Possible strategy:

- Counting on from the larger number: I started at 75 on the number grid and counted up 2 tens and 5 ones. I landed on 100, which is 100¢, or $1.00.

Possible number model:

- $75 + 25 = 100$

Professional Development

The sample strategies listed in this lesson demonstrate children's understanding of one or more of the following: place value, properties of operations, and the relationship between addition and subtraction. These are not the only possible strategies. They are listed as they are to help you connect your class's work to **1.NBT.4** and **1.NBT.6**. Do not expect children to use these phrases.

NOTE If the total cost of the vending machine items is greater than 100¢, remind children that they can use cents or dollars-and-cents as the unit. For example, record the sum of 60¢ + 45¢ as 105¢, or $1.05

Academic Language Development

Scaffold children's strategy discussions by asking probing questions such as: *Why did you count on from _____ and not _____? What would happen if we counted on from the smaller number?* Sample answer: It would take longer to count on.

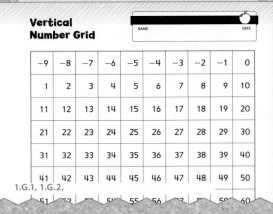

Adjusting the Activity

Differentiate To make it easier for children to decide which strategy to use, present only a few strategies at a time.

Go Online Differentiation Support

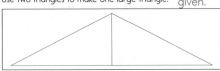

Christopher wants a pack of cheese sticks (25¢), granola (40¢), and crackers (30¢). How much does he need? 95¢

Possible strategy:

- Grouping addends: I added the prices with 0 in the ones place first to get 70. Then I added 25 more, so he needs 95¢.

Possible number model:

- $40 + 30 + 25 = 95$

The following problem lists sample strategies that demonstrate understanding of the relationship between addition and subtraction.

Natalia bought rice cakes (50¢) and one other snack. She spent 80¢. What else did she buy? How much did it cost? crackers; 30¢

Possible strategies:

- Counting up to subtract: I counted up by 10s from 50 to 80 on the number grid. That was 3 rows, or 30¢.
- Changing to easier numbers: I know that $50 + 30 = 80$ because I know that $5 + 3 = 8$. It's like solving facts with dimes or long base-10 blocks. So the other snack cost 30¢.

Possible number models:

- $80 - 30 = 50$
- $30 + 50 = 80$

After practicing with different strategies, ask children to create number stories and to solve them using a variety of strategies.

✓ Assessment Check-In CCSS 1.NBT.4

Observe as children add 2-digit numbers. Expect most children to find the sums using one of the strategies listed above. **GMP6.4** Some children may find the sums mentally while others may count up on number grids. For those who struggle, provide base-10 blocks.

Assessment and Reporting **Go Online** to record student progress and to see trajectories toward mastery for these standards.

► Finding Differences between 2-Digit Numbers

Math Journal 2, p. 197 and inside back cover; Math Masters, p. TA42

WHOLE CLASS	SMALL GROUP	PARTNER	INDEPENDENT

Ask children to compare the prices of rice cakes and sunflower seeds.

- *How much do rice cakes cost?* 50¢ *Sunflower seeds?* 60¢
- *Which one costs more?* Sunflower seeds
- *How much more do the sunflower seeds cost?* 10¢ more
- *How do you know?* **GMP6.1** Sample answer: $60 - 50 = 10$; I looked on the number grid and saw that 50 is right above 60, so they are 10 apart.

Pose number stories in which children find the difference between prices. Encourage them to use various strategies. Write number models, and draw situation diagrams to represent each problem.

The following problem lists a sample strategy that demonstrates understanding of the relationship between addition and subtraction.

How much more do the sunflower seeds (60¢) cost than the granola (40¢)?

Possible strategy:

- Counting up to subtract: I count up from 40 to 60 on the number grid and get 20, so $40 + 20 = 60$, which means $60 - 40 = 20$.

The following problem lists a sample strategy that demonstrates understanding of place value.

How much more do the rice cakes (50¢) cost than the granola (40¢)?

Possible strategy:

- Fact extensions: I know $5 - 4 = 1$, so $50 - 40 = 10$.

Summarize Have children share with their partners their favorite strategy or tool for adding or subtracting 2-digit numbers.

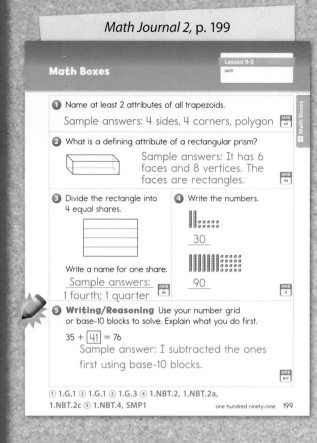

Math Journal 2, p. 199

Math Boxes

Lesson 9-5
DATE

① Name at least 2 attributes of all trapezoids.
Sample answers: 4 sides, 4 corners, polygon

② What is a defining attribute of a rectangular prism?
Sample answers: It has 6 faces and 8 vertices. The faces are rectangles.

③ Divide the rectangle into 4 equal shares.

Write a name for one share.
Sample answers:
1 fourth; 1 quarter

④ Write the numbers.
30
90

⑤ **Writing/Reasoning** Use your number grid or base-10 blocks to solve. Explain what you do first.
$35 + \boxed{41} = 76$
Sample answer: I subtracted the ones first using base-10 blocks.

① 1.G.1 ② 1.G.1 ③ 1.G.3 ④ 1.NBT.2, 1.NBT.2a, 1.NBT.2c ⑤ 1.NBT.4, SMP1   one hundred ninety-nine   199

---

3 Practice   15–20 min

Go Online

ePresentations   eToolkit   Home Connections

▶ # Turning Rectangles into Other Shapes

Math Journal 2, p. 198; Math Masters, p. 269

| WHOLE CLASS | SMALL GROUP | **PARTNER** | **INDEPENDENT** |

Have children practice rearranging composite shapes to create new composite shapes.

▶ # Math Boxes 9-5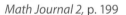

Math Journal 2, p. 199

| WHOLE CLASS | SMALL GROUP | **PARTNER** | **INDEPENDENT** |

Mixed Practice Math Boxes 9-5 are paired with Math Boxes 9-2.

▶ # Home Link 9-5

Math Masters, p. 270

Homework Children practice adding prices.

Math Masters, p. 270

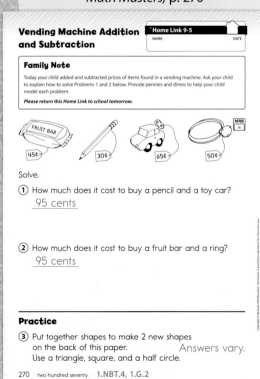

Vending Machine Addition and Subtraction   Home Link 9-5

NAME   DATE

Family Note

Today your child added and subtracted prices of items found in a vending machine. Ask your child to explain how to solve Problems 1 and 2 below. Provide pennies and dimes to help your child model each problem.

Please return this Home Link to school tomorrow.

FRUIT BAR 45¢   30¢   65¢   50¢

Solve.

① How much does it cost to buy a pencil and a toy car?
95 cents

② How much does it cost to buy a fruit bar and a ring?
95 cents

Practice

③ Put together shapes to make 2 new shapes on the back of this paper.   Answers vary.
Use a triangle, square, and a half circle.

270   two hundred seventy   1.NBT.4, 1.G.2

Two-Digit Comparison Number Stories

Overview Children use addition and subtraction strategies to solve comparison number stories.

▶ **Before You Begin**
For Part 1, select and sequence Quick Look Cards 107, 103, and 115.

 Common Core State Standards

Focus Clusters
- Represent and solve problems involving addition and subtraction.
- Use place value understanding and properties of operations to add and subtract.

1 Warm Up 15–20 min

	Materials	
Mental Math and Fluency Children do Quick Looks with double ten frames.	Quick Look Cards 103, 107, 115	**1.OA.6** SMP2, SMP6
Daily Routines Children complete daily routines.	See pages 2-37.	See pages xiv–xvii.

2 Focus 30–35 min

Math Message Children find a difference in vending machine prices.		**1.OA.1, 1.NBT.4, 1.NBT.6** SMP5
Adding and Subtracting New Vending Machine Prices Children add and subtract prices of items found in a vending machine.	*Math Journal 2*, pp. 197 and 200; *Math Masters*, p. TA42; slate; base-10 blocks, pennies, and dimes (optional)	**1.OA.1, 1.OA.2, 1.NBT.4, 1.NBT.6** SMP3, SMP6
Assessment Check-In See page 813.	*Math Journal 2*, p. 200; base-10 blocks (optional)	**1.NBT.6** SMP6
Creating Vending Machine Number Stories Children write and solve vending machine number stories.	*Math Journal 2*, pp. 197 and 201	**1.OA.1, 1.OA.2, 1.NBT.4, 1.NBT.6** SMP6

CCSS 1.NBT.6 **Spiral Snapshot**

GMC Subtract multiples of 10 from multiples of 10.

| 7-5
Practice | 8-7
Warm Up
Practice | 8-11
Focus | 9-2
Focus | 9-5
Focus | 9-6
Focus | 9-9
Practice | 9-11
Warm Up
Practice |

III **Spiral Tracker** **Go Online** to see how mastery develops for all standards within the grade.

3 Practice 15–20 min

Completing Name-Collection Boxes Children use name collection boxes to write equations.	*Math Journal 2*, p. 202	**1.OA.6, 1.OA.7**
Math Boxes 9-6 Children practice and maintain skills.	*Math Journal 2*, p. 203	See page 813.
Home Link 9-6 **Homework** Children use strategies to solve number stories.	*Math Masters*, p. 273	**1.OA.8, 1.NBT.4**

---

connectED.mcgraw-hill.com

Plan your lessons online with these tools.

ePresentations Student Learning Center Facts Workshop Game eToolkit Professional Development Home Connections Spiral Tracker Assessment and Reporting English Learners Support Differentiation Support

Differentiation Options

 RtI

CCSS 1.OA.1, 1.OA.8, 1.NBT.4, SMP6

Readiness
10–15 min

WHOLE CLASS
SMALL GROUP
PARTNER
INDEPENDENT

Practicing Counting Up to Subtract

Math Journal 2, pp. 188–189; slate

To provide additional experience counting up to subtract, have children solve number stories using the School Store Mini-Posters (*Math Journal 2*, pages 188–189). Pose a problem such as: *You have 37¢. You want to buy a pen. How much money will you have left?* Have children write 37 − 31 = ☐ on their slates. Remind children that they can count up in subtraction problems when the two numbers are close together. Write 31 + ☐ = 37, and have children keep track of how many numbers they say as they count up (32, 33, 34, 35, 36, 37). **GMP6.4** Continue with similar examples.

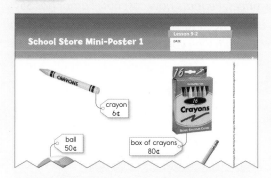

School Store Mini-Poster 1   Lesson 9-2

crayon 6¢

ball 50¢   box of crayons 80¢

School Store Mini-Poster 2   Lesson 9-2

colored paper 38¢ per pack

colored pencil 29¢

CCSS 1.OA.1, 1.OA.2, 1.NBT.4, 1.NBT.6, SMP3

Enrichment
15–20 min

WHOLE CLASS
SMALL GROUP
PARTNER
INDEPENDENT

Make Your Own Vending Machine

Activity Card 115; *Math Masters*, pp. 271–272; crayons

To further explore writing and solving number stories, have children create new vending machines. Encourage them to write comparison stories or stories about buying several items or making change. Children solve each other's stories and try to understand each other's strategies. **GMP3.2**

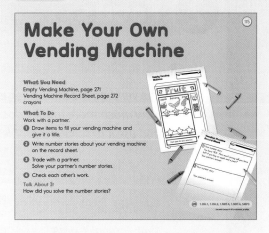

Make Your Own Vending Machine 115

What You Need
Empty Vending Machine, page 271
Vending Machine Record Sheet, page 272
crayons

What To Do
Work with a partner.
1. Draw items to fill your vending machine and give it a title.
2. Write number stories about your vending machine on the record sheet.
3. Trade with a partner. Solve your partner's number stories.
4. Check each other's work.

Talk About It
How did you solve the number stories?

1.OA.1, 1.OA.2, 1.NBT.4, SMP3

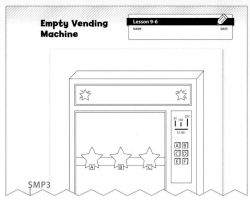

Empty Vending Machine   Lesson 9-6

NAME   DATE

SMP3

CCSS 1.NBT.4, 1.NBT.6, SMP5

Extra Practice
5–15 min

WHOLE CLASS
SMALL GROUP
PARTNER
INDEPENDENT

Playing *Stop and Go*

My Reference Book, pages 164–165 (optional); *Math Masters*, pp. G42 (optional), G43–G45; number grids; counters; base-10 blocks

To provide practice adding and subtracting 2-digit numbers, have children play *Stop and Go*. For detailed instructions, see *My Reference Book*, pages 164–165.

Observe
- What strategies and tools do children use to add and subtract? **GMP5.1**

Discuss
- *Imagine that the STOP player won* Stop and Go *in the fewest turns possible. What cards did each player draw?*

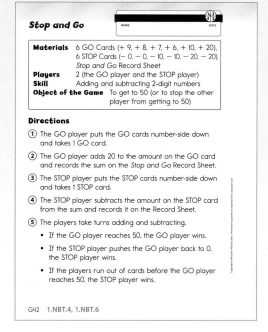

Stop and Go   NAME   DATE

Materials	6 GO Cards (+ 9, + 8, + 7, + 6, + 10, + 20), 6 STOP Cards (− 0, − 0, − 10, − 10, − 20, − 20), *Stop and Go* Record Sheet
Players	2 (the GO player and the STOP player)
Skill	Adding and subtracting 2-digit numbers
Object of the Game	To get to 50 (or to stop the other player from getting to 50)

Directions
1. The GO player puts the GO cards number-side down and takes 1 GO card.
2. The GO player adds 20 to the amount on the GO card and records the sum on the *Stop and Go* Record Sheet.
3. The STOP player puts the STOP cards number-side down and takes 1 STOP card.
4. The STOP player subtracts the amount on the STOP card from the sum and records it on the Record Sheet.
5. The players take turns adding and subtracting.
   - If the GO player reaches 50, the GO player wins.
   - If the STOP player pushes the GO player back to 0, the STOP player wins.
   - If the players run out of cards before the GO player reaches 50, the STOP player wins.

G42   1.NBT.4, 1.NBT.6

English Language Learners Support

Beginning ELL Pair English language learners with English-proficient children when they role play the vending machine activity. Have the English-proficient children tell what they would like to buy and the English language learner make the purchase from the vending machine. Have the pairs work together to solve the number stories and to record their work.

 Go Online   ELL **English Learners Support**

Standards and Goals for
Mathematical Practice

SMP3 Construct viable arguments and critique the reasoning of others.
GMP3.2 Make sense of others' mathematical thinking.

SMP5 Use appropriate tools strategically.
GMP5.1 Choose appropriate tools.

SMP6 Attend to precision.
GMP6.1 Explain your mathematical thinking clearly and precisely.

1 Warm Up 15–20 min Go Online

▶ Mental Math and Fluency

Do Quick Looks with children. Display Quick Look Cards in this order: 107, 103, 115. Show each image for 2–3 seconds. Have children share what they saw and how they saw it. **GMP2.2, GMP6.1** Ask children to share strategies for Card 107, highlighting near doubles. Encourage children to apply this strategy on the next two cards.

▶ Daily Routines

Have children complete the Daily Routines. For detailed instructions, see pages 2–37. For specifics on CCSS coverage, see pages xiv–xvii.

2 Focus 30–35 min Go Online

▶ Math Message

The vending machine has new prices. Peanuts now cost 60¢. Lamar has 55¢. How much more money does Lamar need to buy the peanuts? Choose a tool, and use it to solve the problem. Explain why you chose that tool. **GMP5.1**

▶ Adding and Subtracting New Vending Machine Prices

Math Journal 2, pp. 197 and 200; *Math Masters*, p. TA42

Math Message Follow-Up Have children share their responses to the Math Message. Sample answers: I picked a number grid as my tool and counted up from 55: 56, 57, 58, 59, 60, so I knew that he needed 5 more cents. I knew that when I count by 5s, I say 55, then 60. So I knew that 55 and 60 were 5 apart. As children share their solution strategies, highlight counting up to subtract.

Tell children that all the prices in the vending machine have changed. Today they will solve number stories using the new prices.

Academic Language Development

Different uses of the word *change* with vending machines may confuse children. It can refer to coins, money given back, or making or becoming different. Scaffold children's understandings of *change* as making or becoming different by contrasting the new prices in the vending machine with the old prices. Describe the contrast as *a change from the old price to the new.* Remind children that the *change*, or *difference*, in the new price can be more or less than the old price.

Display *Math Masters,* page TA42, and have children look at journal page 197. Display and update the prices (as listed in the margin) on the master using a pencil or brightly colored marker, and have children follow along, crossing out all the current prices with a light-colored marker so they can still see the original prices. (They will need the original prices to solve some of the number stories.)

Ask: *Which snacks cost more than they did before?* Crackers, peanuts, pretzels, water, granola, cheese sticks Have children record these snacks on the top half of journal page 200. Then have them calculate the price increase for each snack and share their solution strategies. Highlight instances of adding or counting up. For example, to calculate the increase of the price for a package of cheese sticks, children may think about adding 5 to 25 to get 30. Also highlight the efficiency of counting up by 10s. For example, to calculate the granola price increase, children may start from 40 and count up 3 tens: 50, 60, 70. Ask: *Which snack had the greatest difference in price?* The price of a granola bar went up 30¢.

Give children opportunities to practice finding differences with a wide range of 2-digit numbers. Sample problems follow. Select children to play the role of the buyer as the class solves number stories using whatever strategies they wish. Encourage them to record their thinking with number models on their slates. Ask children to state which strategies they used. Encourage them to try to understand each other's strategies and to use them to solve the next number story. GMP3.2

Differentiate | **Adjusting the Activity**

At this point, many children will be able to solve these problems mentally. Encourage mental math whenever possible, but don't expect it from all children. Make sure tools (such as base-10 blocks, number grids, and counters) are available as needed.

Go Online ▸ | Differentiation Support

Sarah has 60¢ to buy sunflower seeds, but the price changed. How much money will she have left after she buys the sunflower seeds? 20¢

Possible strategy:

- Counting up to subtract: I counted up 2 tens from 40 to 60, so she has 20¢ left.

Possible number model:

$60 - 40 = 20$

Old and New Vending Machine Prices

Item	Old Price	New Price
Crackers	30¢	50¢
Peanuts	55¢	60¢
Pretzels	65¢	75¢
Raisins	45¢	35¢
Rice Cakes	50¢	40¢
Water	75¢	85¢
Granola	40¢	70¢
Sunflower Seeds	60¢	40¢
Mints	35¢	20¢
Cheese Sticks	25¢	30¢

Math Journal 2, p. 200

Vending Machine Prices — Lesson 9-6 — DATE

These snacks went up in price. How much did each price go up?

Crackers	20 ¢		Water	10 ¢
Peanuts	5 ¢		Granola	30 ¢
Pretzels	10 ¢		Cheese Sticks	5 ¢

Niko has 30¢. Leela has 35¢. Wei has 50¢.
Niko, Leela, and Wei each buy a package of cheese sticks. Sample number models given.

How much money does each child have left?

Niko will have _0_ ¢ left. Number model: _30 − 30 = 0_

Leela will have _5_ ¢ left. Number model: _30 + 5 = 35_

Wei will have _20_ ¢ left. Number model: _20 = 50 − 30_

Try This

Niko, Leela, and Wei put together all of the money they have left. How much do they have in all? _25_ ¢

Number model: Sample answer: 0 + 5 + 20 = 25

200   two hundred   1.OA.1, 1.OA.2, 1.NBT.4, 1.NBT.6, SMP6

Creating Vending Machine Number Stories

Lesson 9-6

DATE

Write number stories about the vending machine on journal page 197.

Trade with a partner.
Solve the number stories and then write a number model.
Trade again and check your partner's answers. Answers vary.

①

②

③

1.OA.1, 1.OA.2, 1.NBT.4, 1.NBT.6, SMP6

two hundred one 201

Name-Collection Boxes

Lesson 9-6

DATE

① Write four different names for 19.

| 19 | 9 + 9 + 1 |

Answers vary.

Write a true number sentence for each name you added.

$9 + 9 + 1 = 19$

② Cross out the names that do not belong.

| 15 |

7 + 8

25 − 10

⊠

⌸⌸⌸⌸⌸

5 + 5 + 5

20 ⊠ 5

Write a true number sentence for each name that does belong.

$7 + 8 = 15$
$25 − 10 = 15$
$15 = 5 + 5 + 5$

Darrell has 43¢ in his pocket. How much money will he have left if he buys raisins at the new price? 8¢

Possible strategy:

- Counting up: I counted by 5s from 35 and then by 1s on the number line: 35, 40, 41, 42, 43.

Possible number model:

- $35 + 8 = 43$

Miles wants to buy a pack of cheese sticks, mints, and sunflower seeds. How much money does he need? 90¢

Possible strategy:

- Using fact extensions and grouping addends: I used base-10 blocks to add 2 tens, 3 tens, and 4 tens. I added 2 and 3 first; then I added 4 to find $20 + 30 + 40 = 90$, so Miles needs 90¢.

Possible number model:

- $20 + 30 + 40 = 90$

Kiera has 38¢. Maya has 9¢. How much money do they have all together? 47¢
If they buy rice cakes, how much money will they have left? 7¢

Possible strategies:

- Changing to an easier number: I added 10¢ to 38¢ to get 48¢; then I subtracted 1¢ because 9¢ is 1¢ less than 10¢, so all together they have 47¢.
- Counting up: I started at 40 and counted up to 47, so they would have 7¢ left.

Possible number models:

- $38 + 9 = 47$
- $40 + 7 = 47$

Jacob has 35¢. Ritu has 25¢. How much money do they have in all? 60¢
Which one item could they buy and not have any money left over? peanuts

Possible strategy:

- Combining place-value groups and composing a ten: I showed 35 and 25 with base-10 blocks. I added 3 longs and 2 longs to get 5 longs. Then I exchanged the cubes for one more long. 6 longs is 6 tens, or 60¢.

Possible number models:

- $50 + 10 = 60$
- $35 + 25 = 60$

Have children solve the problems at the bottom of journal page 200; then discuss their solution strategies. GMP3.2, GMP6.2

Assessment Check-In CCSS **1.NBT.6**

Math Journal 2, p. 200

As children complete the journal page, ask them to explain how they figured out how much money Wei would have left. GMP6.1 Expect most children to accurately solve the problem. Suggest to children who struggle that they use base-10 blocks. You may wish to have children who excel complete the Enrichment activity.

 Assessment and Reporting · Go Online to record student progress and to see trajectories toward mastery for these standards.

▶ Creating Vending Machine Number Stories

Math Journal 2, pp. 197 and 201

| WHOLE CLASS | SMALL GROUP | **PARTNER** | **INDEPENDENT** |

Have children write and solve their own number stories on journal page 201 based on the new vending machine prices. Pose a few of their number stories for the whole class to solve. Have volunteers explain their solution strategies. GMP6.1

Summarize Have partnerships discuss which strategies they used most often to solve vending machine problems. Then have them share with the class as time permits.

(3) **Practice** 15–20 min · Go Online · ePresentations · eToolkit · Home Connections

▶ Completing Name-Collection Boxes

Math Journal 2, p. 202

| WHOLE CLASS | SMALL GROUP | **PARTNER** | **INDEPENDENT** |

Have children practice identifying equivalent names for numbers by completing name-collection boxes and writing equations.

▶ Math Boxes 9-6

Math Journal 2, p. 203

| WHOLE CLASS | SMALL GROUP | **PARTNER** | **INDEPENDENT** |

Mixed Practice Math Boxes 9-6 are paired with Math Boxes 9-4.

▶ Home Link 9-6

Math Masters, p. 273

Homework Children solve number stories involving addition and subtraction of 2-digit numbers.

Math Journal 2, p. 203

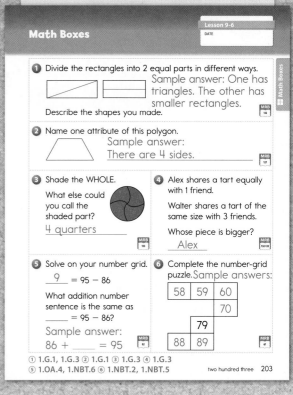

Math Masters, p. 273

Efficient Strategies for 2-Digit Addition and Subtraction

Overview Children choose and use tools and strategies to solve addition and subtraction number stories and explain their choices.

▶ **Before You Begin**
For the first activity in the Focus section, children will need the Animal Cards from their toolkits.

 Common Core State Standards

Focus Clusters
- Represent and solve problems involving addition and subtraction.
- Use place value understanding and properties of operations to add and subtract.

---

1 Warm Up 15–20 min

	Materials	
Mental Math and Fluency Children solve number stories.	slate	1.OA.1, 1.OA.6
Daily Routines Children complete daily routines.	See pages 2–37.	See pages xiv–xvii.

---

2 Focus 30–35 min

Math Message Children explain how to solve an addition number story with tools.		1.NBT.4 SMP5
Creating and Solving Silly Animal Stories Children apply addition and subtraction strategies to make up and solve number stories using Animal Cards.	*Math Journal 2*, p. 204, inside back cover (optional); Animal Cards; base-10 blocks (optional)	1.OA.1, 1.NBT.4, 1.NBT.6 SMP1, SMP5, SMP8
Assessment Check-In See page 819.	*Math Journal 2*, p. 204	1.NBT.4

CCSS 1.NBT.4 Spiral Snapshot

GMC Understand adding 2-digit numbers.

| 7-1
Warm Up
Practice | 7-9
Warm Up | 8-7
Warm Up | 9-2 through 9-6
Focus
Practice | 9-7
Focus
Practice | 9-8
Focus
Practice | 9-9
Practice |

Spiral Tracker   **Go Online**   to see how mastery develops for all standards within the grade.

---

3 Practice 15–20 min

Playing *Tric-Trac* **Game** Children practice addition facts.	*My Reference Book*, pp. 168–169; *Math Masters*, p. G54; per partnership: 2 dot dice, 20 pennies	1.OA.3, 1.OA.6 SMP1
Math Boxes 9-7 Children practice and maintain skills.	*Math Journal 2*, p. 205	See page 819.
Home Link 9-7 **Homework** Children practice adding 2-digit numbers.	*Math Masters*, p. 274	1.OA.1, 1.OA.2, 1.NBT.4, 1.NBT.6

---

connectED.mcgraw-hill.com

Plan your lessons online with these tools.

ePresentations Student Learning Center Facts Workshop Game eToolkit Professional Development Home Connections Spiral Tracker Assessment and Reporting English Learners Support Differentiation Support

Differentiation Options RtI

CCSS 1.NBT.4, 1.NBT.6, SMP5	CCSS 1.NBT.3, 1.NBT.4, SMP1, SMP6	CCSS 1.OA.1, 1.NBT.4, 1.NBT.6, SMP4

Readiness   10–15 min

Adding and Subtracting with the Number Grid

WHOLE CLASS
SMALL GROUP
PARTNER
INDEPENDENT

slate, 2 sets of number cards 0–9, number grid

To provide experience adding and subtracting 2-digit numbers on the number grid, have partners each turn over two number cards to make a 2-digit number. They add the numbers by counting up from the larger number on the number grid. GMP5.2

To provide experience subtracting multiples of 10, have each partner draw one card, which can show how many tens. Using the number grid, children subtract the smaller number of tens from the larger number of tens. Children can record number sentences on their slates.

Enrichment   10–15 min

Playing *Animal Weight Top-It*

WHOLE CLASS
SMALL GROUP
PARTNER
INDEPENDENT

Math Masters, p. G46 (optional); Animal Cards; base-10 blocks; number grid

To further explore addition of 2-digit numbers, children play *Animal Weight Top-It*. For detailed instructions, see Lesson 5-11.

Observe

• Do children use efficient strategies? GMP6.4

Discuss

• *How do you check that your addition makes sense?* GMP1.4

Animal Weight Top-It

NAME   DATE

Materials	2 sets of Animal Cards, base-10 blocks, number grid
Players	2 to 4
Skill	Adding 2-digit numbers
Object of the Game	To collect the most cards

Directions

① Play with a partner.
② Shuffle and place the cards in a stack, weight-side down.
③ Both players take two cards from the top of the deck.
④ Find the total weight of the animals on your two cards.
⑤ The player whose animals weigh more all together takes all the cards.
⑥ The game is over when all the cards have been used.
⑦ The player with more cards wins.

1.NBT.3, 1.NBT.4

Extra Practice   10–15 min

Telling Animal Growth Stories

WHOLE CLASS
SMALL GROUP
PARTNER
INDEPENDENT

Activity Card 116, Animal Cards, slate

To provide practice finding differences, have children tell and solve number stories about animal growth using Animal Cards. They write number models for their stories. GMP4.1

English Language Learners Support

Beginning ELL Ask children to use concrete objects to role play number stories told for the Mental Math and Fluency. For example, give children counters or pictures of a tomato and then read the first problem. Pausing after each sentence, prompt children to show the number of counters for each part of the problem.

Go Online   ELL English Learners Support

SMP1 Make sense of problems and persevere in solving them.

GMP1.6 Compare the strategies you and others use.

SMP5 Use appropriate tools strategically.

GMP5.1 Choose appropriate tools.

SMP8 Look for and express regularity in repeated reasoning.

GMP8.1 Create and justify rules, shortcuts, and generalizations.

---

1 Warm Up   15–20 min   Go Online

ePresentations   eToolkit

▶ Mental Math and Fluency

Have children solve and record their answers on their slates and then share their strategies.

- ●○○ Esther bought 8 tomatoes. She used 4 of them to make salsa. How many tomatoes are left? 4 tomatoes
- ●●○ Akari had 12 books in her room. She took 6 of them back to the library. How many books are still in Akari's room? 6 books
- ●●● Julio has 14 paintbrushes in his art kit. He used 7 of them to paint a picture. How many paintbrushes are still in Julio's art kit?
  7 paintbrushes

▶ Daily Routines

Have children complete the Daily Routines. For detailed instructions, see pages 2–37. For specifics on CCSS coverage, see pages xiv–xvii.

---

2 Focus   30–35 min   Go Online

ePresentations   eToolkit

▶ Math Message

Damien used counters to solve 46 + 38. What other tools could he have used? Record your ideas, and use one of them to solve. **GMP5.1**

▶ Creating and Solving Silly Animal Stories

Math Journal 2, p. 204

Math Message Follow-Up Have children share their ideas about tools Damien could have used. They may suggest using base-10 blocks to model 46 and 38, or using a number grid to count up 38 from 46. Have children share their solutions. 84 You may wish to discuss which tools help them solve the problem the quickest. Tell children that today they will create and solve number stories and compare the tools, as well as strategies, they use to solve them.

Math Journal 2, p. 204

Silly Animal Stories   Lesson 9-7
DATE

Example:

Unit
inches

koala 24 in.
penguin 36 in.

How tall are the koala and penguin all together?

24 + 36 = 60

60 inches

❶ Silly Story

Unit

Answers vary.

❷ Silly Story

Unit

Answers vary.

204 two hundred four 1.OA.1, 1.NBT.4

Have children reexamine their Animal Cards. Remind them that the height or length of an animal is on one side and that the weight is on the other.

Tell animal number stories such as those that follow. Elicit a variety of strategies as children discuss and compare tools and strategies for solving the stories. Be sure to include units when discussing solutions. Have children suggest number models to represent the problems. Keep in mind that more than one number model may be appropriate. You may also wish to display situation diagrams. Focus the discussion on choosing tools and strategies that make sense for each number story. GMP5.1

Suppose the rabbit (11 in.) and the raccoon (23 in.) lie nose to nose. How long will they be in all? 34 inches

Possible strategies:

- Counting up: I know that 11 is 1 more than 10. So, I added 23 and 10, and then I added 1 more. Starting at 23 on the number grid, I moved down one row to 33. Then I moved right one space to 34. So, they are 34 inches long all together.
- Combining place-value groups: I used base-10 blocks to show 11 and 23; then I combined tens with tens and ones with ones to get 34 inches.

Possible number models:

- 23 + 11 = 34

- 30 + 4 = 34

- 11 + 23 = 34

Ask: *What is different about these two strategies? Which do you like better for solving this problem?* Sample answer: I can count on the number grid in my head, but I have to get out blocks and arrange them to add with base-10 blocks. I can find the answer quicker if I count on the number grid, so I like it better. GMP1.6 Explain that many strategies can help children find the right answers but that they should choose tools and strategies that make the most sense to them for a given story.

Encourage children to be flexible in their solution strategies. Whether they use mental computation, number grids, or visual representations, they should carefully select strategies and tools that will allow them to understand and solve problems as efficiently as possible. Emphasize that children can use tools, like the number grid, to check answers as well as to help solve problems.

First-grade girl	43 in.	41 lb
7-year-old boy	50 in.	50 lb
Cheetah	48 in.	120 lb
Porpoise	72 in.	98 lb
Penguin	36 in.	75 lb
Beaver	30 in.	56 lb
Fox	20 in.	14 lb
Cat	12 in.	7 lb
Raccoon	23 in.	23 lb
Koala	24 in.	19 lb
Eagle	35 in.	15 lb
Rabbit	11 in.	6 lb
Starfish	6 in.	10 lb
Blue crab	4 in.	2 lb
Peacock	60 in.	8 lb
Sun bear	54 in.	110 lb
Toad	5 in.	3 lb
Orangutan	48 in.	165 lb
Parrot	31 in.	3 lb
Squirrel	5 in.	4 lb
Skunk	8 in.	9 lb
Owl	20 in.	5 lb
Flamingo	40 in.	9 lb
Octopus	36 in.	20 lb

Animal Card heights/lengths and weights

Academic Language Development

Encourage children to contrast strategies and to argue for using some strategies over others. Scaffold their arguments by providing sentence frames such as this: "I used _____ to get the answer, but when I used _____, I found the answer more quickly."

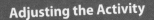
My Reference Book, pp. 168–169

Games

Tric-Trac

Materials	☐ 2 six-sided dice
	☐ 22 pennies
	☐ 1 *Tric-Trac* Game Mat for each player
Players	2
Skill	Addition facts 0–10
Object of the Game	To have the lower sum.

Directions

① Cover the empty circles on your game mat with pennies.

② Take turns. When it is your turn:
- Roll the dice. Find the total number of dots. This is your sum.
- Move 1 of your pennies and cover your sum on your game mat.

OR

- Move 2 or more of your pennies and cover any numbers that can be added together to equal your sum.

168 one hundred sixty-eight

A toad (3 lb) jumped up and landed on top of a sun bear (110 lb). How much do they weigh all together? 113 lb

Possible strategy:

- Counting on from the larger number: I started at 110 and counted up 111, 112, 113; so, they weigh 113 pounds all together.

Possible number model:

- $110 + 3 = 113$

Ask: *If you combined base-10 blocks, would you get the same answer?*
GMP1.6 Sample answer: Yes, but it might take longer to get out and count the blocks.

A beaver (56 lb) and a skunk (9 lb) sit together on a bench. What do they weigh all together? 65 lb

Possible strategy:

- Changing to an easier number: I have a rule I use when I add 9s. Since I know that 9 is 1 less than 10, I add 10 and then take away 1. A beaver and a skunk weigh 65 pounds together. So, $56 + 10 = 66$, and $66 - 1 = 65$. Thus, $56 + 9 = 65$. **GMP8.1**

Possible number model:

- $56 + 9 = 65$

If the cheetah (48 in.) and the octopus (36 in.) sleep head-to-toe in a line, how long would the line be? 84 in.

Possible strategy:

- Composing a 10: I used base-10 blocks to show the lengths of the cheetah and octopus. I added tens to tens and ones to ones to get 7 longs, 14 cubes. Then I exchanged 10 cubes for 1 long, which leaves 4 cubes left over. The length is 8 tens and 4 ones, or 84 inches total.

Possible number model:

- $84 = 48 + 36$

How much longer is the flamingo (40 in.) than the octopus (36 in.)? 4 in.

Possible strategy:

- Counting back: I started at 40 and counted back 4 to get to 36.

Possible number model:

- $40 - 4 = 36$

Have children make up number stories about comparing or finding total animal lengths or weights. Suggest that they imagine two animals on the same weight scale or one animal standing on top of another. Partnerships use their Animal Cards to make up and record two silly stories on journal page 204.

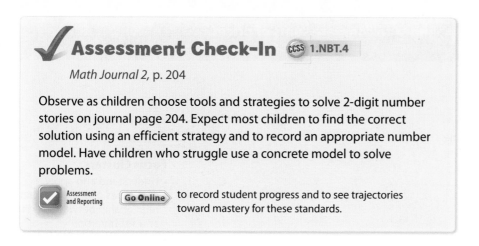

✓ Assessment Check-In **CCSS** 1.NBT.4

Math Journal 2, p. 204

Observe as children choose tools and strategies to solve 2-digit number stories on journal page 204. Expect most children to find the correct solution using an efficient strategy and to record an appropriate number model. Have children who struggle use a concrete model to solve problems.

☑ Assessment and Reporting   **Go Online** to record student progress and to see trajectories toward mastery for these standards.

Summarize Have partnerships discuss two different strategies they could use to find the combined weight of the starfish (10 lb) and the penguin (75 lb) and decide which one makes the most sense.

---

3 Practice 15–20 min   **Go Online** ePresentations eToolkit Home Connections

▶ Playing *Tric-Trac*

My Reference Book, pp. 168–169; *Math Masters,* p. G54

| WHOLE CLASS | SMALL GROUP | **PARTNER** | INDEPENDENT |

To provide practice with addition facts, have children play *Tric-Trac*. For detailed instructions, see *My Reference Book,* pages 168 and 169.

Observe

• Which children use more than two addends to cover their sums?

Discuss

• *How did you use more than two addends to cover your sum?* **GMP1.5**

▶ Math Boxes 9-7

Math Journal 2, p. 205

| WHOLE CLASS | SMALL GROUP | **PARTNER** | **INDEPENDENT** |

Mixed Practice Math Boxes 9-7 are paired with Math Boxes 9-9.

▶ Home Link 9-7

Math Masters, p. 274

Homework Children solve number stories involving 2-digit numbers.

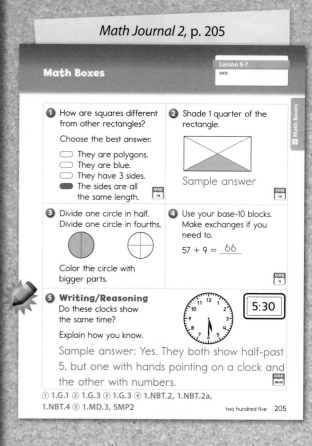

Math Journal 2, p. 205

Math Boxes   Lesson 9-7
                                                       DATE

① How are squares different from other rectangles?

Choose the best answer.
○ They are polygons.
○ They are blue.
○ They have 3 sides.
● The sides are all the same length.

② Shade 1 quarter of the rectangle.

Sample answer

③ Divide one circle in half. Divide one circle in fourths.

Color the circle with bigger parts.

④ Use your base-10 blocks. Make exchanges if you need to.

$57 + 9 =$ __66__

⑤ **Writing/Reasoning** Do these clocks show the same time?

Explain how you know.

5:30

Sample answer: Yes. They both show half-past 5, but one with hands pointing on a clock and the other with numbers.

① 1.G.1 ② 1.G.3 ③ 1.G.3 ④ 1.NBT.2, 1.NBT.2a, 1.NBT.4 ⑤ 1.MD.3, SMP2

two hundred five 205

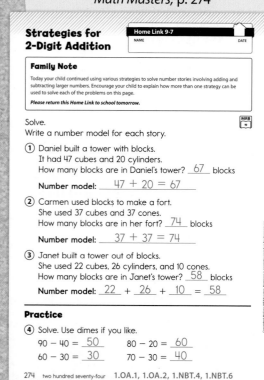

Math Masters, p. 274

Strategies for 2-Digit Addition   **Home Link 9-7**
                                         NAME   DATE

Family Note

Today your child continued using various strategies to solve number stories involving adding and subtracting larger numbers. Encourage your child to explain how more than one strategy can be used to solve each of the problems on this page.

Please return this Home Link to school tomorrow.

Solve.
Write a number model for each story.

① Daniel built a tower with blocks. It had 47 cubes and 20 cylinders. How many blocks are in Daniel's tower? __67__ blocks

Number model: __47 + 20 = 67__

② Carmen used blocks to make a fort. She used 37 cubes and 37 cones. How many blocks are in her fort? __74__ blocks

Number model: __37 + 37 = 74__

③ Janet built a tower out of blocks. She used 22 cubes, 26 cylinders, and 10 cones. How many blocks are in Janet's tower? __58__ blocks

Number model: __22__ + __26__ + __10__ = __58__

Practice

④ Solve. Use dimes if you like.

$90 - 40 =$ __50__ $80 - 20 =$ __60__

$60 - 30 =$ __30__ $70 - 30 =$ __40__

274 two hundred seventy-four 1.OA.1, 1.OA.2, 1.NBT.4, 1.NBT.6

Review: Relations and Equivalence

Overview Children use <, =, and > to compare sums of prices.

▶ **Before You Begin**
Display the name-collection box shown in the Math Message.

Common Core State Standards

Focus Clusters
- Understand and apply properties of operations and the relationship between addition and subtraction.
- Add and subtract within 20.
- Work with addition and subtraction equations.
- Understand place value.
- Use place value understanding and properties of operations to add and subtract.

1 Warm Up 15–20 min

	Materials	
Mental Math and Fluency Children mentally find 10 more or 10 less.	slate	1.NBT.5
Daily Routines Children complete daily routines.	See pages 2–37.	See pages xiv–xvii.

2 Focus 35–40 min

Math Message Children write equivalent names for 13.	slate	1.OA.3, 1.OA.6, 1.OA.7
Reviewing Equivalence Children identify true and false equations.	counters (optional)	1.OA.3, 1.OA.6, 1.OA.7 SMP1
Comparing Sums of Prices Children choose tools and strategies to compare sums.	*Math Journal 2*, pp. 188–189, 206, and inside back cover; base-10 blocks; coins	1.NBT.3, 1.NBT.4 SMP4, SMP5

✓ **Assessment Check-In**   See page 824.   *Math Journal 2*, p. 206   1.NBT.3

CCSS 1.NBT.3 **Spiral Snapshot**

GMC Record comparisons using <, =, or >.

5-5 Focus Practice	5-7 Warm Up Practice	5-9 Focus Practice	5-11 Practice	6-2 Warm Up Practice	6-11 Warm Up Practice	8-10 Warm Up Practice	9-8 Focus Practice

Spiral Tracker   Go Online   to see how mastery develops for all standards within the grade.

3 Practice 10–15 min

Making Smallest and Largest Numbers Children practice writing and comparing numbers.	*Math Journal 2*, p. 207; number cards 1–9	1.NBT.2, 1.NBT.3
Math Boxes 9-8 Children practice and maintain skills.	*Math Journal 2*, p. 208	See page 825.
Home Link 9-8 **Homework** Children write relation models for sums.	*Math Masters*, p. 277	1.NBT.3, 1.NBT.4, 1.G.3

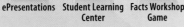

connectED.mcgraw-hill.com

Plan your lessons online with these tools.

ePresentations   Student Learning Center   Facts Workshop Game   eToolkit   Professional Development   Home Connections   Spiral Tracker   Assessment and Reporting   English Learners Support   Differentiation Support

Differentiation Options RtI

<table>
<tr><td>

CCSS 1.OA.7, 1.NBT.3, SMP2

Readiness   10–15 min

</td><td>

CCSS 1.NBT.3, 1.NBT.4, SMP3

Enrichment   15–20 min

</td><td>

CCSS 1.NBT.3, 1.NBT.4, SMP3

Extra Practice   10–15 min

</td></tr>
</table>

Reviewing Relation Symbols

WHOLE CLASS
SMALL GROUP
PARTNER
INDEPENDENT

slate

To provide additional experience with relation symbols, say pairs of 2-digit numbers and have children record the numbers and write $<$, $>$, or $=$ to represent the comparisons on their slates. **GMP2.1** You may wish to review the $<$ and $>$ symbols by reminding children that the "open mouth" of the symbol should "eat" the bigger number. Be sure to include some pairs in which both numbers are the same (such as 53 and 53) to ensure that children understand the use of the equal sign in these situations.

Spending 50¢

WHOLE CLASS
SMALL GROUP
PARTNER
INDEPENDENT

Activity Card 117;
Math Journal 2, pp. 206–207;
Math Masters, p. 275

To extend adding and comparing sums of 2-digit numbers, have children find pairs of School Store items they can buy for 50¢ or less. They share how to quickly tell whether pairs will cost more than 50¢. **GMP3.1** Allow them to use any tools they like.

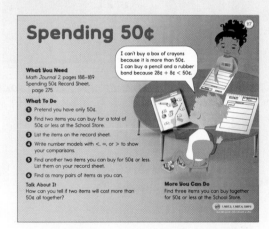

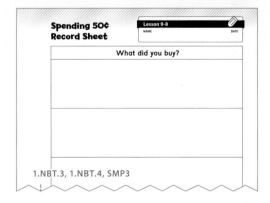

1.NBT.3, 1.NBT.4, SMP3

Playing *Top-It* with School Store Cards

WHOLE CLASS
SMALL GROUP
PARTNER
INDEPENDENT

Math Masters, p. 276; scissors; number grids; number lines; base-10 blocks

To provide practice adding and comparing sums of 2-digit numbers, have children play *Top-It* with items from the School Store Mini-Posters. Have them cut out School Store *Top-It* Cards (*Math Masters,* page 276). Each child turns over two cards and adds the prices on each card. The child with the higher total cost takes all the cards. Have children discuss whether it is possible to compare some sums without actually adding the prices. **GMP3.1**

Observe

- Which children correctly add and compare the sums?

Discuss

- *What strategies did you use to add and compare sums?*

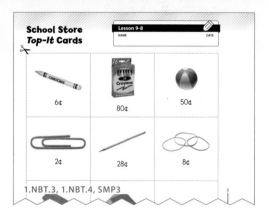

1.NBT.3, 1.NBT.4, SMP3

English Language Learners Support

Beginning ELL Children may think of *tools* as everyday work tools, such as hammers, as well as the tools in their math toolkits. Think aloud using visuals and actions to show using tools to solve problems, such as using a number line or number grid to help you count. Encourage children to name tools they use and explain their actions while using their tools. Ask: *What tools did you use? What tools did you work with? Show me how you used your tools.*

 Go Online   ELL   English Learners Support

Standards and Goals for Mathematical Practice

SMP1 Make sense of problems and persevere in solving them.
GMP1.4 Check whether your answer makes sense.

SMP4 Model with mathematics.
GMP4.1 Model real-world situations using graphs, drawings, tables, symbols, numbers, diagrams, and other representations.

SMP5 Use appropriate tools strategically.
GMP5.1 Choose appropriate tools.

1 Warm Up · 15–20 min · Go Online

ePresentations · eToolkit

▶ Mental Math and Fluency

Ask children to solve mentally and to record their answers on their slates.

◐○○ 10 more than 10? 20; 10 more than 40? 50; 10 more than 70? 80

◐◐○ 10 less than 30? 20; 10 less than 10? 0; 10 less than 100? 90

◐◐◐ 10 more than 25? 35; 10 less than 55? 45; 10 more than 47? 57

▶ Daily Routines

Have children complete the Daily Routines. For detailed instructions, see pages 2–37. For specifics on CCSS coverage, see pages xiv–xvii.

2 Focus · 30–40 min · Go Online

ePresentations · eToolkit

▶ Math Message

This is a name-collection box for 13 using only the numbers 3, 5, 4, and 1 to make names. Some of the names are incorrect. Copy this box onto your slate. Cross out the incorrect names, and add three more correct names.

13
$5 + 5 + 3$
$1 + 3 + 3 + 3$
$1 + 3 + 4$
$1 + 4 + 4 + 4$

▶ Reviewing Equivalence

| **WHOLE CLASS** | SMALL GROUP | PARTNER | INDEPENDENT |

Math Message Follow-Up Have children share the names they think do not belong in the box. $1 + 3 + 4$; $1 + 3 + 3 + 3$ Check children's answers by writing equations for each name in the box. **GMP1.4** Have children give thumbs up or down to indicate whether an equation is true or false. Circle any true equations, and strike through any false equations.

Tell children that today they will review what they have learned about the equal sign. Dictate some equations that are true and some that are false, selecting a different child to display each equation on his or her slate. Then have children give thumbs up or down for each. *Suggestions:*

- $13 = 1 + 3 + 4$ False
- $4 + 5 = 5 + 4 + 1$ False
- $4 + 4 = 3 + 5$ True
- $13 = 4 + 5 + 3 + 3$ False

For false equations, have children discuss how they could change the numbers to make the number sentences true.

Differentiate **Adjusting the Activity**

Provide counters to help children understand equations with sums written on the left, such as $13 = 5 + 5 + 3$. Encourage them to represent the total first. Display the total. Then ask children to separate their counters into two piles of 5 and one pile of 3. Display $5 + 5 + 3$ to represent the piles. Ask children whether any counters were added or taken away. Insert the equal sign to make a number sentence, and ask children to explain why it is true. Sample answer: because when we split 13 into 5, 5, and 3, we still have 13

Go Online Differentiation Support

▶ # Comparing Sums of Prices

Math Journal 2, pp. 188–189 and 206

| WHOLE CLASS | SMALL GROUP | PARTNER | INDEPENDENT |

Have children turn to the School Store Mini-Posters on journal pages 188 and 189. Pose the following problem: *Pedro bought a pencil (28¢) and a paper clip (2¢). Nevaeh bought stickers (25¢) and a crayon (6¢). Who spent more money: Pedro or Nevaeh?* Nevaeh Encourage children to choose appropriate tools (number grids, number lines, base-10 blocks, coins) to solve the problem. GMP5.1

Have the class discuss whether the tools children used were appropriate for solving this particular problem. GMP5.1 For example, counting on fingers may not be appropriate or efficient for solving this problem.

Display the symbols $<$ and $>$. Review how children could use one of these symbols to write a number model comparing the money Pedro and Nevaeh spent. Sample answers: $28 + 2 < 25 + 6$; $30¢ < 31¢$ GMP4.1

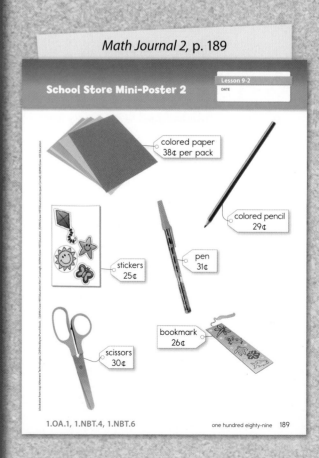

Math Journal 2, p. 188

Math Journal 2, p. 189

Math Journal 2, p. 206

Adding and Comparing Sums

Lesson 9-8
DATE

Solve the number stories using the School Store Mini-Posters.
Write number models for your answers using < or >.

	Unit
	cents

① Natalie buys colored paper and a crayon.
Jeffrey buys scissors and an eraser.
How much did each child spend?

Natalie spent 44 cents. Jeffrey spent 47 cents.

Explain how you know. Answers vary.

Which child spent more? __Jeffrey__

Number model: __Sample answer: 44 < 47__

② Ramzi buys a ball and a colored pencil.
Nikita buys a pen and a pencil.
How much did each child spend?

Ramzi spent 79 cents. Nikita spent 59 cents.

Explain how you know. Answers vary.

Which child spent more? Ramzi

Number model: __Sample answer: 79 > 59__

206   two hundred six   1.NBT.3, 1.NBT.4, SMP4

Math Journal 2, p. 207

Making Smallest and Largest Numbers

Lesson 9-8
DATE

① Shuffle number cards 1–9. Put them facedown on your desk.
② Draw two number cards.
③ Make the smallest number you can.
④ Make the largest number you can.
⑤ Record the numbers.
⑥ Put the cards back and do it again.

Answers vary.

Turn	Cards	Smaller Number	Larger Number
Example	5, 3	35	53
1			
2			
3			

Use the numbers in the table to write a number sentence for
each row using <, >, and =.

① _____
② _____
③ _____

Try This

What is the smallest number you can make with your cards?
__12__ The largest? __98__

1.NBT.2, 1.NBT.3   two hundred seven   207

Pose more problems in which children compare the sums of two School Store prices. Continue to have children share solution strategies and to summarize their answers with number models. Keep in mind that more than one model may be appropriate. *Suggestions:*

- *Rashaad buys a pen (31¢) and an eraser (17¢). Mimi buys a bookmark (26¢) and stickers (25¢). Who spends more money: Rashaad or Mimi?* 31 + 17 = 48; 26 + 25 = 51. 48¢ < 51¢. Mimi spends more money.
- *Ben buys scissors (30¢) and a ball (50¢). Ellen buys a ball (50¢) and a pencil (28¢). Who spends less money?* 30 + 50 = 80; 50 + 28 = 78. 80¢ > 78¢. Ellen spends less money.

Have children complete journal page 206, recording the results of their comparisons with relation symbols. **GMP4.1** You may wish to have children work in partnerships so that one child's journal shows the School Store Mini-Posters and the other child's journal shows the problems. As time permits, discuss the strategies and tools children used to solve the problems, emphasizing tools that are appropriate and efficient for each problem. **GMP5.1**

Academic Language Development Demonstrate with everyday examples and pictures that some tools and strategies are better than others for solving a problem. Ask: *Would this tool be the best for _____? Why? Why not?* For example, show a fork and ask whether it would be the best tool for eating soup.

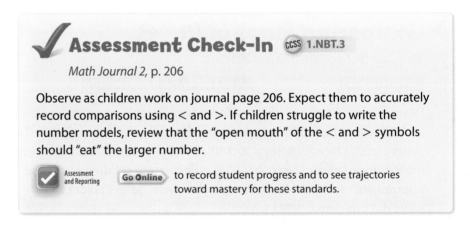

✓ **Assessment Check-In** (CCSS) **1.NBT.3**

Math Journal 2, p. 206

Observe as children work on journal page 206. Expect them to accurately record comparisons using < and >. If children struggle to write the number models, review that the "open mouth" of the < and > symbols should "eat" the larger number.

✓ Assessment and Reporting   Go Online ▷ to record student progress and to see trajectories toward mastery for these standards.

Summarize Discuss different number models children could use to represent their answers to Problem 1 on journal page 206. **GMP4.1**

3 Practice   15–20 min

Go Online

ePresentations   eToolkit   Home Connections

► Making Smallest and Largest Numbers

Math Journal 2, p. 207

| WHOLE CLASS | SMALL GROUP | PARTNER | **INDEPENDENT** |

Have children draw number cards and use the digits to practice making and comparing numbers.

► Math Boxes 9-8

Math Journal 2, p. 208

| WHOLE CLASS | SMALL GROUP | **PARTNER** | INDEPENDENT |

Mixed Practice Math Boxes 9-8 are paired with Math Boxes 9-11.

► Home Link 9-8

Math Masters, p. 277

Homework Children calculate the sums of prices of items in the school store and write number models.

Math Journal 2, p. 208

Math Boxes

Lesson 9-8
DATE

① Draw two different circles.

Answers vary.

What is different?
Sample answer:
The sizes and colors.

② Find a base-10 cube.

How many faces does a cube have? __6__ faces

How many vertices does a cube have? __8__ vertices

③ Color 1 half of each rectangle a different way.

Answers vary.

④ Circle the longer object:
base-10 cube   paper clip

Measure the length of your journal with both. Answers vary.
The length of my journal is
_____ base-10 cubes or
_____ paper clips.

⑤ Circle the numbers that are less than:

⑬   113

23   ⑭

⑥ Fill in the missing numbers.

in	out
95	105
60	70
112	122
76	86

Rule: + 10

208   two hundred eight

① 1.G.1 ② 1.G.1 ③ 1.MD.1, 1.G.3 ④ 1.MD.2
⑤ 1.NBT.2, 1.NBT.2a, 1.NBT.3 ⑥ 1.NBT.5

Math Masters, p. 277

Review: Relations and Equivalence

Home Link 9-8
NAME   DATE

Family Note

Today your child reviewed equivalence by determining whether number sentences were true or false. Children also added prices and wrote comparison number models.
Please return this Home Link to school tomorrow.

Ryan and Janae are choosing things to buy.
Circle the group that costs more money.

Write a number model with < or > to compare the prices.

① stickers 25¢   pen 31¢   ⟨rubber bands 8¢   box of crayons 80¢⟩

Number model: _Sample answer: 56 < 88_

② colored pencil 29¢   pen 31¢   paper clip 2¢   ⟨eraser 17¢   ball 50¢⟩

Number model: _Sample answer: 62 < 67_

Practice

③ Jada and Martin cut a pizza in half to share. Then Min and Julius want to share the pizza, too. So they cut the pizza into fourths.
Are the shares now larger or smaller? _Smaller_

1.NBT.3, 1.NBT.4, 1.G.3, SMP4, SMP7   two hundred seventy-seven   277

Place Value

Overview Children review place value. They apply their understanding of place value to solve number-grid puzzles.

▶ **Before You Begin**
For Part 2, place stick-on notes over the numbers 67, 76, 77, and 78 on the Class Number Grid.

 Common Core State Standards

Focus Clusters
- Understand place value.
- Use place value understanding and properties of operations to add and subtract.

1 Warm Up 15–20 min

Materials

	Materials	
Mental Math and Fluency Children read an analog clock and record times digitally.	demonstration clock, slate	1.MD.3
Daily Routines Children complete daily routines.	See pages 2-37.	See pages xiv–xvii.

2 Focus 30–35 min

Math Message Children show 42 and 60 with base-10 shorthand and coins.	slate	1.NBT.2, 1.NBT.2c SMP2
Reviewing Place Value Children review 2-digit place value through a variety of activities.	base-10 blocks, slate	1.NBT.2, 1.NBT.2a, 1.NBT.2b, 1.NBT.2c, 1.NBT.5 SMP2
✓ **Assessment Check-In** See page 829.	base-10 blocks (optional)	1.NBT.2
Making Up and Solving Number-Grid Puzzles Children use place-value understanding to create and solve number-grid puzzles.	*Math Masters,* p. TA43; Class Number Grid; stick-on notes	1.NBT.2, 1.NBT.5 SMP7

 **1.NBT.2 Spiral Snapshot**

GMC Understand place value.

| 8-8
Practice | 8-10
Warm Up
Focus
Practice | 8-11
Warm Up
Focus | 9-2
Warm Up
Practice | 9-3
Warm Up | 9-5
Warm Up
Practice | 9-8
Practice | 9-9
Focus
Practice |

Spiral Tracker **Go Online** to see how mastery develops for all standards within the grade.

3 Practice 15–20 min

Using Different Strategies Children use different strategies to solve problems.	*Math Journal 2,* p. 209	1.NBT.4 SMP1, SMP5
Math Boxes 9-9 Children practice and maintain skills.	*Math Journal 2,* p. 210	See page 831.
Home Link 9-9 Homework Children complete number-grid puzzles.	*Math Masters,* p. 278	1.OA.6, 1.OA.8, 1.NBT.2, 1.NBT.5, 1.NBT.6 SMP7

connectED.mcgraw-hill.com

Plan your lessons online with these tools.

 ePresentations   Student Learning Center   Facts Workshop Game   eToolkit   Professional Development   Home Connections   Spiral Tracker   Assessment and Reporting   English Learners Support   Differentiation Support

Differentiation Options

RtI

Readiness — 10–15 min

Telling Place-Value Riddles

WHOLE CLASS
SMALL GROUP
PARTNER
INDEPENDENT

slate, base-10 blocks

To review place value with 2-digit numbers, have children tell place-value riddles to each other. One child thinks of a 2-digit number and tells how many tens and ones the number has. The other children represent it with base-10 blocks and write the number on their slates.

GMP2.2

Enrichment — 15–20 min

Writing Numbers in Expanded Form

WHOLE CLASS
SMALL GROUP
PARTNER
INDEPENDENT

Activity Card 118, slate, base-10 blocks

To extend children's understandings of place value, have them write 2- and 3-digit numbers in expanded form. Children use each expanded form in a number sentence. *For example:* For the number 34, the expanded form is 30 + 4, and a number sentence would be 30 + 4 = 34 (or 34 = 30 + 4). **GMP2.1**

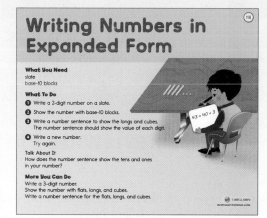

Extra Practice — 10–15 min

Playing *The Digit Game*

WHOLE CLASS
SMALL GROUP
PARTNER
INDEPENDENT

Math Masters, p. G35 (optional); per partnership: 4 each of number cards 0–9

Have children practice comparing 2-digit numbers based on place value by playing *The Digit Game.* For detailed directions, see Lesson 5-1.

Observe

- Which children use place value to explain which number is larger?

Discuss

- *Who wins if both players have a 9 in the 10s place? Explain your reasoning.*
  GMP3.1

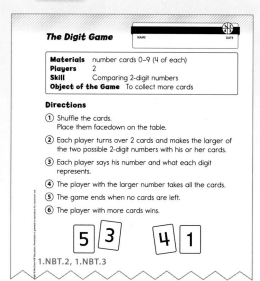

English Language Learners Support

Beginning ELL Teach the terms *most* and *fewest* using money examples and gestures. For example, show 10 pennies, and say: *This is ten cents.* Then show 1 dime, and say: *And, this is ten cents.* Point to the 10 pennies, spread your hands apart, and say: *These are the most coins I can use for 10 cents.* Repeat with the dime, bringing your hands together as you say: *This is the fewest I can use for 10 cents.* Continue showing combinations of pennies and dimes, challenging children to identify which set has the most and which set has the fewest possible coins for the same amounts of money.

Go Online   English Learners Support

Standards and Goals for
Mathematical Practice

SMP2 **Reason abstractly and quantitatively.**

GMP2.1 Create mathematical representations using numbers, words, pictures, symbols, gestures, tables, graphs, and concrete objects.

GMP2.3 Make connections between representations.

SMP7 **Look for and make use of structure.**

GMP7.2 Use structures to solve problems and answer questions.

1 Warm Up 15–20 min   Go Online

ePresentations   eToolkit

▶ Mental Math and Fluency

Show the following times on the demonstration clock, and have children record the times in digital notation on their slates.

● ○ ○ 9:00, 5:00, 8:00

● ● ○ 1:30, 3:30, 9:30

● ● ● 6:00, 5:30, 12:30

▶ Daily Routines

Have children complete the Daily Routines. For detailed instructions, see pages 2–37. For specifics on CCSS coverage, see pages xiv–xvii.

2 Focus 30–35 min   Go Online

ePresentations   eToolkit

▶ Math Message

Use the fewest possible | and • to show 42 and 60 on your slate. Then use the fewest possible ⒟ and ⓟ symbols to show 42¢ and 60¢. How are the representations alike and different? GMP2.3

▶ Reviewing Place Value

| WHOLE CLASS | SMALL GROUP | PARTNER | INDEPENDENT |

Math Message Follow-Up Discuss children's answers to the Math Message. Emphasize that the cube and the penny represent ones and that the long and the dime represent tens, or groups of 10 cubes or 10 pennies. GMP2.3 Ask children to explain what they notice about their drawings of 60 and 60¢. There are no dots and no pennies. Remind children that 2-digit numbers with a 0 in the ones place are special; they have some tens and no ones.

Tell children that today they will review what they know about place value and digits. Choose some of the following activities for review based on the needs of your class. Though these activities review place value of 2-digit numbers, you may wish to expand to 3-digit numbers based on the skill level of your class.

- Have several children demonstrate making a 2-digit number using base-10 blocks. *For example:* For the number 31, 3 children each hold one long and 1 child holds one cube. The class counts the longs and then the cubes; 10, 20, 30, 31. Instruct children to hold up their base-10 blocks as you count each new base-10 block.

- Display a 2-digit number, such as 19. Children name the place for each digit. 1 is in the tens place; 9 is in the ones place. Then ask them to identify the value of the digits. The 1 in 19 represents 10; the 9 in 19 represents 9.

- Specify a 2-digit number for children to write on their slates. *For example,* say: *Write a 2-digit number with 8 in the tens place. Write a 2-digit number with 5 in the ones place.*

- Display two digits. Have children write on their slate the larger or the smaller of the two possible whole numbers that use both digits. *For example:* Ask children to write the larger of the two possible numbers they can make with 2 and 5. 52 After doing this a few times, have children write the largest and smallest 1-digit numbers. 9 and 0 Then have them write the largest and smallest 2-digit numbers. 99 and 10

- Display a collection of base-10 blocks. Ask children to write the number represented by the blocks. First display sets that do not require composing a new ten. *For example:* Show 62 as 6 longs and 2 cubes. Then show sets that require children to exchange ones for a ten; for example, show 75 as 6 longs and 15 cubes.

- Display a number, and have children show base-10 blocks that represent the number. Ask whether they can find other ways to show the number, perhaps by exchanging a long for 10 cubes; for example, children might show 27 as 2 longs and 7 cubes, 1 long and 17 cubes, or 27 cubes. GMP2.1

- Review counting by 10s from any 2-digit number. *For example:* Count up orally by 10s beginning with 56: 56, 66, 76, 86, and so on. Then have children find 10 more and 10 less than a given number; for example, 10 more than 32 is 42, and 10 less than 32 is 22.

 Assessment Check-In (CCSS) 1.NBT.2

Observe whether children identify the number of tens and ones in a 2-digit number and understand the value of each digit in a 2-digit number. Expect children to correctly identify the tens and ones in 2-digit numbers and the value of each digit. For example, children should be able to tell you that the number 24 has 2 tens and 4 ones and that the 2 represents 20. For children who struggle, represent 2-digit numbers using base-10 blocks.

Assessment and Reporting — Go Online to record student progress and to see trajectories toward mastery for these standards.

Academic Language Development

Encourage children to explain their thinking using the terms *digit, value,* and *place* by providing them with a sentence frame, such as: "The digit _____ is in the _____ place, so it has a value of _____." to justify their choices for the largest and smallest possible 2-digit numbers.

Adjusting the Activity

Differentiate As the class counts by 10s from a 2-digit number, direct children who struggle to place their fingers on the number on a number grid and move their fingers down the column.

Go Online Differentiation Support

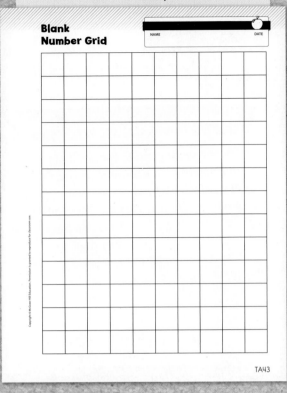

Math Journal 2, p. 209

Solve the number sentences below. Record the tool or strategy you used and explain why you chose it.

- number grid
- counting back
- making 10
- changing to an easier number
- number line
- base-10 blocks
- fact extensions
- counting up
- 10 more or 10 less in my head
- combining place-value groups

Example: 20 + 36 = **56**

I used a number grid because moving down 2 rows is an easy way to add 20. I started at 36 and moved down two rows.

❶ 58 + 3 = **61** Explanations will vary.

❷ 17 + 30 = **47**

❸ 79 + 14 = **93**

❹ 6 + 46 + 4 = **56**

1.NBT.4, SMP1, SMP5   two hundred nine   209

▶ Making Up and Solving Number-Grid Puzzles

Math Masters, p. TA43

| WHOLE CLASS | SMALL GROUP | **PARTNER** | **INDEPENDENT** |

Show children the number grid you covered with stick-on notes before the lesson. Remind children of the number-grid puzzles they have done before. Ask how they could determine the numbers you covered on the grid. Sample answers: count up by 1s from 66 and 75; count up 10 more from 57

Have children fold *Math Masters*, page TA43 into four equal parts. Invite children to make up number-grid puzzles that their partners can solve using their understanding of tens and ones patterns on the number grid. **GMP7.2** Each part of the blank grid can be used for a separate number-grid puzzle. Then have children draw lines between the four parts or cut them apart.

Children draw around some of the grid cells on one part of the sheet to make a puzzle piece and then write a 2-digit number in one of the cells. They trade the puzzles with their partners who then fill in all of the missing numbers.

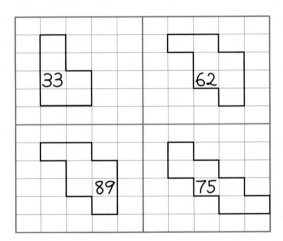

Summarize Have children close their eyes. Ask them to tell you the numbers above and below 26 on the number grid. Then ask them to tell you the numbers to the left and right of 26 on the number grid. Discuss how children determined these numbers. **GMP7.2**

3 Practice 15–20 min Go Online

ePresentations eToolkit Home Connections

▶ Using Different Strategies

Math Journal 2, p. 209

| WHOLE CLASS | SMALL GROUP | PARTNER | **INDEPENDENT** |

Have children practice choosing efficient strategies to add 2-digit numbers. **GMP1.5, GMP5.1**

▶ Math Boxes 9-9

Math Journal 2, p. 210

| WHOLE CLASS | SMALL GROUP | **PARTNER** | **INDEPENDENT** |

Mixed Practice Math Boxes 9-9 are paired with Math Boxes 9-7.

▶ Home Link 9-9

Math Masters, p. 278

Homework Children solve number-grid puzzles.

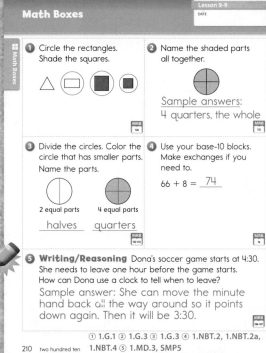

Math Journal 2, p. 210

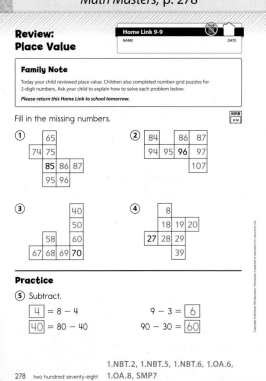

Math Masters, p. 278

Lesson 9-9 **831**

Lesson 9-10

Review: 3-Dimensional Geometry

Overview Children review defining attributes and names of 3-dimensional shapes. They use 3-dimensional shapes to form composite shapes.

▶ **Before You Begin**
For Part 2, copy *Math Masters,* pages 279–282 on cardstock. Consider cutting out some of the templates before the lesson. For the optional Enrichment activity, provide each child with about 30 four-inch straws and about 15 twist ties.

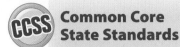

Common Core State Standards

Focus Cluster
Reason with shapes and their attributes.

	Materials	
① Warm Up 15–20 min		
Mental Math and Fluency Children read an analog clock and record times digitally.	demonstration clock, slate	1.MD.3
Daily Routines Children complete daily routines.	See pages 2–37.	See pages xiv–xvii.
② Focus 30–40 min		
Math Message Children find classroom objects that are similar in shape to a cylinder, a cube, and a sphere.		1.G.1 SMP7
Reviewing Attributes of 3-Dimensional Shapes Children review the attributes of 3-dimensional shapes.	*Math Journal 2,* p. 211 (optional); 3-dimensional shape blocks; plastic egg; everyday objects	1.G.1 SMP7
Composing 3-Dimensional Shapes Children construct 3-dimensional shapes from templates. They use the shapes to make new composite shapes.	*Math Masters,* pp. 279–282; tape; scissors	1.G.2 SMP6
✓ **Assessment Check-In** See page 836.	*Math Journal 2,* p. 211 (optional); *Math Masters,* p. TA40 (optional)	1.G.1, 1.G.2 SMP6

CCSS 1.G.1 **Spiral Snapshot**

GMC Distinguish between defining and non-defining attributes.

6-3 Focus Practice	7-5 through 7-7 Focus Practice	8-1 Focus Practice	8-5 Focus Practice	8-6 Focus Practice	9-5 Practice	9-10 Focus Practice

Spiral Tracker 〉 Go Online 〉 to see how mastery develops for all standards within the grade.

③ Practice 10–15 min		
Playing *Time Match* **Game** Children practice telling time to the half hour.	*Math Masters,* pp. G55–G56 and G60–G61, p. G57 (optional)	1.MD.3 SMP2
Math Boxes 9-10 Children practice and maintain skills.	*Math Journal 2,* p. 212	See page 837.
Home Link 9-10 **Homework** Children use descriptions and pictures to name 3-dimensional shapes.	*Math Masters,* p. 283	1.OA.2, 1.OA.3, 1.G.1 SMP7

connectED.mcgraw-hill.com

Plan your lessons online with these tools.

ePresentations | Student Learning Center | Facts Workshop Game | eToolkit | Professional Development | Home Connections | Spiral Tracker | Assessment and Reporting | English Learners Support | Differentiation Support

Differentiation Options

 RtI

Readiness 10–15 min

	WHOLE CLASS
	SMALL GROUP
	PARTNER
	INDEPENDENT

Playing *I Spy*

Math Masters, p. G58 (optional)

To provide additional experience describing and identifying shapes, have children play *I Spy* with 2- and 3-dimensional shapes. Before playing, generate a list of math words used to describe shapes, such as *side, face, vertex, corner, point, edge, length,* and *size.* **GMP3.2** Also generate a list of 2- and 3-dimensional shapes that children know. See Lesson 8-1 for detailed instructions.

Observe

• Which children use defining attributes to correctly describe 3-dimensional shapes?

Discuss

• *How could you use the word* face *to help you describe a cylinder?*

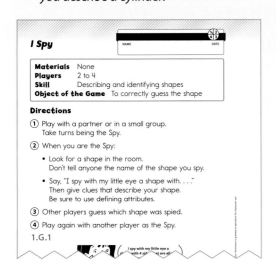

I Spy

Materials	None
Players	2 to 4
Skill	Describing and identifying shapes
Object of the Game	To correctly guess the shape

Directions

① Play with a partner or in a small group. Take turns being the Spy.

② When you are the Spy:

• Look for a shape in the room. Don't tell anyone the name of the shape you spy.

• Say, "I spy with my little eye a shape with. . . ." Then give clues that describe your shape. Be sure to use defining attributes.

③ Other players guess which shape was spied.

④ Play again with another player as the Spy.

1.G.1

Enrichment 15–20 min

	WHOLE CLASS
	SMALL GROUP
	PARTNER
	INDEPENDENT

Constructing Regular Polyhedrons

Activity Card 119;
per child: about 30 four-inch straws and about 15 twist ties

To further explore the properties of 3-dimensional figures, have children build regular polyhedrons with straws and twist ties. **GMP2.1** Provide examples of polyhedrons (such as cubes and tetrahedrons) from your block collection. Prior to the activity, you may wish to discuss the information about polyhedrons found on the Activity Card.

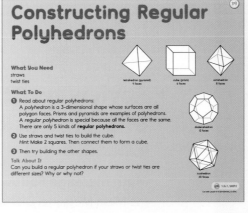

Constructing Regular Polyhedrons

What You Need
straws
twist ties

What To Do

① Read about regular polyhedrons:
A polyhedron is a 3-dimensional shape whose surfaces are all polygon faces. Prisms and pyramids are examples of polyhedrons. A regular polyhedron is special because all the faces are the same. There are only 5 kinds of **regular polyhedrons.**

② Use straws and twist ties to build the cube.
Hint: Make 2 squares. Then connect them to form a cube.

③ Then try building the other shapes.

Talk About It
Can you build a regular polyhedron if your straws or twist ties are different sizes? Why or why not?

1.G.1, SMP2

Extra Practice 10–15 min

	WHOLE CLASS
	SMALL GROUP
	PARTNER
	INDEPENDENT

Playing *Make My Design*

Math Masters, p. G59 (optional); pattern blocks; folder

To provide practice creating composite shapes, have children play *Make My Design.* For detailed instructions, see Lesson 8-5.

Observe

• Which children describe a variety of shapes using language clear enough for their partners to create their designs? **GMP6.3**

Discuss

• *What parts of the design were hardest to describe? Why?*

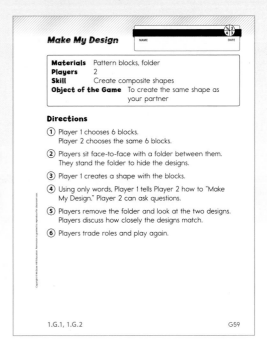

Make My Design

Materials	Pattern blocks, folder
Players	2
Skill	Create composite shapes
Object of the Game	To create the same shape as your partner

Directions

① Player 1 chooses 6 blocks. Player 2 chooses the same 6 blocks.

② Players sit face-to-face with a folder between them. They stand the folder to hide the designs.

③ Player 1 creates a shape with the blocks.

④ Using only words, Player 1 tells Player 2 how to "Make My Design." Player 2 can ask questions.

⑤ Players remove the folder and look at the two designs. Players discuss how closely the designs match.

⑥ Players trade roles and play again.

1.G.1, 1.G.2 G59

English Language Learners Support

Beginning ELL Help children remember and use geometry terms by providing unique visuals that help convey the attributes. For example, write the word *curved* in the shape of the letter C, *vertex* with a point on the vertex of the V, *circle* in the shape of a circle, and so on. Encourage children to make their own shapes and attributes dictionaries using such visuals.

Go Online ELL English Learners Support

1 Warm Up   15–20 min

Go Online   ePresentations   eToolkit

▶ Mental Math and Fluency

Show times on the demonstration clock, and have children record them
in digital notation on their slates.

- ●○○ 10:00, 4:00, 12:00
- ●●○ 5:30, 8:30, 2:30
- ●●● 1:00, 6:30, 11:30

▶ Daily Routines

Have children complete the Daily Routines. For detailed instructions,
see pages 2–37. For specifics on CCSS coverage, see pages xiv–xvii.

2 Focus   30–40 min

Go Online   ePresentations   eToolkit

▶ Math Message

*Find objects that look like a cylinder, a cube, and a sphere. Record what you
found and which attributes make your object look like a cylinder, a cube, or
a sphere.*   **GMP7.1**

▶ Reviewing Attributes
of 3-Dimensional Shapes

| **WHOLE CLASS** | SMALL GROUP | PARTNER | INDEPENDENT |

Math Message Follow-Up Have children share their responses to
the Math Message. Discuss the attributes that make each object look like
a particular 3-dimensional shape.   **GMP7.1**   For example, ask: *How did
you know the can was like a cylinder?* Sample answer: It has two flat faces
that are shaped like circles, and the side of it is curved.

Tell children that today they will review what they have learned about
3-dimensional shapes. Then they will make their own 3-dimensional
shapes.

Do Option 1 or Option 2 as a review activity. If you have at least
one 3-dimensional shape for each child in your class, Option 1 is
appropriate. If you do not, do Option 2.

Math Journal 2, p. 211

**Reviewing Attributes of
3-Dimensional Shapes**

Lesson 9-10
DATE

Write a defining attribute for each shape. Answers vary.

1.G.1, SMP7   two hundred eleven   **211**

Option 1

Distribute a 3-dimensional shape block or everyday object to each child. You may wish to do this by having children stand in a large circle around the objects. Like *Musical Chairs,* have children walk around the shapes until you indicate for them to stop. Children pick up the objects in front of them. Give defining and nondefining attributes as clues for a mystery shape. Have children take one step toward the center of the circle if the clue fits their shape. (*See example in margin.*) Point out that clues about defining attributes generally apply to only 1 or 2 shapes, but clues about nondefining attributes can apply to a lot of different shapes because they might be about any or none of the shapes. Continue providing clues until the mystery shape is revealed and all children in the middle have that shape. Have the children in the middle hold up the mystery shape and describe the shape using defining attributes. **GMP7.1**

Repeat as time permits. Continue to highlight the differences between defining and nondefining attributes.

Option 2

Have children complete journal page 211. Then display a sphere, cube, rectangular prism, cone, pyramid, and cylinder and refer to them while reviewing children's answers. Guide children to think about defining and nondefining attributes; for example:

• *Put your finger on a shape that has no corners and no flat faces.*   Sphere

• *Put your finger on a shape that has at least one face that is a circle.* Cylinder or cone

• *Can you put your finger on a shape that is always small?*   No *Why?* Sample answer: The same shape can sometimes be big and sometimes be small.

• *Put your finger on a shape that always has 8 vertices.*   Cube or rectangular prism

Help children refine their defining attributes of spheres by displaying and comparing a sphere and a plastic egg. Ask children to look at the defining attributes they wrote for spheres. Explain that neither eggs nor spheres have vertices or flat faces, that both can roll, and that eggs are not spheres. Then pass the shapes around, and have children examine them and try to rotate and roll them in all directions. Ask children what differences they see between the egg and the sphere. Sample answers: The egg is longer one way than another. The sphere looks the same no matter which way you turn it. The egg does not roll evenly like the sphere does. Have children add these traits to their list of defining attributes of spheres. Note that other differences, such as the sphere being larger, held higher, or being a different color, describe nondefining attributes.

Example:

• *Take one step toward the center if your shape has edges.* Everyone except those holding spheres steps forward.

• *Take one step toward the center if your shape is made of wood.*

• *Take one step toward the center if your shape has 6 faces.* Anyone with a cube or rectangular prism steps forward.

• *Take one step toward the center if your shape always has its flat face facing down.* Nobody steps forward.

• *Take one step toward the center if all the faces on your shape are squares.* Anyone with a cube steps forward.

The mystery shape is a cube.

Adjusting the Activity

Differentiate To provide additional reinforcement for the terms with which children struggle, point to evidence of the clues on at least two of the shapes in the collection of objects.

Go Online Differentiation Support

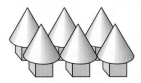

A Composite 3-Dimensional Shape

Math Masters, p. G57

Time Match

Materials	Time Match Cards
Players	2 or more
Skill	Telling time
Object of the Game	To match as many cards as you can

Directions

① Mix the cards, and place them facedown on the table. Take turns.

② Flip over 2 cards.

③ If the cards match, keep them and take another turn. If the cards do not match, flip them back over.

④ When all the cards have been matched, the player with the most matches wins.

1.MD.3   G57

Have children record several defining attributes for shapes of their choice on the bottom portion of the journal page or wherever children have room. **GMP7.1** If time permits, have them read their defining attributes to their partners, and have the partners try to guess the shapes.

► Composing 3-Dimensional Shapes

Math Masters, pp. 279–282

WHOLE CLASS	SMALL GROUP	PARTNER	INDEPENDENT

Divide children into four groups. Assign each group a different 3-dimensional shape found on *Math Masters,* pages 279– 282 to ensure that examples of each solid will be available for the second part of this activity. Then demonstrate how to fold along the lines and tape the tabs in place to create each geometric solid.

Distribute tape, scissors, and copies of the assigned page to each group. Have children write their names on the shapes *before* they begin folding. Have them work in partnerships to cut, fold, and tape paper into the assigned shape. Assist children as they construct the shapes.

After each group has constructed at least 4 examples of their shape, have them work with another group to compose new shapes. Have the cube and cone groups place cones on top of cubes to create composite shapes. Then have children in the group make other cone-cube units and put those together to make a larger composite shape.

Repeat with the cylinder and rectangular prism groups. Encourage children to describe the shapes using terms such as *side, face, vertex, corner, point, edge, length,* and *size.* **GMP6.3** Ask: *Do the new shapes remind you of other shapes or objects?* As time permits, allow children to cut and fold more shapes, combining them with others (including the class collection of blocks and everyday objects) to make new 3-dimensional shapes.

✓ Assessment Check-In CCSS 1.G.1, 1.G.2

Observe children as they create new shapes by combining their 3-dimensional shapes. As they build, ask: *What shapes make up your new shape?* Then choose one of the shapes and ask: *What are some defining attributes of this shape?* **GMP6.3** Expect most children to provide the proper geometric name for the 3-dimensional shapes they used; however, some may struggle with these terms. Have children refer to their 3-dimensional shape cards (*Math Masters,* page TA40) to help them remember the names. To challenge children with more complicated shapes and names, have them complete the Enrichment activity.

 Assessment and Reporting   **Go Online** to record student progress and to see trajectories toward mastery for these standards.

Summarize Partnerships discuss which 3-dimensional shape (or group of shapes) is their favorite and why. Emphasize that 3-dimensional shapes can be put together to make new shapes, which can then be combined to make even more shapes.

③ Practice 15–20 min

Go Online

ePresentations   eToolkit   Home Connections

▶ Playing *Time Match*

Math Masters, pp. G55–G56 and G60–G61

| WHOLE CLASS | SMALL GROUP | **PARTNER** | INDEPENDENT |

Have children practice telling time to the hour and half hour with the full set of *Time Match* Cards. For directions for playing *Time Match,* use *Math Masters,* page G57. For more information, see Lesson 7-11.

Observe

- Which children correctly distinguish between the minute hand and the hour hand?

Discuss

- *How do you know when it is half past the hour on an analog clock? On a digital clock?* **GMP2.3**

▶ Math Boxes 9-10 ✏️

Math Journal 2, p. 212

| WHOLE CLASS | SMALL GROUP | **PARTNER** | **INDEPENDENT** |

Mixed Practice Math Boxes 9-10 are paired with Math Boxes 9-12.

▶ Home Link 9-10

Math Masters, p. 283

Homework Children match characteristics of 3-dimensional shapes to corresponding pictures of the shapes.

Math Journal 2, p. 212

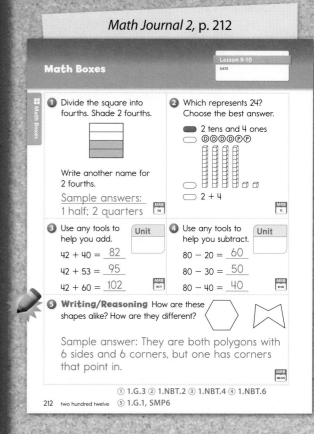

Math Masters, p. 283

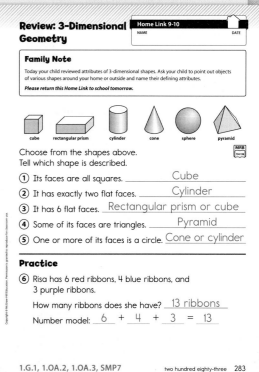

Review: Equal Shares

Overview Children review dividing shapes into 2 and 4 equal shares, naming one share, and naming the whole.

▶ **Before You Begin**
Place copies of *Math Masters*, page 284 near the Math Message. You may wish to have children cut apart the shapes on this page and on *Math Masters*, page 285 before the lesson.

CCSS **Common Core State Standards**

Focus Cluster
Reason with shapes and their attributes.

1 Warm Up 15–20 min

	Materials	
Mental Math and Fluency Children mentally find 10 more or 10 less.	slate	1.NBT.4, 1.NBT.5, 1.NBT.6
Daily Routines Children complete daily routines.	See pages 2–37.	See pages xiv–xvii.

2 Focus 30–35 min

Math Message Children show two different ways they could share a granola square with a friend.	*Math Masters*, p. 284; scissors	1.G.3 SMP1
Partitioning Granola Squares Children show how they could share a granola square among 4 people. They name the shares.	*Math Journal 2*, p. 213; *Math Masters*, p. 284; scissors; tape or glue	1.G.3 SMP1, SMP6
✓ **Assessment Check-In** See page 841.	*Math Journal 2*, p. 213	1.G.3, SMP6
Partitioning Watermelon Slices Children share a circle among 2 or 4 people. They discuss ways to name the shares and compare the sizes of the shares.	*Math Masters*, p. 285; scissors; markers or crayons	1.G.3 SMP2

CCSS 1.G.3 **Spiral Snapshot**

GMC Understand that more equal shares mean smaller equal shares.

 Spiral Tracker   **Go Online** to see how mastery develops for all standards within the grade.

8-3 Focus	8-4 Focus Practice	8-7 Practice	8-9 Practice	9-4 Focus Practice	9-7 Practice	9-9 Practice	9-11 Focus Practice

3 Practice 15–20 min

Playing *Beat the Calculator* **Game** Children practice addition facts.	*Math Masters*, p. G49 (optional); calculator; slate; Fact Triangles	1.OA.3, 1.OA.4, 1.OA.6 SMP5
Math Boxes 9-11 Children practice and maintain skills.	*Math Journal 2*, p. 214	See page 843.
Home Link 9-11 **Homework** Children divide shapes in halves and fourths.	*Math Masters*, p. 289	1.NBT.4, 1.NBT.6, 1.G.3

connectED.mcgraw-hill.com

Plan your lessons online with these tools.

ePresentations   Student Learning Center   Facts Workshop Game   eToolkit   Professional Development   Home Connections   Spiral Tracker   Assessment and Reporting   English Learners Support   Differentiation Support

Differentiation Options RtI

 CCSS 1.G.3, SMP2

Readiness 10–15 min

Reviewing Equal Shares

Math Masters, pp. 286–287

WHOLE CLASS
SMALL GROUP
PARTNER
INDEPENDENT

For additional experience identifying and drawing equal shares, have children discuss the examples on *Math Masters,* page 286. Help them recognize that the first example shows halves because the circle has been divided into 2 equal parts, but the rectangle does not show halves because the 2 parts are not equal in size. GMP2.2 Have children draw several more examples of halves and not-halves on the page. GMP2.1 If time permits, have them discuss and draw examples of fourths and not-fourths on *Math Masters,* page 287.

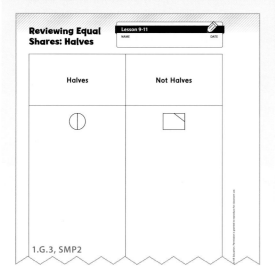

 CCSS 1.G.3, SMP4, SMP5

Enrichment 15–20 min

More Equal Sharing Problems

Activity Card 120, counters

WHOLE CLASS
SMALL GROUP
PARTNER
INDEPENDENT

To extend children's understanding of equal shares, have them use tools to solve real-world problems about sharing collections of objects equally. GMP4.1, GMP5.1

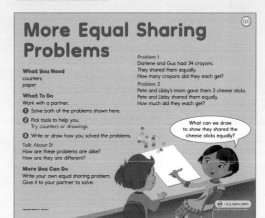

CCSS 1.G.3, SMP2

Extra Practice 10–15 min

Naming Equal Shares

Math Masters, p. 288

WHOLE CLASS
SMALL GROUP
PARTNER
INDEPENDENT

To provide practice naming equal shares, have children choose names to describe drawings of equal shares. GMP2.3

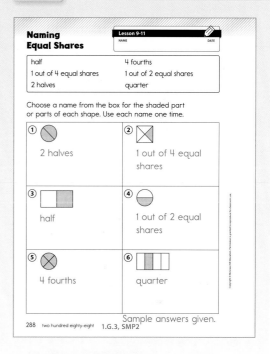

English Language Learners Support

Beginning ELL Teach children different meanings of *share* as used in the classroom. Role play sharing ideas with a partner, as when solving a number story. Demonstrate how to share items, such as apples, by cutting them into as many parts as there are people who want to share. Show children how to count the number of shares after you cut the apple. Use Total Physical Response prompts to provide children with practice using the term in different ways.

Go Online > **ELL** English Learners Support

SMP1 **Make sense of problems and persevere in solving them.**

GMP1.5 Solve problems in more than one way.

SMP2 **Reason abstractly and quantitatively.**

GMP2.3 Make connections between representations.

SMP6 **Attend to precision.**

GMP6.3 Use clear labels, units, and mathematical language.

---

1 Warm Up 15–20 min   Go Online ePresentations eToolkit

▶ Mental Math and Fluency

Ask children to solve mentally and to record their answers on their slates.

●○○ 10 more than 20 30; 10 less than 20 10; 10 more than 50 60

●●○ 30 − 10 20; 50 + 10 60; 100 − 10 90

●●● 35 + 10 45; 65 − 10 55; 28 + 10 38

▶ Daily Routines

Have children complete the Daily Routines. For detailed instructions, see pages 2–37. For specifics on CCSS coverage, see pages xiv–xvii.

---

2 Focus 30–35 min   Go Online ePresentations eToolkit

▶ Math Message

Math Masters, p. 284

Take one page. Cut out 2 granola squares and save the rest for later. Show two different ways you could share a granola square equally with a friend. **GMP1.5** *Explain how you know your shares are equal.*

▶ Partitioning Granola Squares

Math Journal 2, p. 213; *Math Masters,* p. 284

| WHOLE CLASS | SMALL GROUP | PARTNER | INDEPENDENT |

Math Message Follow-Up Ask children to share their two solutions with their partners and to discuss how they know the shares are equal. Have several partnerships share their ideas with the class. Then ask children how they would name one of the two shares. Expect them to list names such as half, 1 half, and 1 out of 2 equal shares. Say: *Imagine you split a granola square into 2 equal shares, and your friend decided she was not hungry. You get to eat both parts. What names could you give for both parts?* Sample answers: 2 halves; 2 out of 2 equal shares; the whole granola square **GMP6.3**

Math Masters, p. 284

Granola Squares

| Lesson 9-11 |
| NAME | DATE |

Have children cut out two more squares. Tell them to imagine that they have to share one granola square among 4 people. Have them show two different ways to share one granola square among 4 people and glue the squares onto journal page 213. **GMP1.5** You may wish to remind children that a name for 1 of 4 equal shares is *quarter*. Then have children answer the questions. Discuss children's answer to Problem 4. Make sure that they understand that increasing the number of equal shares results in smaller shares.

Differentiate **Adjusting the Activity**

Have children who struggle understanding how the number of equal shares affects the size of the shares role-play the activity. Invite two children to cut a piece of paper in half and to compare their shares. Have two more children join them and cut another, same-size piece of paper into quarters. Then have children compare numbers of shares and the size of each share. Ask: *What would happen if you had to share the piece of paper among even more people?* Sample answers: They would get less paper. Their shares would be smaller.

Go Online Differentiation Support

 **Assessment Check-In** **CCSS** **1.G.3**

Math Journal 2, p. 213

Look at the names that children recorded for all 4 shares in Problem 3 of journal page 213. Expect most children to record a name for all of the parts. **GMP6.3** Have children who struggle use the names they recorded for single shares to help them generate names for all 4 shares. For example, if a child recorded 1 fourth for a single share, then point to each of the shares and count with them: *1 fourth, 2 fourths, 3 fourths, 4 fourths.* Encourage children who exceed expectations to generate multiple names for all the shares.

 Assessment and Reporting **Go Online** to record student progress and to see trajectories toward mastery for these standards.

Academic Language Development

Contrast *equal shares* with everyday expressions such as *share your toys with your friend* or *give me the bigger share.* Have children think about whether the shares have to be the same size in everyday examples. Build on their ideas to talk about the special meaning of *equal shares,* emphasizing the importance of making shares equal sizes, particularly in math class.

Math Journal 2, p. 213

Partitioning Granola Squares | Lesson 9-11
DATE

1 Show two different ways to share a granola square among 4 people. Sample answers:

2 Write two different names for 1 of the shares.
Sample answers: 1 fourth; quarter; 1 out of 4 parts

3 Sasha ate all 4 shares of the granola square. Name the share she ate. Sample answers: 4 fourths; 4 quarters; the whole granola square

4 Circle which is bigger.
(1 half of a granola square)   1 fourth of a granola square
Why? Sample answer: 4 equal parts are smaller than 2 equal parts.

1.G.3, SMP1, SMP6   two hundred thirteen 213

Watermelon Slices

Lesson 9-11

NAME   DATE

1.G.3   two hundred eighty-five   285

Beat the Calculator

NAME   DATE

Materials	Fact Triangles
	1 calculator
Players	1 "Caller," 1 "Calculator," and 1 "Brain"
Skill	Mental addition
Object of the Game	To add numbers faster than a player using a calculator

Directions

① Mix the Fact Triangles. Place them facedown on the table.

② The Caller:
  • takes a Fact Triangle from the pile.
  • covers the sum.
  • says the fact *without* the sum.
  For example, "3 + 4 is equal to what?"

③ The Calculator solves the problem *with* a calculator.
  The Brain solves it *without* a calculator.
  The Caller decides who found the sum first.

④ Play about 10 rounds.
  Then trade roles.

Another Way to Play

The Caller covers other numbers on the Fact Triangles. Players find the covered number. They can subtract to find the missing addend.

1.OA.3, 1.OA.4, 1.OA.6   G49

▶ # Partitioning Watermelon Slices

Math Masters, p. 285

| WHOLE CLASS | SMALL GROUP | **PARTNER** | INDEPENDENT |

Tell children to imagine that the circle on *Math Masters,* page 285 is a slice of watermelon that they can color and decorate with seeds. In each partnership, one child should divide a slice into 2 equal shares, and the other child should divide a slice into 4 equal shares. Ask partnerships to name the shares and to compare the sizes of the shares.

When most partnerships have finished, have children hold up their watermelon pieces and list some of the names they came up with. Ask children to explain why a half of a watermelon slice is larger than a quarter of a watermelon slice. Sample answer: A half is 1 of 2 equal shares. A quarter is 1 of 4 equal shares. More shares of the same-size slice mean smaller shares.

Hold up one of the watermelon halves and one of the granola square halves from the Math Message. Ask:

• *We can call each of these parts 1 half. But they do not look alike. Why are both called halves?* Sample answer: Half is 1 of 2 equal shares. They are shares of different things, but both are 1 of 2 equal shares.

• *How are the halves alike?* Sample answer: Both show 1 part of something that was divided into 2 equal shares.

• *How are the halves different?* Sample answer: They are different shapes, and the watermelon slices are bigger.

• *If I divided a banana into 2 equal shares, would it look like the granola square or the watermelon?* No *Would I still call it a half?* Yes; it would be half of a banana. GMP2.3

Use similar questions to guide a discussion about fourths, or quarters. Be sure that children understand that the way a fourth looks is dependent on the whole. However, if something is divided into 4 equal shares, then we can always call 1 of the shares a fourth, or a quarter.

Summarize Ask children to explain what is important to remember about making equal shares. Expect them to mention how shares must be the same size, how a greater number of shares results in smaller shares, and how shares can be named.

3 Practice 15–20 min

Go Online

ePresentations eToolkit Home Connections

▶ Playing *Beat the Calculator*

| WHOLE CLASS | SMALL GROUP | PARTNER | INDEPENDENT |

Have children practice facts by playing *Beat the Calculator*. For directions for playing the game, use *Math Masters,* page G49. For more information, see Lesson 7-2.

Observe

• Which children can solve most addition facts in their heads?

Discuss

• *Why is it sometimes more useful to solve facts in your head rather than on a calculator?* **GMP5.1**

▶ Math Boxes 9-11

Math Journal 2, p. 214

| WHOLE CLASS | SMALL GROUP | **PARTNER** | **INDEPENDENT** |

Mixed Practice Math Boxes 9-11 are paired with Math Boxes 9-8.

▶ Home Link 9-11

Math Masters, p. 289

Homework Children review names for 1 half and 1 fourth.

Math Journal 2, p. 214

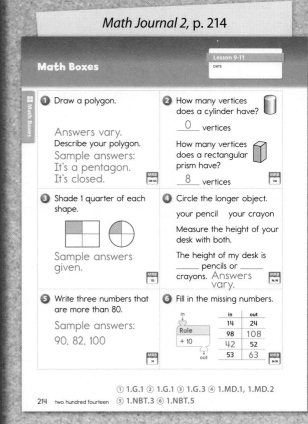

Math Boxes

Lesson 9-11
DATE

① Draw a polygon.

Answers vary.
Describe your polygon.
Sample answers:
It's a pentagon.
It's closed.

② How many vertices does a cylinder have?
___0___ vertices

How many vertices does a rectangular prism have?
___8___ vertices

③ Shade 1 quarter of each shape.

Sample answers given.

④ Circle the longer object.
your pencil   your crayon

Measure the height of your desk with both.

The height of my desk is ___ pencils or ___ crayons. Answers vary.

⑤ Write three numbers that are more than 80.

Sample answers:
90, 82, 100

⑥ Fill in the missing numbers.

in
Rule
+ 10
out

in	out
14	24
98	108
42	52
53	63

214 two hundred fourteen

① 1.G.1 ② 1.G.1 ③ 1.G.3 ④ 1.MD.1, 1.MD.2
⑤ 1.NBT.3 ⑥ 1.NBT.5

Math Masters, p. 289

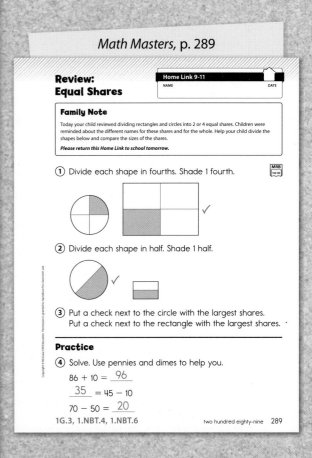

Review: Equal Shares

Home Link 9-11
NAME   DATE

Family Note

Today your child reviewed dividing rectangles and circles into 2 or 4 equal shares. Children were reminded about the different names for these shares and for the whole. Help your child divide the shapes below and compare the sizes of the shares.

Please return this Home Link to school tomorrow.

① Divide each shape in fourths. Shade 1 fourth.

② Divide each shape in half. Shade 1 half.

③ Put a check next to the circle with the largest shares. Put a check next to the rectangle with the largest shares.

Practice

④ Solve. Use pennies and dimes to help you.

86 + 10 = ___96___

___35___ = 45 − 10

70 − 50 = ___20___

1G.3, 1.NBT.4, 1.NBT.6   two hundred eighty-nine 289

Unit 9 Progress Check ✓

2-Day Lesson

Overview　**Day 1:** Administer the Unit Assessments.
　　　　　　　Day 2: Administer the Open Response Assessment.

Student Learning Center
Students may take assessments digitally.

Assessment and Reporting
Record results and track progress toward mastery.

Day 1: Unit Assessments

1 Warm Up　5–10 min

Materials

Self Assessment
Children complete the Self Assessment.

Assessment Handbook, p. 64

2a Assess　35–50 min

★ **Unit 9 Assessment**
These items reflect mastery expectations of Grade 1 content.

Assessment Handbook, pp. 65–69

Unit 9 Challenge (Optional)
Children may demonstrate progress beyond expectations.

Assessment Handbook, pp. 70–71

CCSS

Common Core State Standards	Goals for Mathematical Content (GMC)*	Lessons	Self Assessment	Unit 9 Assessment	Unit 9 Challenge
1.OA.1	Solve number stories by adding and subtracting.	9-2, 9-5 to 9-7		5–7	
	Model parts-and-total, change, and comparison situations.	9-2, 9-5 to 9-7		5–7	
1.OA.6	Add doubles automatically.	9-4			1–5
	Add and subtract within 20 using strategies.	9-4, 9-8			1, 2, 4, 5
1.NBT.2	Understand place value.	9-9		3, 4	
	Represent whole numbers as tens and ones.	9-9		3, 4	
	Understand 10, 20, …, 90 as some tens and no ones.	9-9		4	
1.NBT.3	Compare and order numbers.	9-8		3, 5, 6	
	Record comparisons using >, =, or <.	9-8		5, 6	
1.NBT.4	Understand adding 2-digit numbers and 1-digit numbers.	9-2 to 9-8	3		
	Understand adding 2-digit numbers and multiples of 10.	9-2 to 9-8	3	5, 6	
	Understand adding 2-digit numbers.	9-2 to 9-8	3	5	3–5
1.NBT.6	Subtract multiples of 10 from multiples of 10.	9-2, 9-5 to 9-7	3	7	
1.MD.2	Measure length using same-size units with no gaps or overlaps.	9-1	1	1	
	Express length as a whole number of units.	9-1	1	1	
1.G.1	Distinguish between defining and non-defining attributes.	9-10		2	
1.G.2	Build composite shapes.	9-10	5	8	
1.G.3	Partition shapes into equal shares.	9-4, 9-11	6	9	
	Describe equal shares using fraction words.	9-4, 9-11	6	9	
	Understand that *more equal shares* means *smaller equal shares*.	9-4, 9-11		9	

*All instruction and practice on the Grade 1 content is complete.

 Spiral Tracker　(Go Online) to see how mastery develops for all standards within the grade.

Common Core State Standards	Goals for Mathematical Practice (GMP)	Lessons	Self Assessment	Unit 9 Assessment	Unit 9 Challenge
SMP1	Reflect on your thinking as you solve your problem. **GMP1.2**	9-2, 9-3	2		
	Compare the strategies you and others use. **GMP1.6**	9-2, 9-7	4		
SMP2	Create mathematical representations using numbers, words, pictures, symbols, gestures, tables, graphs, and concrete objects. **GMP2.1**	9-9		4	1–3
	Make connections between representations. **GMP2.3**	9-9, 9-11		4	
SMP4	Model real-world situations using graphs, drawings, tables, symbols, numbers, diagrams, and other representations. **GMP4.1**	9-2, 9-8		5–7	
SMP5	Use tools effectively and make sense of your results. **GMP5.2**	9-1	1	1, 7	
SMP6	Explain your mathematical thinking clearly and precisely. **GMP6.1**	9-3, 9-5, 9-6		6, 7, 9	
	Use clear labels, units, and mathematical language. **GMP6.3**	9-3, 9-4, 9-10, 9-11		9	
SMP7	Look for mathematical structures such as categories, patterns, and properties. **GMP7.1**	9-10		2	4, 5

Spiral Tracker **Go Online** to see how mastery develops for all standards within the grade.

1 Warm Up 5–10 min

▶ Self Assessment

Assessment Handbook, p. 64

| WHOLE CLASS | SMALL GROUP | PARTNER | **INDEPENDENT** |

Children complete the Self Assessment to reflect on their progress in Unit 9.

Remind children of a time when they did each type of problem.

Item	Remind children that they . . .
1	measured using a paper-clip ruler. (Lesson 9-1)
2	reflected on their thinking when they were solving number stories. (Lessons 9-2, 9-3, 9-5, 9-6, and 9-7)
3	added and subtracted numbers to solve number stories. (Lessons 9-2, 9-3, 9-5, 9-6, and 9-7)
4	shared and compared strategies for solving number stories. (Lessons 9-2, 9-3, 9-5, 9-6, and 9-7)
5	composed new shapes from 3-dimensional shapes. (Lesson 9-10)
6	made and named equal shares of granola bars and watermelon slices. (Lesson 9-11)

Assessment Handbook, p. 64

NAME　　　　　　　DATE　　　　　Lesson 9-12 ✓

Unit 9 Self Assessment

Put a check in the box that tells how you do each skill.			
Skills	I can do this by myself. I can explain how to do this.	I can do this by myself.	I can do this with help.
① Measure length.			
② Reflect on your thinking when you are solving a problem.			
③ Add and subtract large numbers.			
④ Compare your strategies to strategies your classmates use.			
⑤ Compose new shapes.			
⑥ Make and name equal shares.			

64　　Assessment Handbook

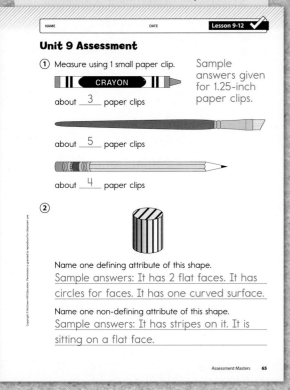

NAME _____ DATE _____ Lesson 9-12 ✓

Unit 9 Assessment

① Measure using 1 small paper clip.

CRAYON

about __3__ paper clips

Sample answers given for 1.25-inch paper clips.

about __5__ paper clips

about __4__ paper clips

②

Name one defining attribute of this shape.
Sample answers: It has 2 flat faces. It has circles for faces. It has one curved surface.

Name one non-defining attribute of this shape.
Sample answers: It has stripes on it. It is sitting on a flat face.

Assessment Masters **65**

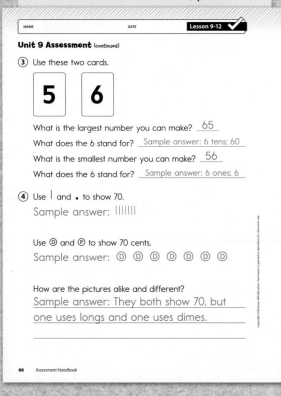

NAME _____ DATE _____ Lesson 9-12 ✓

Unit 9 Assessment (continued)

③ Use these two cards.

| 5 | 6 |

What is the largest number you can make? __65__
What does the 6 stand for? _Sample answer: 6 tens; 60_
What is the smallest number you can make? __56__
What does the 6 stand for? _Sample answer: 6 ones; 6_

④ Use | and ▪ to show 70.
Sample answer: |||||||

Use Ⓓ and Ⓟ to show 70 cents.
Sample answer: Ⓓ Ⓓ Ⓓ Ⓓ Ⓓ Ⓓ Ⓓ

How are the pictures alike and different?
Sample answer: They both show 70, but one uses longs and one uses dimes.

66 Assessment Handbook

2b Assess

35–50 min   Go Online

Differentiation Support   Assessment and Reporting

▶ Unit 9 Assessment

Assessment Handbook, pp. 65–69

| WHOLE CLASS | SMALL GROUP | PARTNER | **INDEPENDENT** |

Children complete the Unit 9 Assessment to demonstrate their progress on the Common Core State Standards covered in this unit.

Generic rubrics in the *Assessment Handbook* appendix can be used to evaluate children's progress on the Mathematical Practices.

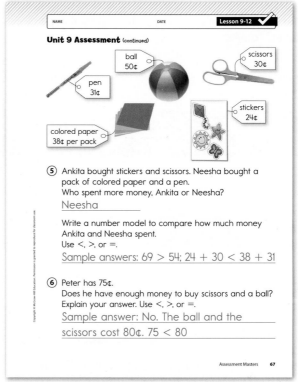

NAME _____ DATE _____ Lesson 9-12 ✓

Unit 9 Assessment (continued)

pen 31¢   ball 50¢   scissors 30¢

stickers 24¢

colored paper 38¢ per pack

⑤ Ankita bought stickers and scissors. Neesha bought a pack of colored paper and a pen.
Who spent more money, Ankita or Neesha?
Neesha

Write a number model to compare how much money Ankita and Neesha spent.
Use <, >, or =.
Sample answers: 69 > 54; 24 + 30 < 38 + 31

⑥ Peter has 75¢.
Does he have enough money to buy scissors and a ball?
Explain your answer. Use <, >, or =.
Sample answer: No. The ball and the scissors cost 80¢. 75 < 80

Assessment Masters **67**

Assessment Handbook, p. 67

Item(s)	Adjustments

Differentiate | **Adjusting the Assessment**

1	To scaffold item 1, provide children with 7 small paper clips. To extend item 1, have children measure in inches or centimeters using a ruler.
2	To scaffold item 2, have children describe everything they notice about the cylinder. Help them determine which of those things can only apply to cylinders and which can apply to non-cylindrical shapes. To extend item 2, have children name more than one defining attribute and more than one non-defining attribute of the cylinder.
3	To scaffold item 3, provide children with base-10 blocks to represent the numbers. To extend item 3, have children make the largest and smallest numbers possible with the digits 5, 6, and 8.
4	To extend item 4, have children show two more representations for 70 using base-10 blocks and coins.
5–7	To scaffold items 5–7, have children use base-10 blocks or a number grid to solve. To extend items 5–7, have children write their own number stories about the school store mini-posters for a partner to solve.
8	To extend item 8, have children add onto the new shape to make another, larger composite shape.
9	To scaffold item 9, have them write a name for one share. Then have them use that name to generate a name for multiple shares. To extend item 9, have children provide more than one name for each of the requested shares.

Advice for Differentiation

All instruction in *First Grade Everyday Mathematics* is complete. If you have concerns about children's progress on this content, see the online differentiation options for support.

NOTE See the Unit Organizer on pages 768–769 or the online Spiral Tracker for details on Unit 9 focus topics and the spiral.

Assessment Handbook, p. 68

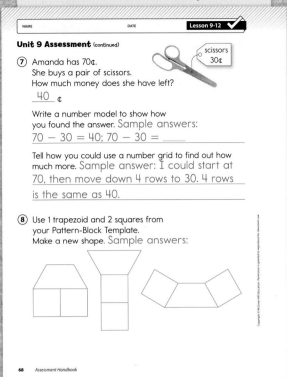

NAME _____ DATE _____ Lesson 9-12 ✓

Unit 9 Assessment (continued)

⑦ Amanda has 70¢. She buys a pair of scissors. How much money does she have left?

__40__ ¢

Write a number model to show how you found the answer. Sample answers:

70 − 30 = 40; 70 − 30 = ____

Tell how you could use a number grid to find out how much more. Sample answer: I could start at 70, then move down 4 rows to 30. 4 rows is the same as 40.

⑧ Use 1 trapezoid and 2 squares from your Pattern-Block Template. Make a new shape. Sample answers:

68 *Assessment Handbook*

Assessment Handbook, p. 69

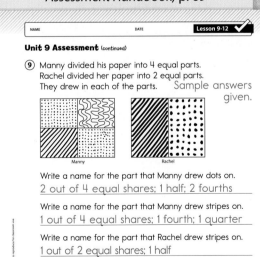

NAME _____ DATE _____ Lesson 9-12 ✓

Unit 9 Assessment (continued)

⑨ Manny divided his paper into 4 equal parts. Rachel divided her paper into 2 equal parts. They drew in each of the parts. Sample answers given.

Manny   Rachel

Write a name for the part that Manny drew dots on.
2 out of 4 equal shares; 1 half; 2 fourths

Write a name for the part that Manny drew stripes on.
1 out of 4 equal shares; 1 fourth; 1 quarter

Write a name for the part that Rachel drew stripes on.
1 out of 2 equal shares; 1 half

Write a name for the part that Rachel drew on.
2 out of 2 equal shares; 2 halves; the whole

Whose striped part is smaller? Manny's

Explain why you think so. Sample answer: Manny's paper is in 4 equal shares. Rachel's paper is in 2 equal shares. So, Manny's parts are smaller.

Assessment Masters 69

NAME   DATE   Lesson 9-12 ✓

Unit 9 Challenge
More Addition Doubles

This table includes some doubles you already know.
Complete the table with sums for the other addition doubles.

double	sum
6 + 6 =	12
7 + 7 =	14
8 + 8 =	16
9 + 9 =	18
10 + 10 =	20
11 + 11 =	22
12 + 12 =	24
13 + 13 =	26
14 + 14 =	28
15 + 15 =	30

These numbers can
be written as a double. ↗

Example: 14 = 7 + 7

You can write 14 as two 7s.

70   Assessment Handbook

▶ Unit 9 Challenge (Optional)

Assessment Handbook, pp. 70–71

| WHOLE CLASS | SMALL GROUP | PARTNER | **INDEPENDENT** |

Children can complete the Unit 9 Challenge after they complete the Unit 9 Assessment.

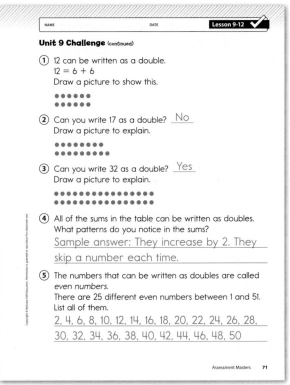

Assessment Handbook, p. 71

Day 2: Open Response Assessment

2b Assess   50–55 min

Materials

Solving the Open Response Problem
After a brief introduction, children solve an open response problem using place-value patterns on a number grid and write an explanation of their thinking.

Assessment Handbook, p. 72

Discussing the Problem
After completing the problem, children share how they used patterns to help them solve the problem.

Assessment Handbook, p. 72

CCSS   Common Core State Standards

	Goal for Mathematical Content (GMC)	Lessons
1.NBT.5	Mentally find 10 more or 10 less than a 2-digit number.	9-9
	Goal for Mathematical Practice (GMP)	
SMP7	Use structures to solve problems and answer questions. GMP7.2	9-9

Spiral Tracker   **Go Online** to see how mastery develops for all standards within the grade.

▶ **Evaluating Children's Responses**
Evaluate children's abilities to mentally find 10 more than a 2-digit number. Use the rubric below to evaluate their work based on **GMP7.2**.

Goal for Mathematical Practice GMP7.2 Use structures to solve problems and answer questions.	Not Meeting Expectations	Partially Meeting Expectations	Meeting Expectations	Exceeding Expectations
	Does not locate or correct Deena's error, and does not explain how patterns are used to fill in the puzzle.	Locates and corrects Deena's error, but provides only a partial explanation of the structure of the number grid. For example, may describe the pattern of increasing by 10 when moving down a row, but not decreasing by 1 when moving left.	Locates and corrects Deena's error, and writes an explanation that reveals the use of structure (pattern) to correct Deena's error by referencing increasing by 10 when moving down a row and decreasing by 1 when moving left.	Meets expectations and explains why Deena's answer was incorrect.

3 Look Ahead   10–15 min

Materials

Math Boxes 9-12
Children practice and maintain skills.

Math Journal 2, p. 215

Home Link 9-12
Children take home the end-of-year Family Letter.

Math Masters, pp. 290–293, TA43

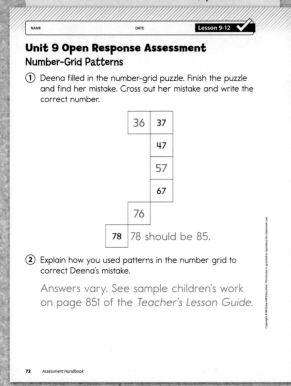

NAME   DATE   Lesson 9-12 ✓

Unit 9 Open Response Assessment
Number-Grid Patterns

① Deena filled in the number-grid puzzle. Finish the puzzle and find her mistake. Cross out her mistake and write the correct number.

36	37
	47
	57
	67
76	
78	78 should be 85.

② Explain how you used patterns in the number grid to correct Deena's mistake.

Answers vary. See sample children's work on page 851 of the *Teacher's Lesson Guide*.

72   Assessment Handbook

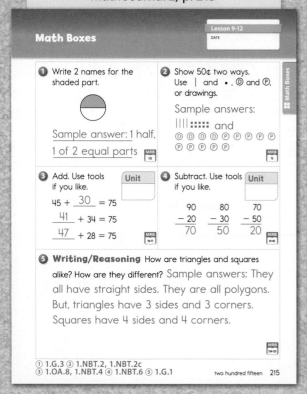

Math Boxes   Lesson 9-12
                                  DATE

① Write 2 names for the shaded part.

Sample answer: 1 half, 1 of 2 equal parts   MRB

② Show 50¢ two ways. Use | and • , Ⓓ and Ⓟ, or drawings.

Sample answers:
||||| ::::: and
Ⓓ Ⓓ Ⓓ Ⓟ Ⓟ Ⓟ Ⓟ Ⓟ
Ⓟ Ⓟ Ⓟ Ⓟ Ⓟ   MRB

③ Add. Use tools if you like.   Unit

$45 + \underline{30} = 75$
$\underline{41} + 34 = 75$
$\underline{47} + 28 = 75$ MRB

④ Subtract. Use tools if you like.   Unit

| 90 | 80 | 70 |
| − 20 | − 30 | − 50 |
| 70 | 50 | 20 | MRB

⑤ **Writing/Reasoning** How are triangles and squares alike? How are they different? Sample answers: They all have straight sides. They are all polygons. But, triangles have 3 sides and 3 corners. Squares have 4 sides and 4 corners.   MRB

① 1.G.3 ② 1.NBT.2, 1.NBT.2c
③ 1.OA.8, 1.NBT.4 ④ 1.NBT.6 ⑤ 1.G.1   two hundred fifteen   215

2b Assess   50–55 min

Assessment and Reporting

▶ Solving the Open Response Problem

Assessment Handbook, p. 72

| WHOLE CLASS | SMALL GROUP | **PARTNER** | **INDEPENDENT** |

The open response problem requires children to apply skills and concepts from Unit 9 to identify a mistake in a number-grid puzzle. The focus of this task is **GMP7.2:** Use structures to solve problems and answer questions.

Distribute *Assessment Handbook*, page 72. Read the directions aloud. Tell children that they should use patterns to help them fill in the number-grid puzzle and correct Deena's mistake.   GMP7.2

Circulate and observe children as they work. You may wish to take notes about children's performance on the task. Expect that children will apply the patterns of increasing by 10 when moving down on the number grid and decreasing by 1 when moving left on the number grid to fill in the puzzle and correct the mistake. However, some may struggle to explain their thinking. Encourage them to explain any patterns they see in the completed puzzle and think about how they used those patterns to help them solve the problem.

> **Differentiate**   **Adjusting the Assessment**
>
> Children who have trouble explaining their thinking in writing should be given the opportunity to describe their thinking orally.

▶ Discussing the Problem

Assessment Handbook, p. 72

| **WHOLE CLASS** | **SMALL GROUP** | PARTNER | INDEPENDENT |

After children complete the page, invite individuals to explain how they filled in the number-grid puzzle and how they knew how to correct Deena's mistake. Encourage children to help each other communicate their thinking by asking them to describe what other children have said and explain why they agree or disagree.

Evaluating Children's Responses (CCSS) 1.NBT.5

Collect children's work. For the content standard, expect most children to correctly fill in the number-grid puzzle and count by 10s, not 1s, to find numbers when moving down on the number grid. You can use the rubric on page 849 to evaluate children's work for **GMP7.2**.

See the sample in the margin. This work meets expectations for the content standard because this child filled in all the correct numbers on the number-grid puzzle. Also, the explanation illustrates that the child added 10 and therefore did not count by 1s to find the numbers when moving down on the number grid. The work meets expectations for the mathematical practice because the explanation accurately describes the pattern of adding 10 when moving down and subtracting 1 when moving left on the number grid and connects the patterns to finding and correcting Deena's mistake. **GMP7.2**

Go Online ✓ Assessment and Reporting

Sample child's work, "Meeting Expectations"

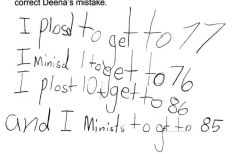

1. Deena filled in the number-grid puzzle. Finish the puzzle and find her mistake. Cross out her mistake and write the correct number next to it.

36	37
	47
	57
	67

76 77

85 ~~87~~ 86

2. Explain how you used patterns in the number grid to correct Deena's mistake.

I plosd to get to 77
I minisd 1 to get to 76
I plost 10 to get to 86
and I minists to get to 85

---

3 Look Ahead 10–15 min

Go Online 🏠 Home Connections

▶ Math Boxes 9-12 ✏️

Math Journal 2, p. 215

| WHOLE CLASS | SMALL GROUP | **PARTNER** | INDEPENDENT |

Mixed Practice Math Boxes 9-12 are paired with Math Boxes 9-10.

▶ Home Link 9-12: End-of-Year Family Letter

Math Masters, pp. 290–293, TA43

Home Connection Give each child one copy of the Family Letter and two copies of *Math Masters*, page TA43, the blank number grid. The End-of-Year Family Letter thanks family members for their participation in *First Grade Everyday Mathematics,* suggests home-based activities for the summer, and provides a preview of *Second Grade Everyday Mathematics.*

NOTE Additional samples of evaluated children's work can be found in the *Assessment Handbook* appendix.

Math Masters, pp. 290–293

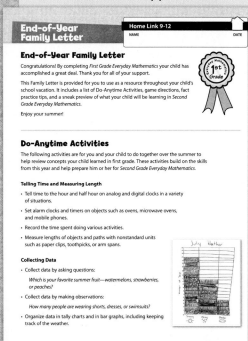

End-of-Year Family Letter

Home Link 9-12

NAME   DATE

End-of-Year Family Letter

Congratulations! By completing *First Grade Everyday Mathematics* your child has accomplished a great deal. Thank you for all of your support.

This Family Letter is provided for you to use as a resource throughout your child's school vacation. It includes a list of Do-Anytime Activities, game directions, fact practice tips, and a sneak preview of what your child will be learning in *Second Grade Everyday Mathematics.*

Enjoy your summer!

Do-Anytime Activities

The following activities are for you and your child to do together over the summer to help review concepts your child learned in first grade. These activities build on the skills from this year and help prepare him or her for *Second Grade Everyday Mathematics.*

Telling Time and Measuring Length

- Tell time to the hour and half hour on analog and digital clocks in a variety of situations.
- Set alarm clocks and timers on objects such as ovens, microwave ovens, and mobile phones.
- Record the time spent doing various activities.
- Measure lengths of objects and paths with nonstandard units such as paper clips, toothpicks, or arm spans.

Collecting Data

- Collect data by asking questions:
    Which is your favorite summer fruit—watermelons, strawberries, or peaches?
- Collect data by making observations:
    How many people are wearing shorts, dresses, or swimsuits?
- Organize data in tally charts and in bar graphs, including keeping track of the weather.

290   two hundred ninety

End-of-Year Assessment

Assessment Handbook, pp. 89–100

You may want to administer the End-of-Year Assessment over a period of several days to check children's mastery of many of the concepts and skills in *First Grade Everyday Mathematics.* See the *Assessment Handbook.*

Glossary

Everyday Mathematics strives to define terms clearly, especially when they can be defined in multiple ways.

This glossary focuses on terms and meanings for elementary school mathematics and omits details and complexities required at higher levels. The definitions here are phrased for teachers. Information for explaining terms and concepts to children can be found within the lessons themselves. Additional information is available online. In a definition, most terms in italics are defined elsewhere in this glossary.

0–9

1-dimensional (1-D) (1) Having *length* but not area or volume; confined to a curve, such as an arc. (2) A figure whose points are all on one *line*. Line segments are 1-dimensional. Compare *2-dimensional* and *3-dimensional*.

2-dimensional (2-D) (1) Having *area* but not volume; confined to a *surface*. A 2-dimensional surface can be flat or curved, such as the surface of a sphere. (2) A figure whose points are all in one *plane* but not all on one line. Examples include polygons and circles. Compare *1-dimensional* and *3-dimensional*.

3-dimensional (3-D) Having *volume*. Solids such as cubes, cones, and spheres are 3-dimensional. Compare *1-dimensional* and *2-dimensional*.

A

accurate (1) As correct as possible for a given context. An answer can be accurate without being very precise if the units are large. For example, the driving time from Chicago to New York City is about 13 hours. See *approximate*. (2) Of a measurement or other quantity, having a high degree of correctness. A more accurate measurement is closer to the true value. Accurate answers must be reasonably *precise*.

add-to situation A situation in which there is a starting quantity, an additional quantity, and an ending quantity. Any of the three quantities may be unknown. See *addition/subtraction use class*.

addend Any one of a set of numbers that are added. For example, in $5 + 3 + 1 = 9$, the addends are 5, 3, and 1.

addition fact Two whole numbers from 0 through 10 and their sum, such as $9 + 7 = 16$. See *arithmetic facts*.

addition/subtraction use class A category of problem situations that can be solved using addition or subtraction or other methods, such as counting or direct modeling. *Everyday Mathematics* distinguishes four addition/subtraction use classes: *parts-and-total, change-to-more, change-to-less,* and *comparison situations*. The table below shows how these use classes correspond to those in the Common Core State Standards.

Everyday Mathematics	CCSS
change-to-more	add to
change-to-less	take from
parts-and-total	put together/take apart
comparison	compare

Additive Identity The number zero (0). The Additive Identity is the number that when added to any other number, yields that other number. See *additive inverses*.

additive inverses Two numbers whose sum is 0. Each number is called the additive inverse, or opposite, of the other. For example, 3 and −3 are additive inverses because $3 + (−3) = 0$. Zero is its own additive inverse: $0 + 0 = 0$. See *Additive Identity*.

adjacent angles Two nonoverlapping *angles* with a common *side* and *vertex*.

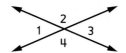 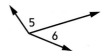

Angles 1 and 2, 2 and 3, 3 and 4, and 4 and 1 are pairs of adjacent angles.

Angle 5 is adjacent to angle 6.

adjacent sides (1) Two sides of a *polygon* with a common *vertex*. (2) Two faces of a *polyhedron* with a common *edge*.

A.M. The abbreviation for *ante meridiem,* meaning "before the middle of the day" in Latin. From midnight to noon.

analog clock (1) A clock that shows the time by the positions of the hour and minute hands. (2) Any device that shows time passing in a continuous manner, such as a sundial. Compare *digital clock*.

anchor chart A classroom display that is cocreated by a teacher and children and focuses on a central concept or skill.

angle (1) A figure formed by two rays or line segments with a common endpoint called the *vertex* of the angle. The rays or segments are called the sides of the angle. Angles can be named after their vertex point alone, as in ∠*A*; or by three points, one on each side and the vertex in the middle, as in ∠*BCD*. One side of an angle is *rotated* about the vertex from the other side through a number of *degrees*. (2) The measure of this rotation in degrees.

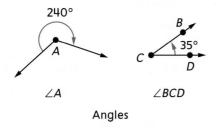

∠*A*   ∠*BCD*

Angles

-angle A suffix meaning angle or corner, for example, triangle and rectangle.

apex (1) In a *pyramid,* the *vertex* opposite the *base.* All the nonbase faces meet at the apex. (2) The point at the tip of a *cone.*

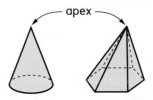

approximate Close to exact. In many situations it is not possible to get an exact answer, but it is important to be close to the exact answer. We might draw an angle that is approximately 60° with a protractor. In this case *approximate* suggests that the angle drawn is within a degree or two of 60°. Compare *precise.*

area The amount of *surface* inside a *2-dimensional* figure. The figure might be a triangle or a rectangle in a *plane,* the curved surface of a cylinder, or a state or a country on Earth's surface. Area can be measured in *square units,* such as square miles or square centimeters, or other units, such as acres.

A triangle with area
21 square units

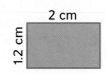

A rectangle with area
1.2 cm * 2 cm = 2.4 square centimeters

arithmetic facts The *addition facts* (*whole-number addends* 10 or less); their inverse subtraction facts; multiplication facts (whole-number factors 10 or less); and their inverse division facts, except there is no division by zero. Facts and their corresponding inverses are organized into *fact families.*

arrow rule In *Everyday Mathematics,* a rule that determines the number that goes into the next *frame* in a *Frames-and-Arrows* diagram. There may be more than one arrow rule per diagram.

arrows In *Everyday Mathematics,* the links representing the *arrow rule(s)* in a *Frames-and-Arrows* diagram.

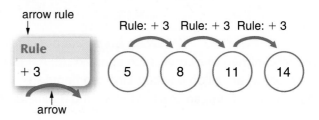

Associative Property of Addition A *property* of addition that for any numbers *a, b,* and *c,* (*a* + *b*) + *c* = *a* + (*b* + *c*). The grouping of the three *addends* can be changed without changing the *sum.* For example, (4 + 3) + 7 = 4 + (3 + 7) because 7 + 7 = 4 + 10. Subtraction is not associative. For example, (4 − 3) + 7 ≠ 4 − (3 + 7) because 8 ≠ −6. Compare *Commutative Property of Addition.*

attribute A characteristic or *property* of an object or a common characteristic of a set of objects. Size, shape, color, and number of sides are attributes.

attribute blocks A set of blocks that vary in four *attributes:* color, size, thickness, and shape. The blocks are used for attribute identification and sorting activities. Compare *pattern blocks.*

automaticity The ability to solve problems with great efficiency either by using recall or applying quick strategies. For example, one might "just know" that 8 + 7 = 15 or quickly think 8 + 2 = 10 and 5 more is 15. Compare *fluency.*

bar graph A graph with horizontal or vertical bars that represent (typically categorical) *data*. The lengths of the bars may be scaled.

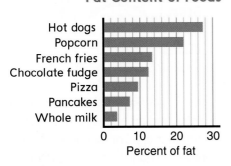

Fat Content of Foods

base angles of a trapezoid Two *angles* that share a *base of a trapezoid*.

base of a number system The foundation number for a *place-value* based numeration system. For example, our usual way of writing numbers uses a base-10 place-value system. In programming computers or other digital devices, bases of 2, 8, 16, or other powers of 2 are more common than base 10.

base of a prism or a cylinder Either of the two *parallel* and *congruent faces* that define the shape of a *prism* or a *cylinder*.

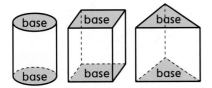

base of a pyramid or a cone The *face* of a *pyramid* or a *cone* that is opposite its *apex*.

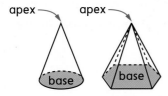

base of a trapezoid (1) Either of a pair of *parallel* sides in a *trapezoid*. (2) The *length* of this *side*. The area of a trapezoid is the average of a pair of bases times the corresponding height.

base ten (1) Related to powers of 10. (2) The most common system for writing numbers, which uses only 10 symbols 0, 1, 2, 3, 4, 5, 6, 7, 8, and 9, called *digits*. One can write any number using one or more of these 10 digits. Each digit has a value that depends on its place in the number (its *place value*). In the base-10 system, each place has a value 10 times that of the place to its right, and one-tenth the value of the place to its left.

base-10 blocks A set of blocks to represent numbers. In *Everyday Mathematics*, the four standard sizes of blocks are called cubes, longs, flats, and big cubes from smallest to largest. Once one of these is designated as the unit whole, the values of the other blocks follow. For example, if the cube is the unit whole, then the long, flat, and big cube represent 10, 100, and 1,000, respectively. However, if the flat is the unit whole, then the cube, long, and big cube represent $\frac{1}{100}$, $\frac{1}{10}$, and 10, respectively. The cube is 1 cm on each edge. The long is 10 cm by 1 cm by 1 cm. The flat is 10 cm by 10 cm by 1 cm. The big cube looks like 1,000 cubes and is 10 cm by 10 cm by 10 cm. See *long*, *flat*, and *big cube* for photos of the blocks. See *base-10 shorthand*.

base-10 shorthand In *Everyday Mathematics*, a written notation for *base-10 blocks*.

Name	Base-10 block	Base-10 shorthand
cube	▫	▪
long			
flat		▢
big cube		▱

big cube In *Everyday Mathematics*, a *base-10 block* cube that measures 10 cm by 10 cm by 10 cm. A big cube is equivalent to one thousand 1-cm cubes.

billion By U.S. custom, 1 billion is 1,000,000,000, or 10^9. By British, French, and German custom, 1 billion is 1,000,000,000,000, or 10^{12}.

C

calendar (1) A *reference frame* to keep track of the passage of time over weeks, months, and years. (2) A practical model of the reference frame, such as the large, reusable Class Calendar in Kindergarten through Second Grade *Everyday Mathematics*. (3) A schedule or listing of events.

cardinality The number of things in a *set* is its cardinality. The cardinality of a set can often be found by counting the objects in the set; the last number counted is the cardinality of the set. The cardinality of the days of a week is 7. See *cardinal number*.

cardinal number A number telling how many things are in a *set*. Compare *ordinal number*.

Celsius A *temperature* scale on which pure water at sea level freezes at 0° and boils at 100°. The Celsius scale is used in the *metric system*. A less common name for this scale is centigrade because there are 100 degrees between the freezing and boiling points of water. Compare *Fahrenheit*.

cent A penny; $\frac{1}{100}$ of a dollar. From the Latin word *centesimus*, which means "a hundredth part."

center of a circle The point in the *plane* of a *circle* that is equally distant from all points on the circle.

center

center of a sphere The point equally distant from all points on a *sphere*.

center

chance The possibility that an *outcome* or an event will occur. For example, in flipping a coin there is an equal chance of getting HEADS or TAILS.

change diagram In *Everyday Mathematics*, a diagram used to model situations in which quantities are increased or decreased. The diagram includes a starting quantity, an ending quantity, and an amount of change. See *situation diagram*, *change-to-less situation*, and *change-to-more situation*.

Change		
Start		End
14	−5	9

A change diagram for $14 - 5 = 9$

change-to-less situation A situation involving a starting quantity, a change, and an ending quantity that is less than the starting quantity. For example, a situation about spending money is a change-to-less situation. Compare *change-to-more situation*. See *addition/subtraction use class*.

change-to-more situation A situation involving a starting quantity, a change, and an ending quantity that is more than the starting quantity. For example, a situation about earning money is a change-to-more situation. Compare *change-to-less situation*. See *addition/subtraction use class*.

circle The set of all points in a *plane* that are equally distant from a fixed point in the plane called the *center*. The distance from the center to the circle is the radius of the circle. The diameter of a circle is twice the radius. A circle together with its interior is called a *disk* or a circular region.

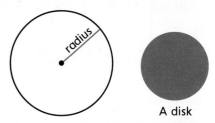
radius

A disk

clockwise rotation A turn in the direction in which the hands move on a typical *analog clock*; a turn to the right.

closed   A *2-dimensional* shape is closed if its boundary divides the *plane* into two regions, the inside and the outside. A *3-dimensional* shape is closed if its boundary divides the space into two regions, the inside and the outside. To pass from the inside to the outside of a closed shape, a line segment must cross the boundary of the shape. Compare *open*.

column   (1) A vertical arrangement of objects or numbers in an array or a table.

column

(2) A vertical selection of cells in a spreadsheet.

combinations of 10   Pairs of *whole numbers* from 0 to 10 that add to 10. Combinations of 10 are key *helper facts* for addition and subtraction fact strategies. For example, $4 + 6$ and $0 + 10$ are both combinations of 10.

Commutative Property of Addition   A *property* of addition that for any numbers a and b, $a + b = b + a$. Two numbers can be added in either order without changing the *sum*. For example, $5 + 10 = 10 + 5$. In *Everyday Mathematics*, this and the Commutative Property of Multiplication are called *turn-around rules*. Subtraction is not commutative. For example, $8 - 5 \neq 5 - 8$ because $3 \neq -3$.

comparison diagram   In *Everyday Mathematics*, a diagram used to model situations in which two quantities are compared. The diagram represents two quantities and their *difference*. See *situation diagram*.

A comparison diagram for $12 = 9 + ?$

comparison situation   A situation involving two quantities and the *difference* between them. See *addition/ subtraction use class*.

compose   To make up or form a number or shape by putting together smaller numbers or shapes. For example, one can compose a 10 by putting together ten 1s:
$1 + 1 + 1 + 1 + 1 + 1 + 1 + 1 + 1 + 1 = 10$.
One can compose a pentagon by putting together an equilateral triangle and a square.

A composed pentagon

concave polygon   A *polygon* on which there are at least two points that can be connected with a *line segment* that passes outside the polygon. For example, segment *AD* is outside the hexagon between *B* and *C*. Informally, at least one *vertex* appears to be "pushed inward." At least one interior *angle* has a measure greater than 180°. Same as *nonconvex polygon*. Compare *convex polygon*.

A concave polygon

cone   A *geometric solid* comprising a circular *base*, an *apex* not in the *plane* of the base, and all *line segments* with one endpoint at the apex and the other endpoint on the circumference of the base.

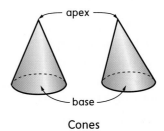

Cones

congruent figures   Figures having the same size and shape. Two figures are congruent if they match exactly when one is placed on top of the other after a combination of isometric transformations (slides, flips, and/or turns). In diagrams of congruent figures, the corresponding congruent sides may be marked with the same number of hash marks. The symbol ≅ means "is congruent to."

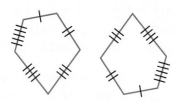

Congruent pentagons

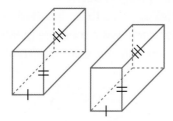

Congruent right rectangular prisms

conjecture   A claim that has not been proved, at least by the person making the conjecture.

consecutive   Following one after another in an uninterrupted order. For example, A, B, C, and D are four consecutive letters of the alphabet; 6, 7, 8, 9, and 10 are five consecutive whole numbers.

convex polygon   A *polygon* on which no two points can be connected with a *line segment* that passes outside the polygon. Informally, all *vertices* appear to be "pushed outward." Each *angle* in the polygon measures less than 180°. Compare *concave polygon*.

A convex polygon

corner   Informal for *vertex* or *angle*.

count   (1) The number of objects in a *set*. (2) For use as a verb, see *rote counting, rational counting,* and *skip counting*.

counterclockwise rotation   Opposite the direction in which the hands move on a typical *analog clock;* a turn to the left.

counting all   An addition strategy that involves counting out each *addend* and then counting the total to find the *sum*. For example, to find the sum of 4 objects and 3 objects, one would count out 4, count out 3, and then count all 7 objects to determine that the sum is 7.

counting numbers   The numbers used to *count* things. The set of counting numbers is {1, 2, 3, 4, . . .}. Sometimes 0 is included but not in *Everyday Mathematics*. Counting numbers are also known as *natural numbers*.

counting on   (1) Counting up or back from a number other than 0, 1, or −1. (2) An addition strategy that involves starting from one *addend* and counting on only the remaining addend to find the *sum*. For example, to solve 4 + 3, one might start from 4 and count 3 more— "4, 5, 6, 7"—to find 7 total.

counting-up subtraction   A subtraction strategy in which a *difference* is found by counting or adding up from the smaller number to the larger number. For example, to calculate 87 − 49, one could start at 49, add 30 to reach 79, and then add 8 more to reach 87. The difference is 30 + 8 = 38.

cube   (1) A *regular polyhedron* with 6 square *faces*. A cube has 8 *vertices* and 12 *edges*.

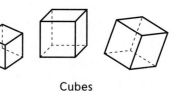

Cubes

(2) In *Everyday Mathematics*, the smaller cube of the *base-10 blocks,* measuring 1 cm on each edge.

curved surface   A *2-dimensional surface* that does not lie in a *plane*. Spheres, cylinders, and cones have curved surfaces.

customary system of measurement   In *Everyday Mathematics*, same as *U.S. customary system* of measurement. See *Tables of Measures*.

cylinder   A *geometric solid* with two *congruent, parallel* circular regions for *bases* and a curved *face* formed by all the *segments* that have an endpoint on each *circle* and that are parallel to the segment with endpoints at the *centers of the circles*. Also called a *circular cylinder*.

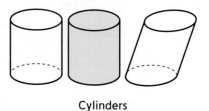

Cylinders

D

data   Information that is gathered by counting, measuring, questioning, or observing. Strictly, data is the plural of datum, but data is often used as a singular word.

decagon   A 10-sided *polygon*.

decimal   (1) A number written in standard *base-10* notation containing a *decimal point*, such as 2.54. (2) Any number written in standard base-10 notation. See *decimal fraction*, *repeating decimal*, and *terminating decimal*.

decimal fraction   (1) A *fraction* or mixed number written in standard *decimal notation*. (2) A fraction $\frac{a}{b}$, where b is a positive power of 10, such as $\frac{84}{100}$.

decimal notation   Same as *standard notation*.

decimal point   A mark used to separate the ones and tenths places in decimals. A decimal point separates dollars from cents in dollars-and-cents notation. The mark is a dot in the U.S. but a comma in Europe and some other countries.

decompose   To separate a number or a shape into smaller numbers or shapes. For example, 14 can be decomposed into 1 ten and 4 ones. A square can be decomposed into two isosceles right triangles. Any even number can be decomposed into two equal counting numbers or integers: $2n = n + n$.

defining attributes of a shape   Characteristics of a shape that are consequences of the definition. The shape will always have those characteristics. For example, four right angles is a defining *attribute* of a square, but being orange or having an area of 4 square inches are not defining attributes of a square. Compare *nondefining attributes of a shape*.

degree (°)   (1) A unit of measure for *angles* based on dividing a *circle* into 360 equal parts. Latitude and longitude are measured in degrees based on angle measures. (2) A unit for measuring *temperature*. See *Celsius* and *Fahrenheit*. The symbol ° means degrees of any type.

difference   (1) The distance between two numbers on a *number line*. The difference between 5 and 12 is 7. (2) The result of subtracting one number from another. For example, in $12 - 5 = 7$ the difference is 7, and in $5 - 12$ the difference is −7. Compare *minuend* and *subtrahend*.

digit   (1) Any of the symbols 0, 1, 2, 3, 4, 5, 6, 7, 8, and 9 in the *base-10 numeration* system. For example, the *numeral* 145 is made up of the digits 1, 4, and 5. (2) Additional symbols in other place-value systems, such as A, B, C, D, E, and F in base-16 numeration.

digital clock   A clock that shows the time with numbers of hours and minutes, usually separated by a colon. This display is discrete, not continuous, meaning that the display jumps to a new time after a minute has elapsed. Compare *analog clock*.

dimension   (1) A measurable extent such as *length*, width, or *height*. Having two measurable extents makes the measured figure 2-dimensional. See *1-*, *2-*, and *3-dimensional*. (2) The measures of those extents. For example, the dimensions of a box might be 24 cm by 20 cm by 10 cm. (3) The number of coordinates necessary to locate a point in a geometric space. A *plane* has two dimensions because an ordered pair of two coordinates uniquely locates any point in the plane.

direct comparison   A measurement comparison made by aligning the objects. Compare *indirect comparison*.

disk   A *circle* and its *interior* region.

dot pattern   A representation of a quantity using dots arranged in a pattern to facilitate quick recognition. For example, dot patterns might mirror patterns on dice or dominos or use a grouping such as two groups of 5.

double ten frame   Two side-by-side *ten frames*. Double ten frames can be used to represent the numbers 0–20 as well as *addition facts* in a variety of ways and are particularly useful for encouraging children to develop addition-fact strategies, such as *near doubles* and *making 10*.

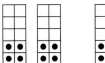

Sequence of double ten frames to elicit near-doubles strategy

doubles fact   The *sum* of a number 0 through 10 added to itself, such as $4 + 4 = 8$, sometimes written as that number multiplied by 2: $2 \times 4 = 8$. These are key *helper facts*.

E

edge   (1) Any *side* of a *polyhedron's faces.*

edges

(2) A *line segment* or curve where two *surfaces* of a *geometric solid* meet.

edge

elapsed time   The amount of time that has passed from one point in time to the next. For example, between 12:45 P.M. and 1:30 P.M., 45 minutes have elapsed.

equal   (1) Identical in number or measure; neither more nor less. (2) *Equivalent.*

equal parts   *Equivalent* parts of a whole. For example, dividing a pizza into 4 equal parts means each part is $\frac{1}{4}$ of the pizza and is equal in size to each of the other 3 parts.

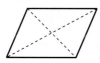

4 equal parts, each $\frac{1}{4}$ of a pizza

4 equal parts, each $\frac{1}{4}$ of the area

equal share   One of several parts of a whole, each of which has the same amount of area, volume, mass, or other measurable or countable quantity. Sometimes called fair share. See *equal parts.*

equation   A *number sentence* that contains an equals sign. For example, $5 + 10 = 15$ and $P = 2l + 2w$ are equations.

equilateral polygon   A *polygon* in which all sides are the same length.

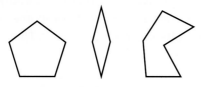

Equilateral polygons

equilateral triangle   A *triangle* with all three sides equal in length. Each angle of an equilateral triangle measures 60°, so it is also called an equiangular triangle. All equilateral triangles are *isosceles triangles.*

An equilateral triangle

equivalent   *Equal* in value but possibly in a different form. For example, $\frac{1}{2}$, 0.5, and 50% are all equivalent.

equivalent names   Different ways of naming the same number. For example, $2 + 6$, $4 + 4$, $12 - 4$, $18 - 10$, $100 - 92$, $5 + 1 + 2$, eight, VIII, and ⊬⊬ /// are all equivalent names for 8. See *name-collection box.*

estimate   (1) An answer close to, or approximating, an exact answer. (2) To make an estimate.

evaluate a numerical expression   To carry out the operations in a numerical expression to find a single value for the *expression.*

even number   (1) A *counting number* that is divisible by 2: 2, 4, 6, 8 . . .. (2) An *integer* that is divisible by 2. Compare *odd number.*

Explorations   In First through Third Grade *Everyday Mathematics*, independent or small-group activities that focus on concept development, manipulatives, data collection, problem solving, games, and skill reviews.

expression   (1) A mathematical phrase made up of numbers, *variables, operation symbols,* and/or grouping symbols. An expression does not contain *relation symbols* such as $=$, $>$, and $\leq$. (2) Either side of an *equation* or an *inequality.*

extended facts   Variations of basic arithmetic facts involving multiples of 10, 100, and so on. For example, $30 + 70 = 100$, $40 \times 5 = 200$, and $560 \div 7 = 80$ are extended facts. See *fact extensions.*

F

face   (1) A flat *surface* on a *closed, 3-dimensional* figure. Some special faces are called *bases.* (2) More generally, any *2-dimensional* surface on a 3-dimensional figure.

a flat face

a curved face

fact extensions Calculations with larger numbers using knowledge of basic *arithmetic facts*. For example, knowing the addition fact $5 + 8 = 13$ makes it easier to solve problems such as $50 + 80 = ?$ and $65 + ? = 73$. Fact extensions apply to all four basic arithmetic operations. See *extended facts*.

fact family A set of related *arithmetic facts* linking two inverse operations. For example,

$$5 + 6 = 11 \qquad 6 + 5 = 11$$
$$11 - 5 = 6 \qquad 11 - 6 = 5$$

are an addition/subtraction fact family. Similarly,

$$5 \times 7 = 35 \qquad 7 \times 5 = 35$$
$$35 \div 7 = 5 \qquad 35 \div 5 = 7$$

are a multiplication/division fact family. Same as *number family*.

Fact Triangle In *Everyday Mathematics*, a triangular flash card labeled with the numbers of a *fact family* that children can use to practice addition/subtraction and multiplication/division facts. The two addends or factors and their *sum* or product (marked with a dot) appear in the corners of each triangle.

facts table A chart showing *arithmetic facts*. An addition/subtraction facts table shows addition and subtraction facts. A multiplication/division facts table shows multiplication and division facts.

Fahrenheit A *temperature* scale on which pure water at sea level freezes at 32° and boils at 212°. The Fahrenheit scale is widely used in the United States but in few other places. Compare *Celsius*.

fair game A game in which every player has the same *chance* of winning.

false number sentence A *number sentence* that is not true. For example, $8 = 5 + 5$ is a false number sentence. Compare *true number sentence*.

flat In *Everyday Mathematics*, the *base-10 block* that is equivalent to one hundred 1-cm cubes.

A flat

fluency The ability to compute using efficient, appropriate, and flexible strategies. Compare *automaticity*.

fraction (1) A number in the form $\frac{a}{b}$ or a/b, where a and b are *integers* and b is not 0. A fraction may be used to name part of an object or part of a collection of objects, compare two quantities, or represent division. For example, $\frac{12}{6}$ might mean 12 eggs divided in groups of 6, a ratio of 12 to 6, or 12 divided by 6. Also called a common fraction. (2) A fraction that satisfies the previous definition and includes a unit in both the numerator and denominator. For example, the rates $\frac{50 \text{ miles}}{1 \text{ gallon}}$ and $\frac{40 \text{ pages}}{10 \text{ minutes}}$ are fractions. (3) A number written using a fraction bar, where the fraction bar is used to indicate division. For example, $\frac{2.3}{6.5}, \frac{\frac{14}{5}}{12}, \frac{\pi}{4}, and \frac{\frac{3}{4}}{\frac{5}{8}}$. Compare *decimal*.

fractional part Part of a whole. Fractions represent fractional parts of numbers, sets, or objects.

frames In *Everyday Mathematics*, the empty shapes in which numbers are written in a *Frames-and-Arrows* diagram.

Frames and Arrows In *Everyday Mathematics*, diagrams consisting of *frames* connected by *arrows*. Frames-and-Arrows diagrams are used to represent number sequences. Each frame contains a number, and each arrow represents a rule that determines which number goes in the next frame. There may be more than one rule, represented by different-color arrows. Frames-and-Arrows diagrams are also called chains.

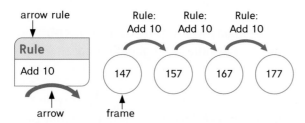

frequency (1) The number of times a value occurs in a set of *data*. (2) A number of repetitions per unit of time, such as the vibrations per second in a sound wave.

frequency graph A graph showing how often each value occurs in a *data* set.

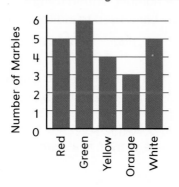

Colors in a Bag of Marbles

frequency table A table in which *data* are *tallied* and organized, often as a first step toward making a frequency graph.

Color	Number of Marbles
red	ⳤ
green	ⳤ I
yellow	IIII
orange	III
white	ⳤ

function (1) A set of *ordered pairs* (*x, y*) in which each value of *x* is paired with exactly one value of *y*. A function is typically represented in a table, by points on a coordinate graph, or by a rule such as an *equation*. (2) A rule that pairs each *input* with exactly one *output*. For example, for a function with the rule "Double," 1 is paired with 2, 2 is paired with 4, 3 is paired with 6, and so on. See *"What's My Rule?"*

function machine An imaginary device that receives *inputs* and pairs them with *outputs* using a rule that is a function. For example, the function machine below pairs an input number with its double. See *function*.

in	out
1	2
2	4
3	6
5	10
20	40
300	600

in
Rule
Double
out

A function machine and function table

G

geoboard (1) A small wooden or plastic board with nails or other posts, usually arranged at equally spaced intervals in a rectangular array. Geoboards and rubber bands are useful for exploring basic concepts in plane geometry.

Geoboard and rubber bands

(2) A digital version of a geoboard.

geometric solid   The *surface* or surfaces that make up a *3-dimensional* figure, such as a *prism, pyramid, cylinder, cone,* or *sphere.* Despite its name, a geometric solid is hollow; that is, it does not include the points in its *interior.* Informally, and in some dictionaries, a solid is defined as both the surface and its interior.

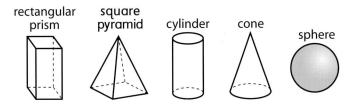

| rectangular prism | square pyramid | cylinder | cone | sphere |

Geometric solids

going through 10   A subtraction fact strategy that involves using 10 as a benchmark to simplify the subtraction. For example, one might solve 16 − 9 by either going up through 10 and thinking 9 + 1 = 10 and 10 + 6 = 16, so the *difference* is 7, or by going back through 10 by thinking 16 take away 6 is 10, and 10 take away 3 more is 7, so the difference is 7. Compare *making 10.*

-gon   A suffix meaning *angle.* For example, a *hexagon* is a plane figure with six angles.

grouping addends   An addition strategy that involves adding three or more numbers in an order that makes the addition simpler, such as recognizing and adding a combination of 10 or a doubles fact first. See *Associative* and *Commutative Properties of Addition.*

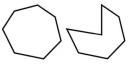

H

height   (1) The length of a perpendicular segment from one *side* of a geometric figure to a *parallel* side or from a *vertex* to the opposite side. (2) The line segment itself.

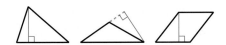

Height of 2-D figures are shown in red.

helper facts   Well-known facts used to derive unknown facts. Doubles and combinations of 10 are key addition/subtraction helper facts. For example, knowing the doubles fact 6 + 6 can help one derive 6 + 7 by thinking 6 + 6 = 12 and 1 more makes 13.

heptagon   A 7-sided *polygon.*

Heptagons

hexagon   A 6-sided *polygon.*

A hexagon

Home Link   In *Everyday Mathematics,* a suggested follow-up or enrichment activity to be done at home.

I

image   A figure that is produced by a transformation of another figure called the *preimage.*

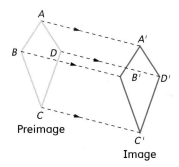

Preimage

Image

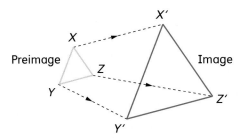

indirect comparison   A measurement comparison between two objects using a third object or *unit.* For example, if one knows that a wall is 8 feet tall and a person is 6 feet tall and one knows that 8 > 6, then one knows that the wall is taller than the person by comparing their heights to the length of a foot. See *transitivity principle of indirect measurement.* Compare *direct comparison.*

indirect measurement The determination, using mathematical relationships, of distances, angle measures, and other quantities that are not measured directly.

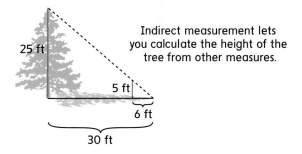

Indirect measurement lets you calculate the height of the tree from other measures.

25 ft

5 ft

6 ft

30 ft

inequality A *number sentence* with a *relation symbol* other than =, such as >, <, ≥, ≤, or ≠. Compare *equation*.

input (1) A number inserted into a *function machine*, which applies a rule to pair the input with an *output*. (2) Numbers or other information entered into a calculator or computer.

integer A number in the set {. . . , −4, −3, −2, −1, 0, 1, 2, 3, 4, . . .}. A *whole number* or its opposite, where 0 is its own opposite. Compare *rational numbers, irrational numbers,* and *real numbers*.

interior of a figure (1) The set of all points in a *plane* bounded by a *closed 2-dimensional* figure, such as a polygon or a circle. (2) The set of all points in space bounded by a closed *3-dimensional* figure, such as a polyhedron or a sphere. The interior is usually not considered to be part of the figure.

irrational numbers Numbers that cannot be written as *fractions,* where both the numerator and denominator are *integers* and the denominator is not zero. For example, $\sqrt{2}$ and π are irrational numbers. In *standard notation,* an irrational number can be written only as a nonterminating, nonrepeating decimal. The number 1.10100100010000. . . is irrational because its pattern does not repeat. Compare *rational numbers*.

isosceles trapezoid A *trapezoid* with a pair of *base angles* with the same measure. See *quadrilateral*.

Isosceles trapezoids

isosceles triangle A *triangle* with at least two *equal-length sides*. *Angles* opposite the equal-length sides are equal in measure.

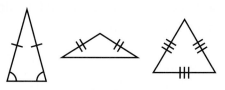

Isosceles triangles

iterate units To repeat a *unit* without gaps or overlaps in order to measure. To measure *length,* units are placed end-to-end along a path. Unit iteration can also be used to measure *area* (by tiling) or *volume* (by filling).

J

join situation A *change-to-more situation*.

K

key sequence The order in which calculator keys are pressed to perform a calculation.

kite A *quadrilateral* that has two nonoverlapping pairs of *adjacent, equal*-length *sides*.

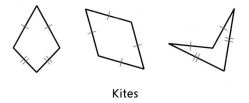

Kites

L

label (1) A descriptive word or phrase used to put a number or numbers in context. Labels encourage children to associate numbers with real objects. (2) In a spreadsheet, table, or graph, words or numbers providing information, such as the title of the spreadsheet, the heading for a row or column, or the variable on an axis.

length The distance between two points on a *1-dimensional* figure. For example, the figure might be a line segment, an arc, or a curve on a map modeling a hiking path. Length is measured in units, such as inches, kilometers, and miles.

line   A *1-dimensional* straight path that extends forever in opposite directions. A line is named using two points on it or with a single, italicized lowercase letter, such as *l*. In formal Euclidean geometry, line is an undefined geometric term.

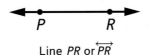

Line *PR* or $\overleftrightarrow{PR}$

line segment   A part of a line between and including two points called the endpoints of the segment. Same as *segment*. A line segment is often named by its endpoints.

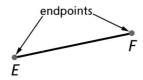

Segment *EF* or $\overline{EF}$

long   In *Everyday Mathematics*, the *base-10 block* that is equivalent to ten 1-cm cubes. Sometimes called a rod.

M

making 10   An addition fact strategy that involves *decomposing* one *addend* and adding part of it to another addend to make 10, then adding the remaining part to find the complete *sum*. For example, 9 + 5 can be solved by decomposing 5 into 1 and 4 and thinking 9 + 1 = 10 and 4 more makes 14. Compare *going through 10*.

Math Boxes   In *Everyday Mathematics*, a collection of problems to practice skills. Math Boxes for each lesson are in the *Math Journal*.

Math Message   In *Everyday Mathematics*, an introduction to the day's lesson designed for children to complete independently.

measurement unit   The reference unit used when measuring. Examples of basic units include inches for length, grams for mass or weight, cubic inches for volume or capacity, seconds for elapsed time, and degrees Celsius for change of temperature. Compound units include square centimeters for area and miles per hour for speed.

memory in a calculator   Where numbers are stored in a calculator for use in later calculations. Most calculators have both short-term memory and long-term memory.

mental arithmetic   Computation done by people "in their heads," either in whole or in part. In *Everyday Mathematics*, children learn a variety of mental-calculation strategies as they develop *automaticity* with basic facts and fact power.

Mental Math and Fluency   In *Everyday Mathematics*, short, leveled exercises presented at the beginning of lessons. Mental Math and Fluency problems prepare children to think about math, warm up skills they need for the lesson, and build mental-arithmetic skills. They also help teachers assess individual strengths and weaknesses.

metric system   The measurement system used in most countries and by virtually all scientists around the world. *Units* within the metric system are related by powers of 10. Units for length include millimeter, centimeter, meter, and kilometer; units for mass and weight include gram and kilogram; units for volume and capacity include milliliter and liter; and the unit for temperature change is degrees Celsius. See *Tables of Measures*.

minuend   In subtraction, the number from which another number is subtracted. For example, in 19 − 5 = 14, the minuend is 19. Compare *subtrahend* and *difference*.

missing addend   An *addend* that is *unknown* within an addition *equation*. A subtraction problem can be represented by an addition number sentence with a missing addend. See *unknown*.

model   A mathematical representation or description of an object or a situation. For example, 60 × 3 can be a model for how much money is needed to buy 3 items that cost 60 cents each. A circle can be a model for the rim of a wheel. See *represent*.

N

name-collection box   In *Everyday Mathematics*, a diagram that is used for collecting *equivalent names* for a number.

25
37 − 12
20 + 5
̶H̶I̶I̶ ̶H̶I̶I̶ ̶H̶I̶I̶ ̶H̶I̶I̶ ̶H̶I̶I̶
twenty-five
veinticinco

natural numbers   Same as *counting numbers*.

near doubles   An addition fact strategy that involves relating a given fact to a nearby *doubles fact* to help solve the fact. For example, 7 + 8 can be solved by thinking 7 + 7 = 14; then 14 + 1 = 15. It can also be solved by thinking 8 + 8 = 16; then 16 − 1 = 15.

negative numbers   Numbers less than 0; the opposites of the *positive numbers,* commonly written as a positive number preceded by a −. Negative numbers are plotted left of 0 on a horizontal number line or below 0 on a vertical number line.

n-gon   A *polygon,* where *n* is the number of *sides*. Polygons that do not have special names are usually named using *n*-gon notation, such as 13-gon or 100-gon.

nonagon   A 9-sided *polygon*.

nonconvex polygon   Same as *concave polygon*.

nondefining attributes of a shape   Characteristics of a shape that are not required by the definition of the shape. Size, color, and orientation are nondefining *attributes* of shapes. Compare *defining attributes of a shape*.

number-and-word notation   A notation consisting of the significant *digits* of a number and words for the *place value*. For example, 27 billion is number-and-word notation for 27,000,000,000. Compare *standard notation*.

number family   Same as *fact family*.

number grid   A table in which *consecutive* numbers are arranged in *rows,* usually 10 *columns* per row. A move from one number to the next within a row is a change of 1; a move from one number to the next within a column is a change of 10.

−9	−8	−7	−6	−5	−4	−3	−2	−1	0
1	2	3	4	5	6	7	8	9	10
11	12	13	14	15	16	17	18	19	20
21	22	23	24	25	26	27	28	29	30
31	32	33	34	35	36	37	38	39	40
41	42	43	44	45	46	47	48	49	50
51	52	53	54	55	56	57	58	59	60
61	62	63	64	65	66	67	68	69	70
71	72	73	74	75	76	77	78	79	80
81	82	83	84	85	86	87	88	89	90
91	92	93	94	95	96	97	98	99	100
101	102	103	104	105	106	107	108	109	110

A number grid

number-grid puzzle   In *Everyday Mathematics*, a piece of a *number grid* in which some but not all of the numbers are missing. Children use number-grid puzzles to practice *place-value* concepts.

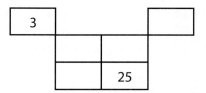

A number-grid puzzle

number line   A *line* on which points are indicated by *tick marks* that are usually at regularly spaced intervals from a starting point called the *origin,* the zero point, or simply 0. Numbers are associated with the tick marks on a *scale* defined by the unit interval from 0 to 1. Every real number locates a point on the line, and every point corresponds to a real number. See *real numbers*.

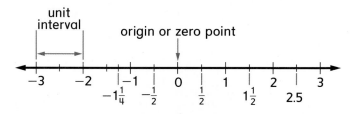

A number line

number model   A *number sentence, expression,* or other representation that fits a *number story* or situation. For example, the number story "Sally had $5, and then she earned $8." can be *modeled* as the number sentence $5 + 8 = 13$, as the expression $5 + 8$, or by

$$\begin{array}{r} 5 \\ + \ 8 \\ \hline 13 \end{array}$$

number scroll   In *Everyday Mathematics,* a series of number grids taped together.

number sentence   Two *expressions* linked by a *relation symbol,* such as =, <, or >.

$$5 + 5 = 10 \qquad 16 \leq a \times b$$
$$2 - ? = 8 \qquad a^2 + b^2 = c^2$$

Number sentences

number sequence   A list of numbers, often generated by a rule. In *Everyday Mathematics,* children explore number sequences using *Frames-and-Arrows* diagrams.

$$1, 2, 3, 4, 5, \ldots \quad 1, 4, 9, 16, 25, \ldots$$

$$1, 2, 1, 2, 1, \ldots \quad 1, 3, 5, 7, 9, \ldots$$

Number sequences

number story   A story that involves numbers and one or more explicit or implicit questions. For example, "I have 7 crayons in my desk; Carrie gave me 8 more crayons" is a number story.

numeral   (1) A combination of *base-10 digits* used to express a number. (2) A word, symbol, or figure that represents a number. For example, six, VI, ̶H̶H̶ /, and 6 are all numerals that represent the same number.

numeration   A method of numbering or of reading and writing numbers. In *Everyday Mathematics,* numeration activities include counting, writing numbers, identifying *equivalent names* for numbers in *name-collection boxes,* exchanging coins such as 5 pennies for 1 nickel, and renaming numbers in computation.

octagon   An 8-sided *polygon.*

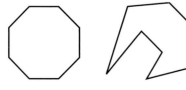

Octagons

octahedron   A *polyhedron* with 8 *faces.* An octahedron with 8 *equilateral triangle* faces is one of the five *regular polyhedrons.*

odd number   (1) A *counting number* that is not divisible by 2. (2) An *integer* that is not divisible by 2. Compare *even number.*

open   Not closed. An open shape has no inside. What might seem like the inside is not separated from what seems like the outside. Compare *closed.*

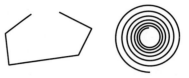

Open figures

open sentence   A *number sentence* with one or more variables that is neither true nor false. For example, $9 + \underline{\quad} = 15$, $? - 24 < 10$, and $7 = x + y$ are open sentences. See *variable* and *unknown.*

operation   An action performed on one or more mathematical objects, such as numbers, *variables,* or *expressions,* to produce another mathematical object. Addition, subtraction, multiplication, and division are the four basic arithmetic operations. Taking a square root, squaring a number, and multiplying both sides of an equation by the same number are also operations. In *Everyday Mathematics,* children learn about many operations along with procedures, or algorithms, for carrying them out.

operation symbol   A symbol used in *expressions* and *number sentences* to stand for a particular mathematical operation. Symbols for common arithmetic operations are addition +; subtraction −; multiplication ×, *, •; division ÷, /; powering ^. See *General Reference.*

order   To arrange according to a specific rule; for example, from smallest to largest, or from largest to smallest. See *sequence.*

ordered pair (1) Two numbers, or coordinates, used to locate a point on a rectangular coordinate grid. The first coordinate *x* gives the position along the horizontal axis of the grid, and the second coordinate *y* gives the position along the vertical axis. The pair is written (*x*, *y*). (2) Any pair of objects or numbers in a particular order, as in letter-number spreadsheet cell names, map coordinates, or functions given as sets of pairs of numbers.

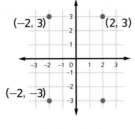

Ordered pairs

ordinal number A number describing the position or order of something in a *sequence,* such as first, third, or tenth. Ordinal numbers are commonly used in dates, as in "May fifth." Compare *cardinal number.*

origin The zero point in a coordinate system. On a *number line,* the origin is the point at 0. On a coordinate grid, the origin is the point (0, 0) where the two axes intersect.

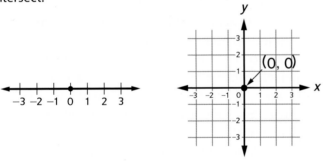

The points at 0 and (0, 0) are origins.

outcome A possible result of a *chance* experiment or situation. For example, HEADS and TAILS are the two possible outcomes of flipping a coin.

output (1) A number paired to an input by an imaginary function machine applying a rule. (2) The values for *y* in a function consisting of ordered pairs (*x*, *y*). (3) Numbers or other information displayed or produced by a calculator or computer.

pairs of numbers that add to 10 Pairs of *whole numbers* that have a *sum* of 10. For example, 4 + 6 and 3 + 7 are both pairs of numbers that add to 10. Pairs that add to 10 are key *helper facts* for addition- and subtraction-fact strategies.

pan balance A measuring device used to weigh objects or compare *weights* or masses. Simple pan balances have two pans suspended at opposite ends of a bar resting on a fulcrum at its midpoint. When the weights or masses of the objects in the pans are *equal,* the bar is level.

Pan balance

parallel *Lines, line segments,* or rays in the same *plane* are parallel if they never cross or meet, no matter how far they are extended. Two planes are parallel if they never cross or meet. A line and a plane are parallel if they never cross or meet. The symbol ∥ means is parallel to.

parallelogram A *trapezoid* that has two pairs of *parallel* sides. See *quadrilateral.*

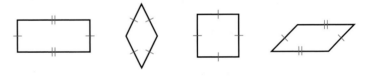

Parallelograms

parts-and-total diagram In *Everyday Mathematics,* a diagram to *model* situations in which two or more quantities (parts) are combined to make a total quantity. See *situation diagram.*

Total	
13	
Part	**Part**
8	?

Parts-and-total diagram for 13 = 8 + ?

parts-and-total situation A situation in which a quantity is made up of two or more distinct parts. For example, the following is a parts-and-total situation: "There are 15 girls and 12 boys in Mrs. Dorn's class. How many children are there in all?" See *addition/subtraction use class.*

pattern   A repetitive order or arrangement. In *Everyday Mathematics*, children mainly explore visual and number patterns in which elements are arranged so that what comes next can be predicted.

pattern blocks   A standard set of *polygon*-shaped blocks used in geometry activities. Compare *attribute blocks*.

Pattern blocks

pentagon   A 5-sided *polygon*.

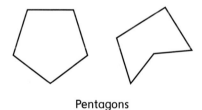

Pentagons

per   For each, as in ten chairs per row or six tickets per family.

picture graph   A graph constructed with icons representing *data* points. They are sometimes called scaled picture graphs when each icon represents more than 1 data point.

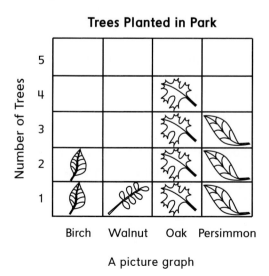

A picture graph

place value   A system that gives a *digit* a value according to its position, or place, in a number. In the standard *base-ten* (decimal) system for writing numbers, each place has a value 10 times that of the place to its right and one-tenth the value of the place to its left.

thousands	hundreds	tens	ones	.	tenths	hundredths

A place-value chart

plane   A *2-dimensional* flat *surface* that extends forever in all directions. In formal Euclidean geometry, plane is an undefined geometric term.

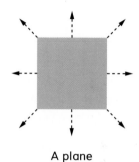

A plane

plane figure   A *set* of points that is entirely contained in a single plane. For example, squares, pentagons, circles, parabolas, lines, and rays are plane figures; cones, cubes, and prisms are not.

P.M.   The abbreviation for *post meridiem,* meaning "after the middle of the day" in Latin. From noon to midnight.

poly-   A prefix meaning many. See *General Reference, Prefixes* for specific numerical prefixes.

polygon   A plane figure formed by three or more *line segments* (*sides*) that meet only at their endpoints (*vertices*) to make a closed path. The sides may not cross one another.

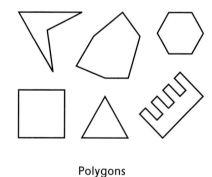

Polygons

Glossary   **G17**

polyhedron A *closed 3-dimensional* figure formed by *polygons* with their *interiors* (*faces*) that do not cross. The plural is polyhedrons or polyhedra.

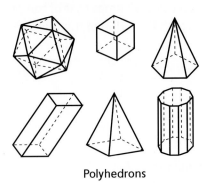

Polyhedrons

positive numbers Numbers greater than 0; the opposites of the *negative numbers*. Positive numbers are plotted to the right of 0 on a horizontal *number line* or above 0 on a vertical number line.

poster In *Everyday Mathematics*, a page displaying a collection of illustrated numerical *data*. A poster may be used as a source of data for developing *number stories*.

precise Of a measurement or other quantity, having a high degree of exactness. A measurement to the nearest inch is more precise than a measurement to the nearest foot. A measurement's precision depends on the *unit* scale of the *tool* used to obtain it. The smaller the unit is, the more precise a measure can be. For instance, a ruler with $\frac{1}{8}$-inch markings can give a more precise measurement than a ruler with $\frac{1}{2}$-inch markings. Compare *accurate*.

preimage The original figure in a transformation. Compare *image*.

prism A polyhedron with two *parallel* and *congruent polygonal* bases and lateral *faces* shaped like *parallelograms*. Right prisms have rectangular lateral faces. Prisms get their names from the shapes of their *bases*.

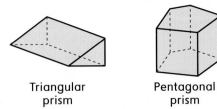

Triangular prism   Pentagonal prism

property (1) A generalized statement about a mathematical relationship, such as the Distributive Property of Multiplication over Addition. (2) A feature of an object or a common feature of a set of objects. Same as *attribute*.

put-together/take-apart situation A situation in which a quantity is made up of two or more distinct parts. See *addition/subtraction use class*.

pyramid A polyhedron with a polygonal *base* and triangular other *faces* that meet at a common *vertex* called the *apex*. Pyramids get their names from the shapes of their bases.

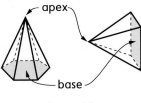

Pyramids

Q

quadrangle Same as *quadrilateral*.

quadrilateral A 4-sided *polygon*. *Squares, rectangles, parallelograms, rhombuses, kites,* and *trapezoids* are organized by *defining attributes* into a hierarchy of shapes.

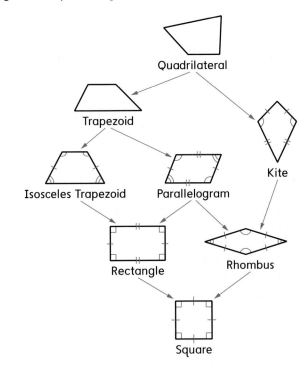

Hierarchy of quadrilaterals

Quick Looks An *Everyday Mathematics* routine in which an image of a quantity is displayed for 2–3 seconds and then removed. Quick Looks encourage children to *subitize, compose,* and *decompose* numbers in flexible ways and also develop strategies for *addition facts*.

R

rational counting   Counting using one-to-one matching. For example, counting a number of chairs, people, or crackers. In rational counting, the last number gives the *cardinality* of the *set*.

rational numbers   Numbers that can be written in the form $\frac{a}{b}$, where *a* and *b* are *integers* and $b \neq 0$. The *decimal* form of a rational number either terminates or repeats. For example, $\frac{2}{3}$, $-\frac{2}{3}$, 0.5, 20.5, and 0.333. . . are rational numbers.

real numbers   All *rational* and *irrational numbers;* all numbers that can be written as *decimals*. For every real number there is a corresponding point on a *number line,* and for every point on the number line, there is a real number.

rectangle   A *parallelogram* with four *right angles*. All rectangles are both parallelograms and *isosceles trapezoids*. See *quadrilateral*.

A rectangle

rectangular prism   A *prism* with rectangular *bases*. The four *faces* that are not bases are formed by rectangles or *parallelograms*. For example, a rectangular prism in which all sides are rectangular models a shoebox.

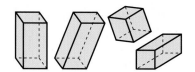

Rectangular prisms

rectangular pyramid   A *pyramid* with a rectangular *base*.

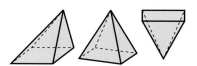

Rectangular pyramids

reference frame   A system for locating numbers within a given context, usually with reference to an *origin* or a zero point. For example, *number lines,* clocks, *calendars,* temperature scales, and maps are reference frames.

regular polygon   A *polygon* in which all *sides* are the same *length* and all interior *angles* have the same measure.

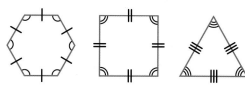

Regular polygons

regular polyhedron   A *polyhedron* whose *faces* are all formed by *congruent regular polygons* and in which the same number of faces meet at each *vertex*. There are only five. They are called the Platonic solids.

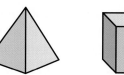

Tetrahedron	Cube	Octahedron
(4 equilateral triangles)	(6 squares)	(8 equilateral triangles)

Dodecahedron	Icosahedron
(12 regular pentagons)	(20 equilateral triangles)

relation symbol   A symbol used to express a relationship between two quantities, figures, or *sets,* such as $\leq$, $\parallel$, or $\subset$. See *General Reference, Symbols*.

repeating decimal   A *decimal* in which one digit or block of digits is repeated without end. For example, 0.3333. . . and $0.\overline{147}$ are repeating decimals. Compare *terminating decimal*.

represent   To show, symbolize, or stand for something. For example, numbers can be represented using base-10 blocks, spoken words, or written numerals. See *model*.

rhombus   A *parallelogram* with four sides of the same length. All rhombuses are both parallelograms and *kites*. See *quadrilateral*.

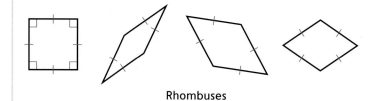

Rhombuses

right angle A 90° *angle*.

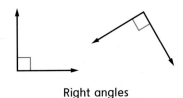

Right angles

rotation (1) A turn about an axis or point. (2) Point *P′* is a rotation *image* of point *P* around a center of rotation *C* if *P′* is on the *circle* with center *C* and radius *CP*. If all the points in one figure are rotation images of all the points in another figure around the same center of rotation and with the same *angle* of rotation, then the figures are rotation images. The center can be inside or outside of the original image. Reflections, rotations, and translations are types of isometry transformations. (3) If all points on the image of a 3-dimensional figure are rotation images through the same angle around a point or a line called the axis of rotation, then the image is a rotation image of the original figure.

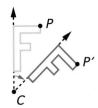

Rotation preimage and image

rote counting Reciting a string of number words by rote, without necessarily understanding their significance. See *skip counting*.

round (1) To approximate a number to make it easier to use or make it better reflect the *precision* of the *data*. "Rounding up" means to approximate larger than the actual value. "Rounding down" means to approximate smaller than the actual value. (2) Circular in shape.

row (1) A horizontal arrangement of objects or numbers in an array or a table. (2) A horizontal section of cells in a spreadsheet.

ruler (1) Traditionally, a wood, metal, or plastic strip partitioned into same-size standard *units* such as inches or centimeters. (2) In *Everyday Mathematics*, a tool for measuring length comprising an end-to-end collection of same-size units.

S

scale of a number line The unit interval on a *number line* or a measuring device. The scales on this ruler are 1 millimeter on the left side and $\frac{1}{16}$ inch on the right side.

Scale of a number line

scalene triangle A *triangle* with *sides* of three different lengths. The three *angles* of a scalene triangle have different measures.

segment Same as line segment.

separate situation A *change-to-less situation* or a *parts-and-total situation*.

sequence An ordered list of numbers, often with an underlying rule that may be used to generate subsequent numbers in the list. *Frames-and-Arrows* diagrams can be used to represent sequences. See *order*.

set A collection or group of objects, numbers, or other items.

short-term memory Memory in a calculator used to store values for immediate calculation. Short-term memory is usually cleared with a ⓒ, ⒶⒸ, Clear, or a similar key.

side (1) One of the *line segments* that make up a *polygon*. (2) One of the rays or segments that form an *angle*. (3) One of the *faces* of a *polyhedron*.

situation diagram In *Everyday Mathematics*, a diagram used to organize information in a problem situation in one of the addition/subtraction or multiplication/division use classes.

Total	
7	
Part	Part
2	5

Susie has 2 pink balloons and 5 yellow balloons.
She has 7 balloons in all.

skip counting Counting by intervals, such as 2s, 5s, or 10s. See *rote counting*.

slate In *Everyday Mathematics,* a lap-size (about 8-inch by 11-inch) chalkboard or whiteboard that children use for recording responses during group exercises and informal group assessments.

solution of a problem (1) The answer to a problem. (2) The answer to a problem together with the method by which that answer was obtained.

solution of an open sentence A value or values for the *variable*(s) in an open sentence that make the sentence true. For example, 7 is a solution of $5 + n = 12$. Although *equations* are not necessarily open sentences, the solution of an open sentence is commonly referred to as a solution of an equation.

sphere The set of all points in space that are an equal distance from a fixed point called the center of the sphere. The distance from the center to the sphere is the radius of the sphere. The diameter of a sphere is twice its radius. Points inside a sphere are not part of the sphere.

A sphere

Spiral Snapshot In *Everyday Mathematics*, an overview of nearby lessons that addresses one of the Goals for Mathematical Content in the Focus part of the lesson. It appears in the Lesson Opener.

Spiral Trace In *Everyday Mathematics*, an overview of work in the current unit and nearby units on selected Standards for Mathematical Content. It appears in the Unit Organizer.

Spiral Tracker In *Everyday Mathematics*, an online database that shows complete details about learning trajectories for all goals and standards.

square A *rectangle* with four sides of equal length. All squares are both rectangles and *rhombuses*. See *quadrilateral*.

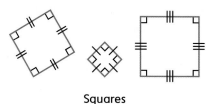

Squares

square corner Same as *right angle*.

square pyramid A *pyramid* with a square *base*.

square unit A *unit* to measure *area*. A model of a square unit is a square with each side a related unit of length. For example, a square inch is the area of a square with 1-inch sides. Square units are often labeled as the length unit squared. For example, 1 cm² is read "1 square centimeter" or "1 centimeter squared."

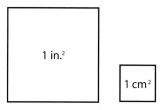

Square units

standard notation The most common way of representing whole numbers, integers, and decimals. Standard notation for real numbers is *base-ten place-value* numeration. For example, standard notation for three hundred fifty-six is 356. Same as *decimal notation*. Compare *number-and-word notation*.

standard unit A *unit* of measure that has been defined by a recognized authority, such as a government or the National Institute of Standards and Technology. For example, inches, meters, miles, seconds, pounds, grams, and acres are all standard units.

subitize To recognize a quantity without needing to count. For example, many children can instantly recognize the number of dots on a rolled die without needing to count them. In *Everyday Mathematics, Quick Looks* help children develop the ability to subitize and use relationships between quantities to find more complex totals.

subtrahend The number being taken away in a subtraction problem. For example, in $15 - 5 = 10$, the subtrahend is 5. Compare *difference* and *minuend*.

sum The result of adding two or more numbers. For example, in $5 + 3 = 8$, the sum is 8.

surface (1) The boundary of a *3-dimensional* object. (2) Any *2-dimensional* layer, such as a *plane* or a *face* of a *polyhedron*.

survey (1) A study that collects *data* by asking people questions. (2) Any study that collects data. Surveys are commonly used to study demographics, such as people's characteristics, behaviors, interests, and opinions.

T

take-apart situation   See *put-together/take-apart situation*.

take-from situation   A *change-to-less situation*.

tally   (1) To keep a record of a count by making a mark for each item as it is counted. (2) The mark used in a count. Also called tally mark.

tally chart   A table to keep track of tallies, typically showing how many times each value appears in a set of data.

Number of Pull-Ups	Number of Children
0	HHT I
1	HHT
2	IIII
3	II

A tally chart

teen number   The teen numbers are 11, 12, 13, 14, 15, 16, 17, 18, and 19. Sometimes 10 is considered a teen number.

temperature   How hot or cold something is relative to another object or as measured on a standardized scale, such as *degrees Celsius* or *degrees Fahrenheit*.

template   In *Everyday Mathematics*, a sheet of plastic with geometric shapes cut out of it, used to draw patterns and designs.

ten frame   A 5-by-2 grid of squares that can be used to represent the numbers 0–10 in a variety of ways. Ten frames are particularly useful for encouraging children to relate given representations to 5 or 10.

Two representations of 6 on ten frames

terminating decimal   A *decimal* that ends. For example, 0.5 and 0.125 are terminating decimals.

thermometer   A tool used to measure *temperature* in *degrees* according to a fixed *scale*. The most common scales are *Celsius* and *Fahrenheit*.

think addition   A subtraction-fact strategy that involves using an addition fact in the same *fact family* to help solve the given subtraction problem. For example, for $10 - 4 = ?$, one might think $4 + ? = 10$ and use $4 + 6 = 10$ to determine that the *difference* is 6. This strategy is particularly useful with doubles or combinations of 10 as *helper facts*.

tick marks   (1) Marks showing the *scale of a number* line or ruler. (2) Marks indicating that two *line segments* have the same *length*. (3) Same as *tally*.

timeline   A number line showing when events took place. In some timelines the origin is based on the context of the events being graphed, such as the birth date of the child's life graphed below. The origin can also come from another reference system, such as the year CE, in which case the scale below might cover the years 2015 through 2020.

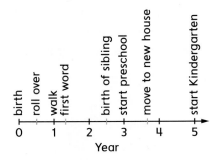

A timeline of a child's milestones

tool   Anything, physical or abstract, that serves as an instrument for performing a task. Physical tools include hammers for hammering, calculators for calculating, and rulers for measuring. Abstract tools include computational algorithms such as partial-sums addition, problem-solving strategies such as "guess and check," and technical drawings such as *situation diagrams*.

Total Physical Response   A teaching technique that facilitates beginning English language learners' acquisition of new English vocabulary through modeling of physical actions or display of visuals as target objects that are named aloud.

transitivity principle of indirect measurement   The property that provides for *indirect comparison* of measures of objects, such as length or weight. Given three measures A, B, and C, if $A > B$ and $B > C$, then $A > C$. See *indirect comparison*.

trapezoid A *quadrilateral* that has at least one pair of *parallel* sides.

Trapezoids

trial-and-error method A systematic method for finding the solution of an equation by trying a sequence of test numbers.

triangle A 3-sided *polygon*.

Triangles

triangular prism A *prism* whose *bases* are triangular.

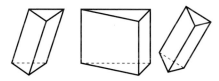

Triangular prisms

triangular pyramid A *pyramid* in which all *faces* are triangular, any one of which is the *base;* also known as a tetrahedron. A regular tetrahedron has four faces formed by *equilateral triangles* and is one of the five regular *polyhedrons.*

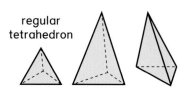

regular
tetrahedron

Triangular pyramids

true number sentence A *number sentence* stating a correct fact. For example, $75 = 25 + 50$ is a true number sentence. Compare *false number sentence.*

turn-around rule A rule for solving addition and multiplication problems based on the *Commutative Properties of Addition and Multiplication.* For example, if one knows that $6 \times 8 = 48$, then, by the turn-around rule, one also knows that $8 \times 6 = 48$.

U

unit A label used to put a number in context. In measuring length, for example, inches and centimeters are units. In a problem about 5 apples, apple is the unit. In *Everyday Mathematics*, children keep track of units in unit boxes.

unit box In *Everyday Mathematics*, a box displaying the unit for the numbers in the problems at hand.

Unit
days

A unit box

unknown A quantity whose value is not known. An unknown is sometimes represented by a _____, a ?, or a letter. See *open sentence* and *variable.*

U.S. customary system The measuring system used most often in the United States. Units for length include inch, foot, yard, and mile; units for weight include ounce and pound; units for volume or capacity include fluid ounce, cup, pint, quart, gallon, and cubic units; and the unit for temperature is degrees Fahrenheit. See *Tables of Measures.*

V

variable (1) A letter or other symbol that can be replaced by any value from a set of possible values. Some values replacing variables in *number sentences* may make them true. For example, to make number sentences true, variables may be replaced by a single number, as in $5 + n = 9$, where $n = 4$ makes the sentence true; many different numbers, as in $x + 2 < 10$, where any number less than 8 makes the sentence true; or any number, as in $a + 3 = 3 + a$, which is true for all numbers. See *open sentence* and *unknown.* (2) A number or data set that can have many values is a variable.

Venn diagram   A picture that uses circles or rings to show relationships between *sets*. In this diagram, $22 + 8 = 30$ girls are on the track team, and 8 are on both the track and the basketball teams.

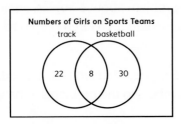

Numbers of Girls on Sports Teams

A Venn diagram

vertex   The point at which the *sides* of an *angle* or a polygon, or the *edges* of a *polyhedron*, meet. The plural is vertexes or vertices. See *corner*.

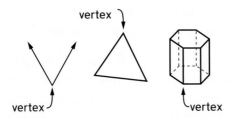

volume   (1) A measure of how much *3-dimensional* space something occupies. Volume is often measured in liquid units, such as gallons or liters, or cubic units, such as cm³ or cubic inches. (2) Less formally, the same as capacity: the amount a container can hold.

W

weight   A measure of how heavy something is. *Everyday Mathematics* does not distinguish weight and mass because the terms and the corresponding units are used interchangeably in everyday life. (2) The force of gravity on an object. An object's mass is constant, but it weighs less in weak gravity than in strong gravity.

"What's My Rule?"   A problem in which two of the three parts of a *function* (input, output, and rule) are known and the third is to be found out. See *function*.

whole   An entire object, collection of objects, or quantity being considered in a problem situation; 100%. Can also be called ONE or unit whole.

whole numbers   The *counting numbers* and 0. The set of whole numbers is {0, 1, 2, 3, …}.

General Reference

Symbols

$+$	plus or positive
$-$	minus or negative
$*, \times$	multiplied by
$\div, /$	divided by
$=$	is equal to
$\neq$	is not equal to
$<$	is less than
$>$	is greater than
$\leq$	is less than or equal to
$\geq$	is greater than or equal to
$\approx$	is approximately equal to
$x^n, x^\wedge n$	nth power of x
$\sqrt{x}$	square root of x
$\%$	percent
$a{:}b, a/b, \frac{a}{b}$	ratio of a to b or a divided by b or the fraction $\frac{a}{b}$
$a\,[bs]$	a groups, b in each group
$n/d \longrightarrow a\ \mathrm{R}b$	n divided by d is a with remainder b
$\{\,\}, (\,), [\,]$	grouping symbols
∞	infinity
$n!$	n factorial
$^\circ$	degree
(a, b)	ordered pair
$\overleftrightarrow{AS}$	line AS
$\overline{AS}$	line segment AS
$\overrightarrow{AS}$	ray AS
$\llcorner$	right angle
$\perp$	is perpendicular to
$\parallel$	is parallel to
$\triangle ABC$	triangle ABC
$\angle ABC$	angle ABC
$\angle B$	angle B
$\cong$	is congruent to
$\sim$	is similar to
$\equiv$	is equivalent to
$|n|$	absolute value of n

Prefixes

Prefix	Meaning	Prefix	Meaning
uni-	one	tera-	trillion (10^{12})
bi-	two	giga-	billion (10^{9})
tri-	three	mega-	million (10^{6})
quad-	four	kilo-	thousand (10^{3})
penta-	five	hecto-	hundred (10^{2})
hexa-	six	deca-	ten (10^{1})
hepta-	seven	uni-	one (10^{0})
octa-	eight	deci-	tenth (10^{-1})
nona-	nine	centi-	hundredth (10^{-2})
deca-	ten	milli-	thousandth (10^{-3})
dodeca-	twelve	micro-	millionth (10^{-6})
icosa-	twenty	nano-	billionth (10^{-9})

Constants

Constants	Approximations
Pi (π)	3.14159 26535 89793
Golden Ratio (ϕ)	1.61803 39887 49894
Radius of Earth at equator	6,378.388 kilometers 3,963.34 miles
Circumference of Earth at equator	40,076.59 kilometers 24,902.44 miles
Velocity of sound in dry air at 0°C	331.36 m/sec 1087.1 ft/sec
Velocity of light in a vacuum	2.997925×10^{10} cm/sec

The Order of Operations

1. Do operations inside grouping symbols following Rules 2–4. Work from the innermost set of grouping symbols outward.
2. Calculate all expressions with exponents or roots.
3. Multiply and divide in order from left to right.
4. Add and subtract in order from left to right.

Tables of Measures

Metric System

Units of Length

1 kilometer (km)	= 1,000 meters (m)
1 meter	= 10 decimeters (dm)
	= 100 centimeters (cm)
	= 1,000 millimeters (mm)
1 decimeter	= 10 centimeters
1 centimeter	= 10 millimeters

Units of Area

1 square meter (m^2)	= 100 square decimeters (dm^2)
	= 10,000 square centimeters (cm^2)
1 square decimeter	= 100 square centimeters
1 are (a)	= 100 square meters
1 hectare (ha)	= 100 ares
1 square kilometer (km^2)	= 100 hectares

Units of Volume and Capacity

1 cubic meter (m^3)	= 1,000 cubic decimeters (dm^3)
	= 1,000,000 cubic centimeters (cm^3)
1 cubic centimeter	= 1,000 cubic millimeters (mm^3)
1 kiloliter (kL)	= 1,000 liters (L)
1 liter	= 1,000 milliliters (mL)

Units of Mass and Weight

1 metric ton (t)	= 1,000 kilograms (kg)
1 kilogram	= 1,000 grams (g)
1 gram	= 1,000 milligrams (mg)

U.S. Customary System

Units of Length

1 mile (mi)	= 1,760 yards (yd)
	= 5,280 feet (ft)
1 yard	= 3 feet
	= 36 inches (in.)
1 foot	= 12 inches

Units of Area

1 square yard (yd^2)	= 9 square feet (ft^2)
	= 1,296 square inches ($in.^2$)
1 square foot	= 144 square inches
1 acre	= 43,560 square feet
1 square mile (mi^2)	= 640 acres

Units of Volume and Capacity

1 cubic yard (yd^3)	= 27 cubic feet (ft^3)
1 cubic foot	= 1,728 cubic inches ($in.^3$)
1 gallon (gal)	= 4 quarts (qt)
1 quart	= 2 pints (pt)
1 pint	= 2 cups (c)
1 cup	= 8 fluid ounces (fl oz)
1 fluid ounce	= 2 tablespoons (tbs)
1 tablespoon	= 3 teaspoons (tsp)

Units of Mass and Weight

1 ton (T)	= 2,000 pounds (lb)
1 pound	= 16 ounces (oz)

System Equivalents (Conversion Factors)

1 inch ≈ 2.5 cm (2.54)	1 liter ≈ 1.1 quarts (1.057)
1 kilometer ≈ 0.6 mile (0.621)	1 ounce ≈ 28 grams (28.350)
1 mile ≈ 1.6 kilometers (1.609)	1 kilogram ≈ 2.2 pounds (2.21)
1 meter ≈ 39 inches (39.37)	1 hectare ≈ 2.5 acres (2.47)

Body Measures

1 *digit* is about the width of a finger.

1 *hand* is about the width of the palm and thumb.

1 *span* is about the distance from the tip of the thumb to the tip of the first (index) finger of an outstretched hand.

1 *cubit* is about the length from the elbow to the tip of the extended middle finger.

1 *yard* is about the distance from the center of the chest to the tip of the extended middle finger of an outstretched arm.

1 *fathom* is about the length from fingertip to fingertip of outstretched arms. Also called an arm span.

Units of Time

1 century	= 100 years
1 decade	= 10 years
1 year (yr)	= 12 months
	= 52 weeks (plus one or two days)
	= 365 days (366 days in a leap year)
1 month (mo)	= 28, 29, 30, or 31 days
1 week (wk)	= 7 days
1 day (d)	= 24 hours
1 hour (hr)	= 60 minutes
1 minute (min)	= 60 seconds (s or sec)

Unpacking the Common Core State Standards

The **Common Core State Standards** include two groups of standards: Standards for Mathematical Content and Standards for Mathematical Practice. The Content Standards define the mathematical content to be mastered at each grade. The Practice Standards define the processes and habits of mind students need to develop as they learn the content for their grade level.

The Content Standards are organized into **Domains**, large groups of related standards. Within each Domain are **Clusters**, smaller groups of related standards. The **Standards** themselves define what students should understand and be able to do by the end of that grade.

The summary page from the Common Core State Standards for Grade 1 is on the following page. This page summarizes the mathematical content that children should learn in Grade 1.

The chart beginning on page EM3 lists the Common Core State Standards for Mathematical Content with corresponding *Everyday Mathematics* Goals for Mathematical Content (GMC).

The chart beginning on page EM6 lists the Common Core State Standards for Mathematical Practice with corresponding *Everyday Mathematics* Goals for Mathematical Practice (GMP).

Common Core State Standards

Standards for Mathematical Content

Domain Operations and Algebraic Thinking 1.OA	*Everyday Mathematics* Goals for Mathematical Content
Cluster Represent and solve problems involving addition and subtraction.	
1.OA.1 Use addition and subtraction within 20 to solve word problems involving situations of adding to, taking from, putting together, taking apart, and comparing, with unknowns in all positions, e.g., by using objects, drawings, and equations with a symbol for the unknown number to represent the problem.	**GMC** Solving number stories by adding and subtracting. **GMC** Model parts-and-total, change, and comparison situations.

"…larger groups of related standards. Standards from different domains may… be… related."

"…groups of related standards. Standards from different clusters may… be… related, because mathematics is a connected subject."

the first standard under this Domain

program goals for finer-grained tracking of student progress

The table on pages EM6–EM9 may be used to trace the *Everyday Mathematics* Goals for Mathematical Practice as they unpack the Common Core Standards for Mathematical Practice.

The Grade 1 Content Standards are introduced in the CCSS document on page 13, as follows:

In Grade 1, instructional time should focus on four critical areas: (1) developing understanding of addition, subtraction, and strategies for addition and subtraction within 20; (2) developing understanding of whole number relationships and place value, including grouping in tens and ones; (3) developing understanding of linear measurement and measuring lengths as iterating length units; and (4) reasoning about attributes of, and composing and decomposing geometric shapes.

(1) Students develop strategies for adding and subtracting whole numbers based on their prior work with small numbers. They use a variety of models, including discrete objects and length-based models (e.g., cubes connected to form lengths), to model add-to, take-from, put-together, take-apart, and compare situations to develop meaning for the operations of addition and subtraction, and to develop strategies to solve arithmetic problems with these operations. Students understand connections between counting and addition and subtraction (e.g., adding two is the same as counting on two). They use properties of addition to add whole numbers and to create and use increasingly sophisticated strategies based on these properties (e.g., "making tens") to solve addition and subtraction problems within 20. By comparing a variety of solution strategies, children build their understanding of the relationship between addition and subtraction.

(2) Students develop, discuss, and use efficient, accurate, and generalizable methods to add within 100 and subtract multiples of 10. They compare whole numbers (at least to 100) to develop understanding of and solve problems involving their relative sizes. They think of whole numbers between 10 and 100 in terms of tens and ones (especially recognizing the numbers 11 to 19 as composed of a ten and some ones). Through activities that build number sense, they understand the order of the counting numbers and their relative magnitudes.

(3) Students develop an understanding of the meaning and processes of measurement, including underlying concepts such as iterating (the mental activity of building up the length of an object with equal-sized units) and the transitivity principle for indirect measurement.[1]

(4) Students compose and decompose plane or solid figures (e.g., put two triangles together to make a quadrilateral) and build understanding of part-whole relationships as well as the properties of the original and composite shapes. As they combine shapes, they recognize them from different perspectives and orientations, describe their geometric attributes, and determine how they are alike and different, to develop the background for measurement and for initial understandings of properties such as congruence and symmetry.

[1] Students should apply the principle of transitivity of measurement to make indirect comparisons, but they need not use this technical term.

Common Core State Standards

Standards for Mathematical Content

Domain Operations and Algebraic Thinking 1.OA	*Everyday Mathematics* **Goals for Mathematical Content**

Cluster Represent and solve problems involving addition and subtraction.

1.OA.1 Use addition and subtraction within 20 to solve word problems involving situations of adding to, taking from, putting together, taking apart, and comparing, with unknowns in all positions, e.g., by using objects, drawings, and equations with a symbol for the unknown number to represent the problem.[2]	**GMC** Solve number stories by adding and subtracting. **GMC** Model parts-and-total, change, and comparison situations.
1.OA.2 Solve word problems that call for addition of three whole numbers whose sum is less than or equal to 20, e.g., by using objects, drawings, and equations with a symbol for the unknown number to represent the problem.	**GMC** Model and solve number stories involving the addition of 3 addends.

Cluster Understand and apply properties of operations and the relationship between addition and subtraction.

1.OA.3 Apply properties of operations as strategies to add and subtract.[3] *Examples: If $8 + 3 = 11$ is known, then $3 + 8 = 11$ is also known. (Commutative property of addition.) To add $2 + 6 + 4$, the second two numbers can be added to make a ten, so $2 + 6 + 4 = 2 + 10 = 12$. (Associative property of addition.)*	**GMC** Apply properties of operations to add or subtract.
1.OA.4 Understand subtraction as an unknown-addend problem. *For example, subtract $10 - 8$ by finding the number that makes 10 when added to 8.*	**GMC** Understand subtraction as an unknown-addend problem.

Cluster Add and subtract within 20.

1.OA.5 Relate counting to addition and subtraction (e.g., by counting on 2 to add 2).	**GMC** Relate counting to addition and subtraction.
1.OA.6 Add and subtract within 20, demonstrating fluency for addition and subtraction within 10. Use strategies such as counting on; making ten (e.g., $8 + 6 = 8 + 2 + 4 = 10 + 4 = 14$); decomposing a number leading to a ten (e.g., $13 - 4 = 13 - 3 - 1 = 10 - 1 = 9$); using the relationship between addition and subtraction (e.g., knowing that $8 + 4 = 12$, one knows $12 - 8 = 4$); and creating equivalent but easier or known sums (e.g., adding $6 + 7$ by creating the known equivalent $6 + 6 + 1 = 12 + 1 = 13$).	**GMC** Recognize and decompose quantities up to 20 using visual patterns. **GMC** Add within 10 fluently. **GMC** Subtract within 10 fluently. **GMC** Add doubles automatically. **GMC** Subtract doubles. **GMC** Add combinations of 10 automatically. **GMC** Subtract combinations of 10. **GMC** Add and subtract within 20 using strategies.

Cluster Work with addition and subtraction equations.

1.OA.7 Understand the meaning of the equal sign, and determine if equations involving addition and subtraction are true or false. *For example, which of the following equations are true and which are false? $6 = 6$, $7 = 8 - 1$, $5 + 2 = 2 + 5$, $4 + 1 = 5 + 2$.*	**GMC** Understand the meaning of the equal sign. **GMC** Determine whether equations involving addition or subtraction are true or false.
1.OA.8 Determine the unknown whole number in an addition or subtraction equation relating three whole numbers. *For example, determine the unknown number that makes the equation true in each of the equations $8 + ? = 11$, $5 = \square - 3$, $6 + 6 = \square$.*	**GMC** Find the unknown in addition and subtraction equations.

[2] See Glossary, Table 1. http://www.corestandards.org/assets/CCSSI_Math Standards.pdf
[3] Students need not use formal terms for these properties.

Common Core State Standards

Standards for Mathematical Content

Domain Number and Operations in Base Ten 1.NBT	*Everyday Mathematics* **Goals for Mathematical Content**

Cluster Extend the counting sequence.

1.NBT.1 Count to 120, starting at any number less than 120. In this range, read and write numerals and represent a number of objects with a written numeral.	GMC Count on from any number. GMC Read and write numbers. GMC Count and represent collections of objects with numerals.

Cluster Understand place value.

1.NBT.2 Understand that the two digits of a two-digit number represent amounts of tens and ones. Understand the following as special cases:	GMC Understand place value. GMC Represent whole numbers as tens and ones.
1.NBT.2a 10 can be thought of as a bundle of ten ones—called a "ten."	GMC Understand exchanging tens and ones.
1.NBT.2b The numbers from 11 to 19 are composed of a ten and one, two, three, four, five, six, seven, eight, or nine ones.	GMC Understand 11 to 19 as a ten and some ones.
1.NBT.2c The numbers 10, 20, 30, 40, 50, 60, 70, 80, 90 refer to one, two, three, four, five, six, seven, eight, or nine tens (and 0 ones).	GMC Understand 10, 20, . . ., 90 as some tens and no ones.
1.NBT.3 Compare two two-digit numbers based on meanings of the tens and ones digits, recording the results of comparisons with the symbols >, =, and <.	GMC Compare and order numbers. GMC Record comparisons using >, =, or <.

Cluster Use place value understanding and properties of operations to add and subtract.

1.NBT.4 Add within 100, including adding a two-digit number and a one-digit number, and adding a two-digit number and a multiple of 10, using concrete models or drawings and strategies based on place value, properties of operations, and/or the relationship between addition and subtraction; relate the strategy to a written method and explain the reasoning used. Understand that in adding two-digit numbers, one adds tens and tens, ones and ones; and sometimes it is necessary to compose a ten.	GMC Understand adding 2-digit numbers and 1-digit numbers. GMC Understand adding 2-digit numbers and multiples of 10. GMC Understand adding 2-digit numbers.
1.NBT.5 Given a two-digit number, mentally find 10 more or 10 less than the number, without having to count; explain the reasoning used.	GMC Mentally find 10 more or 10 less than a 2-digit number.
1.NBT.6 Subtract multiples of 10 in the range 10–90 from multiples of 10 in the range 10–90 (positive or zero differences), using concrete models or drawings and strategies based on place value, properties of operations, and/or the relationship between addition and subtraction; relate the strategy to a written method and explain the reasoning used.	GMC Subtract multiples of 10 from multiples of 10.

Domain Measurement and Data   1.MD

Cluster Measure lengths indirectly and by iterating length units.

1.MD.1 Order three objects by length; compare the lengths of two objects indirectly by using a third object.	GMC Order objects by length. GMC Compare the lengths of objects indirectly.
1.MD.2 Express the length of an object as a whole number of length units, by laying multiple copies of a shorter object (the length unit) end to end; understand that the length measurement of an object is the number of same-size length units that span it with no gaps or overlaps. *Limit to contexts where the object being measured is spanned by a whole number of length units with no gaps or overlaps.*	GMC Measure length using same-size units with no gaps or overlaps. GMC Express length as a whole number of units.

Common Core State Standards

Standards for Mathematical Content

Cluster   Tell and write time.

1.MD.3 Tell and write time in hours and half-hours using analog and digital clocks.	**GMC** Tell and write time using analog clocks.
	GMC Tell and write time using digital clocks.

Cluster   Represent and interpret data.

1.MD.4 Organize, represent, and interpret data with up to three categories; ask and answer questions about the total number of data points, how many in each category, and how many more or less are in one category than in another.	**GMC** Organize and represent data.
	GMC Ask questions about data.
	GMC Answer questions about data.

Domain   Geometry   1.G

Everyday Mathematics
Goals for Mathematical Content

Cluster   Reason with shapes and their attributes.

1.G.1 Distinguish between defining attributes (e.g., triangles are closed and three-sided) versus non-defining attributes (e.g., color, orientation, overall size); build and draw shapes to possess defining attributes.	**GMC** Distinguish between defining and non-defining attributes.
	GMC Build and draw shapes to possess defining attributes.
1.G.2 Compose two-dimensional shapes (rectangles, squares, trapezoids, triangles, half-circles, and quarter-circles) or three-dimensional shapes (cubes, right rectangular prisms, right circular cones, and right circular cylinders) to create a composite shape, and compose new shapes from the composite shape.[4]	**GMC** Build composite shapes.
	GMC Compose new shapes from composite shapes.
1.G.3 Partition circles and rectangles into two and four equal shares, describe the shares using the words *halves*, *fourths*, and *quarters*, and use the phrases *half of*, *fourth of*, and *quarter of*. Describe the whole as two of, or four of the shares. Understand for these examples that decomposing into more equal shares creates smaller shares.	**GMC** Partition shapes into equal shares.
	GMC Describe equal shares using fraction words.
	GMC Describe the whole as a number of shares.
	GMC Understand that *more equal shares* means *smaller equal shares*.

[4] Students do not need to learn formal names such as "right rectangular prism."

Common Core State Standards

Standards for Mathematical Practice	Everyday Mathematics Goals for Mathematical Practice

1 Make sense of problems and persevere in solving them.

Mathematically proficient students start by explaining to themselves the meaning of a problem and looking for entry points to its solution. They analyze givens, constraints, relationships, and goals. They make conjectures about the form and meaning of the solution and plan a solution pathway rather than simply jumping into a solution attempt. They consider analogous problems, and try special cases and simpler forms of the original problem in order to gain insight into its solution. They monitor and evaluate their progress and change course if necessary. Older students might, depending on the context of the problem, transform algebraic expressions or change the viewing window on their graphing calculator to get the information they need. Mathematically proficient students can explain correspondences between equations, verbal descriptions, tables, and graphs or draw diagrams of important features and relationships, graph data, and search for regularity or trends. Younger students might rely on using concrete objects or pictures to help conceptualize and solve a problem. Mathematically proficient students check their answers to problems using a different method, and they continually ask themselves, "Does this make sense?" They can understand the approaches of others to solving complex problems and identify correspondences between different approaches.

GMP1.1 Make sense of your problem.

GMP1.2 Reflect on your thinking as you solve your problem.

GMP1.3 Keep trying when your problem is hard.

GMP1.4 Check whether your answer makes sense.

GMP1.5 Solve problems in more than one way.

GMP1.6 Compare the strategies you and others use.

2 Reason abstractly and quantitatively.

Mathematically proficient students make sense of quantities and their relationships in problem situations. They bring two complementary abilities to bear on problems involving quantitative relationships: the ability to *decontextualize*—to abstract a given situation and represent it symbolically and manipulate the representing symbols as if they have a life of their own, without necessarily attending to their referents—and the ability to *contextualize*, to pause as needed during the manipulation process in order to probe into the referents for the symbols involved. Quantitative reasoning entails habits of creating a coherent representation of the problem at hand; considering the units involved; attending to the meaning of quantities, not just how to compute them; and knowing and flexibly using different properties of operations and objects.

GMP2.1 Create mathematical representations using numbers, words, pictures, symbols, gestures, tables, graphs, and concrete objects.

GMP2.2 Make sense of the representations you and others use.

GMP2.3 Make connections between representations.

Common Core State Standards

Standards for Mathematical Practice	*Everyday Mathematics* Goals for Mathematical Practice

3 Construct viable arguments and critique the reasoning of others.

Mathematically proficient students understand and use stated assumptions, definitions, and previously established results in constructing arguments. They make conjectures and build a logical progression of statements to explore the truth of their conjectures. They are able to analyze situations by breaking them into cases, and can recognize and use counterexamples. They justify their conclusions, communicate them to others, and respond to the arguments of others. They reason inductively about data, making plausible arguments that take into account the context from which the data arose. Mathematically proficient students are also able to compare the effectiveness of two plausible arguments, distinguish correct logic or reasoning from that which is flawed, and—if there is a flaw in an argument—explain what it is. Elementary students can construct arguments using concrete referents such as objects, drawings, diagrams, and actions. Such arguments can make sense and be correct, even though they are not generalized or made formal until later grades. Later, students learn to determine domains to which an argument applies. Students at all grades can listen or read the arguments of others, decide whether they make sense, and ask useful questions to clarify or improve the arguments.

GMP3.1 Make mathematical conjectures and arguments.

GMP3.2 Make sense of others' mathematical thinking.

4 Model with mathematics.

Mathematically proficient students can apply the mathematics they know to solve problems arising in everyday life, society, and the workplace. In early grades, this might be as simple as writing an addition equation to describe a situation. In middle grades, a student might apply proportional reasoning to plan a school event or analyze a problem in the community. By high school, a student might use geometry to solve a design problem or use a function to describe how one quantity of interest depends on another. Mathematically proficient students who can apply what they know are comfortable making assumptions and approximations to simplify a complicated situation, realizing that these may need revision later. They are able to identify important quantities in a practical situation and map their relationships using such tools as diagrams, two-way tables, graphs, flowcharts and formulas. They can analyze those relationships mathematically to draw conclusions. They routinely interpret their mathematical results in the context of the situation and reflect on whether the results make sense, possibly improving the model if it has not served its purpose.

GMP4.1 Model real-world situations using graphs, drawings, tables, symbols, numbers, diagrams, and other representations.

GMP4.2 Use mathematical models to solve problems and answer questions.

Common Core State Standards

| **Standards for Mathematical Practice** | *Everyday Mathematics* **Goals for Mathematical Practice** |

5 Use appropriate tools strategically.

Mathematically proficient students consider the available tools when solving a mathematical problem. These tools might include pencil and paper, concrete models, a ruler, a protractor, a calculator, a spreadsheet, a computer algebra system, a statistical package, or dynamic geometry software. Proficient students are sufficiently familiar with tools appropriate for their grade or course to make sound decisions about when each of these tools might be helpful, recognizing both the insight to be gained and their limitations. For example, mathematically proficient high school students analyze graphs of functions and solutions generated using a graphing calculator. They detect possible errors by strategically using estimation and other mathematical knowledge. When making mathematical models, they know that technology can enable them to visualize the results of varying assumptions, explore consequences, and compare predictions with data. Mathematically proficient students at various grade levels are able to identify relevant external mathematical resources, such as digital content located on a website, and use them to pose or solve problems. They are able to use technological tools to explore and deepen their understanding of concepts.

GMP5.1 Choose appropriate tools.

GMP5.2 Use tools effectively and make sense of your results.

6 Attend to precision.

Mathematically proficient students try to communicate precisely to others. They try to use clear definitions in discussion with others and in their own reasoning. They state the meaning of the symbols they choose, including using the equal sign consistently and appropriately. They are careful about specifying units of measure, and labeling axes to clarify the correspondence with quantities in a problem. They calculate accurately and efficiently, express numerical answers with a degree of precision appropriate for the problem context. In the elementary grades, students give carefully formulated explanations to each other. By the time they reach high school they have learned to examine claims and make explicit use of definitions.

GMP6.1 Explain your mathematical thinking clearly and precisely.

GMP6.2 Use an appropriate level of precision for your problem.

GMP6.3 Use clear labels, units, and mathematical language.

GMP6.4 Think about accuracy and efficiency when you count, measure, and calculate.

Common Core State Standards

Standards for Mathematical Practice	*Everyday Mathematics* Goals for Mathematical Practice

7 Look for and make use of structure.

Mathematically proficient students look closely to discern a pattern or structure. Young students, for example, might notice that three and seven more is the same amount as seven and three more, or they may sort a collection of shapes according to how many sides the shapes have. Later, students will see 7×8 equals the well remembered $7 \times 5 + 7 \times 3$, in preparation for learning about the distributive property. In the expression $x^2 + 9x + 14$, older students can see the 14 as 2×7 and the 9 as $2 + 7$. They recognize the significance of an existing line in a geometric figure and can use the strategy of drawing an auxiliary line for solving problems. They also can step back for an overview and shift perspective. They can see complicated things, such as some algebraic expressions, as single objects or as being composed of several objects. For example, they can see $5 - 3(x - y)^2$ as 5 minus a positive number times a square and use that to realize that its value cannot be more than 5 for any real numbers x and y.

GMP7.1 Look for mathematical structures such as categories, patterns, and properties.

GMP7.2 Use structures to solve problems and answer questions.

8 Look for and express regularity in repeated reasoning.

Mathematically proficient students notice if calculations are repeated, and look both for general methods and for shortcuts. Upper elementary students might notice when dividing 25 by 11 that they are repeating the same calculations over and over again, and conclude they have a repeating decimal. By paying attention to the calculation of slope as they repeatedly check whether points are on the line through (1, 2) with slope 3, middle school students might abstract the equation $(y - 2)/(x - 1) = 3$. Noticing the regularity in the way terms cancel when expanding $(x - 1)(x + 1)$, $(x - 1)(x^2 + x + 1)$, and $(x - 1)(x^3 + x^2 + x + 1)$ might lead them to the general formula for the sum of a geometric series. As they work to solve a problem, mathematically proficient students maintain oversight of the process, while attending to the details. They continually evaluate the reasonableness of their intermediate results.

GMP8.1 Create and justify rules, shortcuts, and generalizations.

Grades K–2 Games Correlation

Game	Grade K Lesson	Grade 1 Lesson	Grade 2 Lesson	Counting and Cardinality***	Operations and Algebraic Thinking	Number and Operations in Base Ten	Measurement and Data	Geometry
Addition Flip-It	9-11				•			
Addition/Subtraction Spin		5-6				•		
Addition Top-It	8-11	5-5	4-3	•	•	•*		
Addition Top-It with Dot Cards	2-2			•	•			
Animal Weight Top-It		5-11				•		
Array Bingo			8-10		•			
Array Concentration			8-10		•	•		
Attribute Spinner	6-10						•	•
Attribute Train Game		7-6						•
Base-10 Exchange (See also The Exchange Game)		5-8				•		
Basketball Addition			7-3			•		
Beat the Calculator		7-2	5-1		•			
Beat the Calculator (Extended Facts)			5-1		•	•		
Beat the Timer	3-9			•				
Before and After		5-6				•		
Bunny Hop		1-5			•	•		
Car Race	8-8				•			
Clear the Board	7-12			•	•			
Count and Sit	1-6			•				
Count and Sit (by Tens)	4-12			•				
Count and Sit (Counting On)	4-12			•				
Dice Addition	7-12			•	•			
Dice Race	3-11			•				
Dice Subtraction	8-5				•			
The Difference Game		5-10	3-5		•	•**		
Digit Game		5-1				•		
Dime-Nickel-Penny Grab			5-2				•	
Disappearing Train	6-9			•	•			
Domino Concentration	7-2				•			
Domino Top-It		3-1			•	•		

*This standard is not covered in the Kindergarten version of the game.
**This standard is not covered in the Grade 1 version of the game.
***Counting and Cardinality is not covered in Grades 1 and 2.

Game	Grade K Lesson	Grade 1 Lesson	Grade 2 Lesson	Counting and Cardinality***	Operations and Algebraic Thinking	Number and Operations in Base Ten	Measurement and Data	Geometry
Evens and Odds			2-9		•			
Find the Block	6-10						•	•
Fishing for 10	9-11	4-9	1-7		•			
Fishing for 100			1-7		•	•		
Frog Hop	7-1			•	•			
Gotcha	1-3			•				
Growing and Disappearing Train	6-12			•	•			
Growing Train	5-11			•	•			
Guess My Number	6-12			•	•	•		
Guess My Shape	6-4							•
Hiding Bears	6-11			•	•			
Hiding Bears (3 Addends)	6-11			•	•			
High Roller		2-6			•	•		
Hit the Target			7-1		•			
How Many Now?	2-6			•				
I Spy (with 2-dimensional shapes)	5-5	8-1						•
I Spy (with 3-dimensional shapes)	6-8	8-6						•
Make My Design	9-1	8-5						•
Making 5	8-11				•			
Match Up with Dot and Number Cards	3-1			•				
Match Up with Dot Cards	2-1			•				
Match Up with Ten Frames and Numbers	4-5			•	•			
Mini Monster Squeeze	3-12			•				
Monster Squeeze	3-12	1-2		•		•*		
Mystery Block	7-13						•	•
Mystery Change	2-5				•			
Name That Number			2-11		•	•		
Number-Grid Cover-Up	4-13			•				
Number-Line Squeeze			1-2		•			
Number Top-It			1-11			•		
Penny-Dice		1-3			•	•		
Penny-Dime-Dollar Exchange		6-11				•		

Game	Grade K Lesson	Grade 1 Lesson	Grade 2 Lesson	Counting and Cardinality***	Operations and Algebraic Thinking	Number and Operations in Base Ten	Measurement and Data	Geometry
Penny-Dime Exchange		5-3				•		
Penny Plate	6-11	2-3	1-7		•	•**		
Quarter-Dime-Nickel-Penny Grab			1-8				•	
Racing Two Trains	5-11			•	•			
Rock, Paper, Scissors		1-8					•	
Rock, Paper, Scissors, Pencil		1-8					•	
Roll and Record	3-11			•				
Roll and Record with Dot Dice	5-2			•	•			
Roll and Record with Doubles	9-10	4-7	2-5		•			
Roll and Record with Numeral Dice	9-6				•			
Roll and Total		2-1			•		•	
Rolling for 50		1-11			•	•		
Salute!		7-3	3-4		•			
Shaker Addition Top-It		7-4			•		•	
Shape Capture			8-2					•
Solid Shapes Match Up	7-4							•
Spin a Number	3-8			•				
Spinning for Money			2-1				•	
Stand Up If…	6-4							•
Stop and Go		5-11				•		
Subtraction Bingo		2-4			•			
Subtraction Roll and Record	9-6				•			
Subtraction Top-It	9-2			•	•			
Target (to 50)			4-7			•		
Target (to 200)			4-7			•		
Teens on Double Ten Frames	5-8			•		•		
Ten Bears on a Bus	5-3			•	•			
Ten-Frame Top-It		2-2			•	•		
The Digit Game (with Symbols)			4-5			•		
The Exchange Game (with Pennies and Nickels)			2-1				•	

*This standard is not covered in the Kindergarten version of the game.

**This standard is not covered in the Grade 1 version of the game.

***Counting and Cardinality is not covered in Grades 1 and 2.

McGraw-Hill Education

Game	Grade K Lesson	Grade 1 Lesson	Grade 2 Lesson	Counting and Cardinality***	Operations and Algebraic Thinking	Number and Operations in Base Ten	Measurement and Data	Geometry
The Exchange Game (with Pennies, Nickels, and Dimes)			1-8				•	
The Exchange Game (with base-10 blocks)			4-10			•		
The Exchange Game (with money)			2-1			•	•	
The Number-Grid Difference Game			3-5			•		
The Number-Grid Game			1-4			•		
Time Match		7-11					•	
Top-It		1-6				•		
Top-It with Dot Cards	2-2			•				
Top-It with Multiple Cards	4-12			•				
Top-It with Number Cards	4-12			•				
Tric-Trac		7-10			•			
Turning Over 10			1-7		•	•		
Two-Fisted Penny Addition			1-6		•	•		
What Changed? Train Game	6-12				•			
"What's My Rule?" Fishing	6-6			•			•	
"What's My Rule?" with Attribute Blocks	7-13						•	•
"What's My Rule?" with Numbers	9-3				•			
"What's My Rule?" with Patterns	8-11							•
What's Your Way		4-11				•		
Which Number Doesn't Belong?	3-7			•				
Who Am I Thinking Of?	6-6						•	

CCSS Domain

Sample Work from Child A

-9	-8	-7	-6	-5	-4	-3	-2	-1	0
1	2	3	④	5	6	7	8	9	10
11	12	13	14	15	16	17	18	19	20
21	22	23	24	25	26	27	28	29	30
31	32	33	34	35	36	37	38	39	40
41	42	43	44	45	46	47	48	49	50
51	52	53	54	55	56	57	58	59	60

③ Use words and pictures to show how you used your tool.

I Stoited at 4 and hopt
41 hos ord it was 44 lbs

Sample Work from Child B

③ Use words and pictures to show how you used your tool.

I Staried at 120
I Cawtied With my Fingrs to
cawt up 20 and my answr
Was 140.

Sample Work from Child C

③ Use words and pictures to show how you used your tool.

OCTOPUS

□□　Blue cɾɑb

Sample Work from Child A

(1) Does Mr. Khan have enough to give each child 1 pencil?

_____15_____

Use words, numbers, or pictures to show how you know.

$$7 + 8 = 15$$

Sample Work from Child B

(1) Does Mr. Khan have enough to give each child 1 pencil?

_____yes_____

Use words, numbers, or pictures to show how you know.

$$18 - 7 - 8 = 3$$

Sample Work from Child C

(1) Does Mr. Khan have enough to give each child 1 pencil?

yes

Use words, numbers, or pictures to show how you know.

becuse 7+8=15 and Mr.Khan has 18 and 18 is more then 15 So he has enough to give each child 1 pencil

Sample Work from Child D

1 Does Mr. Khan have enough to give each child 1 pencil?

yes

Use words, numbers, or pictures to show how you know.

2 How many pencils will Mr. Khan have left over?

_3_ pencils

Show or explain how you found your answer.

I us 8 boysaand 1 Grils then
I cros Them out and mr.Khon ohe
have left Over is ε pencils letr

Sample Work from Child E

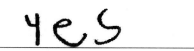   Does Mr. Khan have enough to give each child 1 pencil?

yes

Use words, numbers, or pictures to show how you know.

I put the girls and boys thgethr so I can figrit out and the awnswr was 15. mr. han Has 3 more Pencils.

(2) How many pencils will Mr. Khan have left over?

3 pencils

Show or explain how you found your answer.

Because 7+8=15 so mr.Khan Has 3 more Because if you Have some thing and you give sobode it is sobchracshin.

Sample Work from Child A

① Use pattern blocks to help you fill in the table.

Desks	Chairs
3	5
5	7
4	6
2	3

Example:

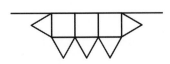

② Write a rule for your table.

Write how you found and checked your rule.

We yousd desks and chairs.

Desks
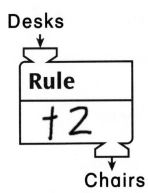

Rule

+2

Chairs

Sample Work from Child B

(1) Use pattern blocks to help you fill in the table.

Desks	Chairs
3	5
7	9
4	6
2	4

Example:

(2) Write a rule for your table.

Write how you found and checked your rule.

I started at 3 and counted to 5 and it was +2

Desks
↓

Rule

+2

↓
Chairs

Sample Work from Child C

① Use pattern blocks to help you fill in the table.

Desks	Chairs
3	5
ten	twelve
six	eight
one	three

Example:

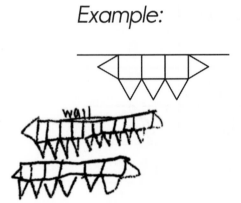

② Write a rule for your table.

Write how you found and checked your rule.

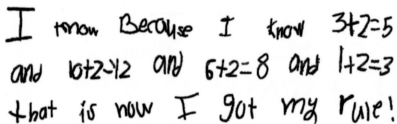

I know Because I know 3+2=5 and 10+2=12 and 6+2=8 and 1+2=3 that is how I got my rule!

Desks
↓
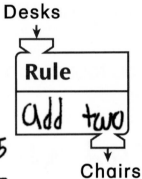
Rule
add two
↓
Chairs

Sample Work from Child D

(1) Use pattern blocks to help you fill in the table.

Desks	Chairs
3	5
5	7
4	6
6	8

Example:

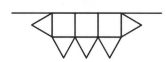

(2) Write a rule for your table.

Write how you found and checked your rule.

Desks

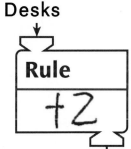

Rule

$+2$

Chairs

I know the Rule is +2. becaze 5+2=7 4+2=6 and 6+2=8 and I f I had 6 desks its just two more chairs for the sides.

(3) How many chairs will Luka need for 18 desks?

20 chairs

Sample Work from Child A

boypaper

a girl would get

girl paper

more paper. a girl would

get a whole

half!

Sample Work from Child B

boys got a Big Paper
because it got more boys

boys

Girls

Girls get a Big Paper
too because
it is 2 Girls

Sample Work from Child C

girls

boys

The boys win because they have more.

Sample Work from Child A

(1) Choose 3 items to buy.

I bought _____a Pencil_____ for 29¢ cents.

I bought _____Some Paper_____ for 38¢ cents.

I bought _____Scissors_____ for 30¢ cents.

(2) How much did you pay for your 3 items in all?
Show or explain how you found your answer.

I STORtd at 29
then I conted
up to 30 I 1ad on 89
then it was 89 cents.

I paid _____89¢_____.

Sample Work from Child B

(1) Choose 3 items to buy.

I bought _____pen_____ for __31__ cents.

I bought _____stickers_____ for __25__ cents.

I bought _____Bookmark_____ for __26__ cents.

(2) How much did you pay for your 3 items in all?
Show or explain how you found your answer.

Me and my parcher
yos are number grid to
make 31+25+26=82

I paid __82__ .

CCSS 1.NBT.4

Work Sample #1 —
Partially Meeting Expectations

This sample work does not meet expectations for the content standard and only partially meets expectations for the mathematical practice. This child did not find the correct total weight of the chosen animals (26 lb). 1.NBT.4 This child's drawing clearly identifies base-10 blocks as the tool used, and the drawing clearly shows the steps of representing the weights of the two animals separately and then putting them together. However, the tool was used incorrectly because the weight of the koala was not correctly represented in the first step. GMP5.2

① Choose two animals.

Animal: _Koala_   Weight: _19_ lb

Animal: _Cat_   Weight: _7_ lb

② What is the total weight of the animals you chose? _17_ lb
Use a tool to help you find your answer.

③ Use words and pictures to show how you used your tool.

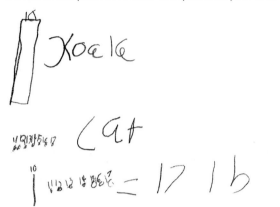

Work Sample #2 —
Partially Meeting Expectations

This work sample meets expectations for the content standard but only partially meets expectations for the mathematical practice. This child found the correct total weight of the chosen animals (65 lb). 1.NBT.4 This child's drawing and written explanation clearly identify base-10 blocks as the tool used. However, the drawing only shows a representation of the first animal weight with base-10 blocks and does not include the remaining steps of showing the second weight and then counting all the blocks together. GMP5.2

① Choose two animals.

Animal: _Peacok_   Weight: _60_ lb

Animal: _Squirrel_   Weight: _5_ lb

② What is the total weight of the animals you chose? _65_ lb
Use a tool to help you find your answer.

③ Use words and pictures to show how you used your tool.

I use blaks And 1S,

1.NBT.4

Work Sample #3 —
Meeting Expectations

This sample work meets expectations for the content standard and for the mathematical practice. This child found the correct total weight of the chosen animals (83 lb). **1.NBT.4** This child's written explanation clearly identifies the number grid as the tool used and describes all the steps for solving the problem. **GMP5.2** Note that this child would have exceeded expectations if a number grid with the hops marked was also included.

(1) Choose two animals.

Animal: _Penguin_   Weight: _75_ lb

Animal: _Pea cock_   Weight: _8_ lb

(2) What is the total weight
of the animals you chose?   _83_ lb
Use a tool to help you find your answer.

(3) Use words and pictures to show how you used your tool.

I Started at 75. And then I
counted up 8. I landed on 83
I used the number grrd.

CCSS 1.OA.1

Work Sample #1 — Meeting Expectations

This sample work does not meet expectations for the content standard but does meet expectations for the mathematical practice. This child found that there were enough pencils, but states that there are 2, not 3, left over. **1.OA.1** This child describes starting at the number of pencils and using fingers to count back according to the number of children. Although the response does not explicitly state that the child started at 18, counting back 8 and ending at 10 implies this. This is evidence of recognizing the need to both compare the number of pencils to the number of children and find out how many more pencils there are. It is not clear why the child ended at 3 and then said that there were 2 pencils left over, but minor errors in computation and interpretation of the final result do not necessarily mean that a child did not make sense of the problem.
GMP1.1

① Does Mr. Khan have enough to give each child 1 pencil?

yes

Use words, numbers, or pictures to show how you know.

I use my fviging. I counted Back eingt that number was 10 then I cainted Back seven that Answen was 3 so he hat two more pencile.

② How many pencils will Mr. Khan have left over?

two pencils

Show or explain how you found your answer.

I used my fingers. I counted back 8 that was 10 the minus 7 that was 3 he had two more penciles.

1.OA.1

Work Sample #2 —
Partially Meeting Expectations

This sample work does not meet expectations for the content standard, but partially meets expectations for the mathematical practice. This child found that there were enough pencils but did not find the correct number that were left over. **1.OA.1** The picture shows 18 children in all, with 3 circled and 15 not circled. This is evidence of recognizing the need to compare the number of children (15) to the number of pencils (18). Although the picture clearly shows that there are 3 more pencils than children, the response to the second question attends to the total number of children, rather than the 3 pencils left over. Thus there is no evidence of recognizing the need to find how many more pencils there are than children. GMP1.1

(1) Does Mr. Khan have enough to give each child 1 pencil?

_____ Yes _____

Use words, numbers, or pictures to show how you know.

(2) How many pencils will Mr. Khan have left over?

_____ 15 _____ pencils

Show or explain how you found your answer.

What I did to get my is I made seven girls and eathit boys and I add them up and I got my answer and I note it down,

Lesson 6-8

CCSS 1.OA.1

Work Sample #3 — Exceeding Expectations

This sample work meets expectations for the content standard and exceeds expectations for the mathematical practice. This child found that there were enough pencils with 3 left over. 1.OA.1 This child's picture shows 18 pencils in all, with sets marked off for the 7 girls and 8 boys and 3 more left over. This is evidence of recognizing the need to compare the number of pencils (18) to the number of children (15). This child's response includes both a picture and subtraction number model that each show there were 3 pencils left over. These representations, combined with a response of 3 pencils to Problem 2, constitute evidence of recognizing the need to find how many more pencils there are than children. Although the written description under Problem 2 could be interpreted as a restatement of how the picture was used, the subtraction model 18 - 15 = 3 suggests a different method of finding the total number of children before taking any pencils away. This paper therefore exceeds expectations.

GMP1.1

---

① Does Mr. Khan have enough to give each child 1 pencil?

Yes

Use words, numbers, or pictures to show how you know.

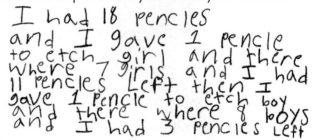

G 🖉🖉🖉🖉🖉🖉 7 Pencles

B 🖉🖉🖉🖉🖉🖉 8 Pencles

L 🖉🖉🖉 3 Left overs

18 - 15 = 3

② How many pencils will Mr. Khan have left over?

___3___ pencils

Show or explain how you found your answer.

I had 18 pencles
and I gave 1 pencle
to etch 7 girl and there
where 7 girls and I had
11 pencles left then I
gave 1 Pencle to etch boy
and there where 8 boys
and I had 3 pencles left

CCSS 1.OA.8

Work Sample #1 — Meeting Expectations

This sample work meets expectations for the content standard and for the mathematical practice. This child found + 2 as the rule and 20 as the number of chairs that Luka will need. 1.OA.8 With revision, this child created a table with four rows that all support the rule of + 2. This child also wrote a justification for the rule in Problem 2 that describes checking all four rows of the table to make sure they work for + 2. This child also wrote "+ 2" between columns in the last three rows of the table, which is further evidence of checking at least three rows against the rule of + 2. GMP8.1

① Use pattern blocks to help you fill in the table.

Desks	Chairs
3	5
4 +2	6
10 +2	12
6 +2	8
18	

Example:

② Write a rule for your table.

Write how you found and checked your rule.

What I did was I checked it a and I pound all my awnsers and all of them were plus two.

 Desks

Rule
+ 2

Chairs

③ How many chairs will Luka need for 18 desks?

____20____ chairs

Work Sample #2 — Partially Meeting Expectations

This sample work meets expectations for the content standard, but partially meets expectations for the mathematical practice. This child found + 2 as the rule and 20 as the number of chairs that Luka will need. 1.OA.8 Although the child added only two new rows to the table, three of the rows support the rule + 2. However, the child's justification of the rule refers to only one row of the table and doesn't show any evidence of checking the rule against other rows. If the teacher had observed this child checking against the other rows, this paper could be evaluated as meeting expectations. GMP8.1

① Use pattern blocks to help you fill in the table.

Desks	Chairs
3	5
6	8
8	10
3	5

Example:

② Write a rule for your table.

Write how you found and checked your rule.

I think it was Plus 2 becuse that is the Rule and It can draw sick to (|||||)#|| 6+a8

Desks

Rule
+2

Chairs

③ How many chairs will Luka need for 18 desks?

____20____ chairs

CCSS 1.OA.8

Work Sample #3 — Exceeding Expectations

This sample work meets expectations for the content standard and exceeds expectations for the mathematical practice. This child found + 2 as the rule and 20 as the number of chairs that Luka will need. **1.OA.8** All four rows of this child's table support the rule + 2. With revision, this child wrote a justification for the rule that refers to two rows of the table (3, 5 and 4, 6). This child also wrote a justification of the answer to Problem 3, explaining that the extra two chairs are the ones on the sides. GMP8.1

① Use pattern blocks to help you fill in the table.

Desks	Chairs
3	5
4	6
2	4
1	3

Example:

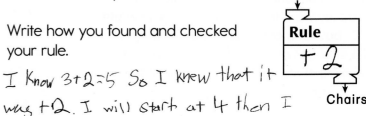

② Write a rule for your table.

Write how you found and checked your rule.

I Know 3+2=5 So I knew that it was +2. I will start at 4 then I cout to 6 then I know the rule is 12.

Desks
↓
Rule
+ 2
↓
Chairs

③ How many chairs will Luka need for 18 desks?

___20___ chairs If there's 18 desks first I set 18 desks then I Just need to set 2 the side.

CCSS 1.G.3

Work Sample #1 —
Partially Meeting Expectations

This sample work does not meet expectations for the content standard, but partially meets expectations for the mathematical practice. This child divided the two squares into equal shares, but did not state that one girl will get a larger share of paper. **1.G.3** With revision, this child created a drawing that shows both paper squares and clearly shows the girls' shares as larger. However, this child did not provide an answer to the question. **GMP4.2**

Work Sample #2 —
Partially Meeting Expectations

This sample work does not meet expectations for the content standard, but partially meets expectations for the mathematical practice. This child determined that one girl will get a larger share of paper, but did not show evidence of dividing the boys' square into equal shares. **1.G.3** This child provided a clear answer to the question and said the girls will get larger pieces. However, although this child attempted to model the situation with a drawing, the drawing does not show the divided square for the boys or the relative sizes of the shares, so the answer is unsupported by the model. **GMP4.2**

Lesson 8-4

CCSS 1.G.3

Work Sample #3 — Exceeding Expectations

This sample work meets expectations for the content standard and exceeds expectations for the mathematical practice. This child divided the two squares into equal shares and stated that the girls will get a larger share. **1.G.3** The drawing shows both paper squares divided into equal shares and clearly shows the girls' shares as larger. The answer is supported by the drawing. This child also stated that fewer people get larger shares, thereby exceeding expectations. **GMP4.2**

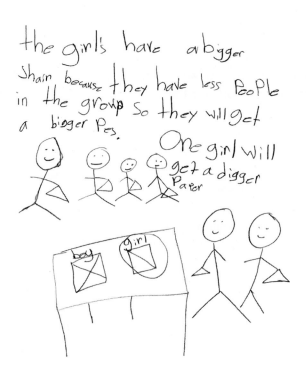

the girl's have a bigger shain because they have less People in the group so they will get a bigger Pes.

One girl will get a digger Paer

boy girl

CCSS 1.NBT.4

Work Sample #1 —
Partially Meeting Expectations

This sample work does not meet expectations for the content standard but partially meets expectations for the mathematical practice. This child did not find the correct total cost of the three items listed in Problem 1 (95 cents). **1.NBT.4** With revision, this child wrote an explanation that describes starting at the first price, counting up to another number by 1s, then counting up again. The explanation says that the child first counted by 1s, but it does not say how many. It is also not clear why the child counted up 8 the second time. GMP6.1

(1) Choose 3 items to buy.

I bought _Colored paper_ for _38_ cents.

I bought _a pen_ for _31_ cents.

I bought _bookmark_ for _26_ cents.

(2) How much did you pay for your 3 items in all? Show or explain how you found your answer.

Colored paper ———————— 38¢

pen ———————————— 31¢

bookmark ——————— 26¢

I started with 38¢. I counted by ones. I landed on 69¢, then I counted up by ones 8 times.

I paid ~~4.60~~ 77¢

Work Sample #2 —
Partially Meeting Expectations

This sample work meets expectations for the content standard but only partially meets expectations for the mathematical practice. This child found the correct total cost for the items listed in Problem 1 (82 cents). **1.NBT.4** The explanation identifies the number grid as the tool used, but does not say how it was used. The revision does not add any clarity or precision to the explanation. GMP6.1

(1) Choose 3 items to buy.

I bought _pen_ for _31_ cents.

I bought _stickers_ for _25_ cents.

I bought _Bookmark_ for _26_ cents.

I sow how muuch is the stickers it cot 25 cents

(2) How much did you pay for your 3 items in all? Show or explain how you found your answer.

Me and my parcher yos are number grid to make 31+ 25+26=82

I paid _82_.

Lesson 9-3

CCSS 1.NBT.4

Work Sample #3 — Meeting Expectations

This sample work meets expectations for the content standard and for the mathematical practice. Although there is an error in one addend (30¢ instead of 31¢) in the number model, this child found the correct total cost of the three items listed in Problem 1 (82 cents). **1.NBT.4** The revised explanation shows that tally marks were used to represent each individual price, and then the tally marks were counted to find the total. This child suggested an additional strategy of counting by 10s first. If the child had explained how to count by 10s and then 1s, the work would exceed expectations.

GMP6.1

① Choose 3 items to buy.

I bought _____ •stickers _____ for __ 25¢ __ cents.

I bought _____ •pen _____ for __ 31¢ __ cents.

I bought _____ •Bookmark _____ for __ 26¢ __ cents.

I should of counted by tens enstead of ones.

② How much did you pay for your 3 items in all? Show or explain how you found your answer.

‖‖‖ ‖‖‖ ‖‖‖ ‖‖‖ ‖‖‖ |
Bookmark 26¢

‖‖‖ ‖‖‖‖‖
Stickers 25¢

all this equals 82¢ = 82¢

‖‖‖ ‖‖ ‖‖‖ ‖‖‖ ‖‖‖
‖‖‖ | 31¢ Pen

25+26+30¢=82¢ ‖‖‖‖‖‖‖..

I paid __ 82¢ __ .

Notes

Notes

Notes

Notes

Notes